Gmelin Handbook of Inorganic Chemistry

8th Edition

Gmelin Handbook
of Inorganic Chemistry

8th Edition

Gmelin Handbuch der Anorganischen Chemie

Achte, völlig neu bearbeitete Auflage

Prepared
and issued by

Gmelin-Institut für Anorganische Chemie
der Max-Planck-Gesellschaft
zur Förderung der Wissenschaften

Director: Ekkehard Fluck

Founded by   Leopold Gmelin

8th Edition   8th Edition begun under the auspices of the
Deutsche Chemische Gesellschaft by R. J. Meyer

Continued by   E. H. E. Pietsch and A. Kotowski, and by
Margot Becke-Goehring

Springer-Verlag Berlin Heidelberg GmbH 1989

Gmelin Handbook
of Inorganic Chemistry

8th Edition

U
Uranium

Supplement Volume B2

Alloys of Uranium with Alkali Metals,
Alkaline Earths,
and Elements of Main Groups III and IV

With 198 illustrations

AUTHORS

Hans Ulrich Borgstedt, Horst Wedemeyer
Kernforschungszentrum Karlsruhe,
Institut für Material- und Festkörperforschung
Karlsruhe, Federal Republic of Germany

CHIEF EDITORS

Karl-Christian Buschbeck, Gmelin-Institut, Frankfurt am Main

Cornelius Keller, Supervising Scientific Coordinator
for the Uranium Supplement Volumes,
Schule für Kerntechnik, Kernforschungszentrum Karlsruhe

System Number 55

Springer-Verlag Berlin Heidelberg GmbH 1989

LITERATURE CLOSING DATE: MID OF 1988
IN SOME CASES MORE RECENT DATA HAVE BEEN CONSIDERED

Library of Congress Catalog Card Number: Agr 25-1383

ISBN 978-3-662-05673-8   ISBN 978-3-662-05671-4 (eBook)
DOI 10.1007/978-3-662-05671-4

© by Springer-Verlag Berlin Heidelberg 1989

Originally published by Springer-Verlag Berlin Heidelberg New York in 1989.

Softcover reprint of the hardcover 8th edition 1989

Preface

The present volume describes the alloys and the intermetallic compounds of uranium with the main group metals — those systems with semimetals such as boron and germanium are to be found in corresponding volumes in uranium C series. In this volume, main emphasis is given to the binary systems. Ternary systems are only dealt with if there is formation of specific compounds or if there are other important characteristics or reactions. In general, the influence of small quantities of a third metal on a system $U-M$, e.g., for mechanical properties or corrosion effects will not be discussed in order not to extend this volume excessively.

The ternary alloys are described immediately after the related binary systems. E.g., $U-Ga-Ni$ is found after $U-Ga$, unless a place after $U-Ni$ could be imaginable from more systematic reasons (system of the last position). The preferred order is also indicated in the sequence of element symbols within the formulas of compounds and systems: The compounds/systems related to $U-Al$ are written as $U-Al-M$, etc.

The behavior of uranium metal towards the main group metals can be divided into two groups:

Practically no solubility in each other and no compound formation. This is the case for the alkali and alkaline earth systems (except beryllium).

Occasional formation of several compounds with limited or extensive solid solubility. This is observed for beryllium and the metals of third and fourth main group.

The most frequently investigated systems are uranium–beryllium and uranium–aluminium because of their special scientific ($U-Be$) and technological ($U-Al$) importance. UBe_{13}, the only compound in the binary $U-Be$ system, is one of the very few heavy-fermion compounds. Therefore, its physical properties especially were extensively and thoroughly studied in recent years. This Gmelin volume gives a complete and very detailed description of this compound, and at the time of publication, the most updated one.

On the other hand, cermets and alloys in the $U-Al$ system are very important as fuel, especially for the high-neutron-flux nuclear research reactors. This fuel is mainly based on ^{235}U-enriched uranium on the aluminium-rich side of the $U-Al$ system, e.g., as UAl_4-Al cermets. In addition to the preparation of these fuels, their physical and chemical properties as well as their irradiation behavior is completely, but also very critically, discussed (see also "Uranium" Suppl. Vol. A3 [1981]).

Therefore, both the reader who is interested in scientific relations and facts, as well as the technologically interested one, will find this volume to be a good source for new and critically updated information.

I want to thank the two competent authors Dr. H. U. Borgstedt and Dr. H. Wedemeyer for their excellent contributions as well as the Literaturabteilung of the Karlsruhe Nuclear Research Center for its help in providing also hard-to-get references and reports.

The very good cooperation with the Gmelin Institute, especially with the editor-in-chief Dr. Karl-Christian Buschbeck and its director Prof. Dr. Ekkehard Fluck is fully appreciated.

Karlsruhe
September 1989

Cornelius Keller

Volumes published on "Radium and Actinides"

U   Uranium

Table of Contents

Uranium Alloys

1 With Alkali Metals

Hans Ulrich Borgstedt,
Institut für Material- und Festkörperforschung,
Kernforschungszentrum Karlsruhe,
Karlsruhe, Federal Republic of Germany

1.1 Rules for the Miscibility of Uranium with Alkali and Alkaline Earth Metals

A rule, based on the work describing the factors governing the alloying behavior of metals [1], allows the prediction of the miscibility or immiscibility of alkali or alkaline earth metals with other metals in the liquid state. Generally, metallic elements are unlikely to dissolve each other if their atomic diameters differ by more than 15% and if there is a high electrochemical tendency to favor the formation of intermetallic compounds. An element is also more likely to dissolve in one of higher valency rather than the converse [2].

The atomic radii of the alkali metals range from 0.156 to 0.27 nm and those of the alkaline earth metals from 0.113 to 0.217 nm. The radius of U in the metallic state is 0.138 nm. Miscibility of the elements in the liquid state should, therefore, be restricted to incomplete miscibility of U with the light alkali and alkaline earth metals. The low valency of the group I a and II a elements compared to U dictate that they are more likely to dissolve in the molten heavy metal [2].

Additionally, a maximum electrochemical factor for solubility, in terms of Pauling's electronegativity concept, was formulated. According to this factor, the difference in electronegativity between the two elements under consideration should be less than 0.4 eV to permit miscibility [3]. One should expect only a limited miscibility of the alkali and alkaline earth metals with U owing to their electrochemical factors. The miscibility in the liquid state is determined by the Hildebrand solubility parameter, δ, which is defined as $\delta = (\Delta E_v/V)^{1/2}$, where ΔE_v is the heat of vaporization and V is the atomic volume [4]. According to Hildebrand's rule, two non-polar liquids should be completely miscible if $\overline{V}(\delta_A - \delta_B)^2 < 2\,RT$. However, this rule neglects the influence of the electronegativity, X. Therefore, Mott introduced the electronegativity difference $(X_A - X_B)$ and n, the number of bonds, which the two elements can form. The expression for the complete miscibility in the modified form is $\overline{V}(\delta_A - \delta_B)^2 - 23060\,n(X_A - X_B)^2 < 2\,RT$. This rule was verified for a large number of alkali metal — heavy metal systems [5].

The temperature coefficients of the solubility in binary metallic systems are dependent on an atomic size effect [6]. According to Mott's rule, the ratios of the atomic radii influence the temperature gradients of the solubility. Experiments showed that the solubilities, which can be expressed by equations of the general form $\log N = B - A/T$, are influenced by the atomic size factor in both constants, B and A. High values of A and B are found in systems with r_B/r_A below 0.8 and around 1.4 [7]. The values of r_B/r_A calculated for the alkali and alkaline earth metals and U atomic radii also predict very poor miscibility of the metals in their liquid state.

The heat of formation of solid alloys of transition metals was calculated on the basis of a cellular model resulting in the equation $\Delta H \approx \{ -P\,e(\Delta\varphi^*)^2 + Q(\Delta n_{ws})^2 \}$. The negative term arises from the difference in the chemical potential, φ^*, of the electrons at the two types of atomic cells. The positive one reflects the discontinuity in the density of electrons, n_{ws}, at the boundary between dissimilar atomic cells [8].

The second term of the equation is preferably written as $Q_0(\Delta n_{ws}^{1/3})^2$. Values for P and Q_0 were derived from basic arguments. The values of the two factors P and Q are nearly equal for widely different alloy systems [9].

The model was tested by comparing calculated values with experimentally measured heats of formation of several alloys, U alloys among them [10]. In the case of alloys such as UBe_{13} (see p. 14), the high coordination number makes the application of this model difficult because a large atom is embedded in a large number of small atoms. The heats of formation are low and the error is, therefore, relatively high.

Calculations of the heats of solution of several liquid metals in liquid metal solvents show that the solution of U in Be has a negative heat of solution in contrast to other alkaline earth metals such as Mg, Ca, or Ba. Hence, there should not be a measurable solubility of U in these three other alkaline earth metals. The heats of solution for the systems U−Li, U−Na, and U−K have markedly positive values. They indicate a very poor miscibility of U with the metals in the liquid state [11].

The same authors also introduced a concentration term in order to evaluate the heats of formation for solid alloys [12]. A computer program in ALGOL 60 was developed to calculate the enthalpy effects for binary alloys containing at least one transition element. The results of the calculated enthalpy of solution are listed in Table 1/1. These values clearly indicate a considerable solubility and possible compound formation only for the U−Be system [13].

Table 1/1
Enthalpy of Solution at Infinite Dilution, $\Delta\overline{H}^\circ$ (in kJ/mol) [13].

U in	Li	Na	K	Rb	Cs	Be	Mg	Ca	Sr	Ba
	+131	+232	+282	+288	+291	−32	+55	+128	+155	+163

References for 1.1:

[1] Hume-Rothery, W. (Inst. Metals Monograph No. 1 [1945]).
[2] Mott, B. W. (Chem. Soc. [London] Spec. Publ. No. 22 [1967] 92/113).
[3] Darken, L. S.; Gurry, R. W. (Physical Chemistry of Metals, McGraw-Hill, New York 1953).
[4] Hildebrand, J. H.; Scott, R. L. (The Solubility of Non-Electrolytes, Reinhold, New York 1950).
[5] Mott, B. W. (Phil. Mag. [8] **2** [1957] 259/83).
[6] Strauss, S. W.; White, J. L.; Brown, B. F. (Acta Met. **6** [1958] 604/6).
[7] Strauss, S. W. (Acta Met. **10** [1962] 171/2).
[8] Miedema, A. R. (J. Less-Common Metals **32** [1973] 117/36).
[9] Miedema, A. R.; Boom, R.; de Boer, F. R. (J. Less-Common Metals **41** [1975] 283/98).
[10] Miedema A. R. (J. Less-Common Metals **46** [1976] 67/83).

[11] Miedema, A. R.; de Boer, F. R.; Boom, R. (CALPHAD **1** [1977] 341/59).
[12] Miedema, A. R.; de Châtel, P. F.; de Boer, F. R. (Physica B + C **100** [1980] 1/28).
[13] Niessen, A. K.; de Boer, F. R.; Boom, R.; de Châtel, P. F.; Mattens, W. C. M.; Miedema, A. R. (CALPHAD **7** [1983] 51/70).

1.2 The Influence of Nonmetals on the U−M Systems

The miscibility of metals in the liquid state can be strongly influenced by the presence of nonmetallic elements, which, depending on their chemical potentials and the temperatures, might cause the formation of compounds. These compounds would be in equilibrium with the solute in the liquid phase instead of the metallic solid phase. The solubility of such compounds is limited by the region of their chemical stability. The free energy of solution of such compounds is often much smaller than the values of the solution of the metals themselves, which causes a crossing of the two saturation curves of the compound and the metallic element, thus defining the border between the regions of existence for the two solid phases in equilibrium with the liquid solutions. The solubility of the metals tends to be extremely low at moderate temperatures.

The nonmetallic elements O, N, and C may interfere with the U−M systems. U forms very stable oxides, nitrides, and carbides, which cannot be reduced by heavier alkali or even alkaline earth metals. On the other hand, these liquid metals often contain the nonmetallic elements as contaminants, which are very difficult to remove by purification techniques. U is a very effective getter and can be applied to remove O and N, and to a lesser degree C, from the liquid metals by chemical reactions. The levels of impurities in liquid metals gettered in this way are very low. Thus, true solubility of metals in the liquid metals can only be observed after such a purification treatment.

The distribution of the nonmetallic elements between the liquid phase and solid U is governed by the thermochemical data of the compounds in the systems and by the solubilities in the two metallic phases. The distribution of O between U and M is given by the distribution coefficient K defined by the equation

$$K = \frac{a_O(s)}{a_O(l)} = \frac{a_O^s(s)}{a_O^s(l)} \exp\left\{\frac{\Delta G_{MO}^{\circ} - \Delta G_{UO_2}^{\circ}}{RT}\right\},$$

in which $a_O(s)$ and $a_O(l)$ are the chemical activities of O in the solid and liquid phases, $a_O^s(s)$ and $a_O^s(l)$ the chemical activities at saturation, and ΔG_{MO}° and $\Delta G_{UO_2}^{\circ}$ the free energies of formation of the stable alkali or alkaline earth oxide and UO_2 [1]. The nitrides or carbides follow the same rule of equilibrium distribution.

Therefore, the free energies of formation and the solubilities of the compounds in the liquid phases M or in U allow the prediction of the possible formation of U compounds when U is in contact with liquid alkali or alkaline earth metals containing nonmetallic contaminants. Values available are listed in Table 1/2. The dissolution of U in the liquid phases M should be restricted to equilibria of the compounds with their solutions in M in systems where the formation of U compounds with the nonmetallic elements is favored.

Table 1/2
Free Energies of Formation of Oxides, Nitrides, and Carbides of U and M (in kJ/g-atom of nonmetal).

compounds	ΔG_f at	800 K	1000 K	Ref.
oxides	UO_2	−473.00	−441.41	[2]
	Li_2O	−468.86	−441.41	[2]
	Na_2O	−302.92	−274.89	[2]
	K_2O	−241.00	−209.20	[2]
	Rb_2O	−215.56	−185.79	[3]

 References for 1.2 on p. 4 1*

Table 1/2 (continued)

compounds	ΔG_f at	800 K	1000 K	Ref.
oxides	Cs_2O	−188.19	−154.78	[3]
	BeO	−520.28	−501.03	[2]
	MgO	−515.89	−493.92	[2]
	CaO	−552.50	−532.20	[2]
	SrO	−511.49	−491.20	[2]
	BaO	−481.16	−464.84	[2]
nitrides	UN	−231.12	−212.37	[4]
	Li_3N	−52.16	−24.09	[5]
	Be_3N_2	−211.29	−197.69	[2]
	Mg_3N_2	−151.46	−130.44	[2]
	Ca_3N_2	−143.83	−	[2]
	Ba_3N_2	−93.51	−69.45	[2]
carbides	UC	−93.97	−90.94	[6]
	Li_2C_2	−21.13	−11.55	[5]
	Na_2C_2	−2.20	8.32	[5]
	CaC_2	−44.35	−40.65	[2]

The free energies of formation of nitrides and oxides indicate that U picks up N and O out of the liquid alkali metals. The alkaline earth oxides, however, are more stable than UO_2. Thus, these metals can reduce UO_2, and even high concentrations of O cannot oxidize U to form UO_2. It is not known if alkaline earth uranates(IV) can be formed under certain conditions in the presence of both metals.

UN (see "Uranium" Suppl. Vol. C7, 1981, p. 6) is more stable than all the alkaline earth nitrides. N contents in these metals, therefore, interfere with the U−M systems. UC (see "Uranium" Suppl. Vol. C12, 1987, p. 9) is the most stable compound among the carbides existent in the U−M alloys. U getters C from the M phases owing to this fact. The formation of compounds containing more than one nonmetallic element, for instance carbide nitrides, may occur if several nonmetallic elements are present.

References for 1.2:

[1] Smith, D. L.; Kassner, T. F. (in: Draley, J. E.; Weeks, J. R., Corrosion by Liquid Metals, Plenum, New York 1970, pp. 137/49).
[2] Wicks, C. E.; Block, F. E. (U.S. Bur. Mines Bull. No. 605 [1963]).
[3] U.S. Bur. Mines Bull. No. 542 [1954].
[4] Shohoji, N.; Katsura, M.; Sano, T. (J. Nucl. Mater. **60** [1976] 52/8).
[5] Rumbaut, N.; Casteels, F.; Brabers, M. (in: Borgstedt, H. U., Material Behavior and Physical Chemistry in Liquid Metal Systems, Plenum, New York 1982, pp. 437/44).
[6] Alcock, C.; Grieveson, P. (Thermodyn. Nucl. Mater. Proc. Symp., Vienna 1963, pp. 563/79).

1.3   Uranium−Lithium

1.3.1   Phase Diagram

A complete phase diagram study of the U−Li system has not yet been done. Some data are available on the Li end of the binary system. The solubility of U in freshly distilled liquid Li was measured by analyses of the filtered samples taken after equilibration of the molten metal with solid uranium [1 to 3]. The graph of the published values is transformed into a log c_{sat} versus reciprocal absolute temperature diagram, which is shown in **Fig.** 1-1. These results were assessed in a general study on the solubility of metallic elements in molten alkali metals. The saturation equation log c_{satU}^{Li} (c in wt%) = $4.529 - 7274.8/T$ (T in K) was calculated from the single points of measured concentrations [4].

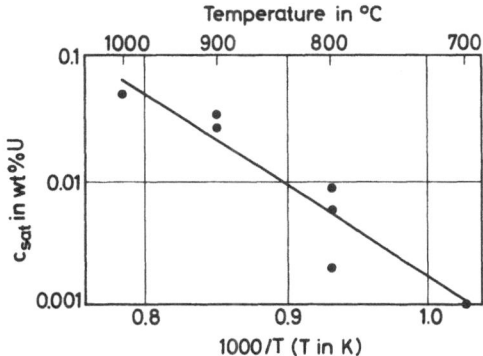

Fig. 1-1. The solubility of U in liquid lithium at 700 to 1000°C [1].

The fact that the Li was purified of nonmetallic impurities by equilibration with U at 700 to 1000°C is emphasized. In accord with thermodynamic considerations, this leads to the conclusion that their data are possibly influenced by the initial presence of impurities, most probably N. It is likely that the solubility of U in very pure Li is considerably lower than predicted by the above equation. The research work on the solubility of metallic elements in Li shows that the concentration of N in Li mainly influences the constant term in the equation. Thus, the free enthalpy of solution remains more or less uninfluenced. The calculation of this value, based on the equation, yields $\Delta H_{sol} = 139.26$ kJ/g-atom. The data on solubility of U in liquid Li are compiled in [2]. The solubility curve suggests that the solubility of U in solid Li should be extremely low, even if the constant is influenced by impurities in Li. At the melting temperature of Li (179°C) the saturation concentration, according to the above equation, should be about 2.7×10^{-11} wt% U. This concentration should be the upper limit of the U solubility in the solid alkali metal [2].

Getter experiments with U foils in molten Li show that the solubility of Li in solid U at temperatures between 700 and 1000 K is also extremely low. Contents of Li in the solid metal were not detected [10].

References for 1.3 on p. 7

1.3.2 Preparation of Alloys

In spite of the poor miscibility of both elements in the liquid and solid state, alloys consisting of two solid phases are known in the U−Li system. They were prepared by co-reduction of UF$_4$ and Li salts or oxides by a metallothermal process applying Ca or Mg [5]. The co-extrusion of U−Li alloys was also reported in the earlier work on nuclear fuel based on metallic U [6]. However, neither method has been used for the U−Li alloy fabrication for fuel elements. Liquid binary U−Li alloys are not mentioned in the literature. The preparation of suspensions of metallic U or the compounds UC, UN, UO$_2$, or U$_3$Si$_2$ as fine grained particles was successfully studied within the development program of liquid nuclear fuels.

1.3.3 Physical Properties

The systems consisting of the two metals as nearly pure phases are mixtures of the ordinary solid structures of the elements.

Physical properties of the solid U−Li alloys and of the suspensions of U or its compounds in liquid Li are not reported nor have the mechanical properties of the U−Li alloys been tested. There is a lack of diffusion and self-diffusion data owing to the very low solubilities of both metals in each other.

Data on the irradiation behavior of the experimentally prepared U−Li alloys are not published.

1.3.4 Chemical Reactions

The U−Li alloys tend to react with the nonmetallic elements C, N, and O owing to the reactivity of both constituents with these elements. The reactions occur even if the nonmetals are only present at very low chemical potentials. They also react with several carbides, nitrides, and oxides at sufficiently high temperatures. As already pointed out (see Section 1.2, p. 3), the compounds UC, UN, and UO$_2$ are more or less stable against Li metal.

The behavior of the alloys against materials such as austenitic or ferritic steels or refractory alloys are not reported. Conclusions can be drawn from the compatibility of the two elements with those materials.

Pure Li metal is sufficiently compatible with cladding materials such as austenitic or ferritic steels, or alloys based upon Nb, V, or Mo, up to temperatures of about 500°C. The solubility of metallic elements in Li, for instance Ni, or the exchange reactions of nonmetallic elements such as C or N cause corrosion at higher temperatures. Only Mo seems to be resistant even at elevated temperatures [11].

Corrosion can be enhanced at temperatures above ca. 700°C in the presence of solid U, which forms intermetallic compounds and eutectic mixtures with several metallic elements. The U metal acts as a sink for Fe, Ni, Cr, etc., thus causing a mass transfer of these metals between the cladding material and the getter. Such reactions were observed in the Li−Zr-steel system in which Zr plays the role of U [7].

1.3.5 Use of U−Li Alloys

The experimentally prepared U−Li alloys were considered to serve as nuclear fuel alloys. Solid or liquid fuels based on U metal are not used in nuclear reactors because of the success of the oxide fuels. Other applications of U−Li alloys are not known.

1.3.6 Ternary Systems

The addition of an alkali metal to liquid Bi raises the solubility of U in this molten metal [8]. Thus, ternary liquid fuel alloys consisting of one Li−Bi−U phase were made.

Liquid fuels containing suspended solid phases such as metallic U, or U carbide, nitride, oxide, or silicide were prepared according to an early patent [8]. The particles were mixed with liquid alloys such as Pb−Li or Bi−Li. The addition of Li is advantageous because it lowers the density of the liquid, resulting in increased stability of the suspension. Natural Li, however, cannot be used in liquid nuclear fuels because of its high cross section for thermal neutrons [9].

References for 1.3:

[1] Bychkov, Yu. F.; Rozanov, A. N.; Yakovleva, V. B. (At. Energiya SSSR **7** [1959] 531/6; Kernenergie **3** [1960] 763/7).

[2] Bychkov, Y. F.; Rozanov, A. N.; Rozanova, V. B. (in: Yemel'yanov, V. S.; Yestynkin, A. I., Metallurgy and Metallography of Pure Metals, Gordon and Breach, New York 1962, pp. 178/89).

[3] Ivanov, O. S.; Badaeva, T. A.; Sofronova, R. M.; Kishenevskii, V. B.; Kushnir, N. P. (Phase Diagrams of Uranium Alloys, Amerind, New Delhi 1983, translated from Diagrammy Sostoyaniya i Fazovye Prevrashcheniya Splavov Urana, Nauka, Moscow 1972).

[4] McKisson, R. L.; Eichelberger, R. L.; Dahleen, R. C.; Scarborough, J. M.; Argue, G. R. (NASA-CR-610 [1966] 1/153; N.S.A. **21** [1967] No. 1891).

[5] Filipovic, M.; Stojsic, I.; Milosavjevic, J. (Tehnica [Belgrade] **22** [1967] 1366/70).

[6] Losco, E. F.; Shapiro, Z. M. (WAPD-PWR-PMM-282 [1955] 1/67; N.S.A. **11** [1957] No. 10901).

[7] Borgstedt, H. U. (J. Nucl. Mater. **103/104** [1981] 693/7).

[8] Wissenschaftlich-Technisches Büro für Reaktorbau (French Patent of Addition No. 75699, Addition to French Patent No. 1225153 [1961]).

[9] Frost, B. R. T. (Nucl. Eng. **1** [1956] 334/9).

[10] Besman, T. M.; Cooper, R. H. (ORNL-TM-9662 [1985] 1/13; INIS Atomindex **17** [1986] No. 007119).

[11] Leavenworth, H. W.; Cleary, R. E. (Acta Met. **9** [1961] 519/20).

1.4 Uranium−Sodium

1.4.1 Phase Diagram

A phase diagram of the U−Na system has not been established. A rough estimation of the solubility of U in liquid Na is based on cryoscopic measurements at the melting point of the alkali metal. The result of this estimation, a saturation concentration $c_s \cong 0.007$ wt% U in Na at 97.8°C is not precise since the cryoscopic method is not specific for U in Na. Other impurities, such as nonmetallic elements, may contribute to the melting temperature depression [1, 2]. The very low solubility of U in liquid sodium suggests that the solubility of U in solid Na is extremely low.

It is obvious from gettering reactions of liquid Na with solid U that the alkali metal does not dissolve in U in measurable concentrations at about 550°C [4].

Corrosion studies by exposure of U sheets to not agitated Na or NaK alloy at 500°C, using alkali metal melts, reduced by gettering procedures with Be or Ca, showed the solubility of U in alkali metals to be much lower than that concluded from the cryoscopic study [3]. The

References for 1.4 on p. 11

solubility is probably about the same as in molten Li. The U $-$ Na system is extremely sensitive to traces of O in the metals. U reduces the Na_2O dissolved in Na to form solid UO_2. O or oxide can dissolve in the U metal and diffuse into the matrix if the amounts of oxide are small enough. In Na with higher contents of oxide, a solid surface layer of UO_2 is formed. The concentrations in Na in equilibrium with U are reduced to very low values in the order of 1 µg/g [4].

Even in Na of gettered purity, U dissolves as an O bearing compound [5]. The formation of complex Na uranate(IV) is not favored under these conditions [6]. The study of the Na corner of the Na $-$ U $-$ O system shows that a new phase, the Na uranate(V), Na_3UO_4, is formed at O concentrations above 2 at%. This compound, with a face-centered cubic structure (a = 4.79 to 4.80 Å) (see "Uranium" Suppl. Vol. C3, 1975, p. 16), might be in equilibrium with the solute in the liquid metal. **Fig.** 1-**2** shows the isothermal section at 850°C of the Na-rich region of the Na $-$ U $-$ O equilibrium diagram composed of the O:U = 2 and O:U = 4 curves and the Na_3UO_4-curve [7].

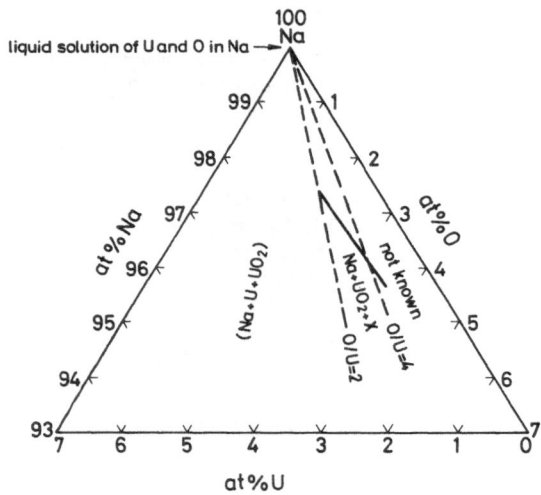

Fig. 1-2. Tentative isothermal section, at 850°C, of the Na-rich region of the Na $-$ U $-$ O equilibrium diagram [7].

1.4.2 Preparation of Alloys

Liquid alloys of U or its compounds with Na can be made by suspending uranium particles of less than 5 µm diameter in the molten alkali metal [8]. The content of the solid particles is limited to 20% of the total volume. The compounds claimed in the cited patent are UC, UN, UO_2, and USi. The particles in these slurries of solid materials in Na or the NaK alloy tend to increase in size with time. The observed growth rates of the grains agree reasonably well with a theory based on the reduction of the interfacial energy [9].

A solid two-phased alloy of the two elements was prepared by infiltration of the liquid alkali metal into a porous body of U metal powder formed by vibrational compacting [10].

1.4.3 Physical Properties

The components U and Na are present in the solid alloys as nearly pure phases owing to their immiscibility. The alloys are mixtures of crystals of U and Na with their ordinary structures [8].

Mechanical properties of the solid U — Na alloys, which were developed for application in the molten state, are not reported. Some tests showed that U — Na slurries with up to 5 vol% solid particles behave as liquids when pumped through circuits of pipes. An addition of solid particles raises the pressure drop of the alkali metal at a given flow rate. The rheological behavior is modified by variations of the concentration of the particles as well as by their shape and size. The changes of the interfacial energy between the solid and liquid influence the properties of the liquid carrier. The slurries pumped through the piping of a loop remain uniform in their composition [11].

Although diffusion data were not reported, the growth of U particles in U — Na slurries containing powdered U indicates some diffusion-controlled transport process in the liquid Na [12].

Suspensions or slurries of U metal in Na, or the NaK eutectic alloy behave under irradiation like their constituents. The molten alkali metals have the capacity, however, to dissolve some of the fission products formed by the irradiation of these fuels. Isotopes of alkali or alkaline earth metals and of halogens or chalcogens are dissolved by Na or NaK to a certain degree [11].

1.4.4 Chemical Reactions

Both elements of the U — Na systems form compounds with nonmetallic elements with which they come into contact. According to the higher stability of U compounds, U getters nonmetals such as C, N, or O out of liquid Na. Additions of Mg to such slurries or liquid ternary alloys prevent the formation of U compounds and improve the wetting behavior of Na. UO_2, UN, and UC are chemically stable towards Na, even highly purified Na. Suspensions of these compounds in Na as slurries containing not more than 5 vol% of solid particles do not undergo chemical reactions [12].

Above 500°C the slurry flocculates and cannot be maintained in a uniform state even at high flow rates. The slurry is resuspended by lowering the temperature to ca. 480°C. The flocculation is due to contaminating Na_2O. The addition of 0.1 vol% U metal reduces it completely to Na. In the purified slurry no settling occurs at temperatures of 500 to 600°C [13].

Hyperstoichiometric UC exposed to liquid Na with less than 1 μg/g O loses C at 700°C. The rate of decarburization of cast carbide specimens is 2×10^{-4} mm/h. It is obvious that the decarburization is controlled by dissolution of C in the UC_2 phase, whereas the U_2C_3 phase is much more stable against decarburization by Na [14].

The compatibility of container materials such as austenitic or ferritic steels or refractory alloys with liquid alloys, suspensions, or slurries of U in Na (or NaK) is similar to that with the alkali metals themselves. In the presence of solid metallic U, however, the slurries are somewhat more aggressive toward iron base alloys owing to the formation of intermetallic compounds of U with Fe, Ni, or Cr. U acts as a sink for these metals, thus causing a mass transfer of these elements through the Na between the steels and the U particles [13].

Some corrosion reaction data taken from [3] for U and its alloys in Na or NaK alloy in static systems are listed in Table 1/3, p. 10.

References for 1.4 on p. 11

1.4.5 Uses of U−Na Alloys

The efforts to prepare U−Na alloys or slurries were directed to their application as nuclear (liquid or solid) fuels. Such fuels, however, were only fabricated for experimental fuel elements. They have not yet been used in practice as reactor fuel.

1.4.6 Ternary Systems

Ternary systems such as liquid alloys U−Bi−Na (NaK) were considered as liquid nuclear fuels. Such molten alloys are advantageously circulated in reactor systems such as graphite-moderated reactors. The compatibility of such liquids and the technology of their reprocessing were studied, however, the development was suspended before a reactor was constructed [13].

The solubility of U in liquid alloys of Na with Bi is lower than in pure Bi and does not show a significant dependence on the content of Na in the solute as is pointed out in Table 1/4 [18].

Table 1/3
Corrosion Data for U and its Alloys in Na and NaK in Static Systems [3].

material	temperature in °C	time in h	purification	weight change in $mg \cdot cm^{-2} \cdot month^{-1}$	
				in Na	in NaK
U	200	144	a		−0.08
U	611	144	a		+3.1
U	405	160	a		+7
U	396	160	a		+1.4
U	500	166	b	−0.1	
U-10Pu	350	160	c		−216
U-10Pu	405	160	c		−288
U-60Al	450	124	b	+30	
U-60Al	500	170	b	+8	
U-60Al	660	164	b	−720	

a) Alkali melt reduced by gettering with U. − b) Alkali metal reduced by gettering with Be or Ca. − c) Alkali metal purified by double filtering.

Table 1/4
Solubility of U in Na−Bi Liquid Alloys Bi78Na22 and Bi68Na32 [18].
The solubility of U in Bi is shown for comparison.

temperature in °C	solubility in at% U in		
	Bi	Bi78Na22	Bi68Na32
500	0.9	−	0.2
550	1.6	−	0.2
600	2.2	0.9	0.6
650	3.1	1.0	0.9
700	4.2	1.2	1.2
750	5.3	1.6	1.7
800	7.1	2.3	−

The addition of alkali metals to the U — Bi alloys with ca. 0.1 wt % U increases the solubility of U in liquid Bi. A phase diagram of this ternary system was never established [15]. For instance, the addition of Na to liquid U — Bi alloys raises the solubility of U in Bi to maximum at 2 wt % Na and decreases it as the Na content is raised to 8 wt % [16]. The addition of Na to Bi — UO_2 stabilizes the slurries by reducing the density of the liquid component of the suspension [17].

References for 1.4:

[1] Douglas, T. B. (AECD-3254 [1951] 1/13; N.S.A. **5** [1951] No. 6756; J. Res. Natl. Bur. Std. **52** [1954] 223/6).
[2] Ivanov, O. S.; Badaeva, T. A.; Sofronova, R. M.; Kishenevskii, V. B.; Kushnir, N. P. (Phase Diagrams of Uranium Alloys, Amerind, New Delhi 1983, translated from: Diagrammy Sostoyaniya i Fazovye Prevrashcheniya Splavov Urana, Nauka, Moscow 1972).
[3] Pearlman, H. (TID-7546-Book 2 [1957] 565; N.S.A. **12** [1958] No. 9417).
[4] Isaacs, H. S. (ANL-7520-Pt-I [1968] 465/70; J. Nucl. Mater. **36** [1970] 322/30).
[5] Caputi, R. W.; Adamson, M. G. (Proc. 2nd Intern. Conf. Liquid Metal Technol. Energy Prod., Springfield, Va., 1980, Vol. 2, pp. 18 — 62/18 — 69; C.A. **94** [1981] No. 164350).
[6] Addison, C. C.; Barker, M. G.; Pulham, R. J. (J. Chem. Soc. **1965** 4483/9).
[7] Tepper, R. T.; Stubbles, J. R.; Tottle, C. R. (Appl. Mater. Res. **3** [1964] 203/7).
[8] Abraham, B. M.; Flotow, H. E. (Nucl. Eng. **4** [1959] 188; Brit. 801288 [1959]).
[9] Greenwood, G. W. (Acta Met. **4** [1956] 243/8).
[10] Kieffer, P.; Sedlatschek, K. (Planseeber. Pulvermet. **5** [1957] 104/19).

[11] Grimes, W. R.; Bohlmann, E. G.; Kirkbride, L. D. (Proc. 3rd Intern. Conf. Peaceful Uses At. Energy, Geneva 1964, Vol. 11, pp. 256/63).
[12] Greenwood, G. W.; Sharpe, B. (AERE-M-R-2250 [1954/57] 1/29; N.S.A. **11** [1957] No. 13486).
[13] Abraham, B. M.; Flotow, H. E.; Carlson, R. D. (Nucl. Sci. Eng. **2** [1957] 501/12).
[14] Watanabe, H.; Kurusawa, T.; Kikuchi, T.; Furukawa, K.; Nihei, I. (J. Nucl. Mater. **40** [1971] 213/20).
[15] Waide, C. H.; Kukacka, L. E.; Meyer, R. A. et al. (BNL-736 [1961] 1/40; N.S.A. **17** [1963] No. 14765).
[16] Brookhaven National Lab.; Upton, N.Y. (BNL-4261 [1959] 1/96; N.S.A. **13** [1959] No. 16621).
[17] Hahn, T. (Ind. Eng. Chem. **51** [1959] 197/9).
[18] Hayes, E. E.; Gordon, P. (TID-2501 [1951] 115/26; N.S.A. **12** [1958] No. 17285).

1.5 Uranium — Potassium

A phase diagram of the U — K system does not exist. Even information on the solubilities of either U in liquid K or K in molten U is not available. Some work on slurries of U metal powder in the eutectic Na — K alloy (with 67.3 at% K) indicates that the system U — K behaves similarly to the system U — Na [1].

Solid U does not dissolve measurable amounts of K even at elevated temperatures. The solid solubility of U in K should be extremely low, as it is in solid Na [see Section 1.4.1, p. 7]. Structural data and physical properties of U — K alloys are not known.

A real solubility of U in liquid and solid K cannot be measured since an oxidation product KUO_3 is formed even at very low oxygen concentration in K. This is shown in tests with solid electrolyte oxygen meters [4].

References for 1.5 to 1.7 on p. 13

Wetting of UO_2 particles by the NaK eutectic alloy was studied by measuring contact angles between droplets of the liquid alloy and UO_2 plaques. This wetting occurs at elevated temperatures. The contact angles between UO_2 and NaK alloy are larger than zero, indicating flocculation of UO_2 in the liquid metal. Oxide contents of the alkali metal enhance the flocculation of slurries [2].

The UO_2-Na-K slurry does not cause erosion or corrosion of stainless steel tubing when circulated in a high-temperature loop. There is also no appreciable comminution of the UO_2 particles [2].

In-pile tests with slurries show that the fission product Cs, which is soluble in NaK alloy, is always found in liquid solution, while Ce and Zr, two insoluble fission products, are fixed on the UO_2 particles. The fission products do not influence the circulation of the slurry, although their formation degrades the UO_2 particles. The addition of small amounts of metallic U stabilizes the UO_2-Na-K slurry by gettering O out of the liquid metal [3].

1.6 Uranium – Rubidium

Although there is no doubt that the U−Rb system behaves as the other U-alkali metals systems, particularly U−Cs, data are not available. Since Rb is one of the fission elements, its alloying behavior with U is of some importance for the formation of phases in irradiated U metal fuel.

1.7 Uranium – Caesium

A phase diagram or even information on the solubility of the two elements in each other is not available. Some knowledge of slurries containing Cs as a fission product, however, indicates that the U−Cs system should be similar to the U−K system. The fission element Cs is always detected in the liquid metal when irradiated U−Na or U−Na−K slurries are analyzed. Thus, one can conclude that Cs is insoluble in solid U as are the other alkali metals [3].

Cs is known as a component of ternary liquid alloys. For instance, additions of Cs to U−Bi liquid alloys raise the solubility of U in Bi. As is shown in **Fig. 1-3**, additions of 1.0 or 1.5 wt% Cs increase the saturation concentrations of U in Bi by 25 to 50% between 400 and 500°C [5].

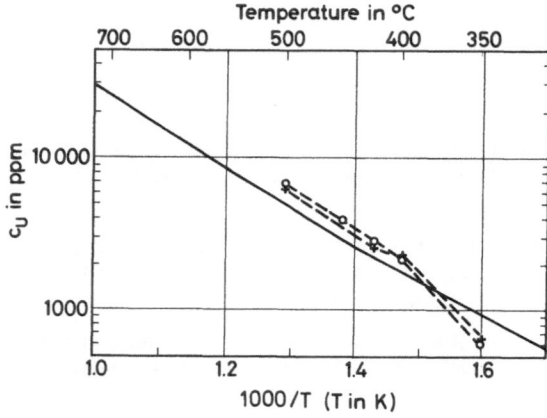

Fig. 1-3. Effects of Cs on the solubility of U in liquid Bi [5].
+ Bi + 1.0 wt% Cs
○ Bi + 1.5 wt% Cs
—— without Cs

References for 1.5 to 1.7:

[1] Grimes, W. R.; Bohlmann, E. G.; Kirkbride, L. D. (Proc. 3rd Intern. Conf. Peaceful Uses At. Energy, Geneva 1964, Vol. 11, pp. 256/63; N.S.A. **18** [1964] No. 4323).

[2] Abraham, B. M.; Flotow, H. E.; Carlson, R. D. (Proc. 2nd Intern. Conf. Peaceful Uses At. Energy, Geneva 1958, Vol. 7, pp. 166/72; N.S.A. **12** [1958] No. 14605).

[3] Carlson, R. D.; Sowa, E. S. (ANL-6644 [1962] 1/69; N.S.A. **17** [1963] No. 11077).

[4] Adamson, M. G.; Aitken, E. A.; Jeter, D. W. (USDE-CONF-760503-P2 [1976] 866/74; C.A. **87** [1977] No. 108079).

[5] Brookhaven National Lab., Upton, N.Y. (BNL-4261 [1959] 1/96; N.S.A. **13** [1959] No. 16621).

1.8 Uranium — Francium

There is nothing published on alloys or intermetallic compounds of U and the unstable alkali metal Fr. The general trends, however, suggest the existence of intermetallic compounds of these elements to be unlikely. The solubility of U in molten Fr or of Fr in molten U should be comparable to the systems of U with other heavy alkali metals. The saturation concentrations are expected to be extremely low. This is in agreement with thermodynamic data of U — Rb and U — Cs systems.

2 With Alkaline Earth Metals

Hans Ulrich Borgstedt,
Institut für Material- und Festkörperforschung,
Kernforschungszentrum Karlsruhe,
Karlsruhe, Federal Republic of Germany

Rules for the miscibility of uranium with alkali and alkaline earth metals and the influence of nonmetals on the U−M systems are discussed in Chapters 1.1 and 1.2 on pp. 1 and 3, respectively.

2.1 Uranium − Beryllium

2.1.1 Phase Diagram

The phase diagram (see **Fig. 2-1**) of the U−Be system was established on the basis of thermal analysis, metallographic examination and X-ray diffraction of as-cast samples [1]. The system is characterized by one intermetallic compound, UBe_{13}, containing 7.14 at% (or 67 wt%) U. The compound has a melting point of ca. 2000°C and has a cubic structure. For lattice constants see on p. 22.

The formula UBe_{13} was checked by determination of the density as well as by chemical analysis. The density at 26°C is D = 4.420 ± 0.002 g/cm³ and the Be content is 32.15 ± 0.05 wt% [1]; both values are close to the computed ones. Thus, the composition UBe_9, as previously reported [2], is not in accordance with the structure and the composition of the intermetallic compound.

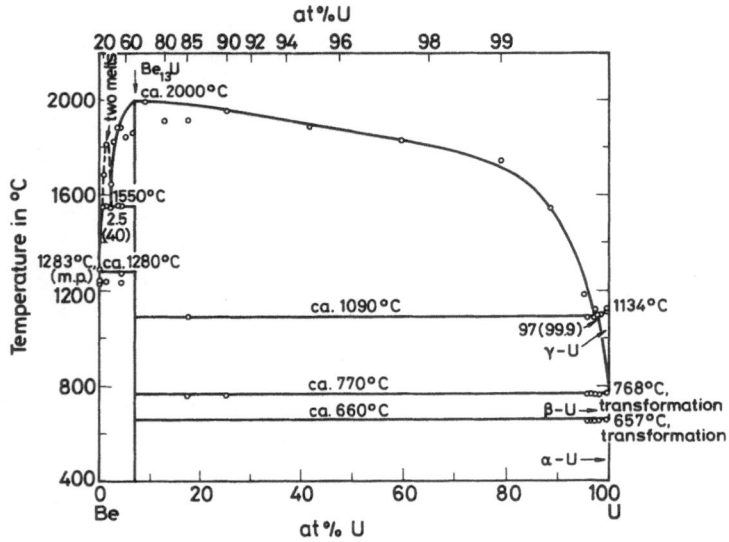

Fig. 2-1. The phase diagram of the U−Be system [3].

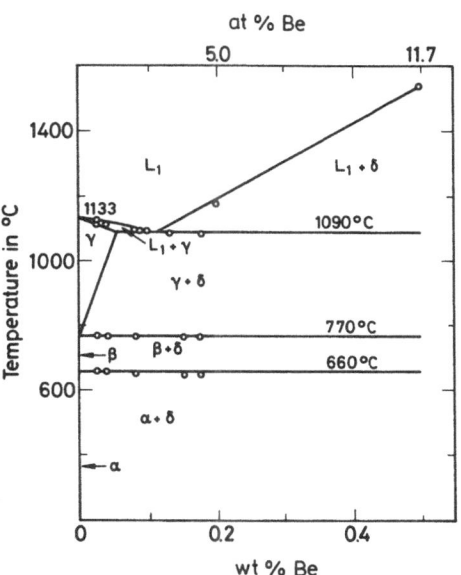

Fig. 2-2. U-rich section of the U—Be phase diagram [1].

A eutectic having a composition of approximately 0.12 wt% of Be occurs between U and UBe₁₃ at 1090°C, and a monotectic is observed at 1550°C and ca. 60 wt% Be [1]. The constitution of the partial system Be—UBe₁₃ seems to be somewhat doubtful, since thermal analysis at temperatures above 1300°C cannot detect effects indicating the formation of two liquids [3], see [4, 5]. The U-rich section of the system is shown in **Fig. 2-2**. The solubility of U in liquid Be in the temperature range between its melting point $t_m = 1287$°C and 1550°C is relatively low [1]. At 1550°C Be dissolves ca. 17.8 at% ($=$ ca. 0.9 wt%) U, at 1290°C, the saturated solution of U in molten Be contains ca. 10 at% ($=$ ca. 0.4 wt%) U. The solubility of U in solid Be should be considerably lower [1]. The solubility of Be in molten U is somewhat higher. At the eutectic temperature of 1090°C the saturated solution contains 0.12 wt% (2.7 at%) Be. γ-U dissolves some Be, about 0.06 wt% (1.6 at%) as determined by an X-ray investigation of quenched alloys at 1075°C [1]. Measurements of the solubility of Be in α- and β-U indicate much lower concentrations than those given above. The presence of Be does not stabilize the β phase of U at lower temperatures as do some other alloying elements [6]. The transformation temperatures of U are not affected by additions of small amounts of Be. The constitution of the U—Be system is also described in [7 to 10].

References for 2.1.1:

[1] Buzzard, R. W. (J. Res. Natl. Bur. Std. **50** [1953] 63/7).
[2] Battelle Memorial Institute, Manhattan Project (CT-1009 [1943]).
[3] Hansen, M.; Anderko, K. (Constitution of Binary Alloys, McGraw-Hill, New York 1958, p. 299).
[4] Bellamy, R. G.; Hill, N. A. (Extraction and Metallurgy of Uranium, Thorium, and Beryllium, Pergamon, Oxford 1963, p. 165).
[5] Rough, F. A.; Bauer, A. A. (Constitutional Diagrams of Uranium and Thorium Alloys, Addison-Wesley, Reading, Mass., 1958, pp. 1/153).

[6] Semenchenkov, A. T.; Ivanov, O. S. (Stronie Svoista Splavov Urana Toriya Tsirkoniya **1963** 22).

[7] Hultgren, R.; Desai, P. D.; Hawkins, D. T.; Gleiser, M.; Kelley, K. K. (Selected Values of Thermodynamic Properties of Binary Alloys, Am. Soc. Metals, Metals Park, Ohio, 1973, pp. 190/1).

[8] Massalski, T. B. (Binary Alloy Phase Diagrams, Vol. 1, Am. Soc. Metals, Metals Park, Ohio, 1985, pp. 482/3).

[9] Ivanov, O. S.; Badaeva, T. A.; Sofronova, R. M.; Kishenevskii, V. B.; Kushnir, N. P. (Phase Diagrams of Uranium Alloys, Amerind, New Delhi 1983, translated from: Diagrammy Sostoyaniya i Fazovye Prevrashcheniya Splavov Urana, Nauka, Moscow 1972).

[10] Vol, A. E. (Handbook of Binary Metallic Systems, Structure and Properties. Vol. I, Physicochemical Properties of the Elements. Systems of Actinium, Aluminium, Americium, Barium, Beryllium, Boron, and Nitrogen, Jerusalem 1966, pp. 453/5, translated from Russian, Israel Program for Scientific Translations).

2.1.2 Preparation of Alloys

Alloys of U with Be were formed by solid state sintering of mixed U and Be powders. U powder (ca. 67 wt%) and Be powder (ca. 33 wt%) of a particle size of ca. 60 µm were cold- or hot-compacted to 80 to 85% of the theoretical density and sintered in vacuum at 950 to 1050°C. The product is UBe_{13} as shown by X-ray diffraction. During the sintering process, the alloy shows an anomalous expansion causing a decrease of its density by more than 60%. Less expansion is observed at lower sintering temperatures (950°C) and with U—Be alloys of different compositions, which contain the compound UBe_{13} mixed with the metals [1]. **Fig. 2-3** shows the effect of the alloy composition on this anomalous expansion [3]. This is due to the development of diffusional porosity in one of the intermetallic phases formed during the homogenization process [2]. This porosity is a consequence of unequal diffusion rates of the two types of constituent atoms in this phase. One method of preventing the expansion problem is to find the phase which causes this porosity, and then to use the appropriate quantities of this phase and one elemental powder to produce the desired alloy [3]. The linear thermal

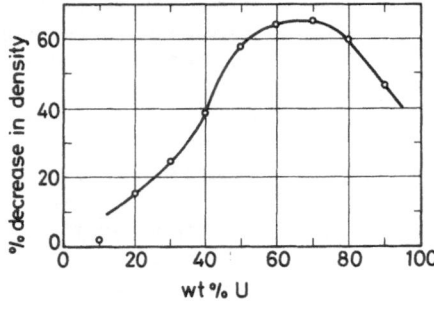

Fig. 2-3. Influence of the composition of U—Be alloys on the decrease in density (due to porosity) during the sintering at 1050°C [3].

expansion of compacted powders of binary metallic mixtures may be applied to detect reactions of the components during the sintering [4].

The production of U−Be alloys is also possible by reactions of a uranium halide with Be at 1300°C to form the U metal, which constitutes an alloy with the excess of the Be metal. Although the U halides are thermodynamically more stable than the Be halides, the reducing reaction proceeds to completion owing to the fact that the Be halide produced is removed by vacuum distillation. [5]. Repeated arc-melting was applied to gain a homogeneous product [9]. The diffusion porosity in the U−Be alloys indicates the higher diffusion rate of Be. This is in contradiction to the rule that the component with the lower melting point should have the higher diffusion rate [8].

In another process, powdered UBe_{13} and Be were blended, compacted and then sintered by rapidly heating to a temperature in the range from 1220 to 1280°C in an inert gas atmosphere in order to get high-density compacts of U−Be alloys [6].

Beryllides of U and their solid solutions were prepared by reaction-sintering, reaction-pressing and arc-melting of the mixed metal powders. The major problems are due to the large swelling during sintering and the losses of Be at temperatures above 1200 to 1300°C. Reaction-pressing avoids these difficulties and yields the purest and most dense products [7].

Despite the high vapor pressure of Be at elevated temperatures, U−Be alloys with compositions ranging from 0 to 100% Be were prepared in high-frequency induction furnaces. The mixtures were melted and then the ingots poured in an atmosphere of purified He. The beryllium oxide crucible and mold assemblies were enclosed in a silica tube. The high-purity beryllium oxide was prefired at approximately 1900°C. The pouring temperatures were determined by sighting through a "sight" tube at the bottom of the stopper (made of beryllium oxide) with an optical pyrometer. The density of the UBe_{13} phase is considerably higher than that of the liquid phase containing Be as the main component. Thus, the compound has a tendency to settle out at the bottom of the crucibles. The alloys were kept in uniform distribution by vigorous stirring before pouring the ingots. The distribution of U in the ingots was checked by chemical analysis as well as by microradiographic examination. Nonuniformity of the ingots was easily detected by examination of radiographic sections. The loss of Be by evaporation must be considered [10].

References for 2.1.2:

[1] Williams, J.; Jones, J. W. S. (AERE-M-R-1974 [1956] 1/27; N.S.A. **10** [1956] No. 10182).
[2] Williams, J.; Jones, J. W. S. (AERE-M-R-1983 [1956]).
[3] Lloyd, H.; Williams, J. (Proc. 2nd Intern. Conf. Peaceful Uses At. Energy, Geneva 1958, Vol. 6, pp. 426/37).
[4] Raub, E.; Plate, W. (Z. Metallk. **42** [1951] 76/82).
[5] Runnalls, O. J. C.; U.S. At. Energy Comm. (U.S. 2875041 [1959]; N.S.A. **13** [1959] No. 22503).
[6] Angier, R. P.; U.S. At. Energy Comm. (U.S. 2979399 [1961]; N.S.A. **15** [1961] No. 13325).
[7] Snyder, M. J.; Tripler, A. B., Jr. (ASTM Spec. Tech. Publ. No. 276 [1959] 293/300).
[8] Kittel, H. (ANL-4937 [1949] 1/29; N.S.A. **10** [1956] No. 5367).
[9] Farkas, M.; Eldridge, E. (J. Nucl. Mater. **27** [1968] 94/6).
[10] Buzzard, R. W. (J. Res. Natl. Bur. Std. **50** [1953] 63/7).

2.1.3 UBe$_{13}$

2.1.3.1 Preparation and Formation

UBe$_{13}$ can be prepared by heating a mixture of UH$_3$ and an excess of Be metal in an Ar atmosphere at 1550°C. The fine powders obtained by crushing and grinding are hydrostatically pressed in order to get dense compacts. These are finally sintered in vacuum at 1550°C [1].

The reaction of Be with UO$_2$ also produces the compound UBe$_{13}$ at temperatures above 600°C. Mixed powders of the components completely react in 28 days at 700°C to form the intermetallic and BeO [2]. Kinetics of the formation of UBe$_{13}$ were measured in the temperature range 600 to 1000°C; the intermetallic phase grows into the Be. The growth of the UBe$_{13}$ phase is diffusion-controlled [3, 4].

The compound UBe$_{13}$ was detected even at the interfaces of contacted plates of both metals at temperatures above 600°C as pointed out in a survey article [5]. The process is diffusion-controlled, the UBe$_{13}$ layer between the two metals grows according to $x^2 = k \cdot t$ with $k = 1.4 \times 10^{-12}$ cm^2/s, indicating volume diffusion (x = layer thickness in cm) [6].

The activation energy of the formation of UBe$_{13}$ at 600 to 650°C is estimated as $\Delta H_{act} = 22.8$ kcal/mol (96 kJ/mol) [6, 7].

UBe$_{13}$ is also formed by reaction of hot-pressed Be with arc-melted UC in the temperature range 700 to 1000°C. A layer of UBe$_{13}$ mixed with a non-identified precipitate grows between discs of Be and UC where they are pressed together. The growth of the layer is controlled by the diffusion of Be into UBe$_{13}$ and occurs at the UBe$_{13}$–UC interface. The activation energy of the reaction is 33.0 kcal/mol (138.9 kJ/mol) [8].

Single phase samples of UBe$_{13}$ with a lattice constant a = 10.254 Å were prepared by fusing the two components in a high-frequency induction furnace. Beryllium oxide was used as crucible material [9]. Pieces of U metal were cleaned by electrolytic polishing in H$_3$PO$_4$. These pieces and electrolytic Be powder were used to prepare the compound (33.09 wt% Be and 66.91 wt% U) in a crucible of BeO in a quartz ampule in pure hydrogen at 1300°C for 1.5 h. Hydrogen was pumped out at 600°C and the cooled product was ground in purified Ar. The beryllide is single-phased and sufficiently pure [10].

UBe$_{13}$ films were prepared by means of vapor codeposition of U and Be onto heated sapphire substrates [11].

References for 2.1.3.1:

[1] Baird, J. D.; West, K. B. C. (A-862 [1958] cited in [54]).
[2] Knapton, A. G.; West, K. B. C. (J. Nucl. Mater. **3** [1961] 239/40).
[3] Hanna, G. L. (Inst. Metals Monograph Rept. Ser. No. 28 [1963] 350).
[4] Röllig, H. E. (Kernenergie **5** [1962] 641/68).
[5] Kittel, H. (ANL-4937 [1949] 1/29; N.S.A. **10** [1956] No. 5367).
[6] Baird, J. D.; Geach, G. A.; Knapton, A. G.; West, K. B. C. (Proc. 2nd Intern. Conf. Peaceful Uses At. Energy, Geneva 1958, Vol. 5, pp. 328/33).
[7] Murdock, J. F. (J. Nucl. Mater. **7** [1962] 192/6).
[8] Lewis, J. R. (J. Metals **13** [1961] 357/62).
[9] Alekseevskii, N. E. (Pis'ma Zh. Eksperim. Teor. Fiz. **40** [1984] 66/9; JETP Letters **40** [1984] 800/3).
[10] Samorukov, O. P.; Kostyukov, V. N.; Kostylev, F. A.; Tumbakov, V. A. (At. Energiya SSSR **37** [1974] 28/31; Soviet At. Energy **37** [1974] 705/8).
[11] Tedrow, P. M.; Quateman, J. H. (Phys. Rev. [3] B **34** [1986] 4595/8).

2.1.3.2 Heat of Formation

The heat of formation of the beryllide UBe_{13} was determined by measuring the heat of solution of UBe_{13} and of the corresponding mixture of the components [1, 3 to 5]. The calorimetric method revealed a value of the standard heat of formation, $\Delta H^\circ_{298} = 39.3 \pm 3.8$ kcal/mol ($= 165.45 \pm 16.0$ kJ/mol). The results are only slightly influenced by impurities present in the intermetallic. Oxygen contents and inaccurate U:Be relationships have the largest influence on the accuracy of the estimation. The standard entropy is calculated from these data, $S_{298} = (41.5 \pm 4)$ e.u. [1]. The values of H and S are included in reviews on thermodynamics of intermetallic systems of actinides [2, 4].

For heat capacity and heat content of UBe_{13} see [6], see Section 2.1.3.3.9, p. 33.

References for 2.1.3.2:

[1] Ivanov, M. I.; Tumbakov, V. A. (At. Energiya SSSR **7** [1959] 33/5; Soviet At. Energy **7** [1959] 559/61).

[2] Johnson, I. (in: Waber, J. T.; Chiotti, P.; Miner, W. N., Compounds of Interest in Nuclear Reactor Technology, Met. Soc. AIME, Warrendale, Pa., 1964, pp. 171/92).

[3] Samorukov, O. P.; Kostyukov, V. N.; Kostylev, F. A.; Tumbakov, V. A. (At. Energiya SSSR **37** [1974] 28/31; Soviet At. Energy **37** [1974] 705/8).

[4] Rand, M. H.; Kubaschewski, O. (The Thermochemical Properties of Uranium Compounds, Oliver and Boyd, Edinburgh — London 1963).

[5] Hultgren, R.; Desai, P. D.; Hawkins, D. T.; Gleiser, M.; Kelley, K. K. (Selected Values of Thermodynamic Properties of Binary Alloys, Am. Soc. Metals, Metals Park, Ohio, 1973, pp. 390/1).

[6] Farkas, M.; Eldridge, E. (J. Nucl. Mater. **27** [1968] 94/6).

2.1.3.3 Physical Properties

2.1.3.3.1 General Remarks

The only compound which exists in the U−Be system, UBe_{13}, exhibits some unusual physical properties. These are related to the fact that the compound is one of the heavy-electron systems in which the conduction electron specific heat is typically some hundred times larger than that found in most metals. Since the f-electron moments become strongly coupled to the conduction electrons and to one another, the effective mass has the very high value of hundred of times the bare electron mass. This causes the unconventional behavior of UBe_{13} at very low temperatures in the normal state, and in the superconducting state as well. Electronic conductivity, magnetic properties, the ultrasound attenuation, and the specific heat are influenced in a typical manner. These properties are highly sensitive to impurities in the compounds, which substitute U in the order of less than 3 at%. The superconductivity, which in ordinary systems is based on the exchange of phonons according to the Bardeen-Cooper-Schrieffer (BCS) theory, is related to an attractive interaction between electrons resulting from a virtual exchange of local moment fluctuations [1 to 6].

The heavy-electron states of U can be formed with elements at the end of the d block and the beginning of the s and p blocks where few states are available for the hybridization with

References for 2.1.3.3.1 on p. 20

the f electrons. The local chemical environment of the f atom is important. Systematic analyses of the heavy-electron compounds showed that superconductivity occurs at the largest specific heat per unit volume ratio. The jump of the electronic specific heat at the superconducting transition indicates that the heavy electrons must be involved in the superconductivity of UBe_{13}. The crystal structure may influence the type of superconducting states [1, 4, 7]. Reviews on the theory of heavy fermions were given in [8, 9]. The most recent state of the theory of heavy-fermion systems was extensively reported by Fulde et al. [10].

References for 2.1.3.3.1:

[1] Stewart, G. R. (Rev. Mod. Phys. **56** [1984] 755/87).

[2] Fisk, Z.; Ott, H. R.; Smith, J. L. (J. Magn. Magn. Mater. **47/48** [1985] 12/6).

[3] Fisk, Z.; Hess, D. W.; Pethick, C. J.; Pines, D.; Smith, J. L.; Thompson, J. D.; Willis, J. O. (Science **239** [1988] 33/42).

[4] Fisk, Z.; Ott, H. R.; Rice, T. M.; Smith, J. L. (Science **230** [1986] 124/9).

[5] Stewart, G. R.; Andraka, B.; Quitmann, C. (Phys. Scr. T **23** [1988] 119/21).

[6] Fritsch, G. (Physik in unserer Zeit **18** [1987] 17/20).

[7] Smith, J. L.; Fisk, Z.; Ott, H. R. (in: Gupta, L. C.; Malik, S. K., Theoretical and Experimental Aspects of Valence Fluctuations and Heavy Fermions, Plenum, New York 1987, pp. 11/5).

[8] Rice, T. M. (Japan. J. Appl. Phys. **26** Suppl. Pt. 3 [1987] 1865/74).

[9] Ott, H. R.; Fisk, Z. (in: Freeman, A. J.; Lander, G. H., Handbook on the Physics and Chemistry of the Actinides, Vol. 5, Elsevier, Amsterdam 1987, pp. 85/224).

[10] Fulde, P.; Keller, J.; Zwicknagl, G. (in: Ehrenreich, H.; Turnbull, D., Solid State Physics, Vol. 41, Academic, San Diego 1988, pp. 1/159).

2.1.3.3.2   Crystal Structure of UBe_{13}

$U-Be$ alloys are mixtures of the intermetallic compound UBe_{13} with either Be or U, depending on their composition. In these mixtures U can be present in different forms, at $<660°C$ as α-U, at 770°C as β-U, or at 1090°C as γ-U. The components mentioned occur in cast alloys as well as in powder-metallurgical products [1, 2], see also [3, 4].

Alloys containing ca. 60 wt% Be show a monotectic at 1550°C. An allotropic transformation in the Be phase possibly occurs at $1250 \pm 10°C$ [1]. A metallographic examination of the alloys in the composition range of 35 to 100 wt% Be indicates that particles of UBe_{13} are embedded in a Be matrix. Thermal treatment at up to 1225°C does not change these structures. In the "as cast" condition, the compound normally appears in an angular form. Globular particles also occur in the composition range of 80 to 92 wt% Be [1].

Alloys with compositions in the vicinity of the $U-UBe_{13}$ eutectic have a definite eutectic structure as shown by metallographic examination of "as cast" alloys. The structure of the compound UBe_{13} was determined by X-ray diffraction. UBe_{13} is isomorphous with $NaZn_{13}$ ($D2_3$ type). The space group is O_h^6-Fm3c (No. 226) and the face-centered cubic cell with $a = 10.2561$ Å contains 8 UBe_{13} per unit cell [5]. The results are confirmed by additional X-ray and neutron diffraction studies [6]. The parameter of UBe_{13} is redetermined as $a = 10.272$ Å. The ratio of electrons to atoms is 30/14 for U^{IV} [7].

A neutron diffraction study of a powder sample of UBe_{13} permitted the determination of the Be positions with the required sensitivity [12]. A more recent study of the intermetallic phase $NpBe_{13}$ by means of X-ray diffraction shows that this compound has the same structure as UBe_{13} and gives X-ray reflections comparable to those of the U beryllide. The results were applied to fix the atom positions [8], which are shown in **Fig.** 2-4. These data are in good agreement with data of other actinide beryllides [9]. According to this survey on the actinide beryllides, the value of a = 10.3698 Å given for UBe_{13} in [1] is abnormally high and should be probably read as 10.2698 Å, which is in better agreement with the other beryllides. **Fig.** 2-5 demonstrates a comparison of the lattice constants of these intermetallics. The upper curve corresponds to the Be-rich compounds, the lower to those with high actinide contents [9].

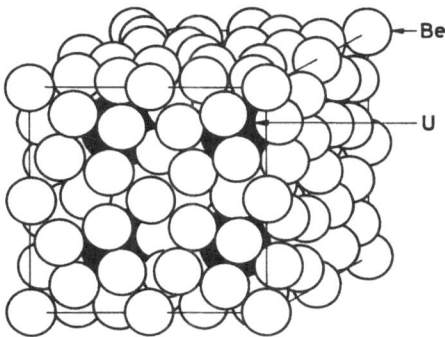

Fig. 2-4. The UBe_{13} unit cell based on the study of the Np beryllide [8].

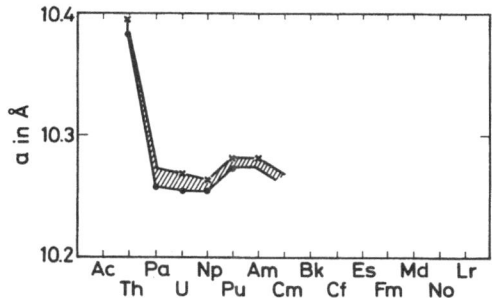

Fig. 2-5. Lattice parameters of the actinide beryllides [9].

The MBe_{13} compounds (examples are $LaBe_{13}$, $ThBe_{13}$) are isodesmic with $NaZn_{13}$, and all 4 peripheral d electrons may take part in the b correlation. This adds valency arguments for the stability and the structure of these intermetallic phases. The valence electron concentration of 2.0 is characteristic of salt-like (FB2) binding [10]. Resonant photoemission measurement using synchrotron radiation with 30 to 120 eV gave, in agreement with the anomalously high γ value of 1.1 $J \cdot mol^{-1} \cdot K^{-1}$ of the superconductor UBe_{13}, a high density of 5f states at the Fermi level E_F (see Section 2.1.3.3.3, p. 23). Zero intensity at E_F in the off-resonance spectrum shows that U 6d electrons are not present. The similarity of this off-resonance curve with a photoemission spectrum of Be suggests that the hybridization of U 5f electrons with the Be sp electrons is small. The UBe_{13} photoemission curves appear to be superpositions of U and Be curves. By using results from UO_2 as a reference, the valence-band emission appears to contain about one 5f electron [11].

References for 2.1.3.3.2 on pp. 22/3

Lattice constants of UBe$_{13}$ were determined using the Brookhaven National Laboratory High-Flux Beam Reactor with 14.7 meV neutrons ($\lambda = 2.359$ Å), a bent pyrolytic graphite (002) monochromator, and a bent pyrolytic graphite (004) analyzer, as a function of temperature as shown in Table 2/1 [12].

Space group $O_h^6 -$ Fm3c (No. 226); positions of atoms: U 8(a): $\pm$ 1/4, 1/4, 1/4; Be(I) 8(b): 0, 0, 0; 1/2, 1/2, 1/2; Be(II) 96(i): $\pm$ 0, y, z, etc. The eight U atoms and eight Be(I) atoms form a CsCl type-like lattice, each U coordinated by 24 Be(II) atoms at the corners of a nearly regular snub cube [12].

Table 2/1
Lattice Constants of UBe$_{13}$ between 10 and 250 K [12].

T in K	a in Å	T in K	a in Å
10	10.24887	175	10.25405
50	10.24987	250	10.26019
100	10.25039		

The distances between nearest neighbors in UBe$_{13}$ are listed in Table 2/2. The distance between the U atoms is $d_{(U-U)} = 5.13$ Å [13].

The bulk modulus c_{bulk} of UBe$_{13}$ was determined by means of Brillouin scattering. A softening of c_{bulk} does not occur in this compound. This behavior indicates an electronic superconducting transition of UBe$_{13}$ due to an interaction different from the usual electron-phonon interaction [14].

Table 2/2
Nearest-neighbor Distances in UBe$_{13}$ at 10 K [13].

bond to atom	number of bonds	d in Å
bonds involving U		
Be(II) (0, y, z)	24	3.01
bonds involving Be(I)		
Be(II) (0, y, z)	12	2.16
bonds involving Be(II)		
Be(I) (0, 0, 0)	1	2.16
U (1/4, 1/4, 1/4)	2	3.01
Be(II) (0, z, 1/2 − y)	2	2.23
Be(II) (z, 0, y)	4	2.25
Be(II) (0, y, z)	1	2.36
Be(II) (Z, 1/2 − y, 0)	2	2.25

References for 2.1.3.3.2:

[1] Buzzard, R. W. (J. Res. Natl. Bur. Std. **50** [1953] 63/7).
[2] Hansen, M.; Anderko, K. (Constitution of Binary Alloys, McGraw-Hill, New York 1958, p. 299).
[3] Bellamy, R. G.; Hill, N. A. (Extraction and Metallurgy of Uranium, Thorium, and Beryllium, Pergamon, Oxford 1963, p. 163).

[4] Rough, F. A.; Bauer, A. A. (Constitutional Diagrams of Uranium and Thorium Alloys, Addison-Wesley, Reading, Mass., 1958, pp. 1/153).

[5] Baenzinger, N. C.; Rundle, R. E. (Acta Cryst. **2** [1949] 258).

[6] Koehler, W. C.; Singer, J.; Coffinberry, A. S. (Acta Cryst. **5** [1952] 394).

[7] Hanna, G. L.; Turner, D. N.; Australian Atomic Energy Comm. (AAEC-E-102 [1963] 1/16; N.S.A. **17** [1963] No. 18667).

[8] Runnals, O. J. C. (Acta Cryst. **7** [1954] 222/3).

[9] Benedict, U.; Buijs, K.; Dufour, C.; Toussaint, J. C. (J. Less-Common Metals **42** [1975] 345/54).

[10] Schubert, K. (Z. Metallk. **73** [1982] 403/8).

[11] Landgren, G.; Jugnet, Y.; Morar, J. F.; Arko, A. J.; Fisk, Z.; Smith, J. L.; Ott, H. R.; Reihl, B. (Phys. Rev. [3] B **29** [1984] 493/6).

[12] Goldman, A. I.; Shapiro, S. M.; Cox, D. E.; Smith, J. L.; Fisk, Z. (Phys. Rev. [3] B **32** [1985] 6042/4).

[13] Pearson, W. B. (Handbook of Lattice Spacings and Structure of Metals, Pergamon, New York 1958, p. 458).

[14] Mock, R.; Hillebrands, B.; Schmidt, H.; Güntherodt, G.; Fisk, Z.; Meyer, A. (J. Magn. Magn. Mater. **47/48** [1985] 312/4).

2.1.3.3.3 Electronic Structure

The electronic properties at low temperatures in the normal and superconducting state of UBe_{13} are highly anomalous; the behavior of the compound seems to be related to the 5f electrons [1]. The research efforts were directed to determine the relationships between the low-temperature properties and the electronic structure of the actinide beryllide. Inelastic neutron scattering measurements in the time-of-flight spectrometer TOF1 of the Melusine reactor at Grenoble were applied to study the electronic structure. Results gained at 8 and 300 K give evidence for the existence of an Einstein-mode-like oscillator around 13 meV, independent of the temperature. The lattice contribution to the low-temperature heat capacity was calculated from phonon densities deduced from the neutron scattering measurements. The lattice contribution is small compared to the specific heat measured for UBe_{13} at low temperature [2]. Within the density-of-states spectrum, the states near the Fermi level E_F are mainly formed from U 5f electrons. States which are at an energy of 0.64 Ry below E_F are formed from Be 2p electrons [3]. These properties are discussed in [4].

The electronic band structure of UBe_{13} has been determined based on the semi-relativistic LAPW (linearized augmented plane wave) method (spin-orbit included by solving the Dirac equation on the final interaction). In the crystal, which has $NaZn_{13}$ structure, there are two inequivalent Be atoms, two of one type and 24 at the other positions. Nearest neighbor Be distances are smaller than in Be metal. The electronic structure separates into two almost independent band systems. The states below 0.64 Ry are mostly Be s and p states with some U d hybridization. Above 0.64 Ry these are mostly U 5f bands hybridized with the Be 2p bands and U 6d states. The density-of-states (DOS) for the type-one atoms is similar to that of Be metal; it has a sharp peak just above the minimum near 0.64 Ry, which falls off near E_F. The electron density near E_F is small, since the other Be atoms have a DOS function which is reduced above the minimum. The hybridized 5f band is about 0.9 eV wide and contains approximately two electrons [3].

The electronic structure of the heavy-fermion superconductor UBe_{13} was discussed in several publications, and some aspects are still under controversial discussion. Bucher et al.

References for 2.1.3.3.3 on pp. 27/8

[5], who discovered the superconducting transition of UBe_{13}, considered the singlet electron state as the most probable ground state. The well-localized $5f^2$ configuration is analogous to that of $PrBe_{13}$. Resonant photoemission using synchrotron radiation with 30 to 120 eV in the IBM two-dimensional spectrometer at Brookhaven was used to get more detailed information on the electronic structure of the superconducting state [6]. A high density of 5f states was found at the Fermi level E_F. U 6d electrons are not present at E_F [6].

The resemblance between the off-resonance curve and a photoemission spectrum of pure Be suggests that the hybridization of U 5f electrons with Be s,p electrons is small; p-wave superconductivity in UBe_{13} is discussed in [7].

Overhauser and Appel [8] relate the unusual properties of the heavy-fermion superconductor UBe_{13} to a s-f hybridization model. They assume the $5f_{5/2} - \Gamma_8$ crystal field level of U in the beryllide is filled and that the Γ_7 (Kramer's doublet) is exactly at the Fermi surface. Resonant hybridization of the Γ_7 state with the Be $(2s) - U$ (7s) conduction band leads to a Lorentzian density-of-states peak at E_F. Its width is 12K and the peak height is ca. 250 times that of the conduction band. The normal-state heat capacity, magnetic susceptibility, and nuclear spin relaxation versus T are explained by this model. It follows that the superconducting energy gap is also anisotropic in $\vec{k}$ space. The heat capacity below T_c is explained with singlet pairing caused by phonon-mediated interactions.

Varma [9] attempted to establish a relationship between the unusual behavior of UBe_{13} and the electronic structure. The heavy-fermion superconductors are not of the conventional s-wave kind. The nuclear spin-lattice relaxation rate and thermal conductivity of UBe_{13} are not consistent with the claim of an axial-like state from specific heat data and by analogy with ^{3}He. It is difficult to reconcile experimental results with group theory. The interpretation of the influence of impurities in UBe_{13} on the value of T_c is also difficult. The measurement of the critical field anisotropy in the cubic UBe_{13} would be very interesting. If the superconducting state is polar-like, there should be a large anisotropy. Varma finds strong evidence that heavy fermions do not arise because of band-structure effects.

As in all heavy-electron U intermetallics, UBe_{13} is beyond the Hill limit in the Hill plot [10], and the metallic radius of U in the heavy-electron compounds is large and close to that of Gd.

The unconventional superconductivity of UBe_{13} is related to the large distances of the nearest-neighbor U atoms ($d_{U-U} = 5.13$ Å). The beryllide is the extreme example of large-distance compounds [11].

Ott [12] suggests, from experimental observations and theoretical models, that the superconducting state of UBe_{13} is due to localized spin fluctuations and therefore is of magnetic origin. It may be characterized by an $1 \neq 0$ pairing of the electrons.

Measurement of the temperature dependence of the magnetic field penetration depth in UBe_{13} at a temperature below T_c provides a different method to study the electron structure of the compound [13, 14]. The observed dependence on T^2 is consistent with an anisotropic gap function for an axial p-wave state. The results support theories of the heavy-fermion normal state, which predict a small Landau parameter [9, 15].

Alekseevskii [16] drew different conclusions from his own experimental results on UBe_{13} and on dissolutions of the isostructural compounds $ZrBe_{13}$ and $CeBe_{13}$ in UBe_{13}. The U beryllides, having strongly localized U 5f electrons, are of $NaZn_{13}$ lattice type and can be regarded as cluster compounds . The introduction of U leads to a decrease of T_c and H_{c2}. In the case of UBe_{13}, a resonant Abrikosov-Suhl state may be possible, which is highly sensitive to a change in the parameters of the system. A covalent instability seems possible, causing H_{c2} to become infinite as J_c approaches zero. Preliminary measurements of J_c do not rule out this possibility.

The self-consistent full potential linearized augmented plane wave method was applied to the properties of the heavy-fermion superconductor UBe_{13} [17]. Total energy calculations in the local density approximation predict an equilibrium lattice constant 2.5% less than the measured value. The calculated f occupation n_f is 2.65 electrons per U atom, reflecting large hybridization of the 4f bands with Be sp bands. An understanding of magnetism in UBe_{13} with respect to the local spin density theory is not yet possible [17].

Delocalization of the 5f states of U can only occur by weak hybridization with the sp wave function tails of the 4 Be atoms located on the cube faces midway between the two nearest U atoms, which are too far apart to form a bond by direct f-f overlap. The outer level spectra reflect, nearly exclusively, the U states which are obviously responsible for the unconventional properties of UBe_{13}. X-ray photoemission and bremsstrahlung isochromat spectroscopies were used to probe the occupied and unoccupied states of the U beryllide. 2 to 3 electrons populate the tail of a broad 5f band (ca. 5 eV) of extended states. The resolution of $\leqq 0.5$ eV is not sufficient to directly observe any peculiarities of the density-of-states at the Fermi energy which would explain the properties of UBe_{13}. Drastic differences between the core level spectra of U and Be indicate that 5f states remain essentially confined around the U atoms and are only weakly hybridized with the sp band states originated from the Be atoms [18]. Auger electron spectra gave evidence of charge transfer of U (5f) electrons. The line shape indicated a weak degree of hybridization [33].

Polarized and unpolarized neutron scattering experiments on U-based heavy-fermion systems have provided a great deal of evidence for the formation of a coherent state involving the f electrons at low temperature. Some results support the assumption that there are additional spin fluctuation modes in UBe_{13} [19]. Several partly differing results on the effect of neutron scattering on the electronic structure of UBe_{13} are discussed in [20].

The entropies of exotic superconductors such as UBe_{13} at low temperatures change by approximately the same amount as those involved in magnetic phase transitions. This is evident by the giant coefficient of electronic specific heat, which is taken to imply the existence of massive fermions. This behavior indicates that a kind of order, different from magnetic order, takes place in the spin system. No dependence of UBe_{13} susceptibility on temperature was observed in high-frequency experiments. Similar results were obtained with other heavy-fermion systems such as UPt_3 and $CeCu_2Si_2$ [21].

It is proposed that the Brinkman-Rice [22] theory of almost-localized Fermi liquids is applicable to the heavy-electron metals. The theory is generalized to finite temperatures in order to describe a smooth transition from a Fermi-liquid regime at low temperatures to a set of random spins at higher temperatures [23]. The theory presented gives a good description of the low temperature entropy (see **Fig. 2-6**, p. 26) and the specific heat in the normal state of the heavy-electron superconductor UBe_{13}. The low-temperature thermodynamics are determined more by the 5f band, owing to its much larger mass, than by the broad conduction band with which it is hybridized. The large specific heat of the almost-localized Fermi liquid arises from the magnetic entropy of the singly occupied (or local moment) sites which appear in the Fermi liquid in a temperature range significantly below the Fermi energy.

Related discussions on the heavy-electron systems as almost-localized Fermi liquids with large spin fluctuations have been published [24, 25]. The model shows a fairly good agreement

                    References for 2.1.3.3.3 on pp. 27/8

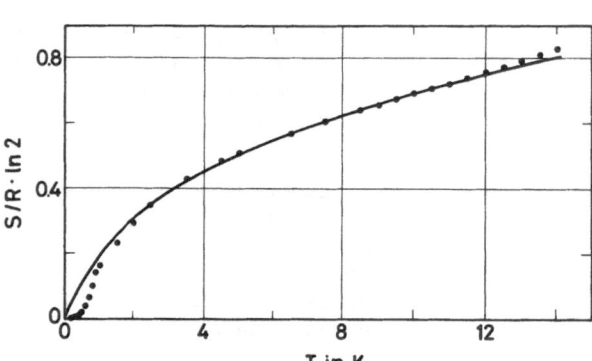

Fig. 2-6. Temperature dependence of the electronic molar entropy of UBe_{13} below 14 K. T_c for superconductivity is 0.9 K. The solid line is the theoretical curve with a mass enhancement of 25 [23].

with experimental values. The axial or Anderson-Brinkman-Morel state should be stable for UBe_{13}. This agrees with the observation of the T^3 dependence of the low-temperature specific heat.

The energy band structure of UBe_{13} was studied by means of a self-consistent APW (augmented plane wave) method with local density approximation [26]. The width of the f bands is ca. 1 eV and the total density of states at the Fermi energy is 19.3 states/Ry per mol. There is evidence that the main properties at the Fermi energy are determined by the f bands.

A comparison of the low-temperature properties of different heavy-fermion compounds of U shows that the properties of U_2PtC_2 are between those of U_6Fe and UBe_{13}. This might be

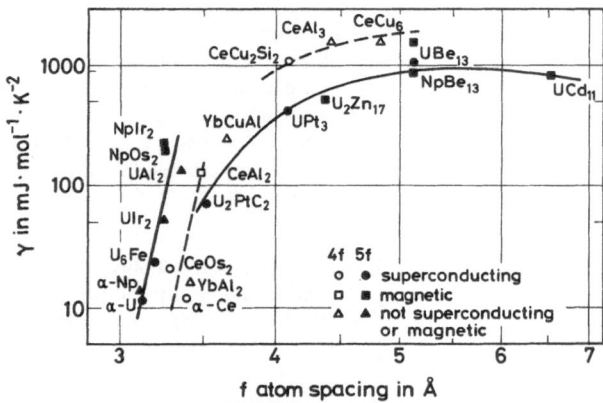

Fig. 2-7. Electronic specific-heat coefficient γ versus f-atom spacing, plotted on logarithmic scales for 4f (open symbols, dashed lines) and 5f (filled symbols, solid lines) in superconducting (circles), magnetic (squares), and neither superconducting nor magnetic (triangles) materials. The values for γ are extrapolated from the superconducting or magnetic transitions except for $NpBe_{13}$, where the upturn in the specific heat accompanying the development of the heavy-fermion state is obscured by magnetic ordering [27].

due to the spacing between f electrons. The electronic specific heat is related to nearly localized f-band formation. These relations between the electronic specific-heat coefficient γ and the f-electron atom spacing is shown for various superconducting, magnetic or non-superconducting or magnetic actinide and lanthanide compounds in **Fig. 2-7** [27].

The electrical quadrupole Kondo effect is proposed as the origin of the heavy-electron behavior observed in UBe_{13}. The possible non-Fermi-liquid character of the quenched quadrupole moment ground state can lead to low-temperature structural instabilities. The question, if this may be a universal behavior of heavy-fermion systems, was discussed [28]. The local moments become compensated via a Kondo-type interaction with the conduction electrons, a consequence of which is that a narrow resonance develops at the Fermi level [29]. A single-order-parameter and a two-gap model (a modified Suhl two-gap model) agrees with experimental data such as NMR and ultrasonic attenuation [30]. The possibility of unconventional pairing on the basis of the Anderson—Morel model was also discussed [31, 32].

References for 2.1.3.3.3:

[1] Ott, H. R.; Rudigier, H.; Fisk, Z.; Smith, J. L. (Phys. Rev. Letters **50** [1983] 1595/8).

[2] Gompf, F.; Renker, B.; Rietschel, H.; Nücker, N.; Beuers, J. (Physica B + C **135** [1985] 41/5).

[3] Boring, A. M.; Albers, R. C.; Mueller, F. M.; Koelling, D.-D. (Physica B + C **130** [1985] 171/4).

[4] Volovik, G. E.; Gor'kov, L. P. (Pis'ma Zh. Eksperim. Teor. Fiz. **39** [1984] 550/3; JETP Letters **39** [1984] 674/7).

[5] Bucher, E.; Maita, J. P.; Hull, G. W.; Fulton, R. C.; Cooper, A. S. (Phys. Rev. [3] B **11** [1975] 440/9).

[6] Landgren, G.; Jugnet, Y.; Morar, J. F.; Arko, A. J.; Fisk, Z.; Smith, J. L.; Ott, H. R.; Reihl, B. (Phys. Rev. [3] B **29** [1984] 493/6).

[7] Ueda, K.; Rice, T. M. (Phys. Rev. [3] B **31** [1985] 7114/9).

[8] Overhauser, A. W.; Appel, J. (Phys. Rev. [3] B **31** [1985] 193/202).

[9] Varma, C. M. (J. Appl. Phys. **57** [1985] 3064/6).

[10] Fisk, Z.; Ott, H. R.; Smith, J. L. (J. Less-Common Metals **133** [1987] 99/106).

[11] Ott, H. R.; Hulliger, F.; Rudigier, H.; Fisk, Z. (Phys. Rev. [3] B **31** [1985] 1329/33).

[12] Ott, H. R. (Physica B + C **130** [1985] 163/7).

[13] Einzel, D., Hirschfeld, P. J.; Gross, F.; Chandrasekhar, B. S.; Andres, K.; Ott, H. R.; Beuers, J.; Fisk, Z.; Smith, J. L. (Phys. Rev. Letters **56** [1986] 2513/6).

[14] Gross, F.; Chandrasekhar, B. S.; Einzel, D.; Andres, K.; Hirschfeld, P. J.; Ott, H. R.; Beuers, J.; Fisk, Z.; Smith, J. L. (Z. Physik B **64** [1986] 175/88).

[15] Rice, T. M.; Ueda, K. (Phys. Rev. Letters **55** [1985] 995/8).

[16] Alekseevskii, N. E. (Pis'ma Zh. Eksperim. Teor. Fiz. **40** [1984] 66/9; JETP Letters **40** [1984] 800/3).

[17] Pickett, W. E.; Krakauer, H.; Wang, C. S. (Physica B + C **135** [1985] 31/3).

[18] Wuilloud, E.; Bear, Y.; Ott, H. R.; Fisk, Z.; Smith, J. L. (Phys. Rev. [3] B **29** [1984] 5228/31).

[19] Neumann, K. U.; Capellmann, H.; Fisk, Z.; Smith, J. L.; Ziebeck, K. R. A. (Solid State Commun. **60** [1986] 641/3).

[20] Goldman, A. I. (Japan. J. Appl. Phys. **26** Suppl. 3 [1987] 1887/92).

[21] Buyers, W. J. L. (Physica B + C **137** [1986] 53/60).

[22] Brinkman, W. F.; Rice, T. M. (Phys. Rev. [3] B **2** [1970] 4302/4).

[23] Rice, T. M.; Ueda, K.; Ott, H. R.; Rudigier, H. (Phys. Rev. [3] B **31** [1985] 594/6).

[24] Valls, O. T.; Tesanovic, Z. (Phys. Rev. Letters **53** [1984] 1497/500).

[25] Béal-Monod, M. T. (Phys. Rev. [3] B **31** [1985] 1647/50).

[26] Takegahara, K.; Harima, H.; Kasuya, T. (J. Magn. Magn. Mater. **47/48** [1985] 263/5).

[27] Meisner, G. P.; Giorgi, A. L.; Lawson, A. C.; Stewart, G. R.; Willis, J. O.; Wire, M. S.; Smith, J. L. (Phys. Rev. Letters **53** [1984] 1829/32).

[28] Cox, D. L. (Physica C **153/155** [1988] 1642/8).

[29] Fisk, Z.; Ott, H. R.; Aeppli, G. (Japan. J. Appl. Phys. **26** Suppl. 3 [1987] 1882/6).

[30] Ji-Hai Xu (J. Low Temp. Phys. **70** [1988] 173/86).

[31] Ott, H. R.; Fisk, Z. (in: Freeman, A. J.; Lander, G. H., Handbook on the Physics and Chemistry of the Actinides, Vol. 5, Elsevier, Amsterdam 1987, pp. 85/225).

[32] Ashauer, B.; Kieselmann, G.; Rainer, D. (J. Low Temp. Phys. **63** [1986] 349/66).

[33] Bevolo, A. J. (J. Less-Common Metals **153** [1989] 101/25).

2.1.3.3.4 Density, Thermal Expansion

The calculated X-ray density of UBe$_{13}$ is $D_{calc} = 4.36$ g/cm^3 [1]. $D_{calc} = 4.23$ g/cm^3 was also based on X-ray data (a = 10.3698 Å) [2]. Compilations of the properties of this compound contain D = 4.373 for X-ray and 4.420 ± 0.002 g/cm^3 for direct measurements [3, 4]. The bulk density of sintered beryllide was D = 4.46 g/cm^3 [5].

Densities of alloys or even the compound prepared by means of powder metallurgical methods are lower than the theoretical density, owing to the formation of voids [6].

The thermal expansion of UBe$_{13}$ is comparable to that of USi or UAl$_2$ (see p. 88). Measurements by means of a vertical quartz-tube dilatometer in vacuum are presented in [7]; the mean linear coefficient of expansion in the temperature range from 20 to 1000°C is $\alpha = 16.7 \times 10^{-6}$ °C^{-1}. The linear thermal expansion coefficients α of UBe$_{13}$ show an anomaly in the sense that they decrease with decreasing temperature down to T$_c$. A sudden decrease of α with falling T occurred below T$_c$ [8].

The low-temperature thermal expansion of single-crystalline UBe$_{13}$ between 0.3 and 10 K in the [100] direction is anomalous in the sense that α increases with decreasing temperature below 7 K. A discontinuous change of α below 1 K is associated with the transition into the superconducting state and is due to the thermodynamic changes at this point [9].

References for 2.1.3.3.4:

[1] Buzzard, R. W. (J. Res. Natl. Bur. Std. **50** [1953] 63/7).

[2] Ferro, R. (Atomic Energy Rev. Spec. Issue **4** [1973] 63/104).

[3] Rough, F. A.; Bauer, A. A. (Constitutional Diagrams of Uranium and Thorium Alloys, Addison-Wesley, Reading, Mass., 1958, pp. 7/9).

[4] Ivanov, O. S.; Badaeva, T. A.; Sofronova, R. M.; Kishenevskii, V. B.; Kushnir, N. P. (Phase Diagrams of Uranium Alloys, Amerind, New Delhi 1983, pp. 27/30 translated from: Diagrammy Sostoyaniya i Fazovye Prevrashcheniya Splavov Urana, Nauka, Moscow 1972).

[5] Albrecht, W. M.; Koehl, B. G. (Proc. 2nd Intern. Conf. Peaceful Uses At. Energy, Geneva 1958, Vol. 6, pp. 116/21).

[6] Williams, J.; Jones, J. W. S. (Powder Met. **1960** No. 5, pp. 45/63).

[7] Snyder, M. J.; Tripler, A. B., Jr. (ASTM Spec. Tech. Publ. No. 276 [1959] 293/300).

[8] Ott, H. R. (Physica B + C **126** [1984] 100/6).

[9] Ott, H. R.; Fisk, Z. (in: Freeman, A. J.; Lander, G. H., Handbook on the Physics and Chemistry of the Actinides, Vol. 5, Elsevier, Amsterdam 1987, pp. 85/225).

2.1.3.3.5 Elastic Constants. Compressibility

The elastic constants determined by means of inelastic neutron scattering at room temperature and at 10 K were consistent with elastic constants of pure Be metal; the angular variation was almost isotropic. The results were used to calculate the low temperature specific heat. The elastic constants of UBe_{13} are listed in Table 2/3 [1].

Table 2/3
Elastic Constants of UBe_{13} in 10^9 N/m^2 [1].

constant	at 298 K	at 10 K
c_{11}	265 ± 16	310 ± 9
c_{12}	7 ± 22	-1 ± 9
c_{44}	134 ± 3	161 ± 5

The average atomic volume of UBe_{13} is 19% larger than that of pure Be metal, so that the UBe_{13} sound velocities are 0.631 of the Be sound velocities. Values of $V_T = 5360$ to 6000 m/s and $V_L = 8018$ to 8580 m/s are predicted, in fair agreement with the elastic constants [1].

The compressibility of UBe_{13} was studied using X-ray diffraction in a diamond anvil cell at room temperature. The pressure − volume relationship of the compound resulted in a bulk modulus of $B_0 = 108$ GPa with an initial pressure derivate B_0' of 5.8. The effect of pressure on the 5f electrons is not sufficient to induce a phase transition; the electronic configuration of UBe_{13} appears to be stable at room temperature [2].

References for 2.1.3.3.5:

[1] Robinson, R. A.; Axe, J. D.; Goldman, A. I.; Fisk, Z.; Smith, J. L.; Ott, H. R. (Phys. Rev. [3] B **33** [1986] 6488/90).
[2] Benedict, U.; Dabos, S.; Gerward, L.; Staun Olsen, J.; Beuers, J.; Spirlet, J. C.; Dufour, C. (J. Magn. Magn. Mater. **63/64** [1987] 403/5).

2.1.3.3.6 Hardness

The Knoop hardness of UBe_{13} at room temperature is about 1300 kg/mm^2 [1, 2]. A value of 1140 measured with a load of 100 g is reported in [3], or 1035 in [4]. The hardness seems to depend on the method of preparation of the beryllide. The value of 960 kg/mm^2 is given for arc-melted UBe_{13}, whereas sintered material has a somethat higher hardness of 1155 kg/mm^2 [5]. Therefore, UBe_{13}has a rather low resistance to thermal shock, and fracture frequently occurs during cooling after hot pressing [1].

References for 2.1.3.3.6:

[1] Alder, K. F. (referred in [2] as private communication).
[2] Lewis, J. F. (J. Metals **13** [1961] 357/62).
[3] Badaeva, T. A.; Dashevskaya, L. I. (in: Ivanov, O. S., Physical Chemistry of Alloys and Refractory Compounds of Thorium and Uranium, Jerusalem 1972, pp. 123/9).

[4] Badaeva, T. A.; Alekseenko, G. K.; Kuznetsova, R. I. (in: Ivanov, O. S., Physical Chemistry of Alloys and Refractory Compounds of Thorium and Uranium, Jerusalem 1972, pp. 148/57).

[5] Snyder, M. J.; Tripler, A. B., Jr. (ASTM Spec. Tech. Publ. No. 276 [1959] 293/300).

2.1.3.3.7 Ultrasonic Attenuation

A peak in the acoustic attenuation of UBe$_{13}$ was observed in measurements of the sound propagation at frequencies of 0.9 to 2.4 GHz and temperatures from 0.01 to 100 K. Instead of a rapid drop, which might be expected for a BCS (Bardeen-Cooper-Schrieffer) superconductor, the peak appeared at 0.9 K. The absorption into a collective mode of a nonsinglet, anisotropic superconductor was proposed [1, 2].

The unconventional properties of the superconducting ground state of the beryllide were also studied by means of the ultrasound attenuation in single crystals in which the sound was propagated in directions parallel to the c axis and the basal plane. At frequencies of 50 to 500 MHz, values of ca. 0.1 dB $\cdot$ cm$^{-1} \cdot$ MHz^{-1} were measured. They were within the range measured in various crystallographic directions in several non-heavy-fermion metals. The results were not enhanced by 10^4 to 10^5, as might be expected from the large effective mass. The mass enhancement of the quasiparticles seemed to be compensated by a reduction of the coupling strength [3, 4].

The ultrasound attenuation and its temperature dependence below T$_c$ were measured at 0.5 to 2.5 GHz. The attenuation in UBe$_{13}$ single crystals was very different from all known superconductors. Instead of the expected abrupt decrease at T$_c$, or even no change as in UPt$_3$, a narrow attenuation peak was observed at T$_c$. Its magnitude was comparable to the electronic attenuation in the normal state. In an applied magnetic field of 20 kOe no shift of the maximum with respect to T$_c$ could be detected. The attenuation peak might be caused by the coupling of sound to collective modes of the order parameter, in analogy to similar effects in superfluid ^{3}He. Partial substitution of U by Th induced a second transition below T$_c$, as will be discussed in the section on ternary compounds (see p. 60) [5].

The normalized ultrasonic attenuation α_s/α_n for UBe$_{13}$ varies as T^2 with temperature, and attempts were made to relate this behavior to thermodynamic quantities calculated from a two-gap model [6].

The longitudinal ultrasonic attenuation constants, obtained from a model for the superconducting behavior of UBe$_{13}$ based on a hydrodynamic theory, were close to values measured in the temperature range 0.75 to 0.90 K. The variables which were used in this model followed a Hamiltonian dynamics which resulted in an internal Josephson effect between the population of spin-up and spin-down Cooper pairs. The effect is similar to the longitudinal spin-resonance behavior which occurred in the superfluid phase of ^{3}He [7].

References for 2.1.3.3.7:

[1] Golding, B.; Bishop, D. J.; Batlogg, B.; Haemmerle, W. H.; Fisk, Z.; Smith, J. L.; Ott, H. R. (Phys. Rev. Letters **55** [1985] 2479/82).

[2] Fisk, Z.; Ott, H. R.; Rice, T. M.; Smith, J. L. (Science **230** [1986] 124/9).

[3] Batlogg, B.; Bishop, D. J.; Golding, B.; Bucher, E.; Hufnagl, J.; Fisk, Z.; Smith, J. L.; Ott, H. R. (Phys. Rev. [3] B **33** [1986] 5906/9).

[4] Ott, H. R.; Fisk, Z. (in: Freeman, A. J.; Lander, G. H., Handbook on the Physics and Chemistry of the Actinides, Vol. 5, Elsevier, Amsterdam 1987, pp. 85/225).

[5] Batlogg, B.; Bishop, D. J.; Bucher, E.; Golding, B.; Varma, C. M.; Fisk, Z.; Smith, J. L.; Ott, H. R. (Physica B + C **135** [1985] 23/6).
[6] Ji-Hai Xu (J. Low Temp. Phys. **70** [1988] 173/86).
[7] Rodriguez, J. P. (Phys. Rev. [3] B **36** [1987] 168/79).

2.1.3.3.8 Phonon Dispersion

The spatial distribution and temperature dependence of the magnetization induced in UBe_{13} single crystals was studied by means of polarized neutron scattering techniques. The induced magnetization in the normal low-temperature state is predominantly of f-electronic character. The generalized electronic susceptibility at Q = 0 (Q is a reciprocal-lattice vector) (see Section 2.1.3.3.11, p. 37) is temperature-independent in the range 4.2 to 0.1 K. The results imply that the superconduction in UBe_{13} is unconventional or that impurities cause spin-orbit scattering. In spite of this, restrictions on the superconducting states are seen [1].

Calculation of the electron-phonon interaction strength, λ_{ep}, based on the conventional theory for metals, which should be applicable in the normal regime, results in the very small value of 0.035 for the heavy-fermion superconductor UBe_{13}. The interactions responsible for the large observed mass enhancement in the heavy-fermion regime will tend to further decrease λ_{ep}, supporting the assumption that either unconventional electron-phonon coupling or electronic interactions are responsible for pairing [2].

Inelastic neutron scattering measurements for UBe_{13} were applied to determine generalized phonon density-of-states (DOS) which are compared with model calculation [3, 4]. The measured phonon DOS are in fair agreement with a model calculation from a fit of a Born-von Karman model. An Einstein-mode-like resonance occurs at 13 meV [5], which may be important in the explanation of the low-temperature heat capacity C(T) of UBe_{13} [6]. The calculated lattice contribution to the specific heat was used to get the purely electronic part of C(T). This is perfectly fitted by $C/T = \gamma + \delta T^2 \cdot \ln T/T_{SF}$ with $\delta = -1.78$ and $T_{SF} = 29$ K in agreement with [7].

The X-ray photoemission intensity for UBe_{13} was calculated on the basis of relativistic formulas. The results support the experimental finding of very low Be intensity at the Fermi energy. The agreement between calculated results and the experimental data suggests that the highly correlated 5f states, detected below 20 K, condense out of the normal one-electron band states. The conclusion is inconsistent with the interpretation of localized 5f states at high temperatures which yield Curie-Weiss magnetic behavior [8].

Inelastic neutron scattering experiments were made to examine a possible relation between the exceptional electronic properties of UBe_{13} and the lattice dynamics. The results were used to calculate the lattice contribution to the specific heat. Thus, the purely electronic part of the specific heat can be estimated. The generalized phonon DOS $G(\hbar\omega)$, as a result of time-of-flight measurements at 300 K, is compared to results of a calculation from a fit of a Born-von Karman model. A large peak at 13 meV is due to resonance-like vibration of the heavy U atoms inside a cage of Be atoms. The frequency does not shift with temperature [3].

Inelastic polarized and unpolarized neutron scattering measurements at 10 K give information on spin fluctuations in UBe_{13} because they couple to the response of the f electrons. A triple axis spectrometer with a pyrolytic graphite monochromator and analyzer was used

 References for 2.1.3.3.8 on pp. 32/3

for the unpolarized neutron-beam measurements. The results indicate magnetic and phonon contributions [9].

For neutron energy loss between 0 and 13 meV, the scattering at $Q = 6.0 \text{ Å}^{-1}$ is significantly smaller than at $Q = 20 \text{ Å}^{-1}$. This is expected for the electronic form factor Q dependence of magnetic scattering. The higher value at 13 meV at $Q = 6.0 \text{ Å}^{-1}$ is caused by phonons. The paramagnetic scattering by polycrystalline UBe$_{13}$ in the unpolarized-neutron measurements might be isolated by subtraction of the phonon contribution to the inelastic spectrum. This can be done by a comparison of the scattering of the nonmagnetic compound ThBe$_{13}$ which has the same structure and nearly the same phonon density of states. Thus, the magnetic scattering contribution for UBe$_{13}$ is calculated from $I_{mag}(UBe_{13}) = I(UBe_{13}) - N(ThBe_{13})$ with N as a normalization constant [9].

The measurements of inelastic polarized-neutron scattering may also be used to eliminate the magnetic scattering. The lower signal rates of this method, however, require some compromises in energy resolution and longer counting times. The measurements were made on a modified triple-axis spectrometer using vertically magnetized Heussler transmission crystals at the monochromator and analyzer positions and magnetic guide fields to maintain the polarization of the neutrons. The measurements at $Q = 2.0 \text{ Å}^{-1}$ are in fair agreement with the unpolarized spectra [9].

The energy spectrum of the magnetic scattering is characterized by a broad quasi-elastic Lorentzian line shape with a width of 13 ± 2 meV. There is no evidence of a narrow f-level resonance predicted from the electronic specific heat coefficient. The energy-integrated magnetic scattering yields a susceptibility which is consistent with bulk measurements (see Section 2.1.3.3.11, p. 37), indicating that any additional response should have a small spectral width [9].

The low-energy phonon dispersion of UBe$_{13}$ was also determined in inelastic neutron scattering experiments at room temperature and 10 K on the H7 triple-axis spectrometer at the Brookhaven National Laboratory High-Flux Beam Reactor [10] (see also Section 2.1.3.3.5, p. 29).

Electron-phonon effects in the superconducting state of heavy-fermion systems as UBe$_{13}$ are partly contrary to those in BCS superconductors. At the superconducting transition temperature T_c an ultrasonic attenuation peak for longitudinal waves appears, which is markedly influenced by the magnetic fields [11].

References for 2.1.3.3.8:

[1] Stassis, C.; Arthur, J.; Majkrzak, C. F.; Axe, J. D.; Batlogg, B.; Remeika, J.; Fisk, Z.; Smith, J. L.; Edelstein, A. S. (Phys. Rev. [3] B **34** [1986] 4382/5).

[2] Pickett, W. E.; Krakauer, H.; Wang, C. S. (Phys. Rev. [3] B **34** [1986] 6546/9).

[3] Renker, B.; Gompf, F.; Suck, J. B.; Rietschel, H.; Frings, P. (Physica B + C **136** [1986] 376/8).

[4] Gompf, F.; Renker, B.; Rietschel, H.; Nücker, N.; Beuers, J. (Physica B + C **135** [1985] 41/5).

[5] Renker, B.; Gompf, F.; Reichardt, W.; Rietschel, H.; Suck, J. B.; Beuers, J. (Phys. Rev. [3] B **32** [1985] 1859/61).

[6] Overhauser, A. W.; Appel, J. (Phys. Rev. [3] B **31** [1985] 193/202).

[7] Ott, H. R. (private communication, cited in [5]).

[8] Boring, A. M.; Albers, R. C.; Schadler, G.; Marksteiner, P.; Weinberger, P. (Phys. Rev. [3] B **35** [1987] 2447/50).

[9] Goldman, A. I.; Shapiro, S. M.; Shirane, G.; Smith, J. L.; Fisk, Z. (Phys. Rev. [3] B **33** [1986] 1627/33).

[10] Robinson, R. A.; Axe, J. D.; Goldman, A. I.; Fisk, Z.; Smith, J. L.; Ott, R. H. (Phys. Rev. [3] B **33** [1986] 6488/90).

[11] Lühti, B. (Japan. J. Appl. Phys. **26** Suppl. 3 [1987] 1893/4).

2.1.3.3.9   Heat Capacity, Enthalpy, Entropy

Measurements of the specific heat of UBe_{13} were performed in a vacuum adiabatic calorimeter in the temperature range 13.5 to 303.4 K [2]. The measured and calculated values are in excellent agreement. The contributions of the lattice and of the electrons, as given by the equation $C_p = \gamma T + 3\,D(\Theta/T)$, are shown in **Fig.** 2-8. The thermodynamic functions, which were calculated from the measured values, are listed in Table 2/4, p. 34 [2].

The temperature dependence of the specific heat c_p (in $cal \cdot g^{-1} \cdot K^{-1}$), between 0 and 106°C, is given by $c_p = 0.7755 - 9.652 \times 10^{-4} \cdot T - 2.926 \times 10^{-4} \cdot T^2$ [6].

The high-temperature heat capacity of UBe_{13}, as measured by a microcalorimetric technique from room temperature to 1010 K, results in the equation $c_p = 1003 + 0.0855 \cdot T - 29.6 \times 10^6/T$ (in $J \cdot kg^{-1} \cdot K^{-1}$) [7].

The Einstein temperature $\Theta_E = 610$ K and the Debye temperature $\Theta_D = 735$ K were calculated from high-temperature specific heat measurements [7].

Significant anomalies of the specific heat were observed in UBe_{13} single crystals below 1.2 K. A peak of the specific heat occurs at 0.8 K, which is evidence for bulk superconductivity of the compound (see Section 2.1.3.3.15.1, p. 43) [8]. From the experimental $c_p(T)$ values an entropy change of about $1\ J \cdot mol^{-1} \cdot K^{-1}$ between 0 and 1 K was calculated. The same anomaly

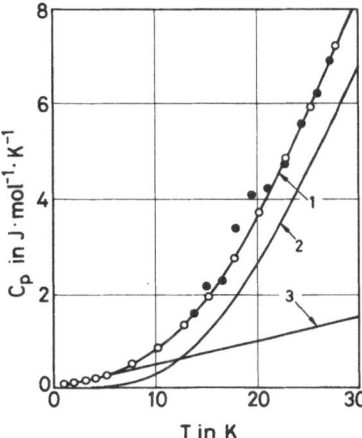

Fig. 2-8: Specific heat of UBe_{13} in the temperature range 1 to 30 K compared with calculated values. Line 2 is the lattice component and line 3 the electronic component of C_p [2].

References for 2.1.3.3.9 on pp. 35/6

of the specific heat was observed in samples of polycrystalline UBe$_{13}$. The $c_p(T)$ values were used to calculate the superconducting transition temperature, $T_c = 0.88$ K for single crystals and $T_c = 0.92$ K for the polycrystalline material [9].

Table 2/4
Thermodynamic Functions and Specific Heat of UBe$_{13}$ [2].

T in K	S_T in $J \cdot mol^{-1} \cdot K^{-1}$	$H_T - H_0$ in J/mol	$H_T - H_0/T$ in $J \cdot mol^{-1} \cdot K^{-1}$	C_p in $J \cdot mol^{-1} \cdot K^{-1}$
0	0	0	0	0
5	0.268	0.686	0.138	0.297
10	0.632	3.473	0.347	0.862
20	1.950	24.10	1.205	3.690
30	3.690	67.61	2.255	8.054
40	6.703	173.05	4.326	13.01
50	10.19	329.95	6.598	18.37
60	13.96	537.48	8.953	23.58
70	17.99	798.98	11.414	29.08
80	22.24	1117.8	13.975	35.31
90	26.77	1503.1	16.702	42.55
100	31.64	1965.1	19.652	50.54
110	36.87	2514.8	22.861	59.89
120	42.51	3163.3	26.363	70.25
130	48.58	3922.3	30.171	81.50
140	55.03	4793.6	34.242	93.05
150	61.86	5783.5	38.556	104.9
160	69.00	6889.8	43.062	116.6
170	76.40	8111.4	47.714	128.1
180	84.04	9447.0	52.484	139.5
190	91.84	10891	57.321	150.5
200	99.81	12446	62.229	160.9
210	107.89	14101	67.149	170.8
220	116.04	15854	72.065	180.2
230	124.24	17700	76.952	189.2
240	132.48	19635	81.814	197.7
250	140.72	21653	86.613	206.0
260	148.95	23753	91.358	213.8
270	157.16	25930	96.035	221.5
280	165.37	28182	100.65	228.8
290	173.47	30506	105.19	236.2
298.15	180.12	32452	108.84	242.1
300	181.63	32902	109.67	243.3

The heat content of UBe$_{13}$ was determined between 0 and 106°C. The results are expressed by the equation $H_T - H_{298} = 0.7755 \cdot T - 4.826 \times 10^{-4} \cdot T^2 + 2.926 \times 10^{-4} \cdot 1/T - 286.4$ in the temperature range 298 to 373 K (with H_T as the heat content in cal/g at the temperature T in K) [6].

The standard entropy was calculated from calorimetric data, $S_{298} = (41.5 \pm 4)$ e. u. [1]. The values of H and S are included in reviews on the thermodynamics of intermetallic systems of actinides [3, 5].

The specific heat was determined using a transient pulse technique, which allowed measurement down to 0.06 K. At this very low temperature a change of the specific heat behavior occurs; a linear temperature term $\gamma_0 = 110$ mJ $\cdot$ mol$^{-1} \cdot$ K^{-2} is extrapolated at T$\rightarrow$0 K [10]. The f-electron contribution to the specific heat of UBe_{13} shows a Schottky anomaly between 50 and 250 K. This behavior is related to the Kondo-demagnetization of the lowest-lying doublet [11].

The ratios of the specific heats in the superconducting and normal states and the respective ratios of the nuclear relaxation rates taken from the literature are in fair agreement with an s-wave model of the superconductivity of $CeCu_2Si_2$, which is applied to UBe_{13} [12].

An increase of the specific heat occurs with a maximum of +9% just above 3 K [13, 14]. Results of measurements of $c_p(T)$ in the temperature range 250 to <50 mK are in fair agreement with earlier findings. The importance of the purity of the compound for the values of c_p is shown [15]. At $T < T_c$, the specific heat c_p deviates significantly from the Bardeen-Cooper-Schrieffer (BCS) theory and obeys a T^3 law rather than an exponential law [16, 17].

Increase of the pressure to 15 kbar causes a shift and a broadening of the maximum of the $C_p(T)$ curve, which occurs at 2 K at atmospheric pressure in UBe_{13}. The rate of this shift is $dT_{max}/dP = 0.23$ K/kbar [18].

Specific heat measurements in magnetic fields up to 8 T indicate drastic changes below the superconducting transition in contrast to the magnetic-field-independent values of C/T in the normal phase [19].

References for 2.1.3.3.9:

[1] Ivanov, M. I.; Tumbakov, V. A. (At. Energiya SSSR **7** [1959] 33/5; Soviet At. Energy **7** [1959] 559/61).

[2] Samorukov, O. P.; Kostyukov, V. N.; Kostylev, F. A.; Tumbakov, V. A. (At. Energiya SSSR **37** [1974] 28/31; Soviet At. Energy **37** [1974] 705/8).

[3] Rand, M. H.; Kubaschewski, O. (The Thermochemical Properties of Uranium Compounds, Oliver & Boyd, Edinburgh — London 1963).

[4] Hultgren, R.; Desai, P. D.; Hawkins, D. T.; Gleiser, M.; Kelley, K. K. (Selected Values of Thermodynamic Properties of Binary Alloys, ASM, Metals Park, Ohio, 1973, pp. 390/1).

[5] Johnson, I.; Waber, J. T.; Chiotti, P.; Miner, W. N. (Compounds of Interest in Nuclear Reactor Technology, Metall. Soc. AIME, Warrendale, Pa., 1964, pp. 171/92).

[6] Farkas, M.; Eldridge, E. (J. Nucl. Mater. **27** [1968] 94/6).

[7] Chipaux, R.; Cecilia, G.; Beauvy, M.; Troc, R. (J. Less-Common Metals **121** [1986] 347/51).

[8] Ott, H. R.; Rudigier, H.; Fisk, Z.; Smith, J. L. (Phys. Rev. Letters **50** [1983] 1595/8).

[9] Ott, H. R.; Rudigier, H.; Fisk, Z.; Smith, J. L. (in: Buyers, W. J. L., Moment Formation in Solids, Plenum, New York 1984, pp. 305/11).

[10] Ravex, A.; Flouquet, J.; Tholence, J. L.; Jaccard, D.; Meyer, A. (J. Magn. Magn. Mater. **63/64** [1987] 400/2).

[11] Felten, R.; Weber, G.; Rietschel, H. (J. Magn. Magn. Mater. **63/64** [1987] 383/5).

[12] Chen Chang-Feng; Zhang Li-Yuan (J. Magn. Magn. Mater. **63/64** [1987] 426/8).

[13] Ott, H. R.; Rudigier, H.; Fisk, Z.; Smith, J. L. (Physica B + C **127** [1984] 359/65).

[14] Ott, H. R.; Fisk, Z. (in: Freeman, A. J.; Lander, G. H., Handbook on the Physics and Chemistry of the Actinides, Vol. 5, Elsevier, Amsterdam 1987, pp. 85/225).

[15] Brison, J. P.; Lasjanias, J. C.; Ravex, A.; Flouquet, J.; Jaccard, D.; Fisk, Z.; Smith, J. L. (Physica B + C **153/155** [1988] 437/8).

[16] Ott, H. R.; Rudigier, H.; Rice, T. M.; Ueda, K.; Fisk, Z.; Smith, J. L. (Phys. Rev. Letters **52** [1984] 1915/8).

[17] Ott, H. R.; Felder, E.; Bernasconi, A.; Fisk, Z.; Smith, J. L.; Taillefer, L.; Lonzarich, G. G. (Japan. J. Appl. Phys. **26** Suppl. 3 [1987] 1217/8).

[18] Thompson, J. D. (J. Magn. Magn. Mater. **63/64** [1987] 358/64).

[19] Brison, J. P.; Ravex, A.; Flouquet, J.; Fisk, Z.; Smith, J. L. (J. Magn. Magn. Mater. **76/77** [1988] 525/6).

2.1.3.3.10 Thermal Conductivity

The thermal conductivity of UBe$_{13}$ depends on the temperature as shown in **Fig.** 2-9 [1]. The thermal conductivity λ of single crystalline UBe$_{13}$ around the transition temperature T$_c$ depends on the strength of applied magnetic fields [2].

The temperature dependence of λ is linear in a field of 75 kOe, but in a field of 1 kOe a nonlinear behavior occurs below 0.79 K. The temperature dependence of the "Lorentz number", $\lambda \cdot D/T$, is quasi-linear in both cases. The temperature function of the thermal conductivity of single crystals of the U beryllide in the superconductive state differs from that in the normal state. Measurements between 0.3 and 4.4 K indicate a dependency $\lambda = \alpha T^2$. In the normal state, however, the dependency is linear with a change of the factor at a temperature of 2.5 K [3].

The thermal conductivity λ was again measured between 0.1 and 3.0 K under the influence of an external magnetic field H of up to 75 kOe. The purpose of applying magnetic fields was to suppress superconductivity to get low-temperature values in the normal state of UBe$_{13}$. The

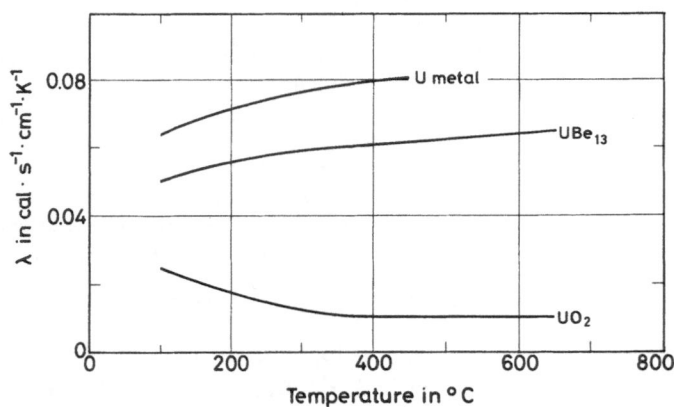

Fig. 2-9. Thermal conductivity λ of UBe$_{13}$ as a function of temperature (compared with U and UO$_2$) [1].

values of λ showed a tendency like that in the zero-field case; a field of 75 kOe caused an additional increase of λ in the order of 0.1 mW·cm^{-1}·K^{-1} [4, 5, 10]. A simple assignment of a temperature dependence is not possible. The resulting Lorentz number λ·D/T in the normal state of UBe$_{13}$ is temperature-dependent and exceeds the classical value L$_o$ = 2.45 × 10^{-8} W·Ω·K^{-2} [4].

Measurements of the thermal conductivity between 0.05 and 1 K indicated the existence of large linear temperature terms. The relationship $\lambda = \alpha \cdot T$ with α = 0.03 mW·cm^{-1}·K^{-1} is valid below T = 0.1 K. A quasi-linear dependence of ln $(\lambda - \alpha T)$ versus $\lambda(T_c)$ was reported [6]. The thermal conductivity in a sample of high purity did not show any signature of a residual linear T term, which is in agreement with a theoretical calculation of the $\lambda(T)$ [7]. Results at very low temperatures differed in their slope with temperature compared to earlier ones, though their magnitudes did not show significant differences [8, 9].

References for 2.1.3.3.10:

[1] Snyder, M. J.; Tripler, A. B., Jr. (ASTM Spec. Tech. Publ. No. 276 [1959] 293/300).

[2] Jaccard, D.; Flouquet, J.; Fisk, Z.; Smith, J. L.; Ott, H. R. (J. Phys. Lettres [Paris] **46** [1985] 811/7).

[3] Alekseevskii, N. E.; Mitin, A. V.; Rudenko, A. S.; Sorokin, A. A. (Pis'ma Zh. Eksperim. Teor. Fiz. **43** [1986] 533/5; JETP Letters **43** [1986] 690/3).

[4] Jaccard, D.; Flouquet, J. (J. Magn. Magn. Mater. **47/48** [1985] 45/50).

[5] Ott, H. R.; Fisk, Z. (in: Freeman, A. J.; Lander, G. H., Handbook on the Physics and Chemistry of the Actinides, Vol. 5, Elsevier, Amsterdam 1987, pp. 85/225).

[6] Ravex, A.; Flouquet, J.; Tholence, J. L.; Jaccard, D.; Meyer, A. (J. Magn. Magn. Mater. **63/64** [1987] 400/2).

[7] Brison, J. P.; Lasjanias, J. C.; Ravex, A.; Flouquet, J.; Jaccard, D.; Fisk, Z.; Smith, J. L. (Physica C **153/155** [1988] 437/8).

[8] Ott, H. R.; Felder, E.; Bernasconi, A.; Fisk, Z.; Smith, J. L.; Taillefer, L.; Lonzarich, G. G. (Japan. J. Appl. Phys. **26** Suppl. 3 [1987] 1217/8).

[9] Sulpice, A.; Gandit, P.; Chaussy, J.; Flouquet, J.; Jaccard, D.; Lejay, P.; Tholence, J. (Low Temp. Phys. **62** [1986] 39/54).

[10] Jaccard, D.; Flouquet, J. (Helv. Physica Acta **60** [1987] 108/21).

2.1.3.3.11 Magnetic Susceptibility

The eight heavy-fermion systems known up to the present time (CeCu$_2$Si$_2$, UBe$_{13}$, UPt$_3$, NpBe$_{13}$, U$_2$Zn$_{17}$, UCd$_{11}$, CeAl$_3$, CeCu$_6$) are similar in their magnetic susceptibilities [1]. These show a large temperature dependence which quantitatively obeys the Curie-Weiss law at near room temperature [2]. Results of measurements between 0 and 300 K are shown in **Fig. 2-10**, p. 38 [3]. A temperature-independent Van Vleck-type susceptibility that would have provided evidence for a 5f^2 configuration of the U ions could not be detected [4]. The temperature dependence becomes Pauli-like at lower temperatures [5].

Curie-Weiss behavior of the U beryllide was observed above ca. 120 K; μ_{eff} = 3.08 μ_B/U ion and Θ_p = −53 K [7]. These results confirmed earlier similar measurements. **Fig. 2-11**, p. 38, shows the temperature dependence of the reciprocal susceptibility χ^{-1} up to 1000 K.

 References for 2.1.3.3.11 on p. 39

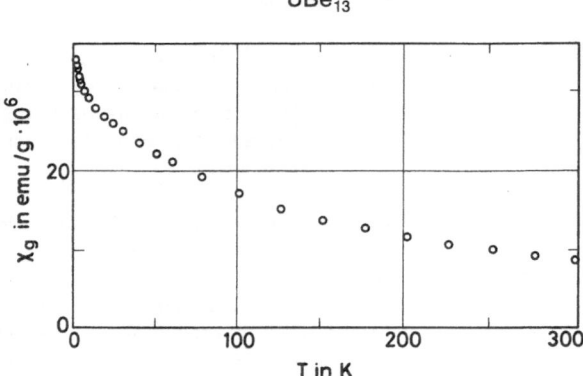

Fig. 2-10. Susceptibility of UBe$_{13}$ as a function of temperature up to 300 K [3].

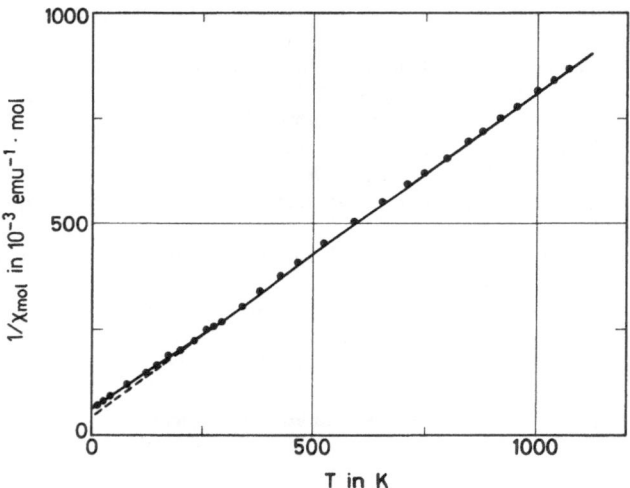

Fig. 2-11. Reciprocal susceptibility of UBe$_{13}$ as a function of temperature between 4.2 and 1000 K [8].

There is evidence for local moment paramagnetism, as expected from the large U-atom spacing. The effective paramagnetic moment agrees with the 5f^3 configuration of UBe$_{13}$. Thus, the U ion was found to be trivalent if the f electrons are considered to be localized [8]. The ratio χ/γ is comparably small as for all heavy-mass — heavy-fermion superconductors [9].

The measurements of the magnetic susceptibility of UBe$_{13}$ were summarized and discussed in [10]. The exhaustion of the bulk susceptibility at 10 K, revealed by inelastic neutron scattering, is due to a broad quasi-elastic response, and the linewidth grows with increasing temperature, saturating at $\Gamma \approx 40$ meV [11].

The magnetic parameters, as measured by different authors, are in fair agreement as is shown in Table 2/5.

Table 2/5
Magnetic Parameters of UBe_{13}.

μ_{eff}	χ (T = 0) in 10^{-3} emu/mol	Curie-Weiss temperature Θ in K	Ref.
3.4	13.5	−70	[8]
2.99	12.1	−68	[3]
3.52	16.6 at 1.4 K	−98	[4]
2.6 *)	15 at T_c	−53	[6]
3.08	15	−53	[7]

*) 3.08 is considered to be the correct value [126], 2.6 seems to be due to an error.

References for 2.1.3.3.11:

[1] Stewart, G. R. (Rev. Mod. Phys. **56** [1984] 755/87).

[2] Fritsch, G. (Physik Unserer Zeit **18** [1987] 17/20).

[3] Brodsky, M. B.; Friddle, R. J. (in: Graham, C. D.; Lander, G. H.; Rhyne, J. J., Magnetism and Magnetic Materials — 1974, Am. Inst. Phys., New York 1975, pp. 353/4).

[4] Bucher, E.; Maita, J. P.; Hull, G. W.; Fulton, R. C.; Cooper, A. S. (Phys. Rev. [3] B **11** [1975] 440/9).

[5] Ma, Y. P.; Brooks, J. S.; Schmiedeshoff, G. M.; Maple, M. B.; Fisk, Z.; Smith, J. L.; Ott, H. R. (Japan. J. Appl. Phys. **26** Suppl. 3 [1987] 1235/6).

[6] Ott, H. R.; Rudigier, H.; Fisk, Z.; Smith, J. L. (Phys. Rev. Letters **50** [1983] 1595/8).

[7] Ott, H. R.; Rudigier, H.; Fisk, Z.; Smith, J. L. (in: Buyers, W. J. L., Moment Formation in Solids, Plenum, New York 1984, pp. 305/11).

[8] Troc, R.; Trzebiatowski, W.; Pirrek, K. (Bull. Acad. Polon. Sci. Ser. Sci. Chim. **19** [1971] 427/32).

[9] Newns, D. M. (Phys. Scr. T **23** [1988] 113/5).

[10] Ott, H. R.; Fisk, Z. (in: Freeman, A. J.; Lander, G. H., Handbook on the Physics and Chemistry of the Actinides, Vol. 5, Elsevier, Amsterdam 1987, pp. 85/225).

[11] Goldmann, A. I. (Japan. J. Appl. Phys. **26** Suppl. 3 [1987] 1887/92).

2.1.3.3.12　Electrical Resistivity

The electrical resistivity of sintered UBe_{13} is 113 $\mu\Omega \cdot$ cm at 27.6°C [1]. The electrical resistivity shows a temperature dependence between 0.5 and 290 K, which is anomalous for an intermetallic compound [2]. The value in single crystalline UBe_{13} is $\varrho = 137$ $\mu\Omega \cdot$ cm at room temperature. At lower temperature, ϱ rises steadily with decreasing temperature, finally

References for 2.1.3.3.12 on p. 40

reaching a flat maximum of 228 $\mu\Omega \cdot$ cm at around 10 K. With further cooling, ϱ increases again and reaches a maximum of 234 $\mu\Omega \cdot$ cm at around 2.35 K. Below this maximum, ϱ decreases to the superconducting transition at 0.86 K. The transition into the superconducting state has a very high gradient (see Section 2.1.3.3.15, p. 43).

The resistivity ϱ at constant magnetic field is composed of two terms, $\varrho(T) = \varrho_0(H) + A_H T^2$. They represent a residual contribution characteristic of impurity effects and a quadratic temperature dependence, which may have an intrinsic origin. The extrapolated high-field residual resistivity $\varrho_0(H) = 12 \mu\Omega \cdot$ cm is the lowest value yet reported [3]. A survey of the resistivity of UBe₁₃ at low temperatures is given in [4]. The resistivity is compared with values of other heavy-fermion compounds in [5].

The electrical resistivity of UBe₁₃ is influenced by an applied hydrostatic pressure of 15 kbar. The maximum of ϱ, which occurs at atmospheric pressure at 2 K, is shifted and broadened. The rate of temperature shift is $dT_{max}/dP = 0.23$ K/kbar [6].

The electrical resistivity of UBe₁₃ was measured between 1.2 and 300 K at pressures up to 67 kbar and in magnetic fields up to 6 T. The resistivity reflects the partial delocalization of the f resonance [7].

References for 2.1.3.3.12:

[1] Tripler, A. B., Jr.; Snyder, M. J.; Duckworth, W. H. (BMI-1313 [1959] 1/55; N.S.A. **13** [1959] No. 8908).
[2] Ott, H. R.; Rudigier, H.; Fisk, Z.; Smith, J. L. (Phys. Rev. Letters **50** [1983] 1595/8).
[3] Brison, J. P.; Lasjanias, J. C.; Ravex, A.; Flouquet, J.; Jaccard, D.; Fisk, Z.; Smith, J. L. (Physica C **153/155** [1988] 437/8).
[4] Ott, H. R.; Fisk, Z. (in: Freeman, A. J.; Lander, G. H., Handbook on the Physics and Chemistry of the Actinides, Vol. 5, Elsevier, Amsterdam 1987, pp. 85/225).
[5] Fritsch, G. (Physik Unserer Zeit **18** [1987] 17/20).
[6] Thompson, J. D. (J. Magn. Magn. Mater. **63/64** [1987] 358/64).
[7] Mao, S. Y.; Jaccard, D.; Sierro, J.; Fisk, Z.; Smith, J. L. (J. Magn. Magn. Mater. **76/77** [1988] 241/2).

2.1.3.3.13   Magnetoresistance and Hall Effect

The very large magnetoresistance of UBe₁₃ at low temperature was described in 1985. The discovery resulted from measurements on a single crystal in the temperature range 0.15 to 1.2 K and magnetic fields up to 60 kOe. The residual resistivity seems to depend on the magnetic field [1].

Magnetoresistivity experiments at very low temperatures exhibit a simple quadratic dependence of the resistivity on T just above the transition temperature T_c. The extrapolated residual resistivity depends on the magnetic field. The behavior might be due to the presence of light and heavy particles [2].

The anisotropy of the magnetoresistance $\varrho(H)$ is relatively slight in UBe_{13} single crystals above T_c. Measurements were made at 1.9 and 4.2 K in field orientations $H\|c_2$ and $H\|c_4$. In a field of 75 kOe the field anisotropy $\varrho(H)$ does not exceed the order of 2%. This is an order of magnitude lower than in the case of the heavy-fermion system $CeCu_6$ [3].

The experimental results on the magnetic properties of UBe_{13} in the normal state (suscepti-bility and magnetoresistance) and in the superconducting state (Meissner effect, upper and lower critical field) are discussed in relation to the nature of the heavy-fermion superconductor. The compounds behave as a dense array of independent Kondo impurities in the normal state. At $T < 1$ K the onset of coherent scattering leads to new phenomena which are characteristic of the Kondo lattice. A superconducting order parameter at T_a is only found on a part of the Fermi surface. In pure and Th-doped UBe_{13} samples the remainder of the Fermi surfaces becomes superconducting at $0.5\,T_c$. The results do not permit conclusions concerning the nature of the superconducting state of the compound [4]. The low-temperature magneto-resistivity of UBe_{13} shows evidence for a field dependence of interactions between the heavy fermions [14].

The superconducting transition temperature T_c and the upper-critical magnetic field of UBe_{13} decrease under hydrostatic pressure. The low-temperature magnetoresistance remains large and negative under pressure. The region in which the resistivity has a dependence on T^2 increases with the magnetic field and with pressure [5].

The Hall voltages of single phase samples of UBe_{13} with a lattice constant a = 10.254 Å were measured between 1.9 and 4.2 K in magnetic fields of 20 to 140 kOe. The results are shown in **Fig. 2-12**. The Hall constant increases with decreasing temperature; these results of Alekseevskii et al. [6, 7] are confirmed in the more recent measurements of the Hall coefficients of UBe_{13} single crystals. The Hall coefficient reaches a maximum at 1.5 K [8] indicating the transition into the heavy-fermion state. Below this maximum a decrease by 20% occurs before the transition temperature T_c is reached. Similar results are reported in [9].

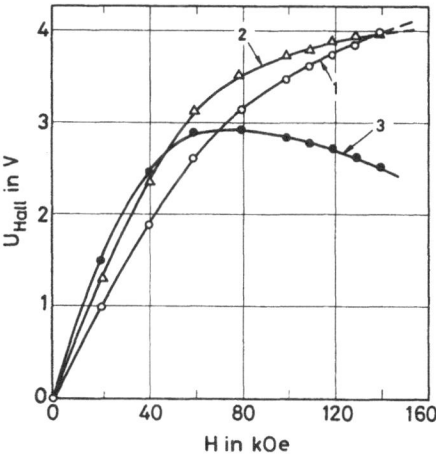

Fig. 2-12. The influence of magnetic field and temperature on the Hall voltage of a sample of UBe_{13}; 1) T = 4.2 K, 2) T = 3 K, 3) T = 1.9 K. The measurement current is 10 mA [6].

References for 2.1.3.3.13 on p. 42

The very low temperature necessary for this type of experiment can be reached in crucibles of Er^{3+}- or Nd^{3+}-substituted yttrium—aluminium garnet single crystals, which are treated in a simple adiabatic demagnetization apparatus in order to cool the UBe$_{13}$ samples. The method allows investigations in zero or high magnetic fields [10].

The Hall coefficient R_H of UBe$_{13}$, as measured by a dc method, decreases below T = 1 K and the coefficient becomes negative below 0.7 K [11]. This R_H(T) behavior is characteristic of the transition in the coherent scattering regime, which sets in just below the superconducting transition temperature T_c. The Hall voltage of the compound is also dependent on the magnetic field strength [12]. At T = 1.7 K, the Hall voltage passes a maximum at a magnetic field of H ≈ 60 kOe. The maximum of the coefficient disappears owing to the effect of rising impurity concentration in UBe$_{13}$ [13]. The Hall voltage is strongly nonlinear in a magnetic field [14]. The knowledge on the Hall effect in the U beryllide was summarized in [13].

Measurements under hydrostatic pressure up to 4 kbar show a small change of the Hall coefficient by ca. 5%. At T = 100 K the Hall coefficient rises with pressure but at T = 4.2 K the value decreases [7].

References for 2.1.3.3.13:

[1] Maple, M. B.; Chen, J. W.; Lambert, S. E.; Fisk, Z.; Smith, J. L.; Ott, H. R.; Brooks, J. S.; Naughton, M. J. (Phys. Rev. Letters **54** [1985] 477/80).

[2] Remenyi, G.; Jaccard, D.; Flouquet, J.; Briggs, A.; Fisk, Z.; Smith, J. L.; Ott, H. R. (J. Phys. [Paris] **47** [1986] 367/72).

[3] Alekseevskii, N. E.; Mitin, A. V.; Nizhankovskii, V. I.; Firsov, V. I.; Khlybov, E. P. (Pis'ma Zh. Eksperim. Teor. Fiz. **41** [1985] 335/7; JETP Letters **41** [1986] 410/3).

[4] Rauchschwalbe, U. (Physica B + C **147** [1987] 1/80).

[5] Willis, J. O.; Thompson, J. D.; Smith, J. L.; Fisk, Z. (J. Magn. Magn. Mater. **63/64** [1987] 461/3).

[6] Alekseevskii, N. E.; Narozhnyi, V. N.; Nizhankovskii, V. I.; Nikolaev, E. G.; Khlybov, E. P. (Pis'ma Zh. Eksperim. Teor. Fiz. **40** [1984] 421/3; JETP Letters **40** [1984] 1241/4).

[7] Alekseevskii, N. E.; Nizhankovskii, V. I.; Narozhnyi, V. N.; Khlybov, E. P.; Mitin, A. V. (J. Low Temp. Phys. **64** [1986] 87/104).

[8] Penney, T.; Stankiewicz, J.; von Molnar, S.; Fisk, Z.; Smith, J. L.; Ott, H. R. (J. Magn. Magn. Mater. **54/57** [1986] 370/2).

[9] Alekseevskii, N. E.; Aliev, F. G.; Brandt, N. B.; Zalyalyutdinov, M. K.; Kovachik, V., Moshchalkov, V. V.; Mitin, A. V. (Pis'ma Zh. Eksperim. Teor. Fiz. **43** [1986] 482/4; JETP Letters **43** [1986] 622/5).

[10] Alekseevskii, N. E. (Pis'ma Zh. Eksperim. Teor. Fiz. **40** [1984] 66/9; JETP Letters **40** [1984] 5598/600).

[11] Aliev, F. G.; Brandt, N. B.; Moshchalkov, V. V.; Zalyaljutdinov, M. K.; Alekseevskii, N. E.; Khlybov, E. P. (J. Magn. Magn. Mater. **63/64** [1987] 458/60).

[12] Alekseevskii, N. E.; Mitin, A. V.; Narozhnyi, V. N.; Nizhankovskii, V. I.; Khlybov, E. P. (J. Magn. Magn. Mater. **63/64** [1987] 467/8).

[13] Ott, H. R.; Fisk, Z. (in: Freeman, A. J.; Lander, G. H., Handbook on the Physics and Chemistry of the Actinides, Vol. 5, Elsevier, Amsterdam 1987, pp. 85/225).

[14] Brison, J. P.; Briggs, A.; Flouquet, J.; Lapierre, F.; Fisk, Z.; Smith, J. L. (J. Magn. Magn. Mater. **76/77** [1988] 243/4).

2.1.3.3.14 Thermoelectric Power

The thermoelectric power Q shows a sharp change at ≈ 0.8 K if the measurements are made in magnetic fields up to 20 kOe; the changes are in the order of ≈ -24 µV/K. The proportionality of Q to temperature, that is usual for a metal at low temperature, does not occur with UBe_{13}. Such behavior occurs at a field of 75 kOe. The thermoelectric power is about -20 µV/K above the transition temperature; its value decreases with increasing temperature [1]. The thermoelectric power was measured between 1.2 and 300 K at pressures up to 67 kbar and in magnetic fields up to 6 T. The marked oscillation of the low-temperature thermopower strongly suggests that pressure drives a magnetic order above the transition temperature T_c [2]. The data are compiled in [3].

References for 2.1.3.3.14:

[1] Jaccard, D.; Flouquet, J.; Fisk, Z.; Smith, J. L.; Ott, H. R. (J. Phys. Lettres [Paris] **46** [1985] L811/L817).

[2] Mao, S. Y.; Jaccard, D.; Sierro, J.; Fisk, Z.; Smith, J. L. (J. Magn. Magn. Mater. **76/77** [1988] 241/2).

[3] Ott, H. R.; Fisk, Z. (in: Freeman, A. J.; Lander, G. H., Handbook on the Physics and Chemistry of the Actinides, Vol. 5, Elsevier, Amsterdam 1987, pp. 85/225).

2.1.3.3.15 Superconductivity

2.1.3.3.15.1 Heavy-Fermion Superconductivity

The superconductivity in the high-effective-mass (ca. 200 m_e) electrons of intermetallics of actinides was discovered in 1979, and since that time, several such "heavy-fermion" systems have been discovered, among them the compound UBe_{13}. This unusual behavior of UBe_{13} was first observed by Ott et al. [1] in 1983. The discovery of the heavy-fermion superconductivity of UBe_{13} has an interesting history. Bucher et al. [2] found a sharp superconduction transition in the compound at 0.97 K, which was only slowly depressed by the applied magnetic field. The authors only measured the specific heat of this material down to 1.8 K and made an interpretation of the phenomenon which related the superconductivity to U filaments. Had they continued the measurements of the specific heat to a temperature below the transition temperature, they would have opened the field of heavy-fermion superconductivity of the compound five years earlier.

One year after the publication of the discovery of the heavy-fermion superconductor UBe_{13}, Stewart [3] published a first survey on this class of superconductors and their low-temperature physical properties. UBe_{13} is among eight heavy-fermions, which include superconductors (UBe_{13}, UPt_3, $CeCu_2Si_2$), magnets ($NpBe_{13}$, UZn_{17}, UCd_{11}), and materials in which no ordering was observed ($CeAl_3$, $CeCu_6$). This group of materials was studied intensively since 1979.

A second survey on the superconductivity and magnetism in heavy-electron U intermetallics was published in 1985 [4]. According to this survey, the group of intermetallics might be best described as a Fermi liquid of quasi-particles with very large effective masses. They all show a large electronic specific heat, a temperature-independent magnetic susceptibility and a temperature dependence of the electrical resistivity proportional to T^2. At higher temperatures a strongly temperature-dependent C_p/T ratio, a Curie-Weiss type magnetic susceptibility and an electrical resistivity occurs, increasing with decreasing temperature. After passing a maximum, the resistivity varies with a positive slope $\partial\varrho/\partial T$. All these relation-

References for 2.1.3.3.15.1 on p. 47

ships indicate that an unusual electronic state occurs. However, the discovery of bulk super-conductivity of UBe$_{13}$ below 0.9 K [1] was rather surprising. One might, therefore, understand that first magnetic indications for this behavior were misunderstood and ascribed to precipitated U filaments. The large anomaly of the specific heat at the transition temperature confirms the bulk character of the superconductivity and demonstrates the electronic nature of the large specific heat and the participation of the "heavy" electrons in the formation of the superconducting state. Another unusual behavior of UBe$_{13}$ at the transition temperature is the very uncommon temperature dependence of the upper critical field of the compound (see also [5 to 8]).

The superconducting transition temperature T_c in a thin film of UBe$_{13}$ of about 30 μm, prepared by means of sputtering, is very close to that of bulk material. The critical fields are also in the same order [9].

A Ginsburg-Landau model of even-parity superconducting order parameters with transition temperatures T_0 and T_2 was discussed in relation to the ternary compound $(U_xTh_{1-x})Be_{13}$ (see p. 59). The critical temperatures for the pure system were assumed such that $T_2 > T_0$. It was suggested that the impurity scattering strongly suppresses T_2, whereas T_0 remains unaffected. The inequality is thus reversed for an impurity concentration $x > x_c$ [10].

The unusual superconducting properties of this compound were also treated on the basis of a phenomenological model [11] which incorporated the effect of fluctuations on the interactions between superconducting and coherent Kondo screening in heavy-fermion superconductors. It was pointed out that the appearance of a pseudo-gap near the Fermi energy leads to a temperature-dependent effective BCS interaction between the electrons forming the Cooper pairs [12].

The low-temperature specific heat of UBe$_{13}$ single crystals was measured below 1.2 K in zero magnetic field [1]. As shown in **Fig. 2-13**, an anomaly of the specific heat occurs at 0.8 K, which is evidence for bulk superconductivity in the compound. The anomaly cannot be related to magnetic ordering. From the experimental $c_p(T)$ values, an entropy change of about $1 \text{ J} \cdot \text{mol}^{-1} \cdot \text{K}^{-1}$ between 0 and 1 K was calculated. The same anomaly of the specific heat was observed in polycrystalline UBe$_{13}$ by the same authors [13].

Fig. 2-14 shows the $c_p(T)$ values between 0 and 13 K. The $T_{c \text{ [onset]}}$ was deduced from c_p measurements; the value for single crystals is $T_c = 0.88$ K, for the polycrystalline material 0.92 K [13].

The specific heat of another sample of the beryllide was determined using a transient heat pulse technique [14]. The lowest temperature in this series of measurements was $T = 0.06$ K. At this very low temperature a change of the specific heat behavior occurred, and a linear temperature term $\gamma_o = 110 \text{ mJ} \cdot \text{mol}^{-1} \cdot \text{K}^{-2}$ was extrapolated at $T \rightarrow 0$ K.

The f-electron contribution to the specific heat of UBe$_{13}$ shows a Schottky anomaly at temperatures between 50 and 250 K [15]. This behavior is connected with the Kondo demagnetization of the lowest-lying doublet.

The magnetization and magnetoresistance of the Kondo lattice of UBe$_{13}$ indicate a break-down of a description in terms of a single energy scale T_k. T_k becomes temperature-dependent below ca. 10 K, and the canonical relationship between R(H) and M(H) does not hold. The possibilities of non-cubic strain, induced by anisotropic superconductivity in UBe$_{13}$, was investigated by means of ultrasound experiments [16]. The suggestion of a tetragonal strain in Th-doped UBe$_{13}$ was found to be not applicable. The observed frequency- and amplitude-dependent absorption anomalies at $T \ll T_c$ might reflect domain wall motion.

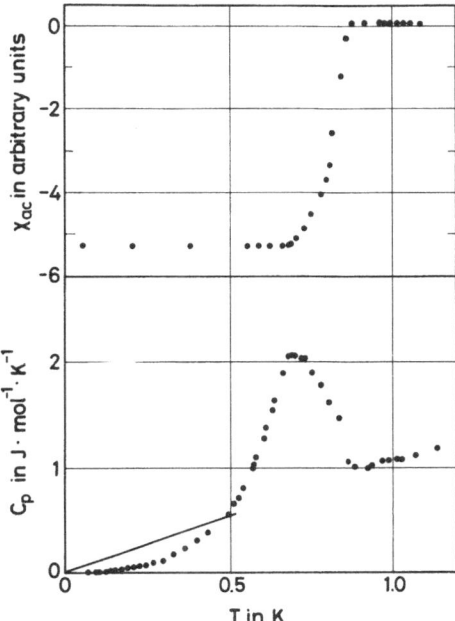

Fig. 2-13. Temperature dependence of the low-frequency ac magnetic susceptibility (upper diagram) and the specific heat of UBe_{13} single crystals below 1.2 K in zero magnetic field ($H_{ac} < 0.3$ Oe) [1].

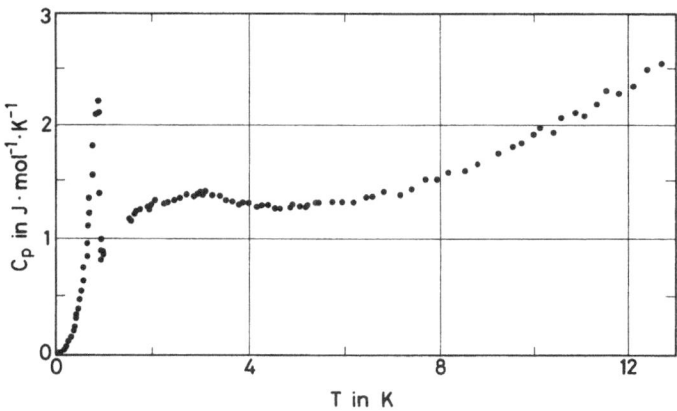

Fig. 2-14. Low-temperature specific heat of polycrystalline UBe_{13} [13].

Using UBe_{13} as an example of heavy-electron metals, normal state, and superconducting state properties of this class of materials were discussed in [17]. Concerning the experimental data of the specific heat and the theoretical data, the superconducting state is the electronic analogue of the A phase of superfluid ^3He. This idea is supported by the effects of impurities

References for 2.1.3.3.15.1 on p. 47

in the compound and by properties of similar compounds which are not superconductors, but are magnetically ordered. An increase of the specific heat at a temperature below 3 K occurs with a maximum of +9%, just above 3 K. Above 9 K the effect of the strong magnetic field disappears. This might be explained by means of a very-narrow-band model.

The specific heat of UBe$_{13}$ at $t < T_c$ significantly deviates from the Bardeen-Cooper-Schrieffer (BCS) theory and obeys a T^3, rather than an exponential law [18]. A good description is obtained by the assumption of an Anderson-Brinkman-Morel p-wave superconducting state at all temperatures. **Fig.** 2-**15** shows the low-temperature specific heat of UBe$_{13}$ in comparison to the Anderson-Brinkman-Morel p-wave superconduction model. The close analogy in the physical properties between the heavy-Fermi liquids UBe$_{13}$ and ^{3}He confirms the assumption of p-wave superconductivity according to this model. The spin-fluctuation parameter is large and consistent with the stability of such a state.

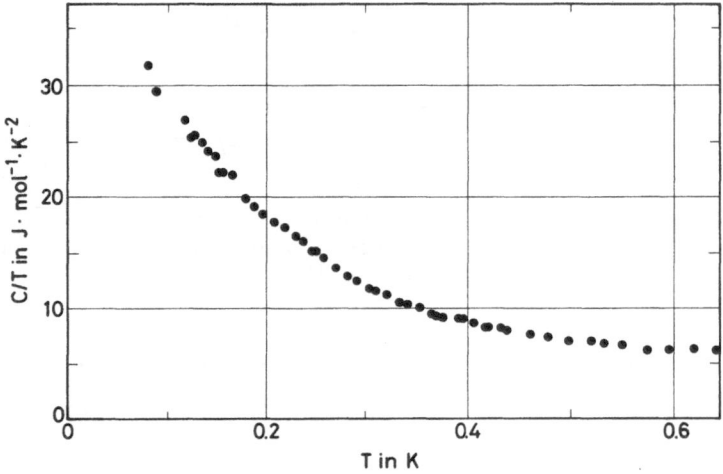

Fig. 2-15. Low-temperature specific heat C$_p$/T of UBe$_{13}$ [18].

A theoretical approach relates the p-wave superconducting behavior of UBe$_{13}$ to a hydrodynamical model. The variables which are used in this model follow a Hamiltonian dynamics that results in an internal Josephson effect between the population of spin-up and spin-down Cooper pairs. The effect is similar to the longitudinal spin-resonance behavior, which is also observed in the superfluid phase of ^{3}He. The longitudinal ultrasonic attenuation constants, which are gained with this theory, are close to measured values in the temperature range 0.75 to 0.90 K [19]. The available data indicate an unconventional pairing, strong Fermi-liquid effects, resonant impurity scattering and multiband effects that need support and explanation by theory [20].

It was suggested that at least part of the Fermi surface, which does not participate in superconductivity at $T_c = 0.9$ K, develops an order parameter in the accessible temperature range around 0.55 K. The effect can be sharply enhanced by means of doping UBe_{13} with small amounts of Th [21].

The superconducting transition temperature T_c of UBe_{13} varies with the hydrostatic pressure up to 12 kbar. A small depression of T_c occurs in polycrystalline UBe_{13}, as well as in single crystals of the beryllide, with a slope of ca. 0.01 K/kbar [22]. The temperature coefficient of the specific heat of this compound shows a decrease of ca. 30% at a pressure of the order of 9 kbar. This decrease appears in all highly-correlated superconductors [23, 24].

References for 2.1.3.3.15.1:

[1] Ott, H. R.; Rudigier, H.; Fisk, Z.; Smith, J. L. (Phys. Rev. Letters **50** [1983] 1595/8).

[2] Bucher, E.; Maita, J. P.; Hull, G. W.; Fulton, R. C.; Cooper, A. S. (Phys. Rev. [3] B **11** [1975] 440/9).

[3] Stewart, G. R. (Rev. Mod. Phys. **56** [1984] 755/87).

[4] Ott, H. R.; Rudigier, H.; Fisk, Z.; Smith, J. L. (J. Appl. Phys. **57** [1985] 3044/8).

[5] Ott, H. R. (Physica B + C **130** [1985] 163/7).

[6] Fisk, Z.; Ott, H. R.; Smith, J. L. (Physica B + C **130** [1985] 159/62).

[7] Fisk, Z.; Ott, H. R.; Smith, J. L. (J. Magn. Magn. Mater. **47/48** [1985] 12/6).

[8] Lee, P. A.; Rice, T. M.; Serene, J. W.; Sham, J. L.; Wilkins, J. W. (Comments Condens. Matter. Phys. **12** No. 3 [1986] 99/161).

[9] Kang, J. H.; Maps, J.; Goldman, A. M.; Brooks, J. S.; Fisk, Z.; Smith, J. L. (Japan. J. Appl. Phys. **26** Suppl. 3 [1987] 1233/4).

[10] Kumar, P.; Wolfle, P. (Phys. Rev. Letters **59** [1987] 1954/7).

[11] Moshchalkov, V. V. (Pis'ma Zh. Eksperim. Teor. Fiz. **45** [1987] 181/4; JETP Letters **45** [1987] 223/7).

[12] Chen Chang-fen; Zhang Li-yuan (Physica B + C **144** [1987] 193/9).

[13] Ott, H. R.; Rudigier, H.; Fisk, Z.; Smith, J. L. (in: Buyers, W. J. L., Moment Formation in Solids, Plenum, New York 1984, pp. 305/11).

[14] Ravex, A.; Flouquet, J.; Tholence, J. L.; Jaccard, D.; Meyer, A. (J. Magn. Magn. Mater. **63/64** [1987] 400/2).

[15] Felten, R.; Weber, G.; Rietschel, H. (J. Magn. Magn. Mater. **63/64** [1987] 282/5).

[16] Batlogg, B.; Bishop, D. J.; Bucher, E.; Golding, B., Jr.; Ramirez, A. P.; Fisk, Z.; Smith, J. L.; Ott, H. R. (J. Magn. Magn. Mater. **63/64** [1987] 441/6).

[17] Ott, H. R.; Rudigier, H.; Fisk, Z.; Smith, J. L. (Physica B + C **127** [1984] 359/65).

[18] Ott, H. R.; Rudigier, H.; Rice, T. M.; Ueda, K.; Fisk, Z.; Smith, J. L. (Phys. Rev. Letters **52** [1984] 1915/8).

[19] Rodriguez, J. P. (Phys. Rev. [3] B **36** [1987] 168/79).

[20] Rainer, D. (Phys. Scr. T **23** [1988] 106/12).

[21] Rauchschwalbe, U.; Steglich, F.; Sparn, G.; Bredl, C. D.; Fulde, P.; Maki, K. (Japan. J. Appl. Phys. **26** Suppl. 3 [1987] 1225/6).

[22] Chen, J. W.; Lambert, S. E.; Maple, M. B.; Fisk, Z.; Ott, H. R. (Proc. 17th Intern. Conf. Low Temp. Phys., Karlsruhe 1984, Pt. I, pp. 325/6).

[23] Olsen, J. A.; Fisher, R. A.; Phillips, N. E.; Giorgi, A. L.; Stewart, G. R. (Physica B + C **144** [1986] 54/5).

[24] Olsen, J. A.; Fisher, R. A.; Phillips, N. E.; Stewart, G. R.; Giorgi, A. L. (Bull. Am. Phys. Soc. [2] **31** [1986] 648).

2.1.3.3.15.2 Josephson Effect

The Josephson effect was used to check spin-pairing states between contacted superconductors. Josephson tunneling experiments between UBe$_{13}$ and niobium tips have, therefore, been performed. Anomalous proximity-induced s-wave superconductivity has been observed via the ac Josephson effect and the magnetic field and temperature dependences of the Josephson current in Nb/UBe$_{13}$ junctions. The effect occurs in UBe$_{13}$ at much higher relative temperatures T/T_c^* than in indium or tin. The observation of s-wave superconductivity in U beryllide favors the possibility of an s-wave ground state for this intermetallic. It can be, however, that the influence of the Nb pairs may be so strong that it overcomes the preference of UBe$_{13}$ for a triplet state below 0.85 K [1].

The oscillatory variation of the critical current with magnetic field is a fundamental property of the Josephson effect. The dependence of the phase difference Φ between the two pair-states wave functions is shown in a $I_c(H)$ plot for a UBe$_{13}$ — Nb contact. A minimum, which occurs at ca. 43 Oe, is consistent with a single contact of a dimension of 5 to 10 µm [2].

The Josephson tunneling supercurrent between the two superconductors is predicted to depend on the relative parity of the Cooper pairs in the two electrodes. In the singlet-triplet case the supercurrent J_0 is predicted to be on the fourth order [1]. Experiments using a Nb point and K-band microwaves on the heavy fermion superconductor UBe$_{13}$ were performed in a vacuum can, which could be cooled to below 0.7 K [3]. The step structure, weakly present in the I-V curve, is made more clearly evident in the derivative spectrum dV/dI. The spacing of the Shapiro step structure is consistent with J_0 tunneling between two s-wave superconductors. The s-wave superconducting state of UBe$_{13}$ is induced by proximity to Nb. The effect is surprisingly strong, with s-wave pairs in UBe$_{13}$ persisting above 7 K. The anomalously strong s-wave proximity effect might be due to the electronic properties (see Section 2.1.3.3.3, p. 23) of UBe$_{13}$. The observations appear to favor an s-wave ground state for UBe$_{13}$ as proposed by Overhauser, Appel [4].

Josephson current measurements were also performed between electropolished UBe$_{13}$ surfaces and Ta wire probes in order to support the suggestion that the pairing is an odd-parity spin triplet. The results give evidence of the suppression of the induced singlet super-conductivity by means of the bulk UBe$_{13}$ superconductivity. This indeed indicates a triplet superconducting state of odd-parity in UBe$_{13}$ [5].

The unusual superconductivity of the heavy-fermion system UBe$_{13}$ was again studied by means of point contact and tunneling effects. Normal metals such as W or Pt (metallic point contacts), or GdAs (Schottky-barrier tunneling contacts) were used at temperatures 50 mK to 1 K. The metal-point-contact characteristics (dV/dI versus V) show zero-bias minima of width 2 below the transition temperature T_c. The ratio $2/k_B T_c$ is close to the BCS value [6].

The p-wave superconducting behavior of UBe$_{13}$ is related to a hydrodynamical theory. The variables, which are used in this model, follow a Hamiltonian dynamics that results in an internal Josephson effect between the population of spin-up and spin-down Cooper pairs. The effect is similar to the longitudinal spin-resonance behavior which is also observed in the superfluid phases of ^{3}He. The longitudinal ultrasonic attenuation constants (see Section 2.1.3.3.7, p. 30), which are gained with this theory, are close to measured values in the temperature range 0.75 to 0.90 K [7].

References for 2.1.3.3.15.2:

[1] Pals, J. A.; van Haeringen, W.; van Maaren, M. H. (Phys. Rev. [3] B **15** [1977] 2592/9).

[2] Siyuan Han; Ng, K. W.; Wolf, E. L.; Braun, H. F.; Tanner, L.; Fisk, Z.; Smith, J. L.; Beasley, M. R. (Phys. Rev. [3] B **32** [1985] 7567/70).

[3] Wolf, E. L.; Noer, R. J.; Han, S.; Ng, K. W.; Chen, T. P.; Finnemore, D. K.; Tanner, L. (Physica B + C **135** [1985] 65/8).

[4] Overhauser, A. W.; Appel, J. (Phys. Rev. [3] B **31** [1985] 193/202).

[5] Siyuan Han; Ng, K. W.; Wolf, E. L.; Millis, A.; Smith, J. L.; Fisk, Z. (Phys. Rev. Letters **57** [1986] 238/41).

[6] Nowack, A.; Heinz, A.; Oster, F.; Wohlleben, D.; Güntherodt, G.; Fisk, Z.; Menovsky, A. (Phys. Rev. [3] B **36** [1987] 2436/9).

[7] Rodriguez, J. P. (Phys. Rev. [3] B **36** [1987] 168/79).

2.1.3.3.15.3 The Influence of Magnetic Fields on the Superconductivity

The magnetic field is expected to change the specific heat, either by effecting the interactions between electrons (if λ is large), or by broadening the very narrow electronic density-of-states (if λ is small), or by both mechanisms in the intermediate case. The effect of a magnetic field of 11 T on the specific heat of UBe_{13} at temperatures below 120 K is shown in **Fig. 2-16** [1].

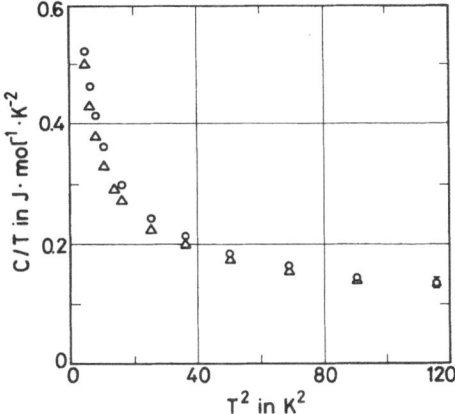

Fig. 2-16. The influence of a magnetic field of 11 T ($\bigcirc$) on the specific heat data of a flux-grown single crystal of UBe_{13}; zero-field data: ($\triangle$). The increase of the specific heat with 11 T disappears above 9 K and decreases below 3 K [1].

The experimental results on the magnetic properties of UBe_{13} in the normal state (susceptibility and magnetoresistance) and in the superconducting state (Meissner effect, upper and lower critical field) are discussed in relation to the nature of the heavy-fermion superconductor. The compound behaves as a dense array of independent Kondo impurities in the normal state. At T < 1 K the onset of coherent scattering leads to new phenomena, which are charac-

References for 2.1.3.3.15.3 on pp. 51/2
4

teristic of the Kondo lattice. A superconducting order parameter at T_c is only found on a part of the Fermi surface. In pure and Th-doped UBe$_{13}$ samples, the remaining part of the Fermi surfaces becomes superconducting at 0.5 T_c. The results do not permit a conclusion concerning the nature of the superconducting state of the compound [2].

Impurities such as Zr or Fe (see also the ternary Zr compounds in Section 2.1.7.4, p. 65), or a thermomechanical treatment, influence the magnetoresistivity of UBe$_{13}$ in the normal state at temperatures in the range 1.8 to 4.2 K [3]. Impurities such as Lu, Gd, La, and Th depress the transition temperature T_c and influence the critical magnetic fields and the influence they have on T_c [4].

The superconducting transition temperature T_c and the upper critical magnetic field of UBe$_{13}$ decrease under hydrostatic pressure. The low temperature magnetoresistance remains large and negative under pressure. The region in which the resistivity has a dependence on T^2 increases with the magnetic field and with pressure [5].

UBe$_{13}$, with substitutions of Th, shows a rich behavior in the superconducting state. The depression of the transition temperature T_c with growing Th content is not monotonic. A second specific heat anomaly below T_c is observed at low Th concentrations. The unconventional character of the superconducting compound UBe$_{13}$ (high purity) is indicated by the proximity effect tunneling [6], see also Section 2.1.3.3.15.2, p. 48.

The normal-state electronic specific heat $C_n(T_c)/T_c$ of UBe$_{13}$ at 0.2 to 18 K remains constant at applied magnetic fields of magnetic flux densities B = 1 to 8 T as seen from the values listed in Table 2/6 [7].

Table 2/6
Transition Temperature T_c^* (Onset) and Normal-State Specific Heat $C_n(T_c)/T_c$ in UBe$_{13}$ for Various Fields with Magnetic Flux Densities B [7].

B in T	T_c^* in K	$C_n(T_c)/T_c$ in $J \cdot mol^{-1} \cdot K^{-2}$
0	0.937	0.72
1	0.903	0.71
2	0.860	0.72
4	0.677	0.73
8	0.330	0.72

The pronounced maxima in resistivity, $\varrho(T)$, and specific heat, $C(T)$, in UBe$_{13}$ are partially suppressed by magnetic fields of flux densities B = 8 T [8]. Doping of the compound with Th on U sites, or with Cu on Be sites also changes the resistivity and specific heat peaks. Such measurements were made in a wide range of parameters to get information on the low-temperature and low-field behavior of the magnetoresistivity. As T approaches zero, $\Delta\varrho$ becomes very small; at T < 0.1 K a change of sign of $\Delta\varrho$ from − to + takes place. This behavior is similar to the Ce-based heavy-fermion compounds CeAl$_3$ and CeCu$_2$Si$_2$ [9].

The upper critical magnetic field, $B_{c2}(T)$, of single crystal and polycrystalline samples of UBe$_{13}$ was measured and the results compared with those for a conventional superconductor. The data are also compared with the weak coupling theory of superconductivity (BCS-theory). The preliminary analysis is based on the "dirty-limit model" and the "p-wave model" which represents two simple limits of the theory. The first model is representative of a conventional superconductor. The second is the simplest realization of a triplet superconductor. The results of the study suggest that triplet superconductivity is unlikely. UBe_{13} shows an anomalous magnetoresistivity at fields of the size of B_{c2} [10] (see also [11]).

The critical magnetic field of a UBe_{13} film of 100 to 500 nm thickness is similar to that of bulk material. The magnetoresistance of such films was measured in the temperature range from 0.4 to 40 K in fields up to 20 T. Plots of isotherms of the magnetoresistance $|R(O) - R(H)|/R(O)$ versus magnetic field H show a universal behavior for $T > 2$ K [12].

Measurements of the thermal and electrical conductivity at magnetic fields of $H = 1$ and 75 kOe, corresponding to critical transition temperatures $T_{cp} = 854$ and 220 mK, respectively, revealed in the normal phase, at $H = 75$ kOe, a quasi-linear temperature dependence of the Lorentz number L, the thermal conductivity λ, and the electrical resistivity ϱ. In the superconducting phase ($H = 1$ kOe, $T < 800$ mK) λ follows a T^2 relationship. This suggests that the heat might be carried by boson-like particles, which have a continuum of excitations and are scattered by free electrons [13].

The structure in the lower- and upper-critical magnetic fields, $B_{c1}(T)$ and $B_{c2}(T)$, of UBe_{13} appears to be related to a bump in the normal state specific heat, $C_s(T)$, around $T = 0.5\ T_c$. This may be a precursor of the two calorimetric transitions which occur in the impure beryllide $U_{0.97}Th_{0.03}Be_{13}$ [14], see Section 2.1.7.2, p. 59.

The resistivity below 900 mK at constant magnetic field is composed of two terms according to $\varrho(T) = \varrho_{O(H)} + A_H T^2$. They represent, respectively, a residual contribution characteristic of impurity effects and a quadratic temperature dependence which may have an intrinsic origin [15]. Specific heat measurements in magnetic fields up to 8 T show drastic changes below the superconducting transition in contrast to the field-independent values of C/T in the normal phase. It may be concluded that the normal phase undergoes magnetic order at $T \approx 150$ mK [16].

References for 2.1.3.3.15.3:

[1] Stewart, G. R.; Fisk, Z.; Smith, J. L.; Ott, H. R.; Mueller, F. M. (Proc. 17th Intern. Conf. Low Temp. Phys., Karlsruhe 1984, pp. 321/2).
[2] Rauchschwalbe, U. (Physica B + C **147** [1987] 1/80).
[3] Alekseevskii, N. E.; Mitin, A. V.; Khlybov, E. P.; Gilewski, A.; Gren, B. (Phys. Status Solidi A **100** [1987] K1-65/K1-68).
[4] Smith, J. L.; Fisk, Z.; Willis, J. O.; Ott, H. R.; Lambert, S. E.; Dalichaouch, Y.; Maple, M. B. (J. Magn. Magn. Mater. **63/64** [1987] 464/6).
[5] Willis, J. O.; Thompson, J. D.; Smith, J. L.; Fisk, Z. (J. Magn. Magn. Mater. **63/64** [1987] 461/3).
[6] Willis, J. O. (J. Less-Common Metals **133** [1987] 107/10).
[7] Mayer, H. M.; Rauchschwalbe, U.; Bredl, C. D.; Steglich, F.; Rietschel, H.; Schmidt, H.; Wühl, H.; Beuers, J. (Phys. Rev. [3] B **33** [1986] 3168/71).
[8] Mayer, H. M.; Rauchschwalbe, U.; Steglich, F.; Stewart, G. R.; Giorgy, A. A. (Z. Physik B **64** [1986] 299/304).
[9] Rauchschwalbe, U.; Steglich, F.; Rietschel, H. (Europhys. Letters **1** [1986] 71/6).
[10] Rauchschwalbe, U.; Ahlheim, U.; Steglich, F.; Rainer, D.; Franse, J. J. M. (Z. Physik B **60** [1985] 379/87).
[11] Maple, M. B.; Chen, J. W.; Lambert, S. E.; Fisk, Z.; Smith, J. L.; Ott, H. R.; Brooks, J. S.; Naughton, M. J. (Phys. Rev. Letters **54** [1985] 477/80).

[12] Tedrow, P. M.; Quateman, J. H. (Phys. Rev. [3] B **34** [1986] 4595/8).

[13] Jaccard, D.; Flouquet, J. (J. Magn. Magn. Mater. **47/48** [1985] 45/50).

[14] Rauchschwalbe, U.; Ahlheim, U.; Bredl, C. D.; Mayer, H. M.; Steglich, F. (J. Magn. Magn. Mater. **63/64** [1987] 447/54).

[15] Brison, J. P.; Lasjanias, J. C.; Ravex, A.; Flouquet, J.; Jaccard, D.; Fisk, Z.; Smith, J. L. (Physica C **153/155** [1988] 437/8).

[16] Brison, J. P.; Ravex, A.; Flouquet, J.; Fisk, Z.; Smith, J. L. (J. Magn. Magn. Mater. **76/77** [1988] 525/6).

2.1.3.3.16 Photoelectron Emission

Valence-band photoemission studies were made on UBe$_{13}$ at the National Synchroton Light Source. Fano resonances were employed to isolate 5f-derived features. Oxygen contamination could be avoided, thus the results were different from those reported in [2]. There is significant hybridization between 5f electrons and the nearest-neighbor ligands, which may be essential to the phenomenon of heavy-fermion superconductivity [1].

High-resolution photoemission spectra and systematic resonant data on a series of U narrow-band compounds indicate that the "two-peaked" f-electron spectrum, which is commonly found in Ce compounds, is also observed in UBe$_{13}$. The 5f spectra consist of a superposition of features consistent with the band structure ground state data, plus a 5f satellite due to poorly-screened final state effects [3].

The photoemission curves of UBe$_{13}$ look like superpositions of U and Be curves. The valence band emission contains one 5f electron; the hybridization of U 5f electrons with Be s, p electrons seems to be small [4]; similar results are presented in [5]. Calculations of the X-ray photoemission intensity of the compound, based upon relativistic formulas, support the experimental finding of very low Be intensity at the Fermi energy and suggest that the highly correlated 5f states, which occur below 20 K, may condense out of normal one-electron band states [6].

The high-resolution photoemission spectra of UBe$_{13}$ at low temperature indicate that the 5f density-of-states is broadened by 0.3 eV wide gaussian, and additionally by an energy-dependent Lorentzian to simulate life-time broadening. The difference curves (between experimental results and theory) suggest a localized 5f satellite; there appears to be more structure on the high binding side than expected from a 5f density of states [3].

A survey on the valence-band photoemission spectra of UBe$_{13}$ is presented in [7] together with a discussion.

References for 2.1.3.3.16:

[1] Parks, R. D.; den Boer, M. L.; Raaen, S.; Smith, J. L.; Williams, G. P. (Phys. Rev. [3] B **30** [1984] 1580/2).

[2] Landgren, G.; Jugnet, Y.; Morar, J. F.; Arko, A. J.; Fisk, Z.; Smith, J. L.; Ott, H. R.; Reihl, B. (Phys. Rev. [3] B **29** [1984] 493/6).

[3] Arko, A. J.; Yates, B. W.; Dunlap, B. D.; Koelling, D. D.; Mitchell, A. W.; Lam, D. J.; Zolnierek, Z.; Olson, C. G.; Fisk, Z.; Smith, J. L.; del Giudice, M. (J. Less-Common Metals **133** [1987] 87/97).

[4] Ueda, K.; Rice, T. M. (Phys. Rev. [3] B **31** [1985] 7114/9).

[5] Wuilloud, E.; Baer, Y.; Ott, H. R.; Fisk, Z.; Smith, J. L. (Phys. Rev. [3] B **29** [1984] 5228/31).

[6] Boring, A. M.; Albers, R. C.; Schadler, G.; Marksteiner, P.; Weinberger, P. (Phys. Rev. [3] B **35** [1987] 2447/50).

[7] Ott, H. R.; Fisk, Z. (in: Freeman, A. J.; Lander, G. H., Handbook on the Physics and Chemistry of the Actinides, Elsevier, Amsterdam 1987, Vol. 5, pp. 85/225).

2.1.3.3.17 Raman Scattering

Raman scattering experiments on single-crystal UBe_{13} were performed between 2 and 300 K. The observed Raman-active phonons are consistent with those allowed by the space- and site-group symmetries of UBe_{13} ($O_h^6 - Fm\overline{3}c$). Certain phonon modes have anomalous behavior with decreasing temperature. Additionally, inelastic scattering from magnetic excitations was observed, which was related to spin fluctuations [1].

The results of another Raman scattering study of single-crystal UBe_{13} were compared with those for isostructural compounds MBe_{13} (M = La, Ce, Th) at 3 to 350 K. Ten phonons were observed in the intermetallic U compound, displaying symmetries $2 A_{1g} + 4 E_g + 4 T_{2g}$ which are consistent with predictions based on the space- and site-group symmetries $O_h^6(Fm\overline{3}c)$. Electronic scattering is also evident in these spectra; this exhibits the symmetry of the purely antisymmetric representation, T_{1g}. This scattering seems to be caused by the located excitations of the 5f electrons [2]. These results are discussed and compiled in [3].

References for 2.1.3.3.17:

[1] Cooper, S. L.; Demers, R. T.; Klein, M. V.; Fisk, Z.; Smith, J. L. (Physica B + C **135** [1985] 49/52).

[2] Cooper, S. L.; Klein, M. V.; Fisk, Z.; Smith, J. L.; Ott, H. R. (Phys. Rev. [3] B **35** [1987] 2615/8).

[3] Ott, H. R.; Fisk, Z. (in: Freeman, A. J.; Lander, G. H., Handbook on the Physics and Chemistry of the Actinides, Elsevier, Amsterdam 1987, Vol. 5, pp. 85/225).

2.1.3.3.18 NMR Spectra

The nuclear magnetic resonance spectrum of ^9Be in the compound UBe_{13} was measured at a frequency of 19.06 MHz at 4.2 K. Tests made at various magnetic fields show that the broad wings of both sides of the central components result from quadrupole interaction. The two peaks in the spectrum are due to the existence of two non-equivalent positions for Be in the UBe_{13} lattice. The difference might be caused by different signs of the spin polarization in the positions BeI and BeII [1] (see Section 2.1.3.3.2, p. 22).

The ^9Be nuclear magnetic resonance was used to study heavy-fermion superconductivity in UBe_{13} (and also in $(U, Th)Be_{13}$, see p. 61). The nuclear spin-lattice relaxation rate $1/T_1$, which yields information on thermal excitations in the superconducting state, varies more slowly with temperature than expected for a conventional BCS superconductor with non-zero energy gap Δ. This indicates an enhanced density of excitations for low energies $E \ll \Delta$. At intermediate temperatures the value of $1/T_1 \cdot T^3$ is nearly constant, which is consistent with highly anisotropic pairing. At $T < 0.2$ K, $1/T_1 \cdot T$ is approximately constant in UBe_{13}. This behavior is not due to direct relaxation by paramagnetic impurities [2].

The ^9Be nuclear spin-lattice relaxation time (T_1) and spin-phase memory time (T_2) were measured in the superconducting and normal state of UBe_{13}. Below the transition temperature

References for 2.1.3.3.18 on p. 54

T_c, a rise in $1/T_1$ occurs, which is anomalously large for a type II superconductor, along with a sharp increase of $1/T_2$ as well. This might be evidence for slow magnetic fluctuations in the superconducting state. Above T_c, $1/T_1$ shows a behavior expected for conduction electrons moving in a very narrow band [3].

⁹Be NMR spectra of UBe_{13} single crystals in the normal state at $T = 4.2$ K were used to derive Knight shifts, hyperfine fields, and quadrupolar couplings. The ⁹Be Knight shift at BeI and the three BeII sites and the hyperfine fields H_{hf} are listed in Table 2/7 [4]. The literature concerning NMR spectra of UBe_{13} was reviewed in [5].

Table 2/7
Data Derived from ⁹Be NMR Spectra of UBe_{13} at 4.2 K [4].

site	K_i in %	$\omega_{oi}/2\pi$ in kHz	$(H_{hf})_i$ in Oe/μ_B
BeI	-0.08 (1)	—	-118
(BeII)A	0.07 (1)	± 164 (2)	414
(BeII)B	0.01 (1)	± 97 (2)	147
(BeII)C	0.19 (1)	± 64 (2)	874
(BeII)$_{avg}$	0.09 (1)	± 1 (1)	478

References for 2.1.3.3.18:

[1] Alekseevskii, N. E.; Narozhuyi, V. N.; Nizhankowskii, V. I.; Nikolaev, E. G.; Khlybov, E. P. (Zh. Eksperim. Teor. Fiz. **40** [1984] 421/3; JETP Letters **40** [1984] 1241/4).
[2] Cheng Tien; MacLaughlin, D. E.; Lan, M. D.; Clark, W. G.; Fisk, Z.; Smith, J. L.; Ott, H. R. (Physica B + C **135** [1985] 14/21).
[3] Clark, W. G.; Fisk, Z.; Glover, K.; Lan, M. D.; MacLaughlin, D. E.; Smith, J. L.; Cheng Tien (Proc. 17th Intern. Conf. Low Temp. Phys., Karlsruhe 1984, pp. 227/8).
[4] Clark, W. G.; Lan, M. D.; van Kalkeren, G.; Wong, W. H.; Cheng Tien; MacLaughlin, D. E.; Smith, J. L.; Fisk, Z.; Ott, H. R. (J. Magn. Magn. Mater. **63/64** [1987] 396/9).
[5] Ott, H. R.; Fisk, Z. (in: Freeman, A. J.; Lander, G. H., Handbook on the Physics and Chemistry of the Actinides, Elsevier, Amsterdam 1987, Vol. 5, pp. 85/225).

2.1.3.3.19 Optical Spectra, Reflectance

The reflectance in the range 0.050 to 2 eV of the heavy-fermion superconductor UBe_{13} has a sharp structure at low frequencies, which is superimposed on a smooth decrease down to 50% at 1.6 eV [1]. The Kramers-Kronig analysis of reflectance data yields a sharp interband structure in the compound in the 0.1 eV region. UBe_{13} exhibits normal Drude behavior in the far IR, above 100 K: the optical conductivity agrees well with the dc conductivity. The free carrier plasma frequency $\omega_p(UBe_{13}) = 5.9$ eV is obtained from the evaluation of reflectivity measurements [2].

The reflectance was measured at frequencies between 15 and 27000 cm^{-1} at various temperatures between 2 and 300 K [3]. The data at low frequencies and low temperatures, which reveal an increasing absorption with increasing temperature, are compatible with the temperature dependence of the dc resistivity. Anomalous, however, is the non-monotonic behavior [4].

References for 2.1.3.3.19:

[1] Klassen, R. J.; Bonn, D. A.; Timusk, T.; Smith, J. L.; Fisk, Z. (J. Less-Common Metals **127** [1987] 293/7).

[2] Eklund, P. C.; Hoffman, D. M.; Delong, L. E.; Arakawa, E. T.; Smith, J. L.; Fisk, Z. (Phys. Rev. [3] B **35** [1987] 4250/7).

[3] Bonn, D. A.; Klassen, R. J.; Timusk, T.; Smith, J. L.; Fisk, Z. (unpublished work, referred in [4]).

[4] Ott, H. R.; Fisk, Z. (in: Freeman, A. J.; Lander, G. H., Handbook on the Physics and Chemistry of the Actinides, Elsevier, Amsterdam 1987, Vol. 5, pp. 85/225).

2.1.4 Chemical Reactions of UBe_{13} and U–Be Alloys

The reaction of massive samples of UBe_{13}, prepared by powder metallurgical techniques, with N_2 of 1 bar pressure, follows a parabolic rate law in the temperature range from 500 to 800 °C. The parabolic rate law is caused by the formation of a protective nitride film. The reaction rate of the nitride formation at 500 °C is given by $\Delta W = 110 \cdot t^{0.46}$, with W in $\mu g/cm^2$ and t in s. The temperature dependence of the reaction rate follows the equation $k = 5.10 \times 10^4$ $\exp[-(9450 \pm 550)/RT]$ in the temperature range 500 to 800 °C [1].

UBe_{13} resists oxidation by dry air only up to a temperature of 200 °C [2]. The oxidation of UBe_{13} by O_2 of 1 bar pressure is governed by a linear rate law. Thus, one can conclude that a protective layer does not exist. The oxidation rate is $\Delta W = 8.0 \cdot t$, its dependence on the temperature is according to the equation $k = 1.98 \times 10^8 \exp[-(26200 \pm 750)/RT]$ [1]. The activation energy of the oxide formation is considerably higher than that of the nitride formation. Oxidation of UBe_{13} also occurs in water vapor of 0.039 bar pressure. The rate of oxidation is much lower than in oxygen. The rate law is also linear: $\Delta W = 0.016 \cdot t^{0.75}$. The activation energy of this reaction is of the same order as the value given for oxygen, thus indicating the same type of reaction. The reactivity of UBe_{13} with these gases is compared with reactions of other U compounds in Table 2/8 [1].

Table 2/8
Relative Reactivity of U Compounds [1].
The arrows indicate the direction of increase.

nitrogen	water vapor	oxygen
UB_2	UBe_{13}	UBe_{13}
UBe_{13}	UB_2	U
U	U	UB_2
UC_2	UC_2	UC_2

Sintered UBe_{13} compacts lose some material in boiling water under atmospheric pressure. The weight losses after 72 h exposure are in the order of 2.2 mg/cm^2 [3].

Nuclear fuels have to be dissolved for their reprocessing after reaching the desired burnup. There is no information available on the dissolution of dispersion fuel containing UBe_{13}. The components, U and Be, can be dissolved in boiling concentrated HNO_3.

                    References for 2.1.4 on p. 56

Diffusion of Be in UBe$_{13}$ was observed in reactions of Be metal with UC as well as with UO$_2$. Diffusion coefficients, however, are not calculated from the experimental results [4, 5]. The data collection in [6] concerning diffusion of Be and its alloys does not contain any information on the diffusion in the U−Be alloys.

References for 2.1.4:

[1] Albrecht, W. M.; Koehl, B. G. (Proc. 2nd Intern. Conf. Peaceful Uses At. Energy, Geneva 1958, Vol. 6, pp. 116/21).

[2] Badajeva, T. A.; Dashevskaya, L. I. (in: Ivanov, O. S., Physical Chemistry of Alloys and Refractory Compounds of Thorium and Uranium, Jerusalem 1972, pp. 123/9).

[3] Tripler, A. B., Jr.; Snyder, M. J.; Duckworth, W. H. (BMI-1313 [1959] 1/55; N.S.A. **13** [1959] No. 8908).

[4] Knapton, A. G.; West, K. B. C. (J. Nucl. Mater. **3** [1961] 239/40).

[5] Murdock, J. F. (J. Nucl. Mater. **7** [1962] 192/6).

[6] Dragoo, A. L. (At. Energy Rev. Spec. Issue No. 4 [1973] 173/5).

2.1.5 Nuclear Properties and Irradiation Behavior

The nuclear properties of U−Be alloys, applied as refractory fuel materials, are characterized by the fission behavior of U and the excellent moderating properties of Be. (Both properties make the alloys attractive as nuclear fuels.)

The irradiation damage of UBe$_{13}$ as dispersed fuel alloy should be less severe than in metallic U [1]. The application as dispersed fuel stabilizes the beryllide against irradiation due to the effects of the metallic matrix [2].

Alloys containing 0.5, 1.0, and 3.0 wt% U show an increase in the electrical resistivity in the order of 1.5 to 9%, depending on the U content, after 0.06% burnup. Changes in dimensions and density are observed. In-pile measurements of the thermal conductivity of a Be−20% U sample do not show significant changes at a temperature of 100°C [3].

The superconducting transition temperature T_c of UBe$_{13}$ decreases by 40% upon neutron irradiation by 10^{18} n/cm^2, E > 1 MeV. This decrease is three times more rapid than in other superconductors and comparable to that of UPt$_3$. The sensitivity of heavy-fermion superconductors to irradiation-induced defects does not serve as evidence for unconventional pairing [4].

References for 2.1.5:

[1] Hanle, W. (Metall **11** [1957] 91/9).

[2] Palme, R. (Metall **13** [1959] 386/9).

[3] Billington, D. S. (Proc. 1st. Intern. Conf. Peaceful Uses At. Energy, Geneva 1955, Vol. 7, pp. 421/32).

[4] Andraka, B.; Meisel, M. W.; Kim, J. S.; Wölfle, P.; Stewart, G. R.; Snead, C. L., Jr.; Giorgi, A. L.; Wire, M. S. (Phys. Rev. [3] B **38** [1988] 6402/6).

2.1.6 Uses of U — Be Alloys

The alloys in the system U—Be, particularly the intermetallic compound UBe_{13}, were considered to be applicable as a nuclear fuel. Several properties of the alloys and the compound were evaluated. The high-temperature stability of intermetallic UBe_{13} makes it particularly attractive for this application [1]. Another favorable property of the beryllide fuel is its content of the moderating element Be [6]. The compound has been discussed, therefore, as a dispersed fuel for application in high burn-up fuel elements [2]. The brittleness of the U beryllide is one reason for its preferred application in the dispersed form, the other is the necessity to fabricate it by powder-metallurgical methods. The compound has a fairly good resistance to irradiation damage [3, 4]. The success of the oxide fuels in commercial reactors, however, resulted in a stop in the development of fuels based on U alloys in the sixties. An application of U beryllides as a fuel for nuclear reactors has never been reported.

A possible application of the compound UBe_{13} may be based on its superconductivity which occurs below 0.85 K. The intermetallic is the first example of a so-called exotic superconductor in an actinide-containing system [5].

References for 2.1.6:

[1] Snyder, M. J.; Tripler, A. B., Jr. (ASTM Spec. Tech. Publ. No. 276 [1959] 293/300).
[2] Lewis, J. R. (J. Metals **13** [1961] 357/62).
[3] Weber, C. E. (J. Metals **8** [1956] 561/659).
[4] Hanle, W. (Metall **11** [1957] 91/9).
[5] Ott, H. R.; Rudigier, H.; Fisk, Z.; Smith, J. L. (Phys. Rev. Letters **50** [1983] 1595/8).
[6] Wisnyi, L. G. (Ceram. Ind. [Chicago] **74** No. 2 [1960] 56/75, **74** No. 3 [1960] 77/9).

2.1.7 Ternary Systems

2.1.7.1 Introduction

Several ternary compounds of the types $(U_xM_{1-x})Be_{13}$ or $U(Be_{13-x}M_x)$ were studied because of an interest in the effect of substitution of U or Be on the low-temperature physical properties. Substitution compounds of U were made using the elements Np, Th, Zr, Ce, Lu, Sc, Gd, La, and Ba. Compounds of this type were prepared by melting mixtures of the MBe_{13} compounds in an argon arc furnace. Compounds in which small amounts of Cu, Ga, or B substitute for Be in the Be sublattice of U beryllide were prepared by means of a melting procedure in a purified argon atmosphere. UBe_{13} and the substitute were mixed and heated up to melting [1].

A survey on the influence of ternary elements in UBe_{13} is given in [2]. Changes of the lattice parameters are listed in Table 2/9, p. 59 [1].

The influence of certain ternary elements as Th (**Fig.** 2-**17**, p. 58), Sc, Ce, and Lu (**Fig.** 2-**18**, p. 58) or Y, Zr, and La (**Fig.** 2-**19**, p. 58) on T_c is shown in [2].

References for 2.1.7.1 on p. 59

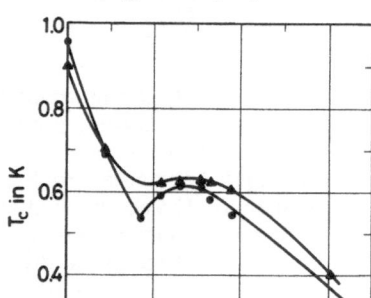

Fig. 2-17. Superconducting transition temperature T$_c$ versus Th concentration. ● χ$_{ac}$, ▲ C$_p$ [2].

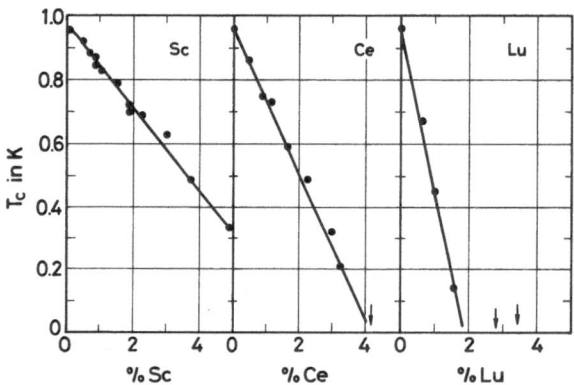

Fig. 2-18. Superconducting transition temperature T$_c$ versus impurity concentration in UBe$_{13}$ (Sc, Ce, Lu) [2].

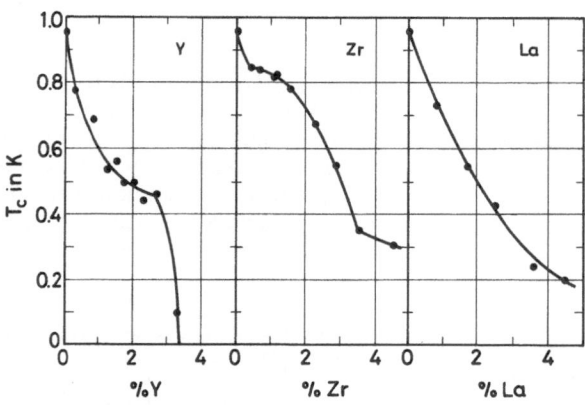

Fig. 2-19. Superconducting transition temperature T$_c$ versus impurity concentration in UBe$_{13}$ (Y, Zr, La) [2].

Table 2/9

Lattice Parameters for Various Ternary Compounds Based on UBe_{13} [2].

The ternary compounds of this table have the general composition $M_xU_{1-x}Be_{13}$. The complete formula, such as $Th_{0.0172}U_{0.9828}Be_{13}$, is only given for the heading compound in a block of compounds. For the following compounds only the part Th_x is given, the remaining parts $U_{1-x}Be_{13}$ are not printed.

compound	a in Å	compound	a in Å
UBe_{13} (several samples)	10.2545 to 10.2550	$Y_{0.0030}U_{0.997}Be_{13}$	10.2543
UBe_{13} (single crystal)	10.2656	$Y_{0.0081}$	10.2541
		$Y_{0.0124}$	10.2545
$Th_{0.0172}U_{0.9828}Be_{13}$	10.2575	$Y_{0.0126}$	10.2545
$Th_{0.0216}$	10.2579	$Y_{0.0176}$	10.2540
$Th_{0.026}$	10.2591	$Y_{0.020}$	10.2544
$Th_{0.0308}$	10.2591	$Y_{0.0332}$	10.2540
$Th_{0.0378}$	10.2605	YBe_{13}	10.2398
$Th_{0.0598}$	10.2635		
$Th_{0.0603}$	10.2642	$Zr_{0.0108}U_{0.9892}Be_{13}$	10.2533
$Sc_{0.0068}U_{0.9932}Be_{13}$	10.2539	$La_{0.008}U_{0.992}Be_{13}$	10.2562
$Sc_{0.0103}$	10.2531	$La_{0.025}$	10.2586
$Sc_{0.030}$	10.2523		
$Sc_{0.0484}$	10.2508	$UBe_{12.99}Al_{0.01}$	10.2564
		$UBe_{12.99}Al_{0.01}$ (annealed)	10.2556
$Lu_{0.0062}U_{0.9928}Be_{13}$	10.2540	$UBe_{12.97}Al_{0.03}$	10.2571
$Lu_{0.016}$	10.2536	$UBe_{11.15}$	10.2543
$LuBe_{13}$	10.1693	$UBe_{15.0}$	10.2545

References for 2.1.7.1:

[1] Smith, J. L.; Fisk, Z.; Willis, J. O.; Batlogg, B.; Ott, H. R. (J. Appl. Phys. **55** [1984] 1996/2000).

[2] Smith, J. L.; Fisk, Z.; Willis, J. O.; Giorgi, A. L.; Roof, R. B.; Ott, H. R.; Rudigier, H.; Felder, E. (Physica B + C **135** [1985] 3/8).

2.1.7.2 U−Be−Th

The structures of Th−U−Be alloys were studied within the composition region between $ThBe_{13}$ and UBe_{13} [1]. Data on the microstructure of annealed specimens prove the occurrence of a continuous series of solid solutions corresponding to the identical fcc lattices of the compounds. The values of the lattice constants a of the solid solutions vary linearly from the value of $ThBe_{13}$ (a = 10.363 Å) to the value of UBe_{13} (a = 10.226 Å). The hardness ($H_v \approx 900$ kg/mm^2) only differs insignificantly in the whole range of compositions. The lattice constants of $(U_{1-x}Th_x)Be_{13}$ with $0 \leqq x \leqq 0.06$ are shown in **Fig.** 2-20, p. 60 [2]. The early work on U−Th−Be alloys is compiled in [3].

Properties of $(U, Th)Be_{13}$ with 10 at% U have been studied regarding use of the alloys as thermal breeding reactor fuels. Preparative methods for the compounds were investigated as well. The Knoop hardness (100 g load) of $(Th_{0.9}U_{0.1})Be_{13}$ is 1114 kg/mm^2 (compared to $ThBe_{13}$ with 1343). The corrosion resistance of this ternary compound against the liquid NaK alloy of 1200 F (649°C) is nearly as good as that of $ThBe_{13}$, which shows excellent compatibility. The density of $(Th_{0.9}U_{0.1})Be_{13}$ is D = 4.28 g/cm^3 [4].

                    References for 2.1.7.2 on p. 64

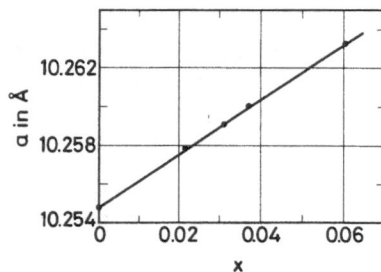

Fig. 2-20. Room-temperature lattice constants of $(U_{1-x}Th_x)Be_{13}$ ternary compounds for x between 0 and 0.06 [2].

Measurement of the low-temperature specific heat of the ternary compound $U_{1-x}Th_xBe_{13}$, with x below 0.06, give evidence for the transition into a new phase in the superconducting state. An increasing content of Th depresses the temperature of the superconducting transition T_c and causes a broadening of the transition. The maximum values of the electronic specific heat, given as c_p/T, shift to lower temperatures. In compounds with x = 0.026 and 0.03 the maximum is split into two peaks at ~ 0.6 and ~ 0.35 K, which indicates a continuous phase transition from one superconducting state below T_{c1} to another one below T_{c2}. The observation is in accordance with the non-conventional character of the superconductivity of UBe_{13} [2].

A Ginsburg-Landau model of even-parity superconducting order parameters with transition temperatures T_0 and T_2 is discussed in relation to $(U_xTh_{1-x})Be_{13}$. The critical temperatures for the pure system are assumed such that $T_2 > T_0$. It is suggested that the impurity scattering strongly suppresses T_2 while T_0 remains unaffected. The inequality is thus reversed for an impurity concentration above the threshold concentration [5].

The unusual superconducting properties of this compound are also treated on the basis of a phenomenological model [6] which incorporates the effect of fluctuations on the interactions between superconducting and coherent Kondo screening in heavy-fermion superconductors. It is pointed out that the appearance of a pseudo-gap near the Fermi energy leads to a temperature-dependent effective BCS interaction between the electrons forming the Cooper pairs [7].

A λ-shaped attenuation peak occurs in $(U,Th)Be_{13}$, which is two orders of magnitude larger than the total contribution from particle-hole scattering, close to the lower transition temperature T_{c2} [8]. The results are consistent with those expected in the neighborhood of a magnetic transition. A state of coexisting anisotropic superconductivity and antiferromagnetic order is suggested below T_{c2}. The sound velocity shows a minimum at the large attenuation peak near T_{c2} [9].

An alternative explanation of this ultrasonic attenuation at the lower transition, T_{c2}, in superconducting $U_{1-x}Th_xBe_{13}$, which is described in [8], is given in [10]. It is assumed in this discussion that the transition might be between two different anisotropic superconducting phases and that the low-temperature phase may be tetragonally distorted. The domain-wall energy at T_{c2} is very small and the attenuation due to the motion of the walls is very large. This attenuation strongly depends on the direction and polarization.

Ultrasound studies of $(U,Th)Be_{13}$ in the 50 to 250 MHz range indicate that the ternary compound has an additional transition below T_c with a very strong ultrasound absorption maximum. At this lower transition temperature $T_{c1} = 0.42$ to 0.43 K the sound velocity decreases by ~27 ppm. The electronic contribution to the attenuation is only in the order 10^{-3} dB/cm. An attenuation peak two orders of magnitude larger than the electronic contribution occurs at further cooling with its maximum at 0.377 K. Such an attenuation might be associated with magnetic transitions. At T_{c2} a small dip of 5 to 8 ppm in the sound velocity is detected. The results suggest a coexistence of superconductivity and magnetism in (U, Th)Be$_{13}$ [11].

Superconducting states coexist for Th concentrations $0 \leqq x \leqq 0.06$ in the compound $U_{1-x}Th_xBe_{13}$. Assuming s-wave and d-wave symmetries for these states, a Ginzburg-Landau free-energy expression is derived, which couples s- and d-wave states and is rotationally invariant. The existence of two eigenfrequencies associated with the dynamics of phase oscillations (internal Josephson effect) is predicted, which are characteristic of the s-wave and d-wave states [12].

Another theoretical model relates the possible superconducting pairing states with the spin-density wave. The basis is an imperfect-nesting-band model in two dimensions. A Hamiltonian, including the on-site repulsive interaction in addition to the attractive interaction within a mean-field approximation, is investigated. Less-competitive and competitive states, according to the combined symmetry of the superconductivity order parameter, are the two classes to be distinguished. The theory shows experimental implication to the heavy-fermion superconductor (U, Th)Be$_{13}$, which exhibits the Fermi-surface-related spin-density-wave instability [13, 14].

Part of the Fermi surface does not participate in superconductivity of UBe$_{13}$ at $T_c \approx 0.9$ K and develops an order parameter in the accessible temperature range (about 0.55 K). This effect is sharply enhanced by doping with a few at% Th, which leads to a second phase transition within the superconducting state [15].

^{9}Be nuclear magnetic resonance (NMR) and spin lattice relaxation experiments in the normal and superconducting states of UBe$_{13}$ and $U_{1-x}Th_xBe_{13}$ with $x = 0.033$ are made in regard to the aspects of pairing [16]. In the ternary compound, they might help to identify T_{c2} as a second superconducting transition temperature. The results are consistent with a class of anisotropic pairing models for which the gap vanishes along the Fermi surface. NMR spectra give no indication of magnetic, structural, or charge density ordering at T_{c2} for $(U_{1-x}Th_x)Be_{13}$ with $x = 0.033$. **Fig. 2-21**, p. 62, shows the influence of the doping on the resistivity of $(U_{1-x}Th_x)Be_{13}$ compounds [17].

The addition of Th results in a non-monotonic depression of T_c, which is extremely unusual for a non-magnetic doping. This might be due to an interplay between the lowest temperature resistivity peak and the transition temperature as the peak is depressed. It indicates that heavy-fermion superconductivity is related to one of the possible ground states for heavy mass electron systems. The effect of Th doping on the magnetic susceptibility χ of such ternary compounds is shown in **Fig. 2-22**, p. 62 [17].

A magnetic field of $H \approx 4$ kOe has no effect on the resistivity of the ternary compound $(U_{1-x}Th_x)Be_{13}$ with $x = 0.026$. However, for $x = 0.0175$, the low-temperature peak in resistivity moves to lower temperatures with raising field strength. The low-temperature decrease of resistivity is inhibited by the superimposure of a magnetic field. Similar influence of the magnetic field of the same order is shown for $x = 0.0378$ [18].

The resistivity and specific heat of the compound $(U_{0.97}Th_{0.03})Be_{13}$ are measured at magnetic fields with $B = 0$ and $B = 8$ T. The maxima near the transition temperature, T_c, are suppressed by the fields. The second transition at the lower temperature, T_{c2}, is observed again. Both transitions are shifted to lower temperatures by means of magnetic fields [19].

References for 2.1.7.2 on p. 64

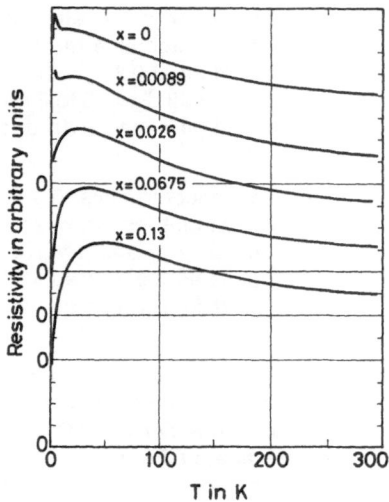

Fig. 2-21. Electrical resistivity of $(U_{1-x}Th_x)Be_{13}$ between 1.4 and 300 K. The curves are offset vertically for clarity with their zeros indicated. The vertical scale for each sample varies, but the height of the high-temperature value indicates the proper normalization [17].

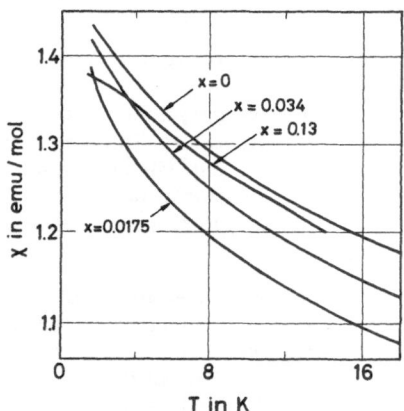

Fig. 2-22. Magnetic susceptibility of $(U_{1-x}Th_x)Be_{13}$ between 1.4 and 18 K [17].

The influence of pressure p on the superconducting transition temperature T_c was determined for $(U_{1-x}Th_x)Be_{13}$ [20]. For $x = 0.0089$ to 0.026 increasing values of p have an increasing effect on T_c. Isobars of $T_c(x)$ indicate a maximum depression of T_c at x_{min}; the slope dT_c/dP becomes larger with increasing x. Pressure and concentration x of Th have a strongest combined effect at $x < x_{min}$. An increase of T_c for $x > x_{min}$ occurs with a maximum and a subsequent decrease for higher values of x. The data may be evidence for two different superconducting states of $(U_{1-x}Th_x)Be_{13}$, see [21]. The splitting caused by the substitution of Th for U is related to the symmetry of the uncommon class of superconductors as UBe_{13} [21].

Measurement of the specific heat, C, of $U_{0.97}Th_{0.03}Be_{13}$ in the temperature range 0.1 to 1.0 K at a pressure of 1.6 to 7.7 kbar, and up to 20 K with p = 0 revealed that peaks which occur in C(T) at 0.33 and 0.54 K are suppressed and shifted to lower temperature by increased pressure. At temperatures above 8 K the compound has the same C as UBe_{13}. Anomalies in C(T) can be correlated to rapid changes in the magnetic susceptibility. The suppression of the peaks and shift of T_c to lower values is in contrast to the behavior of pure UBe_{13}. The broad maximum of C at above 2 K is completely suppressed in the Th-substituted sample [22].

Measurements of the resistivity under pressure up to 18.4 kbar indicate that pressure modifies the temperature dependence of the resistivity of $U_{1-x}Th_xBe_{13}$ (x = 0.0172); this may provide a possible explanation for the unusual behavior of these compounds [23].

The temperature dependence of the upper critical magnetic field $H_{c2}(T)$ was determined from measurements of the ac electrical resistance R for four compositions in the $(U_{1-x}Th_x)Be_{13}$ system. **Fig. 2-23** shows the effect of a magnetic field of 0 to 6 T on the resistance/temperature curves. T_c decreases rapidly as Th is substituted for U. The overall shape of the $H_{c2}(T)$ curves, however, remains nearly the same (see Fig. 2-23), indicating that the nature of super-conductivity of heavy fermions remains unchanged despite a decrease of ca. 30% of T_c at a field of H = 0. The data are characterized by very large values of the initial slope $(-dH_{c2}/dT)T_c$ and strong negative magnetoresistance in the normal state [24].

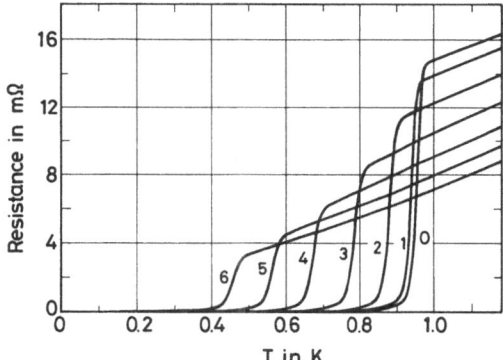

Fig. 2-23. ac electronic resistance versus temperature in various applied magnetic fields in T for a polycrystalline sample of $(U_{1-x}Th_x)Be_{13}$ [24].

Further studies of the influence of Th impurities in UBe_{13} demonstrated that a few % of Th dramatically alter not only the dependence of the electrical resistivity at low temperature but also the superconducting state [25, 26]. In the range 0.01 < x < 0.06, T_c is almost independent of x and is about 0.6 K. At 0.4 K, another superconducting phase was observed, which is confirmed by measurements of the thermal expansion and ultrasonic attenuation (see also [8]). The thermal expansion for the compound $(U_{1-x}Th_x)Be_{13}$ with x = 0.0331 below 1 K shows a discontinuity at $T_{c1} \approx 0.6$ K and a much larger volume effect at the lower transition temperature T_{c2}, thus giving further evidence that the second transition cannot be based on trivial phenomena. The influence of magnetic fields of H = 0 to 50 kG on the specific heat of the same $(U, Th)Be_{13}$ compound causes a shift of the lower transition temperature with increasing field strength. The behavior differs completely from that of pure UBe_{13} in which the magnetic field only shifts the value of T_{c1} [25].

                 References for 2.1.7.2 on p. 64

References for 2.1.7.2:

[1] Ivanov, V. E.; Badajeva, T. A. (Proc. 2nd Intern. Conf. Peaceful Uses At. Energy, Geneva 1958, Vol. 5, pp. 139/55).

[2] Ott, H. R.; Rudigier, H.; Fisk, Z.; Smith, J. L. (Phys. Rev. [3] B **31** [1985] 1651/3).

[3] Ivanov, O. S.; Badaeva, T. A.; Sofronova, R. M.; Kishenevskii, V. B.; Kushnir, N. P. (Phase Diagrams of Uranium Alloys, Amerind, New Delhi 1983, translated from: Diagrammy Sostoyaniya i Fazovye Prevrashcheniya Splavov Urana, Nauka, Moscow 1972).

[4] Farkas, M. S.; Bauer, A. A.; Dickinson, R. F. (BMI-1568 [1962] 1/20; N.S.A. **16** [1962] No. 12651).

[5] Kumar, P.; Wolfle, P. (Phys. Rev. Letters **59** [1987] 1954/7).

[6] Moshchalkov, V. V. (Pis'ma Zh. Eksperim. Teor. Fiz. **45** [1987] 181/4; JETP Letters **45** [1987] 223/7).

[7] Chen Chang-feng; Zhang Li-yuan (Physica **144** [1987] 193/9).

[8] Batlogg, B.; Bishop, D.; Golding, B.; Varma, C. M.; Fisk, Z.; Smith, J. L.; Ott, H. R. (Phys. Rev. Letters **55** [1985] 1319/22).

[9] Fisk, Z.; Hess, D. W.; Pethick, C. J.; Pines, D.; Smith, J. L.; Thompson, J. D.; Willis, J. O. (Science **239** [1988] 33/42).

[10] Joynt, R.; Rice, T. M.; Ueda, K. (Phys. Rev. Letters **56** [1986] 1412/5).

[11] Batlogg, B.; Bishop, D. J.; Bucher, E.; Golding, B.; Varma, C. M.; Fisk, Z.; Smith, J. L.; Ott, H. R. (Physica B + C **135** [1985] 23/6).

[12] Langner, A.; Sahu, D.; George, Th. F. (Phys. Rev. [3] B **38** [1988] 9187/90).

[13] Machida, K.; Kato, M. (Japan. J. Appl. Phys. **26** Suppl. 3 [1987] 1237/8).

[14] Kato, M.; Machida, K. (Phys. Rev. B **37** [1988] 1510/9).

[15] Rauchschwalbe, U.; Steglich, F.; Sparn, G.; Bredl, C. D.; Fulde, P.; Maki, K. (Japan. J. Appl. Phys. **26** Suppl. 3 [1987] 1225/6).

[16] MacLaughlin, D. E.; Tien, C.; Clark, W. C.; Lan, M. D.; Fisk, Z.; Smith, J. L.; Ott, H. R. (Phys. Rev. Letters **53** [1984] 1833/6).

[17] Smith, J. L.; Fisk, Z.; Willis, J. O.; Batlogg, B.; Ott, H. R. (J. Appl. Phys. **55** [1984] 1996/2000).

[18] Willis, J. O.; Smith, J. L.; Fisk, Z.; Ott, H. R. (J. Appl. Phys. **57** [1985] 3079/81).

[19] Mayer, H. M.; Rauchschwalbe, U.; Steglich, F.; Stewart, G. R.; Giorgy, A. A. (Z. Physik B **64** [1986] 299/304).

[20] Lambert, S. E.; Dalichaouch, Y.; Maple, M. B.; Smith, J. L.; Fisk, Z. (Phys. Rev. Letters **57** [1986] 1619/22).

[21] Volovik, G. E.; Khmel'nitskii, D. E. (Pis'ma Zh. Eksperim. Teor. Fiz. **40** [1984] 469/72; JETP Letters **40** [1985] 1299/302).

[22] Fisher, R. A.; Lacy, S. E.; Marcenat, C.; Olsen, J. A.; Phillips, N. E.; Fisk, Z.; Smith, J. L. (Japan. J. Appl. Phys. **26** Suppl. 3 [1987] 1219/20).

[23] Borges, H. A.; Thompson, J. D.; Aronson, M. C.; Smith, J. L.; Fisk, Z. (J. Magn. Magn. Mater. **76/77** [1988] 235/7).

[24] Chen, J. W.; Lambert, S. E.; Maple, M. B.; Naughton, M. J.; Brooks, J. S.; Fisk, Z.; Smith, J. L.; Ott, H. R. (J. Appl. Phys. **57** [1985] 3076/8).

[25] Ott, H. R.; Rudigier, H.; Felder, E.; Fisk, Z.; Smith, J. L. (Phys. Rev. [3] B **33** [1986] 126/31).

[26] Ott, H. R. (Physica B **126** [1984] 100/6).

2.1.7.3   U — Be — Np

Since $NpBe_{13}$ has a lattice parameter (a = 10.276 Å) similar to that of UBe_{13} and $ThBe_{13}$, the effect of doping UBe_{13} with Np on the value of T_c is similar to that with Th [1, 2]. The transition temperature of the compound $(U_{1-x}Np_x)Be_{13}$ with x = 0.011 is $T_c = 0.62$ K. Superconductivity does not occur in the compound with x = 0.68. Thus, the addition of one more f electron by substituting Np for U suppresses superconductivity, whereas the temperature-dependent heavy-fermion contribution γ remains, even in pure $NpBe_{13}$. Superconductivity is apparently replaced by some form of itinerant-electron magnetism. The continuous change of properties in $(U_{1-x}Th_x)Be_{13}$ requires a large replacement of U by Np before this magnetic behavior is apparent, even though superconductivity is rapidly depressed [2].

The effect of an additional f electron is also evident in the low-temperature specific heat data, which show an increase as Np is doped to UBe_{13}. The C/T values for $(U_{0.32}Np_{0.68})Be_{13}$ are larger than those of UBe_{13} by a factor of ~ 1.7 (above T = 3 K). Below 3 K the ratio decreases and at 1 K the C/T values of both materials are approximately the same. Below this temperature UBe_{13} becomes superconducting while C/T for $(U_{0.32}Np_{0.68})Be_{13}$ rises sharply. This might be due to stronger heavy-fermion behavior or to a magnetic transition similar to that in $NpBe_{13}$ at lower temperatures [2].

References for 2.1.7.3:

[1] Smith, J. L.; Fisk, Z.; Willis, J. O.; Batlogg, B.; Ott, H. R. (J. Appl. Phys. **55** [1984] 1996/2000).
[2] Stewart, G. R.; Fisk, Z.; Smith, J. L.; Willis, J. O.; Wire, M. S. (Phys. Rev. [3] B **30** [1984] 1249/52).

2.1.7.4   U — Be — Zr

T_c decreases sharply with increasing x in $(U_{1-x}Zr_x)Be_{13}$ compounds; at x = 0.1, T_c is ≈ 0.07 K, and $\partial H_{c2}/\partial T \approx 15$ kOe/K [1]. For large values of x (x $\approx$ 0.5), T_c drops below 0.02 K, and then increases to $T_c = 1.3$ K as x approaches 1 ($ZrBe_{13}$). The effect of the dissolution of $ZrBe_{13}$ in UBe_{13} is large, although there is only a slight change in the lattice parameters. The properties of the $(U, Zr)Be_{13}$ compounds do not support the assumption that the 5f electrons of the U atoms are primarily responsible for the unusual low-temperature behavior of this class of intermetallics, as pointed out in [1].

The $UBe_{13} - Zr$ alloys belong to a complex section of the U — Be — Zr system. The alloy contains four intermetallic phases, namely $ZrBe_{13}$, Zr_2Be_{17}, $ZrBe_5$, and $ZrBe_2$. A UBe_{13}-rich eutectic is detected which also contains $ZrBe_{13}$ and U-based solid solutions [3]. The alloys were prepared from pure components (99.8% U, 99.8% Be, and 99.9% Zr). The alloys are homogeneous at 500 to 800°C [2].

Ternary systems U — Be — Zr, Nb, or Mo were investigated in order to evaluate the compatibility of UBe_{13} fuel with metallic cladding materials. The beryllide does not react with the three metals at temperatures below 550 to 710°C. Powdered and pressed mixtures of UBe_{13} with Zr, Nb, or Mo of less than 500 μm particle size show a reaction above this temperature limit [3].

References for 2.1.7.4 on p. 66

References for 2.1.7.4:

[1] Alekseevskii, N. E. (Pis'ma Zh. Eksperim. Teor. Fiz. **40** [1984] 66/9; JETP Letters **40** [1984] 800/3).

[2] Ivanov, O. S.; Badaeva, T. A.; Sofronova, R. M.; Kishenevskii, V. B.; Kushnir, N. P. (Phase Diagrams of Uranium Alloys, Amerind, New Delhi 1983, translated from: Diagrammy Sostoyaniya i Fazovye Prevrashcheniya Splavov Urana, Nauka, Moscow 1972).

[3] Badaeva, T. A.; Alekseenko, G. K.; Kuznetsova, R. I. (in: Ivanov, O. S., Physical Chemistry of Alloys and Refractory Compounds of Thorium and Uranium, Israel Program for Scientific Translation, Jerusalem 1972, pp. 148/57).

2.1.7.5   U−Be−Nb

The tetragonal $NbBe_{12}$, UBe_{13}, and U are observed in the U−Be−Nb system. The compound $NbBe_3$ occurs in alloys containing more than ∼20 at% Nb [1]. This system is surveyed in [2].

References for 2.1.7.5:

[1] Badaeva, T. A.; Alekseenko, G. K.; Kuznetsova, R. I. (in: Ivanov, O. S., Physical Chemistry of Alloys and Refractory Compounds of Thorium and Uranium, Israel Program for Scientific Translation, Jerusalem 1972, pp. 148/57).

[2] Ivanov, O. S.; Badaeva, T. A.; Sofronova, R. M.; Kishenevskii, V. B.; Kushnir, N. P. (Phase Diagrams of Uranium Alloys, Amerind, New Delhi 1983, translated from: Diagrammy Sostoyaniya i Fazovye Prevrashcheniya Splavov Urana, Nauka, Moscow 1972).

2.1.7.6   U−Be−RE

$(U, Ce)Be_{13}$ compounds were prepared by melting together the two MBe_{13} compounds in an argon arc furnace [1]. The substitution of Ce for U in UBe_{13} causes a suppression of the transition temperature T_c of the beryllide. The compound $(U_{1-x}Ce_x)Be_{13}$ with x = 0.0158 has a transition temperature T_c = 0.55 K [1]. The depression proceeds with higher amounts of Ce present in the intermetallic. Values of T_c = 0.12 K and $\partial H_{c2}/\partial T \approx 20$ kOe/K are measured in the compound with x = 0.15. The value of the electronic heat capacity coefficient γ of the latter compound, $\gamma = 1.5$ mJ·mol^{-1}·K^{-2}, differs widely from the high value of the pure U beryllide [2]. Further compounds in which electronic specific heats, c_p/T, were determined as a function of the temperature in the range 1 to 14 K are of the compositions $(U_{1-x}Lu_x)Be_{13}$ (x = 0.016 and 0.034) [3] and $(U_{1-x}Sc_x)Be_{13}$ (x = 0.0152) [3]. Lu has a similar effect on the specific-heat anomaly of the U beryllide at the superconducting transition, as has Th. Table 2/10 surveys the influence of some lanthanides on the transition temperature T_c of UBe_{13} [1].

Table 2/10
Superconducting Transition Temperatures of Beryllides $(U_{1-x}M_x)Be_{13}$ (M = lanthanide) [1].

lanthanide	lattice parameter of MBe_{13} in Å (from [8])	concentration x	transition temperature T_c in K
La	10.44	0.017	0.53
Ce	10.376	0.0158	0.55
Gd	10.27	0.0147	0.42
Lu	10.173	0.016; 0.034	<0.045

Doping of UBe_{13} with Gd provides information on the nature of its superconductivity. $(U_{0.999}Gd_{0.001})Be_{13}$ does not show any decrease of the Knight shift of the local Gd moments below the normal-state values in ESR measurements down to 0.4 K, well below T_c. The data support the assumption of even-parity superconductivity of UBe_{13} [5].

The ESR properties of the heavy-fermion compound UBe_{13} doped with Er, Dy, or Gd were determined in the temperature range where the variation in enhanced specific heat is large. The doping with these elements, up to 10000 ppm Er, 2500 ppm Dy, or 1000 ppm Gd, does not generate detectable local element ESR [5]. Thus, it is concluded that the local moments substituting at the U sites are not significantly coupled to the heavy-fermion system [4, 5]. For lower fields the ^9Be spin-lattice relaxation rate is dominated by longitudinal fluctuations of the Gd^{3+} moment [6]. Measurements with Gd-doped UBe_{13} below the transition temperature down to 0.4 K indicate that the Knight shift of the local Gd moments does not decrease below the normal-state value [7].

The transition temperature of $(U_{1-x}Ba_x)Be_{13}$ with $x = 0.022$ is $T_c = 0.80$ K, and of $(U_{1-x}Sc_x)Be_{13}$ with $x = 0.0152$ at $T_c = 0.70$ K [1].

References for 2.1.7.6:

[1] Smith, J. L.; Fisk, Z.; Willis, J. O.; Batlogg, B.; Ott, H. R. (J. Appl. Phys. **55** [1984] 1996/2000).
[2] Alekseevskii, N. E. (Pis'ma Zh. Eksperim. Teor. Fiz. **40** [1984] 66/9; JETP Letters **40** [1984] 800/3).
[3] Ott, H. R.; Rudigier, H.; Felder, E.; Fisk, Z.; Smith, J. L. (Phys. Rev. [3] B **33** [1986] 126/31).
[4] Gandra, F.; Schultz, S.; Oseroff, S. B.; Fisk, Z.; Smith, J. L. (Phys. Rev. Letters **55** [1985] 2719/22).
[5] Bloch, J. M.; Davidov, D.; Felner, I.; Shaltiel, D. (J. Phys. F **6** [1976] 1979/88).
[6] Moore, J. M.; Wong, W. H.; Lan, M. D.; Clark, W. G.; MacLaughlin, D. E.; Fisk, Z.; Smith, J. L.; Ott, H. R. (J. Magn. Magn. Mater. **76/77** [1988] 530).
[7] Hijmans, T. W.; Taleb, S.; Clark, W. G.; Fisk, Z.; Smith, J. L.; Ott, H. R. (Solid State Commun. **60** [1986] 343/6).
[8] Pearson, W. B. (A Handbook of Lattice Spacings and Structures of Metals and Alloys, Vol. 2, Pergamon, Oxford 1967).

2.1.7.7   U − Be − M

Lattice parameters and transition temperatures of ternary intermetallics of the $U(Be_{13-x}M_x)$type are listed in Table 2/11, p. 68 [1]. Additions of Ga cause an expansion of the lattice and a slight depression of T_c. The two effects are somewhat decreased by thermal annealing, possibly because Ga is forced out of the lattice, since the appearance of diffraction lines of UGa_3 indicates such an effect. The two other impurities do not change the lattice constants. The beryllides doped with Cu or B do not show a superconductive transition even at low concentrations of the impurities. The low-temperature specific heat of $U(Be_{12.94}Cu_{0.06})$ indicates that the heavy-fermion state also exists in the doped beryllide. It is not known why the ternary compound does not show superconductivity, but it is postulated that the reason might be similar to that for the heavy-fermion compound $CeAl_3$ [2].

References for 2.1.7.7 on p. 68

Table 2/11

Lattice Parameters and Superconducting Transition Temperatures T$_c$ of Some Ternary Compounds of the Type U(Be$_{13-x}$M$_x$) [1].

M	x	as cast condition		tempered 5 d at 950°C	
		a in Å	T$_c$ in K	a in Å	T$_c$ in K
—	0	10.254	0.81	10.253	0.83
Ga	0.12	10.262	0.60	10.253	0.83
Ga	0.35	10.277	0.39	10.264	0.64
Cu	0.05	10.257		10.257	<0.015
Cu	0.06	10.259		10.259	<0.015
Cu	0.11	10.261		10.261	
Cu	0.51	10.277		10.277	
B	0.10	10.246		10.246	<0.020
B	0.20	10.236		10.236	<0.020

The influence of the substitution of a part of Be by Cu on the low-temperature specific heat of UBe$_{13}$ is shown in **Fig. 2-24** [1]. The resistivity and specific heat peaks of UBe$_{13}$ are suppressed by doping with Cu on Be sites [3]. The ternary compound U (Be$_{12.94}$Cu$_{0.06}$) does not show any indication for a superconducting transition [3].

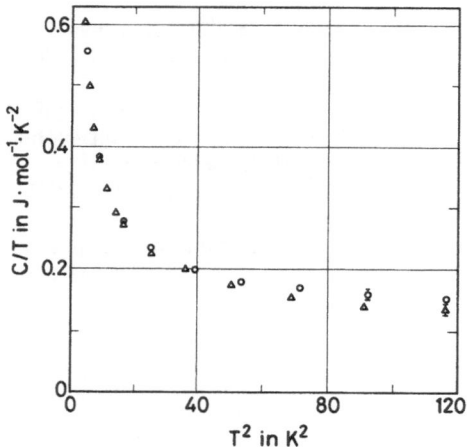

Fig. 2-24. The influence of the substitution of a part of Be by Cu on the low-temperature specific heat of UBe$_{13}$ (triangles), compared with UBe$_{12.94}$Cu$_{0.06}$ (circles) [1].

References for 2.1.7.7:

[1] Giorgi, A. L.; Fisk, Z.; Willis, J. O.; Stewart, G. R.; Smith, J. L. (Proc. 17th Intern. Conf. Low Temp. Phys. LT-17, Karlsruhe 1984, pp. 229/30).

[2] Stewart, G. R.; Giorgi, A. L. (J. Low Temp. Phys. **56** [1984] 379/81).

[3] Mayer, H. M.; Rauchschwalbe, U.; Steglich, F.; Stewart, G. R.; Giorgy, A. A. (Z. Physik B **64** [1986] 299/304).

2.1.7.8   U − Be − Mo

A eutectic mixture is also observed in the U − Be − Mo system, containing $MoBe_{13}$ as well as UBe_{13}. The hexagonal compound $MoBe_2$ is detected in alloys with more than 30 at% Mo. The alloys were prepared in the same way as U − Be − Zr alloys, using 99.99% purity Mo [1, 2].

References for 2.1.7.8:

[1] Badaeva, T. A.; Alekseenko, G. K.; Kuznetsova, R. I. (in: Ivanov, O. S., Physical Chemistry of Alloys and Refractory Compounds of Thorium and Uranium, Israel Program for Scientific Translation, Jerusalem 1972, pp. 148/57).

[2] Ivanov, O. S.; Badaeva, T. A.; Sofronova, R. M.; Kishenevskii, V. B.; Kushnir, N. P. (Phase Diagrams of Uranium Alloys, Amerind, New Delhi 1983, translated from: Diagrammy Sostoyaniya i Fazovye Prevrashcheniya Splavov Urana, Nauka, Moscow 1972).

2.1.7.9   U − Be − Si

The isothermal section at 900°C of the U − Si − Be phase diagram shows the existence of three significant solid solutions. One of them is a substitutional solid solution of up to 4.1 at% Si in UBe_{13}. A second one is the solution of Be in the $USi_{1.88}$ phase extending to 17 at%. A solubility of Be also exists in the hexagonal U_3Si_5 with a saturation concentration of 26 at% Be. The phase $U(Be, Si)_{13}$ has a lattice constant a = 10.312 Å at saturation (compared to UBe_{13}, a = 10.255 Å) [1]. The capacity of U_3Si_5 to take up large amounts of Be is due to the nearly identical ionic radii of Si (2.219 Å in U_3Si_5) and Be (2.18 to 2.22 Å in $MeBe_{13}$). Some two-phase eutectics $(U(Be, Si)_{13} − Si$; $U(Be, Si)_{13} − U_3(Si, Be)_5)$ and a ternary eutectic $(U(Be, Si)_{13} − USi_{13} − Si)$ were discovered by phase analyses. The melting point of the saturated phase $U_3(Si, Be)_5$ is $1830 \pm 20°C$ [2].

References for 2.1.7.9:

[1] Molho, S. (J. Nucl. Mater. **42** [1972] 65/72).
[2] de Tournemine, R.; Molho, S. (J. Nucl. Mater. **37** [1970] 345/6).

2.1.7.10   U − Be − Na

U beryllide suspensions in liquid Na were considered for application as liquid fuel slurries. Some studies concerning the fluid flow behavior of the slurry were performed [1, 2]. The compound UBe_{13} is compatible with the liquid alkali metal [1, 2].

References for 2.1.7.10:

[1] Cairns, R. C. (A Conf. 15 P 1092; N.S.A. **13** [1959] No. 7126).
[2] Cairns, R. C.; Turner, K. S. (AAEC E-16 [1958] 1/31; N.S.A. **13** [1959] No. 20868).

2.1.7.11   U − Be − C

Reactions of Be and UC were examined in the temperature range from 700 to 1000°C. A reaction occurs throughout the whole temperature range, however, an alloy of the three

References for 2.1.7.11 on p. 70

elements has not been identified. The reaction products at the phase boundaries are UBe_{13} and uniformly dispersed carbon [1].

The presence of three pseudobinary sections in the ternary phase diagram between the compounds $UC-UBe_{13}$, UC_2-Be_2C, and Be_2C-UBe_{13} and the eutectic nature of crystallization of alloys in the section $UC-Be_2C$ and $UC-UBe_{13}$ is reported, and a possible ternary phase diagram at 1700°C in the range of alloys limited by the compounds $UC-UC_2-Be_2C$ is suggested [2].

References for 2.1.7.11:

[1] Murdock, J. F. (J. Nucl. Mater. **7** [1962] 192/6).
[2] Ivanov, O. S.; Badaeva, T. A.; Sofronova, R. M.; Kishenevskii, V. B.; Kushnir, N. P. (Phase Diagrams of Uranium Alloys, Amerind, New Delhi 1983, translated from: Diagrammy Sostoyaniya i Fazovye Prevrashcheniya Splavov Urana, Nauka, Moscow 1972).

2.2   Uranium — Magnesium

2.2.1   Phase Diagram

The U—Mg phase diagram, based on analytical, X-ray, thermal, and metallographic methods, shows almost complete immiscibility of the liquids up to 1225°C. U—Mg compounds do not exist. The liquids which coexist at 1135°C and 3 bar contain approximately 0.14 wt% (= 0.014 at%) U in Mg and 0.004 wt% (= 0.04 at%) Mg in U. The solubility of U in Mg decreases to ca. 0.05 wt% (0.005 at%) at 675°C. U has a negligible effect on the melting point of Mg, and Mg does not influence the transformation temperatures of U [1], see also [2 to 4]. The data are also compiled in [5, 6].

Fig. 2-25 [1] shows the U—Mg phase diagram. A revised phase diagram is presented in a more recent survey of the constitution of the U—Mg system [7]. The revised version gives a more precise representation of the solubility of U in liquid Mg in the temperature range from 649 to 1135°C [8]. The Mg-rich part of the phase diagram is shown in **Fig. 2-26** according to [7, 8]. The solubility of U in Mg can be expressed by the relation $\log x_U = -1.031 - 3846/T$ (T in K) which is valid in the temperature range 776 to 1135°C. Here, x_U is the atom fraction of U. Below 776°C, the solubility of U in molten Mg is lower than described by this equation. Saturation values are given in Table 2/12. These values are lower than expected from the solubility equation and from the known transformation enthalpies of γ- and β-U [7].

Table 2/12
Solubility of Uranium in Magnesium [7].
x_U = atom fraction of U, c_U = concentration of U in wt%.

temperature in °C	U content in liquid Mg	
	x_U	c_U
700	7×10^{-6}	7×10^{-3}
650	2×10^{-6}	2×10^{-3}

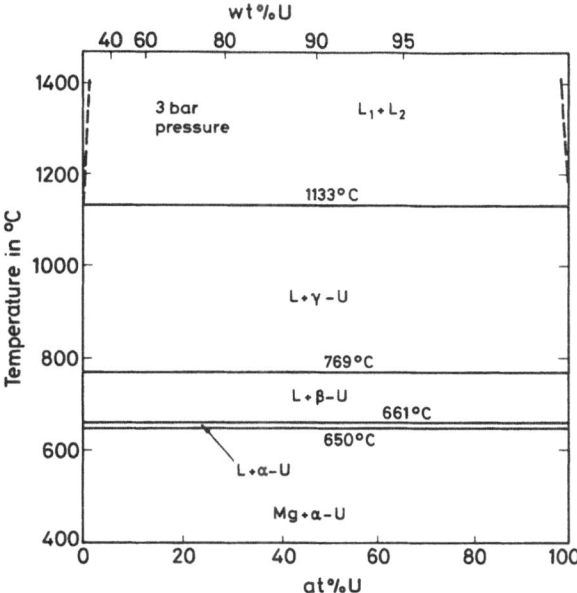

Fig. 2-25. The U—Mg phase diagram [1].

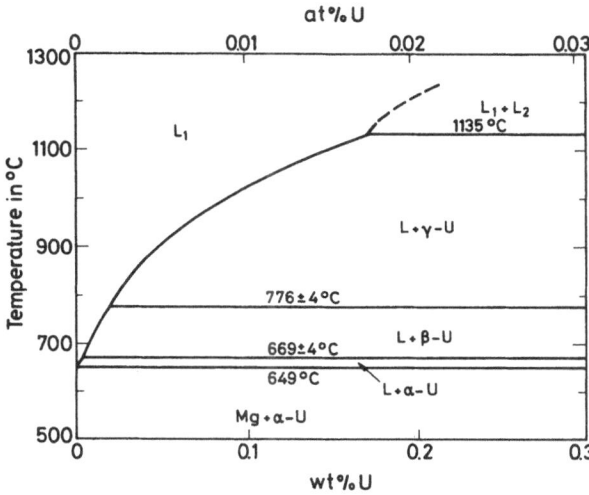

Fig. 2-26. The Mg-rich part of the revised U—Mg phase diagram [7].

The solubility of Mg in solid U at its freezing temperature, ca. 1135°C, is 0.0042 wt% [9].

The activity coefficient and the excess molar free energy of solutions of U in liquid Mg at 800°C (1073 K) are calculated on the basis of the partitioning coefficient of U between a

References for 2.2 on pp. 75/6

melt of $MgCl_2$ and a molten Mg−Zn alloy at different Mg concentrations. The relation $\Delta G_U^{sol} =$ RT ln $\gamma_U = 17600 \pm 4.17$ T (T in K) was deduced from the solubility equation for solid γ-U. The calculated value $\Delta G_U^{sol} = 20700$ cal/mol at 800°C (1073 K) agrees well with the value obtained from the equation [10].

2.2.2 Preparation of Alloys

Sintering of fine powders of U and Mg does not result in alloy formation because of the poor miscibility of both metals [11]. Even the powdered metals are compatible with each other up to 1015°C (= 1288 K) and do not react [12]. The compatibility of the two metals is caused partly by the lack of wetting of U by molten Mg. The addition of small amounts of Al enhances the wetting and allows a diffusion bonding of U with Mg. This diffusion is an indication of a weak alloy formation, a solid solution of Mg in U [13]. The poor wetting of U by molten Mg causes problems in the powder-metallurgical infiltration process. The addition of some Al improves the wetting behavior as well as the alloy preparation. Such alloys consist of a network of sintered U grains filled by the Mg−Al phase. The effect of the Al addition results from the reduction of surface oxides on the U grains [14, 15].

The attempt to prepare alloys by means of the reduction of U halides with molten Mg was unsuccessful owing to the very low solubility of U in the molten Mg. At 1150°C and a pressure of 3 bar, two liquid layers exist in equilibrium: One is rich in U and contains 0.004 at% Mg; the other is Mg-based and contains 0.04 at% U [9].

2.2.3 Structural Data

U−Mg alloys are very dilute solid solutions of either U in Mg or Mg in U. The very low concentrations of the solutes hardly influence the physical properties of the solid solvents. Mg contents in U do not change the transformation temperatures of the U modifications as shown in the phase diagram (see Fig. 2-25). Thus, the structure of U with low Mg contents depends on the temperature. α-U is stable at room temperature up to 650°C, whereas β- and γ-U are stable between 650 and 769°C. The solid solution of U in Mg contains α-U as the stable modification. Because of the very poor miscibility in the solid state, alloys prepared in the molten state should contain precipitated solutes [7].

Density and thermal expansion data of the dilute solid solutions are not yet reported.

2.2.4 Physical Properties

The melting points of the solid solution alloys of U in Mg and Mg in U are not markedly changed compared with those of their constituents. The influence of the solutes on the vapor pressure and the boiling point is not published. It is probably small [7].

Electron microprobe analyses of diffusion couples of U and Mg at 400 and 500°C gave diffusion constants of ca. 3×10^{-11} cm²/s at 500°C and ca. 1×10^{-11} cm²/s at 400°C. However, the estimated concentrations of U in Mg disagree with the solubility measurements. The reported diffusion constants contradict the corrosion experience [16].

The electric, magnetic, optic, and elastic properties of the alloys are not yet known, but the properties of the solvent metals are probably not significantly changed.

Nuclear properties and irradiation effects have not yet been studied because U−Mg alloys cannot be used as nuclear fuel materials.

2.2.5 Chemical Reactions

Mg and Mg-based alloys are compatible with U and can be applied as cladding materials up to high temperatures [12]. In some cases, the two metals react with other metals to form ternary alloys (see ternary systems, below). U and Mg do not interact chemically. The solid solutions of U in Mg and of Mg in U react with acid or alkaline solutions as if they were pure metals [9].

The oxidation and corrosion behavior of the alloys have not yet been studied.

Alloys and diffusion couples of U and Mg can be etched to expose the precipitated phase with a solution consisting of 1 part saturated aqueous sodium fluorosilicate and potassium tartrate with 1 part nitric acid [9].

2.2.6 Uses of U — Mg Alloys

The only application of the system U — Mg is based on the very poor miscibility of the constituents. Molten Mg can be used to extract Th metal from U — Th alloys or from solid mixtures of the two metals since Mg can dissolve considerable amounts of Th, while U remains undissolved [8]. The solubility of U in Mg — Th liquid alloys is only slightly increased by the presence of the dissolved Th [8].

The possible use of the formation of the ternary liquid U — Mg — Zn alloys as a step in reprocessing breeder reactor fuels was studied [18].

2.2.7 Ternary Systems

The solubility of U in liquid Mg — Al alloys is considerably lower than in molten Al and decreases with increasing content of Mg as shown in Table 2/13 [17].

Table 2/13
Solubility of U in Liquid Mg — Al Alloys [17].

temperature in °C	solubility in at% U in		
	Al	$Al_{62}Mg_{38}$	$Al_{28}Mg_{72}$
650	1.7	0.17	0.016
700	2.3	0.21	0.020
750	2.8	0.29	

The solubility of U in liquid Mg is raised slightly by additions of large amounts of Th. **Fig. 2-27**, p. 74, shows the solubility of U in Mg compared with the solubility in Mg — Th containing 35 wt% Th [8]. The addition of only 16 wt% of Th has a much smaller effect on the solubility of U in Mg — Th liquid mixtures [8].

The addition of Mg to molten Zn affects the solubility of U in the liquid alloy [18]. As shown in **Fig. 2-28**, p. 74, the solubility of U is raised by Mg present in concentrations up to ca. 0.5 (atom fraction). Larger amounts of Mg cause a drastic decrease in the U solubility. The effect of additions of Mg is more pronounced at relatively low temperature (600°C), whereas the

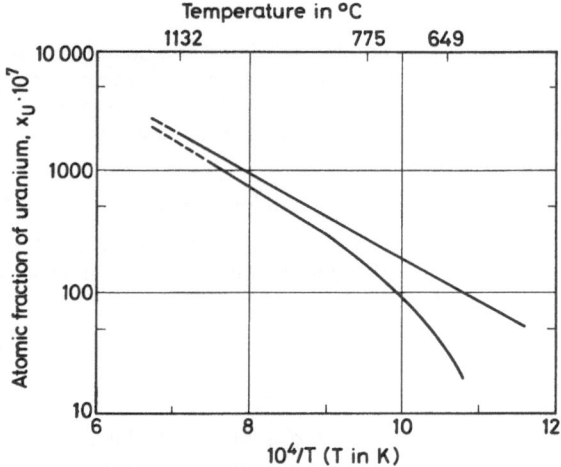

Fig. 2-27. Solubility of U in Mg and in the liquid mixture Mg — 35 wt% Th (upper curve) [8].

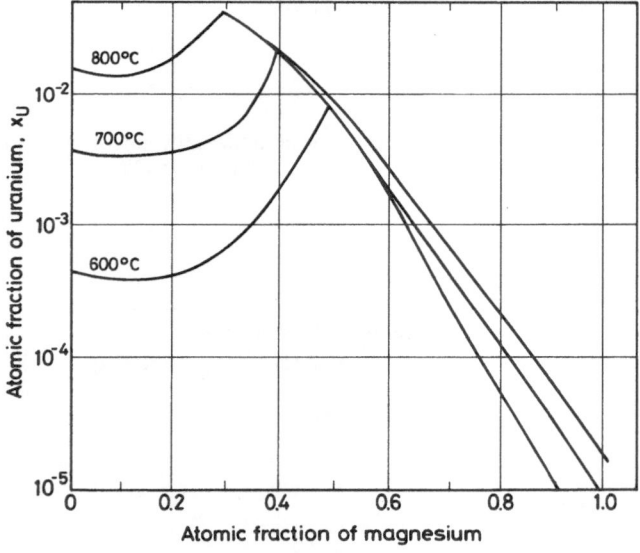

Fig. 2-28. Solubility of U in liquid U — Mg — Zn alloys [18].

effect becomes insignificant at higher temperatures. The maximum in the solubility of U moves toward smaller values of the Mg concentration as the temperature is increased. The behavior is influenced by the formation and decomposition of an intermetallic compound according to UZn_n (s) $\rightleftarrows U_s$ + n Zn (dissolved). A similar effect of Mg was observed in solutions of U in liquid Bi or Bi — Zr alloys. **Fig. 2-29** indicates that an addition of 1 wt% Mg to Bi — Zr alloys

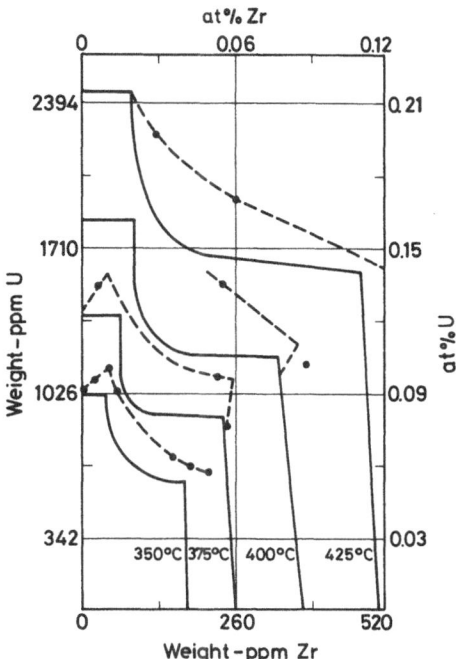

Fig. 2-29. The effect of the addition of 1 wt% Mg on the liquidus curves of the U — Zr — Bi ternary systems (points and dotted lines indicate the effect of Mg) [19].

(with up to 0.12 at% Zr) increases the saturation concentration of U. The effect seems to be small at the lowest and highest regions of Zr concentrations in the alloy. The effect does not markedly change with rising temperature in the range from 350 to 425°C [19].

References for 2.2:

[1] Hansen, M.; Anderko, K. (Constitution of Binary Alloys, McGraw-Hill, New York 1958, p. 299).

[2] Bellamy, R. G.; Hill, N. A. (Extraction and Metallurgy of Uranium, Thorium, and Beryllium, Pergamon, Oxford 1963, p. 165).

[3] Rough, F. A.; Bauer, A. A. (Constitutional Diagrams of Uranium and Thorium Alloys, Addison-Wesley, Reading, Mass., 1958, pp. 1/153).

[4] Saller, H. A.; Rough, F. A. (BMI-1000 [1955] 1/141; N.S.A. **9** [1955] No. 5349).

[5] Massalski, T. B. (Binary Alloy Phase Diagrams, Vol. 1, Am. Soc. Metals, Metals Park, Ohio, 1985, pp. 482/3).

[6] Ivanov, O. S.; Badaeva, T. A.; Sofronova, R. M.; Kishenevskii, V. B.; Kushnir, N. P. (Phase Diagrams of Uranium Alloys, Amerind, New Delhi 1983, translated from: Diagrammy Sostoyaniya i Fazovye Prevrashcheniya Splavov Urana, Nauka, Moscow 1972).

[7] Chiotti, P. (Bull. Alloy Phase Diagrams **1** [1980] 108/9).

[8] Chiotti, P.; Schoenmaker, H. E. (Ind. Eng. Chem. **50** No. 2 [1958] 137/49).

[9] Chiotti, P.; Tracy, G. A.; Wilhelm, H. A. (J. Metals **8** [1956] 562/7).

[10] Bayanov, A. P. (Zh. Fiz. Khim. **43** [1969] 2231/3; Russian J. Phys. Chem. **43** [1969] 1250/1).

[11] Raub, E.; Plate, W. (Z. Metallk. **42** [1951] 76/82).

[12] Röllig, H. E. (Kernenergie **5** [1962] 641/68).

[13] Kraus, V. (Proc. 3rd Intern. Conf. Peaceful Uses At. Energy, Geneva 1964, Vol. 11, pp. 131/6).

[14] Kieffer, P.; Sedlatschek, K. (Planseeber. Pulvermet. **5** [1957] 104/19).

[15] Kieffer, R.; Sedlatschek, K. (Proc. 2nd Intern. Conf. Peaceful Uses At. Energy, Geneva 1958, Vol. 6, pp. 96/103).

[16] Calais, D.; Beyeler, M.; Mouchnino, M.; van Craynest, A.; Adda, Y. (Compt. Rend. **257** [1963] 1285/7).

[17] Hayes, E. E.; Gordon, P. (TID-2501 [1951] 115/26; N.S.A. **12** [1958] No. 17285).

[18] Johnson, I. (J. Nucl. Mater. **51** [1974] 163/77).

[19] Weeks, J.; Minardi, A. (BNL-4261 [1959] 1/96; N.S.A. **13** [1959] No. 16621).

2.3   Uranium – Calcium

2.3.1   Phase Diagram

A phase diagram U–Ca is not established yet. Attempts to prepare U–Ca alloys in 24 h at 800 °C show no reaction between the two metals. Intermetallic compounds are not known and they are unlikely to exist. Saturation concentrations of U in liquid or solid Ca are not known, nor is the solubility of Ca in liquid or solid U. Diffusion of Ca in U indicates a small but measurable solubility of Ca in solid U at high temperature [1] (see also [2 to 4]).

Ca metal is used as a reducing agent to prepare U metal from UO_2. The U powder formed in this process has a higher stability against oxidation by air than that produced by the hydriding of massive U. It is a very fine powder containing more than 90 wt% of a particle size below 60 μm [10].

2.3.2   Preparation and Fabrication of Alloys

U–Ca alloys were prepared by powder-metallurgical methods. The powdered components with 20 to 42 vol% U powder were heated in a stainless steel tube to a temperature just above the melting point of the matrix material and pressed at 1500 lb/in^2 (∼10.6 MPa). After a release of pressure, the compact was heat-treated at 650 °C, during which the pressure was raised to 2000 lb/in^2 (∼15.9 MPa) [5]. The melting of U–Ca mixtures can be performed in Ta crucibles coated with TaB or Ta_2B, if thermal cycling is avoided [6]. The formation of U–Ca alloys is beneficial in the sintering of UO_2 powder [7].

2.3.3   Physical Properties

The solubility of U in Ca and, conversely, Ca in U is limited, the solutes in these very dilute solid solutions do not influence the microstructure. Alloys prepared in the molten state by saturating Ca with U may form supersaturated solid solutions containing a precipitated U-rich phase [5].

Data on the density and thermal expansion of such alloys are not yet published.

The solutes do not significantly affect the melting and boiling temperature or the vapor pressure of solid solutions of U in Ca and Ca in U, in analogy to the U — Mg system (see p. 70).

Data concerning the electrical, magnetic, optical, and elastic properties are not available.

An estimation of diffusion constants and the activation energy of the diffusion process was based on the determination of the growth of a layer in which the Ca-rich precipitates in the U — Ca alloy dissolve, owing to the diffusion of Ca into unalloyed U. The determination uses metallographic techniques. The method, however, has not been used to study the U — Ca system [8].

The nuclear properties and the irradiation behavior of solid U — Ca alloys have never been studied because they have not been used as fuel alloys.

2.3.4   Chemical Reactions

The U — Ca alloys are aggressive against container materials at high temperatures. Ta crucibles are attacked by the liquid mixtures of the two metals at ~1200°C [6]. Liquid Zn extracts Ca from U — Ca mixtures (see ternary systems) [9].

The solution behavior of dilute solid solutions of Ca in U is the same as that of U. Alloys of U dissolved in solid Ca react like the alkaline earth metal.

2.3.5   Uses of U — Ca Alloys

U — Ca alloys were considered to be suitable fuel alloys for an intermediate or fast reactor. They are, however, never used for this purpose owing to the success of the oxide fuel [5]. U — Ca alloys may play a role as intermediate products in processes in which UF_4 is reduced by Ca to receive metallic U [5].

2.3.6   Ternary Systems

Studies on the system U — Ca — Zn (U — Ca — Zn — Mg) have shown that Ca has a similar, though weaker effect than Mg on the solubility of U in Zn — Ca mixtures. The U concentrations in U — Ca — Zn — Mg systems are between the values of the two ternary systems. Solubilities in the U — Ca — Zn system, in dependence on the temperature, are shown in Table 2/14 [9].

Table 2/14
The Solubility of U in Liquid Zn — Ca Alloys [9].

| temperature | composition of ternary alloys (atomic fraction) | | |
in °C	Zn	Ca	U
700	0.79	0.21	0.0033
800	0.80	0.18	0.014

                      References for 2.3 on p. 78

References for 2.3:

[1] Hansen, M.; Anderko, K. (Constitution of Binary Alloys, McGraw-Hill, New York 1958, p. 299).
[2] Bellamy, R. G.; Hill, N. A. (Extraction and Metallurgy of Uranium, Thorium and Beryllium, Pergamon, Oxford 1963, p. 165).
[3] Rough, F. A.; Bauer, A. A. (Constitutional Diagrams of Uranium and Thorium Alloys, Addison-Wesley, Reading, Mass., 1958, pp. 1/153).
[4] Ivanov, O. S.; Badaeva, T. A.; Sofronova, R. M.; Kishenevskii, V. B.; Kushnir, N. P. (Phase Diagrams of Uranium Alloys, Amerind, New Delhi 1983, translated from: Diagrammy Sostoyaniya i Fazovye Prevrashcheniya Splavov Urana, Nauka, Moscow 1972, p. 31).
[5] Brooks, H.; U.S. At. Energy Comm. (U.S. 2934482 [1960]; C.A. **1960** 17224).
[6] Jenkins, I. L.; Keen, N. J. (J. Less-Common Metals **4** [1962] 387/9).
[7] Roake, W. E.; U.S. At. Energy Comm. (U.S. 2952535 [1960]).
[8] Tournier, J. (CEA-R-2446 [1964] 1/40; N.S.A. **19** [1965] No. 16062).
[9] Johnson, I. (J. Nucl. Mater. **51** [1974] 163/77).
[10] Lloyd H.; Williams, J. (Proc. 2nd. Intern. Conf. Peaceful Uses At. Energy, Geneva 1958, Vol. 6, pp. 426/37).

2.4   Uranium — Strontium

2.4.1   Phase Diagram

A phase diagram of the U—Sr system is not yet established [1]. There is some evidence that the solutions of Sr in solid U are similar to solutions of Ca in U [2]. Information on solutions of U in liquid or solid Sr is not available.

Intermetallic compounds are not known and are unlikely to exist.

2.4.2   Preparation and Fabrication of Alloys

The preparation of U—Sr alloys is not described, but their preparation would probably be similar to that of U—Ca alloys.

2.4.3   Structural Data

The solutes in very dilute solutions of Sr in U and of U in Sr do not markedly influence their structures. U—Sr alloys prepared at high temperature or in the molten state precipitate small particles of a Sr-rich phase [2], the composition and structure of which are, however, unknown. Data on the density and thermal expansion of U—Sr alloys are not published.

2.4.4   Physical Properties

The solutes probably do not change the melting and boiling temperature or the vapor pressure of solid and liquid solutions of Sr in U and of U in Sr, but data are not published. Data concerning the electric, magnetic, optical, and elastic properties are not available.

A method of estimating diffusion coefficients of fission elements in U metal in which a dilute solution of Sr in U can be made visible by dissolution of precipitated particles has been applied to U−Sr alloys [3]. The diffusion constants are shown in Table 2/15 [2].

Table 2/15
Diffusion Constants D in cm²/s of Sr in Solid U [2].

temperature in °C	800	850	900	950	1000
D	8×10^{-10}	1.8×10^{-9}	5×10^{-9}	1.1×10^{-8}	2.3×10^{-8}

The temperature dependence of the diffusion constant is given by $D = 2.38 \exp(-47000/R \cdot T)$. In this equation, $D_e = 2.38$ is in cm²/s, and $Q_{act} = 47000$ in cal/g-atom [2].

The nuclear properties of the U metal containing dissolved or precipitated Sr, even as a fission product, are not evaluated.

2.4.5   Chemical Reactions

Information on the chemical properties of U−Sr alloys is not available. There is no doubt that the chemical behavior of solid solutions of Sr in U is similar to the chemistry of the U−Ca alloys.

2.4.6   Uses of U−Sr Alloys

Solutions of Sr in U may play a role in irradiated U metal fuel containing fission products [5]. The heavy alkaline earth metals belong to the group of predominant fission products. Such alloys are, however, not simple U−Sr systems, since several other fission elements are also present. There has been no interest in the direct application of U−Sr alloys as nuclear fuel.

2.4.7   Ternary Systems

The U−Sr−C system, which is of importance for U carbide fuel containing alkaline earth metals as fission products, was examined briefly, and is similar to the U−Ba−C system, studied at 1400°C [4].

References for 2.4:

[1] Elliott, R. P. (Constitution of Binary Alloys, First Suppl. McGraw-Hill, New York 1965).
[2] Adda, Y.; Lévy, V.; Hadari, Z.; Tournier, J. (Compt. Rend. **250** [1960] 536/8).
[3] Tournier, J. (CEA-R-2446 [1964] 1/40; N.S.A. **19** [1965] No. 16062).
[4] Peatfield, M.; Brett, N. H.; Potter, P. E. (J. Nucl. Mater. **89** [1980] 35/40).
[5] Hahn, O.; Strassmann, F. (Naturwiss. **27** [1939] 89/95).

2.5 Uranium – Barium

2.5.1 Phase Diagram

A phase diagram of the U – Ba system is not established, nor are the solubilities of either Ba in liquid or solid U, or U in liquid or solid Ba known. Solid U probably has a capacity to dissolve Ba in amounts similar to the amounts of Sr.

Intermetallic compounds are not known, and it is unlikely that they exist.

2.5.2 Preparation and Fabrication of Alloys

Methods of preparation or fabrication of U – Ba alloys are not published. Their preparation may be possible by the same techniques applied to alloys of U with Ca [1]. Ba might also be used to reduce U compounds, namely UO_2 (see Chapter 1.2, p. 3).

U – Ba alloys, in very impure state, can be formed by irradiation of U metal fuel, in which Ba is formed as one of the most abundant fission elements [4].

2.5.3 Physical Properties

Solutes, not exceeding the weight-ppm level, do not significantly influence the structure of the solid solutions of Ba in U or of U in Ba. Structural data are not published. The structure of precipitated Ba-rich phases in U – Ba alloys is unknown. The poor miscibility of the two metals, however, favor the precipitation of more or less pure Ba.

The melting and boiling temperature, as well as the vapor pressure of solid or liquid solutions of Ba in U or of U in Ba have not been measured. The data should not be markedly influenced by the low concentration of solute.

Electric, magnetic, optical, and elastic properties of the alloys are not published.

Diffusion data in the U – Ba system are not published. The similarity of solutions of Ba in solid U to U – Sr alloys allows the assumption that the diffusion rates of Ba in U are in the same order of magnitude as given in Table 2/15, p. 79.

The nuclear properties and the irradiation behavior of U – Ba alloys have not been studied. Ba-containing U metal fuel has not been studied with respect to its content of this fission element [4].

2.5.4 Chemical Reactions

Chemical properties of U – Ba alloys are not reported. It seems that the two reactive constituents tend to react individually (see ternary systems). The molten alloys are aggressive against metallic containments; even Ta is severely affected [2].

2.5.5 Uses of U – Ba Alloys

As U – Sr alloys, U – Ba alloys are also present in irradiated U metal fuel [4]. The alloys are impure and contain several other fission elements. The technical application of U – Ba alloys has not found any interest.

2.5.6 Ternary Systems

The ternary system U – Ba – C was examined at 1400 °C and the solid state compatibility lines were established [3]. Compositions along the UC – Ba line were produced from starting materials consisting of elemental constituents as well as the UC−Ba mixtures. X-ray examination indicated the presence of fcc UC with a = 4.959 Å (see "Uranium" Suppl. Vol. C12, 1987, p. 15). Ba reflections are not found in the X-ray patterns. Solubility of Ba in UC or of UC in Ba was not detected. The presence of UC in the Ba-rich corner (see **Fig.** 2-30) indicates that a tie line between UC and Ba, BaC_2 does not occur. The system U−Ba−C is of importance for the compatibility of UC fuel with fission products.

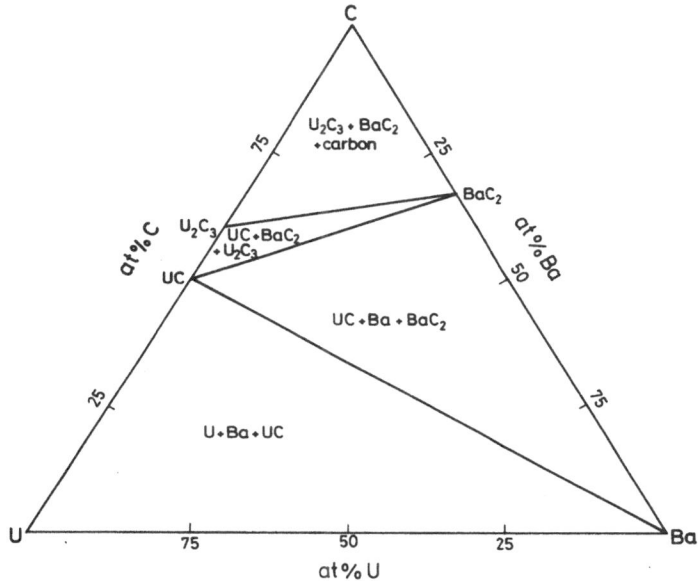

Fig. 2-30. Compatibility relationships at 1400°C in the system U−Ba−C [3].

References for 2.5:

[1] Brooks, H.; U.S. At. Energy Comm. (U.S. 2934482 [1960]; C.A. **1960** 17224).
[2] Jenkins, I. L.; Keen, N. J. (J. Less-Common Metals **4** [1962] 387/9).
[3] Peatfield, M.; Brett, N. H.; Potter, P. E. (J. Nucl. Mater. **89** [1980] 35/40).
[4] Hahn, O.; Strassmann, F. (Naturwiss. **27** [1939] 11/5, 89/95).

2.6 Uranium − Radium

Nothing is known on phase relations and alloys in the U−Ra system. This system seems to behave similarly to systems of U with heavy alkaline earth metals (Sr and Ba). Thermochemical calculations on those systems indicate that the miscibility of the elements in the liquid state should be limited. The formation of intermetallic compounds is very unlikely. The high values of the free energies of formation of the compounds of both metals with nonmetals (O, C, N) permit greater stability for ternary compounds [1, 2].

References for 2.6:

[1] Miedema, A. R.; de Châtel, P. F.; de Boer, F. R. (Physica B + C **100** [1980] 1/28).
[2] Niessen, A. K.; de Boer, F. R.; Boom, R.; de Châtel, P. F.; Mattens, W. C. M.; Miedema, A. R. (CALPHAD **7** [1983] 51/70).

3 With Metals of 3rd Main Group

Horst Wedemeyer,
Institut für Material- und Festkörperforschung,
Kernforschungszentrum Karlsruhe,
Karlsruhe, Federal Republic of Germany

3.1 Uranium — Aluminium

3.1.1 Phase Diagram

In the older literature some methods for the preparation of uranium — aluminium alloys are reported but no methods for the preparation of single-phase U — Al compounds are presented (see "Aluminium" A 5, 1937, pp. 885/6). In the present volume, the newer preparation and characterization work since 1948 is reported.

The binary compounds UAl_2, UAl_3, and UAl_4 (first reported as UAl_5 [1]) are reported to exist in the uranium — aluminium system, based on X-ray diffraction, thermal measurements, and metallographic determinations.

The solubility of aluminium in uranium is very limited (4 to 5 at% at 1105°C) [1], see also [2]. More details of the solubility are reported in [3]: 0.5 wt% Al at 980°C, 0.11 wt% Al at 700°C, 0.2 wt% Al at 800°C, <0.1 wt% Al at 650°C.

The solubility of aluminium in uranium is reported to be less than 1 at% in α-U, 2 at% in β-U, and 3 at% in γ-U at 665, 750, and 1105°C, respectively [4], see also [5], or less than 80 ppm in α-U [6], 34 to 44 ppm [7] at 550 to 665°C, see also [8, 9]. In the presence of the dissolved aluminium, the solid transformations and melting point of uranium, which are for α-U to β-U at 655°C [1], or 667°C [10], for β-U to γ-U at 765°C [1] or 771°C [10], and for γ-U to liquid at 1125°C [1] or 1132°C [10] persist without any larger changes (α-U to β-U at 655°C [1], see also [2, 3], or 665°C [6], see also [8, 10], 668°C [11], see also [12], β-U to γ-U at 750°C [1], see also [3], 758°C [6], see also [10, 12], or about 770°C [2]) in all uranium — aluminium alloys up to the first compound, UAl_2, see also [13]. The solidus temperature of the uranium-rich alloys is lowered to 1105°C [1], see also [3], or 1125°C [10], and this temperature holds with increasing amounts of aluminium up to the phase boundary of UAl_2. A eutectic reaction was assumed at 1105°C with alloys containing 4 to 5 at% aluminium [1], or at 1123°C as reported in [3, 12], or at 1125°C placed at 99.5 at% uranium [10]. The (possible) existence of a further eutectic composition at the β-U to γ-U transformation was assumed on the basis of a slight lowering of the transformation temperature, to 750°C [1], see also [3], or to 757°C as reported in [3, 12], or to 758°C [6], see also [8, 10, 11, 14] at 1.41 at% Al [6, 7], see also [8], or about 770°C at about 1.1 at% Al [2], see also [8]. Small additions of aluminium to uranium increase the hot and cold mechanical characteristics of pure uranium (see Section 3.1.6.1, p. 156) [32 to 38]. The cubic ($MgCu_2$ type) uranium dialuminide has no homogeneity range. The melting point of UAl_2 is 1590°C [1] or 1620°C [15]. No additional phase was found between uranium and UAl_2. The next phase on the aluminium side of UAl_2 is the compound UAl_3. The cubic ($AuCu_3$ type) uranium trialuminide has no homogeneity range [1], but the solubility of uranium in UAl_3 should be at least 2.6 at%, as observed by interdiffusion experiments [16]. This compound is formed by a peritectic reaction of UAl_2 and a uranium — aluminium alloy liquid containing 60 wt% uranium [17], see also [16] at 1350°C [1]. The next phase on the aluminium side of UAl_3 is the compound UAl_4, which is formed by a peritectic reaction of UAl_3 and an aluminium-rich liquid containing 18 wt% uranium [17], see also [16], at 730°C [1], see also [3], or 732°C [18], or at 745°C [10, 12]. The orthorhombic (space group is C_{2v}^{22} — I2ma (No. 46)

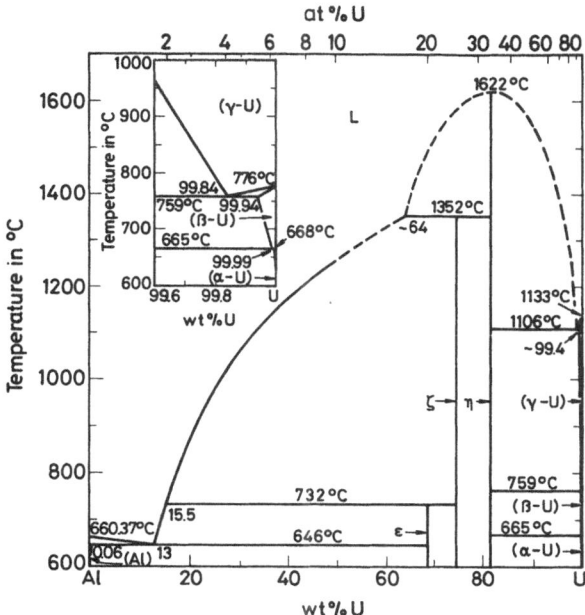

Fig. 3-1. Revised uranium — aluminium phase diagram [31]. The inset shows the U-rich area, taken from [6].

or D_{2h}^{28} — Imma (No. 74) [19]) uranium tetraaluminide, originally referred as UAl_5 [1], exists in a wide homogeneity range, ranging from $UAl_{4.0}$ (i. e., 65.0 wt % U) to $UAl_{4.8}$ (i. e., 68.82 wt % U) [20], or from 64.2 to 66.3 wt % uranium, reported in [21], see also [12], due to a lattice defect structure with unoccupied uranium positions [19], but this domain may also be much narrower [21 to 23]. Eutectic reactions occur with alloys on the aluminium side of UAl_4. The eutectic composition is placed at 1.7 at % uranium (i. e., 13.2 ± 0.2 wt % U) [18, 24] with a eutectic temperature of 640°C [1], see also [3, 18], or 646 ± 0.5°C [10, 25, 26], 642°C [27], see also [12], or 672°C as reported in [3, 12]. Slightly higher values than given in [1] are reported in [24], see also [3, 18, 28] for the aluminium-rich liquidus with UAl_3, based on thermal analysis techniques: 17 wt % U at 755°C, 30.9 wt % U at 1068°C, 20 wt % U at 855°C, 41.7 wt % U at 1190°C, 24.5 wt % U at 954°C, 51.2 wt % U at 1265°C. On heating or cooling aluminium-rich alloys (Al — 20 wt % U) a transformation of β-UAl_4 to α-UAl_4 was observed near the α-Al — UAl_4 eutectic temperature of 646°C. The high-temperature phase, β-UAl_4, precipitated during solidi-fication. After an induction period of 5 to 20 min the nucleation and growth of the low-temperature phase, α-UAl_4, was observed by evolution of energy from which the alloy tempera-ture increases by 1 to 2°C [26].

The solid solubility of uranium in aluminium (if any) is very small [1], see also [10]. The maximum solid solubility is 0.06 ± 0.02 wt % uranium at the eutectic temperature of 646°C and decreases to less than 0.04 wt % uranium at 350°C [25], see also [8].

The first established (1950) phase diagram of the uranium — aluminium system is given in [1]. It is based on thermal measurements, chemical analysis of liquid in equilibrium with solid, X-ray and microscopic determinations, also reported in [3], see also [29, 30]. A re-investigation concerning the aluminium-rich liquidus with UAl_3 led to the phase diagram (1961) given in

[24]. A slightly revised phase diagram (1971) is given in [10]. A further revised phase diagram was published by the American Society for Metals in 1973, especially including the uranium-rich area [31], see **Fig.** 3-1, p. 83. This phase diagram, which is also reported in [32], is based on reviewed literature data presented in [33], the uranium-rich area (given in the inset of Fig. 3-1) was investigated by [6].

References for 3.1.1:

[1] Gordon, P.; Kaufmann, A. R. (J. Metals **2** [1950] 182/4).
[2] Khakimova, D. K.; Virgil'ev, Yu. S.; Ivanov, O. S. (Stroenie Svoistva Splavov Urana Toriya Tsirkoniya Sb. Statei **1963** 5/8; C.A. **1959** 7214).
[3] Rough, F. A.; Bauer, A. A. (BMI-1300 [1959] 1/138; N.S.A. **12** [1958] No. 13935).
[4] Petzow, G.; Tank, R. (Z. Metallk. **54** [1963] 91/8).
[5] Chamberlain, J.; Johnson, M. P.; Kench, J. R.; Young, A. G. (J. Nucl. Mater. **19** [1966] 121/32).
[6] Straatmann, J. A.; Neumann, N. F. (MCW-1488 [1984] 1/22; C.A. **63** [1965] 6675).
[7] Russell, R. B. (NMI-2813 [1964] 1/44; C.A. **62** [1965] 7460).
[8] Shunk, F. A. (Constitution of Binary Alloys, 2nd Suppl., McGraw-Hill, New York — Toronto — London 1969, p. 46).
[9] Nolan, M. F. (MCW-1506 [1966] 1/32; C.A. **67** [1967] No. 6250).
[10] Jesse, A.; Ondracek, G.; Thümmler F. (Powder Met. **14** [1971] 289/97).

[11] Bellot, J.; Henry, J. M.; Cabane G. (Mem. Sci. Rev. Met. **56** [1959] 301/6; UCRL-Trans-714-L [1959] 1/13; N.S.A. **15** [1961] No. 31218).
[12] Elliott, R. P. (Constitution of Binary Alloys, 1st Suppl., McGraw-Hill, New York — Toronto — London 1965, pp. 61/2).
[13] Schierding, R. G.; Fergason, L. A. (MCW-1503 [1966] 1/20; C.A. **67** [1967] No. 6251).
[14] Bellot, J.; Henry, J. M.; Cabane, G. (CEA-765 [1958] 1/16; C.A. **1959** 16897).
[15] Petzow, G.; Steeb, S.; Ellinghaus, I. (J. Nucl. Mater. **4** [1961] 316/21).
[16] Castleman, L. S. (J. Nucl. Mater. **3** [1961] 1/15).
[17] Saller, H. A. (in: Finneston, H. M.; Howe, J. P.; Progress in Nuclear Energy, Ser. 5, Metallurgy and Fuels, Vol. 1, Pergamon, London 1956, pp. 535/43).
[18] Saller, H. A.; Dickerson, R. F.; Rough, F. A.; Foster, E. L.; Bauer, A. A.; Lulay, J. R. (BMI-1066 [1956] 1/52; C.A. **1961** 17433).
[19] Borie, B. S., Jr. (J. Metals **3** [1951] 800/2; ORNL-810 [1950] 1/14).
[20] Jesse, A. (J. Nucl. Mater. **37** [1970] 340/2).

[21] Boucher, R. (J. Nucl. Mater. **1** [1959] 13/27).
[22] Boucher, R. (Symp. Solid State Diffusion, Saclay, France 1958, pp. 111/7; N.S.A. **15** [1961] No. 26564; AEC-tr-4768 [1960] 1/11; N.S.A. **14** [1960] No. 7826).
[23] Boucher, R. (CEA-1298 [1959] 111/7; C.A. **1960** 22247).
[24] Storhok, V. W.; Bauer, A. A.; Dickerson, R. F. (Trans. Am. Soc. Metals **53** [1961] 837/42).
[25] Roy, P. R. (J. Nucl. Mater. **11** [1964] 59/66).
[26] Runnalls, O. J. C.; Boucher, R. (Trans. AIME **233** [1965] 1726/32).
[27] Abramson, R.; Boucher, R.; Fabre, R.; Monti, H. (Proc. 2nd Intern. Conf. Peaceful Uses At. Energy, Geneva 1958, Vol. 6, pp. 174/83).
[28] Storhok, V. W.; Bauer, A. A.; Dickerson, R. F. (BMI-1264 [1958] 1/13; C.A. **1958** 16162).
[29] Wright, E. H.; Willey, L. A. (NP-10411 [1960] 1/50; N.S.A. **15** [1961] No. 25226).
[30] Gulyaeva, B. B. (Povysh. Prochn. Otlivok Mashinostr. **1981** 6/16; C.A. **97** [1982] No. 77039).

[31] Anonymous (in: Lyman, T.; Metals Handbook, 8th Ed., Vol. 8, Am. Soc. Metals, Metals Park, Ohio, 1973).

[32] Boudouresques, B.; Englander, M. (in: Finneston, H. M.; Howe, J. P., Progress in Nuclear Energy, Metallurgy and Fuels, Ser. 5, Vol. 2, Pergamon, London − New York − Paris − Los Angeles 1959, pp. 621/31).

[33] Lehmann, J.; Aubert, H. (Symp. Special Metallurgy, Saclay 1957, pp. 41/6; N.S.A. **13** [1959] No. 2244).

[34] Badaeva, T. A.; Alekseenko, G. K.; Aleksandrova, L. N. (Str. Svoistva Splavov At. Energ. **1973** 45/52; C.A. **81** [1974] No. 174511).

[35] Englander, M. (CEA-776 [1958] 1/79; C.A. **1959** 16728).

[36] Allen, N. P.; Grogan, J. D. (Brit. 811841 [1959]; C.A. **1959** 13034).

[37] Laniesse, J.; Aubert, H. (CEA-R-2584 [1964] 1/24; C.A. **62** [1965] 15710).

[38] Althaus, W. A.; Cook, M. M. (NLCO-978 [1966] 1/13; C.A. **66** [1967] No. 68276).

[39] Massalski, T. B. (Binary Alloy Phase Diagrams, Vol. 1, p. 178, American Society for Metals 1986).

[40] Willey, L. A. (in: van Horn, K. R., Aluminium, Vol. 1, pp. 359/81, American Society for Metals 1967).

3.1.2 Interdiffusion of Uranium and Aluminium

The solid state reaction between uranium and aluminium metal starts with the formation of nuclei of UAl_2 formed near the boundary uranium − aluminium, but always in the uranium phase. There is a close contact between UAl_2 and the uranium phase without any pores at the surface. The next step seems to be a diffusion of uranium and aluminium through the UAl_2 phase into the aluminium and uranium phase, respectively. On the aluminium side two further phases, UAl_3 and UAl_4, are formed, whereas the UAl_2 grains grow into the uranium phase. The UAl_3 phase grows in close contact to UAl_2 on the uranium side of the initial boundary. The UAl_4 phase grows into the aluminium phase with the formation of a coarse and porous structure [1], see also [2 to 8].

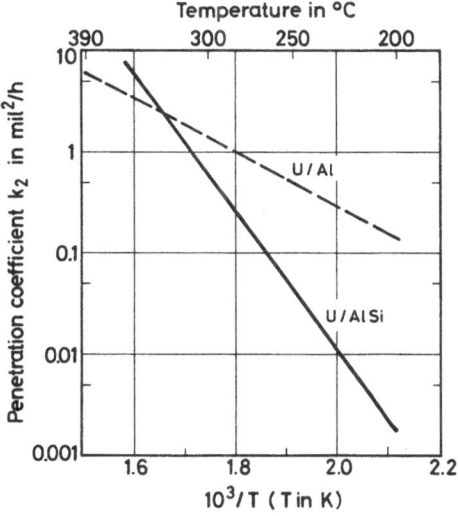

Fig. 3-2. Maximum anticipated penetration coefficient, $k_2 = k_0 \cdot \exp(-Q/RT) = x^2/t$ in mil²/h (1 mil = 10^{-3} in.), for the diffusion of uranium into aluminium and aluminium − silicon alloy from the original U/Al or U/AlSi interface [7].

References for 3.1.2 on p. 88

The measured rates of diffusion in uranium — aluminium couples in the temperature range 200 to 390°C are summarized in **Fig.** 3-2, p. 85, were the maximum penetration coefficient $k_2 = k_0 \cdot \exp(-Q/RT) = x^2/t$ is plotted as a function of the reciprocal absolute temperature [7, 8], see also [9, 10].

Experiments carried out at 400 to 600°C under a pressure of 2 to 15.7 kg/mm^2 showed an increase of the UAl$_3$ layer depending on the applied pressure. It is proposed that the effect of pressure is primarily mechanical, repressing the formation of macroscopic voids and thus increasing the cross-section area available for interdiffusion [2].

Table 3/1
Maximum Penetration of Uranium into Aluminium at a Pressure of 5 t/in.^{2} [12].

time in h	temperature in °C	maximum penetration $x = (K \cdot t)^{1/2}$ in 10^{-3} in.	maximum penetration coefficient K in 10^{-4} in.^{2}/h
119	390	27	6.1
119	390	26	6.1
287	250	12	0.5
329	250	9.3	0.25
314	200	4.8	0.074
330	200	3.0	0.027
317	200	4.3	0.058

Table 3/2
Variation with Time of the Extent of Diffusion $x = a \cdot t^{1/n}$ at Fixed Temperature and Pressure [11].

temperature in °C	pressure in t/in.^{2}	a	1/n
620	10	1.29	0.487
610	10	1.23	0.487
590	10	1.14	0.487
580	10	1.08	0.487
560	10	1.02	0.487
540	10	1.09	0.487
520	10	1.10	0.487
500	10	0.97	0.487
480	10	0.85	0.487
620	5	1.22	0.51
590	5	1.04	0.51
580	5	0.995	0.51
570	5	0.96	0.51
560	5	0.98	0.51
530	5	1.06	0.51
510	5	0.98	0.51
480	5	0.80	0.51
620	1.25	1.12	0.55
590	1.25	0.94	0.55
560	1.25	0.79	0.55
540	1.25	0.69	0.55

Table 3/3
Activation Energies Q According to $a^n = K \cdot \exp(-Q/RT)$ [11].

temperature in °C	pressure in t/in^2	Q in cal/mol
all temperatures	1.25	16000
570	5	15300
	10	14300
525	all pressures	16000
525 to 570	5 and 10	Q effectively negative

The effect of pressure on the rate of diffusion is most evident at temperatures below 540°C. The penetration of uranium into aluminium is about 2.5 times faster than that of aluminium into uranium [11, 12], see also [13, 14]. To understand this phenomenon, the formation of a UAl phase was assumed. The formation of UAl_2 and UAl_3 is accompanied by an appreciable increase in volume of about 7% for UAl_2 and 9% for UAl_3, so that the effect of pressure renders these compounds unstable [40]. In Tables 3/1 to 3/3 the measured extent of diffusion (x = $a \cdot t^{1/n}$) and the derived activation energies from $a^n = K \cdot \exp(-Q/RT)$ are summarized. The effect of pressure on the composition of the uranium−aluminium diffusion zone at 400°C and applied pressures of 300 and 1000 kg/cm^2 (see **Fig. 3-3**) showed a small variation of the uranium concentration after 3 h, ranging from 70 to 80 wt% uranium, which corresponds to a change of composition from $UAl_{2.2}$ to $UAl_{3.8}$ (at 300 kg/cm^2) or a constant concentration of 74.6 wt% uranium, corresponding to the compound UAl_3 (at 1000 kg/cm^2). The UAl_3 lattice constant showed no significant change: a = 4.262 ± 0.002 Å [15, 16].

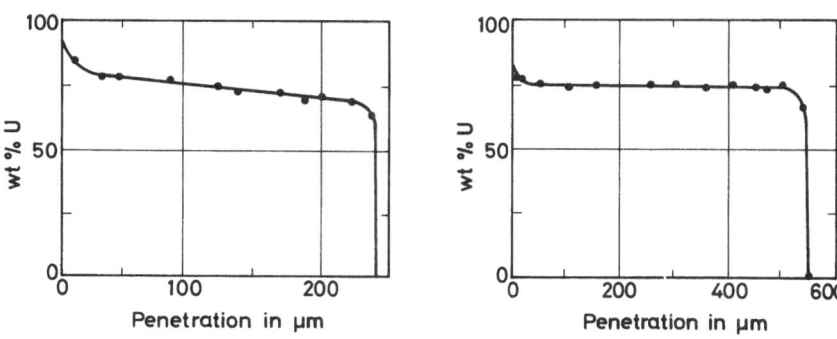

Fig. 3-3. Uranium−aluminium diffusion under applied pressures of 300 kg/cm^2 (left) and 1000 kg/cm^2 (right) for 3 h at 400°C [13].

Slightly in conflict with this observation, an increase in UAl_3-layer growth was observed with an increase in pressure of up to 138 MPa at 550°C after diffusion times of 1 and 2.5 h, but no effect was evident at 600°C after 9000 s. The pressure effect was attributed to mechanical closure of Kirkendahl voids generated during diffusion. Nonparabolic growth rates were obtained for the growth of the intermediate phase in uranium−aluminium couples [17].

References for 3.1.2 on p. 88

References for 3.1.2.:

[1] Kiessling, R. (Proc. 1st Intern. Conf. Peaceful Uses At. Energy, Geneva 1955, Vol. 9, pp. 69/73).
[2] Castleman, L. S. (J. Nucl. Mater. **3** [1961] 1/15).
[3] Boucher, R. (J. Nucl. Mater. **1** [1959] 13/27).
[4] Subramanyam, D.; Notis, M. R.; Goldstein, J. I. (Met. Trans. A **16** [1985] 589/95).
[5] Lebedev, V. A.; Sel'nikov, V. I.; Cherkezov, V. A. (Zh. Fiz. Khim. **49** [1975] 1593; Russ. J. Phys. Chem. **49** [1975] 945; C.A. **83** [1975] No. 138849).
[6] Castleman, V. L. S.; Froot, H. A.; Seigle, L. (SEP-258 [1961] 1/35; C.A. **1962** 13911).
[7] Cunningham, J. E.; Adams, R. E. (TID-7546 [1958] 102/19; C.A. **1958** 9904).
[8] Green, D. R. (HW-49697 [1957] 1/51; N.S.A. **11** [1957] No. 9324).
[9] Green, D. R. (HW-53486 [1957] 1/16; N.S.A. **12** [1958] No. 6607).
[10] Kraus, V. (Jad. Energ. [Prague] **7** [1961] 48/53; C.A. **1961** 10254).

[11] LeClaire, A. D.; Bear, I. J. (J. Nucl. Energy **2** [1955/56] 229/42).
[12] Bierlein, T. K.; Green, D. R. (Nucl. Sci. Eng. **2** [1957] 778/86).
[13] de Luca, L. S.; Sumsion, H. T. (KAPL-1747 [1957] 34; C.A. **1959** 171).
[14] Storchheim, S.; Zambrow, J. (SEP-102 [1952] 1/29; C.A. **57** [1962] 5678).
[15] Adda, Y.; Beyeler, M.; Kirianenko, A.; Pernot, B. (Mem. Sci. Rev. Met. **57** [1960] 423/34; Compt. Rend. **253** [1961] 2967/9; C.A. **1960** 20782).
[16] Adda, Y.; Beyeler, M.; Kirianenko, A.; Pernot, B. (CEA-1664 [1961] 512/24; C.A. **57** [1962] 6987).
[17] Subramanyam, D.; Notis, M. R.; Goldstein, J. I. (Met. Trans. A **16** [1985] 605/11).

3.1.3   Uranium Dialuminide, UAl_2

3.1.3.1   Formation and Preparation

3.1.3.1.1   Preparation

Preparation of UAl_2 by melting stoichiometric amounts of the elements in an induction furnace yielded metallurgical recoveries of 85 to 90% using graphite crucibles coated with magnesium zirconate. During the reaction at 700°C for 18 h uranium soaked into the crucible and aluminium distilled off. Therefore arc melting of the elements was preferred for preparation of small buttons of pure UAl_2 [1 to 3]. An average UAl_2 content of 98.2 wt% (remainder impurities, such as O_2, N_2, H_2, C, ...) was obtained by induction melting using the following method: The aluminium metal was placed at the bottom of an aluminium oxide crucible (vacuum-annealed at 1500°C and 10^{-5} Torr) with the uranium metal on top as a precaution against segregation. The melting conditions were: 400 Torr helium atmosphere, heating rate 50°C per min, melting temperature 1620°C, and optional cooling rate [4].

Usually UAl_2 was prepared by arc melting of stoichiometric amounts of uranium and aluminium metal in an inert atmosphere (argon) using tungsten electrodes and water cooled copper hearths [5, 6]. The compound obtained was remelted several times [7 to 11], see also [12 to 18] and homogenized after remelting at 1650°F for at least 10 h [6], at 1000°C for 72 h [19] in evacuated quartz ampoules. The resulting UAl_2 contained only small amounts of impurities; metallurgical recoveries yielded 97 to 99% [6].

Very pure UAl_2 was prepared by high temperature interdiffusion of aluminium pieces and dispersed uranium obtained by decomposition of uranium hydride. Uranium and the aluminium metal (in stoichiometric amounts) were placed in a beryllium oxide crucible loaded in a quartz ampoule. The crucible and the quartz tube were degassed before loading by heating in vacuum (10^{-2} Torr) at 900°C. After loading, the ampoule was filled with hydrogen and heated to 250 to 270°C to convert the uranium to UH$_3$, which was subsequently decomposed at 600°C. The hydrogen evolved at this stage was not completely removed so the ampoule was heated in an electric furnace with argon at 1000°C for 3 h, during which the hydrogen pressure reached 400 Torr. After further heating to 1300°C for 1.5 h, the ampoule was cooled to 600°C, the hydrogen pumped off, and then the sample was cooled to room temperature. The friable product was unloaded under argon, ground in a jasperite mortar, and again heated under hydrogen to 1300°C for 1.5 h. After cooling to 600°C the hydrogen was completely removed. The chemical analysis of the UAl$_2$ powder resulted in 18.45 ± 0.09 wt% Al (18.47 wt% theoretical), the stoichiometry was UAl$_{1.997}$ [65], see also [6, 21, 22].

The reaction of uranium hydride, UH$_3$, with aluminium is somewhat more complicated than melting of the metals. The hydride and aluminium powders were mixed and slowly heated under vacuum to 540°F to decompose the hydride, then the exothermic reaction started. The UAl$_2$ was single-phase with high purity, sometimes containing minor amounts of UAl$_3$. Good metallurgical recoveries resulted [6, 21], see also [2, 22].

A four-step fluidized-bed technique was developed for manufacturing uranium — aluminium compounds starting with UO$_2$(NO$_3$)$_2$ solution and aluminium powder. This process was mainly developed for the manufacture of UAl$_3$ [23], see also [4, 24] and UAl$_4$ [25], but batches of UAl$_2$ were also processed [25]. A flow sheet for the fluidized-bed production is given in Fig. 3-28, p. 120. The four steps of the process are summarized by the following chemical reactions [23, 25, 26]:

$$UO_2(NO_3)_2 \; (l) \xrightarrow{400°C} UO_3 \; (s) + 2 \; NO_2 \; (g) + 1/2 \; O_2 \; (g)$$

$$UO_3 \; (s) + CH_3OH \; (g) \xrightarrow{360°C} UO_2 \; (s) + COH_2 \; (g) + H_2O \; (g)$$

$$UO_2 \; (s) + CCl_4 \; (g) \xrightarrow{Cl_2 \; (g), \; 360°C} UCl_4 \; (s) + CO_2 \; (g)$$

$$3 \; UCl_4 \; (s) + (4 + 3x) \; Al \; (l) \xrightarrow{680°C} 3 \; UAl_x \; (s) + 4 \; AlCl_3 \; (g)$$

During the first reaction step, uranyl nitrate solution was sprayed into a fluidized-bed of aluminium powder (200 to 250 mesh in size, generally spherical) at a temperature of 400°C to form a layer of UO$_3$ on the aluminium particles using air as the fluidizing gas. In the second step, the coating of UO$_3$ was reduced by methanol to UO$_2$ at 360°C with argon as the fluidizing gas. The reduction proceeded readily when the methanol was injected and vaporized into the fluidizing argon upstream. During the third step the UO$_2$ coating was chlorinated by a mixture of 75 mol% CCl$_4$ and 25 mol% Cl$_2$ at 330°C injected into the fluidizing argon stream, forming UCl$_4$ layers on the aluminium particles. In the fourth step the reaction temperature was raised to 660°C (melting point of aluminium) and the reaction to UAl$_x$ proceeded very rapidly. Intermediately, most of the UCl$_4$ was reduced to UCl$_3$ at about 500°C by the reaction 3 UCl$_4$ + Al → 3 UCl$_3$ + AlCl$_3$. Any small amount of UO$_2$ that was not converted to UCl$_4$ formed Al$_2$O$_3$ and UAl$_x$ by the reaction 3 UO$_2$ + (4 + 3x)Al → 2 Al$_2$O$_3$ + 3 UAl$_x$. Batches of UAl$_2$ were produced by this process [25].

Violent exothermic reaction with ignition was observed when compacts of U$_3$O$_8$ and aluminium powder in stoichiometric amount were heated above the melting point of aluminium. The ignition temperature depends on the U$_3$O$_8$ particle size (see Table 3/4, p. 90). X-ray analysis of the products showed the presence of α-Al$_2$O$_3$, UO$_2$, UAl$_2$, and UAl$_3$. The reaction was the same in vacuum, argon, or air [27, 28], see also [29, 30].

 References for 3.1.3.1 on pp. 90/1

Table 3/4
Effect of U$_3$O$_8$ Particle Size on Ignition (54.4 wt % U$_3$O$_8$) [24].

U$_3$O$_8$ particle size in μm	ignition temperature in °F
> 149 (+ 100 mesh)	1950
105 to 149 (− 100 to + 140 mesh)	1920
74 to 105 (− 140 to + 200 mesh)	1860
53 to 74 (− 200 to + 270 mesh)	1820
44 to 53 (− 270 to + 325 mesh)	1780
< 44 (− 325 mesh)	1780

3.1.3.1.2 Sintering

The production of specified powders from solid UAl$_2$ presents two problems: Solid UAl$_2$ is extremely friable so that powders are easily overground. Hammer-milling led to better results than ball-milling. UAl$_2$ powders are pyrophoric. Controlled atmosphere enclosure is necessary and grinding must be carried out in argon atmosphere [6].

UAl$_2$ powders were sintered in vacuum to 6.13 g/cm^3 (75.2% th.d.) [31]. Sintered bodies had a metallic glint [20]. Sintered densities of 7.8 g/cm^3 (95% th.d.) were achieved after vacuum sintering (10^{-4} to 10^{-5} Torr) with bodies pressed to 77% th.d. The sintering started at 1493 K, i.e., 79% of the melting temperature (1893 K) [10]. UAl$_2$ powder obtained from arc-melting, then broken, wet-ground, and pressed into pellets at 100000 p.s.i., yielded, after vacuum sintering, 75.2% th.d. at 1175°C and 88.0% th.d. at 1300°C [22]. Sintering of UAl$_2$ powders in reducing (hydrogen) or inert (argon) atmosphere was not successful, as X-ray patterns of sintered bodies showed many unidentified reflexion maxima (only partly identified as UO$_2$) [10].

3.1.3.1.3 Single Crystals

Single crystals of UAl$_2$ were grown from the constituents and were initially melted in an arc furnace. The button was then crushed and melted into a rod in a silver boat with induction heating. The single crystals were grown from these rods by an induction-melting floating-zone technique [15]. Large single crystals were obtained by slowly cooling the UAl$_2$ melt [32].

References for 3.1.3.1.:

[1] Gibson, G. W.; Graber, M. J.; Francis, W. C. (IDO-16934 [1963] 1/7; N.S.A. **18** [1964] No. 8723).
[2] Gibson, G. W.; Zukor, M. (IDO-17154 [1966] 3/6; N.S.A. **20** [1966] No. 19178).
[3] Saller, H. A. (U.S. 2917383 [1959]; C.A. **1960** 18313).
[4] Thümmler, F.; Lilienthal, H. E.; Nazare, S. (Powder Met. **12** [1969] 1/22).
[5] Chamberlain, J.; Johnson, M. P.; Kench, J. R.; Young, A. G. (J. Nucl. Mater. **19** [1966] 121/32).
[6] Gibson, G. W.; de Boisblanc, D. R. (Proc. 2nd Intern. Powder Met. Conf., New York 1965 [1966], Vol. 3, pp. 26/35; C.A. **67** [1967] No. 17002).
[7] Petzow, G.; Kvernes, I. (Z. Metallk. **52** [1961] 693/5).
[8] Petzow, G.; Steeb, S.; Tank, R. (Z. Metallk. **53** [1962] 526/9).

[9] Petzow, G.; Steeb, S.; Kiessler, G. (J. Nucl. Mater. **12** [1964] 271/6).

[10] Ondracek, G.; Petzow, G. (Z. Metallk. **56** [1965] 498/502).

[11] Katz, G.; Jacobs, A. J. (J. Nucl. Mater. **5** [1962] 338/40).

[12] Hammond, J. P.; Adamson, G. M. (Carbides Nucl. Energy Proc. Symp., Harwell, Engl., 1963 [1964], Vol. 2, pp. 648/67; C.A. **63** [1965] 1433).

[13] Badaeva, T. A.; Dashevskaya, L. I. (Fiz. Khim. Splavov Tugoplavkikh Soedin Toriem Uranom **1968** 107/13; C.A. **71** [1969] No. 24297).

[14] Franse, J. J. M.; Frings, P. H.; de Boer, F. R.; Menovsky, A. (Phys. Solids High Pressure Proc. Intern. Symp., Bad Honnef, FRG, 1981, pp. 181/91; C.A. **96** [1982] No. 134510).

[15] Franse, J. J. M.; Frings, P. H.; de Boer, F. R.; Menovsky, A.; Beers, C. J.; van Deursen, A. P. J.; Myron, H. W.; Arko, A. J. (Phys. Rev. Letters **48** [1982] 1749/52).

[16] Wire, M. S.; Stewart, G. R.; Johanson, W. R.; Fisk, Z.; Smith, J. L. (Phys. Rev. [3] B **27** [1983] 6518/21).

[17] Wire, M. S.; Thompson, J. D.; Fisk, Z. (Phys. Rev. [3] B **30** [1984] 5591/5).

[18] Wire, M. S.; Stewart, G. R.; Roof, R. B. (J. Magn. Magn. Mater. **53** [1985] 283/9).

[19] Mochalov, G. A.; Tagirova, R. Kh.; Terekhov, G. I.; Ivanov, O. S. (Fiz. Khim. Splavov Tugoplavkikh Soedin Toriem Uranom **1968** 53/62; C.A. **71** [1969] No. 24226).

[20] Ivanov, M. I.; Tumbakov, V. A.; Podol'skaya, N. S. (At. Energiya SSSR **5** [1958] 166/70; Soviet J. At. Energy **5** [1958] 1007/11; C.A. **1960** 9481).

[21] Eding, H. J.; Carr, E. M. (ANL-6339 [1961] 1/39; C.A. **1961** 16243).

[22] Snyder, M. J.; Duckworth, W. H. (BMI-1223 [1957] 1/35; C.A. **1958** 3489).

[23] Grimmet, E. S.; Ballard, R. K.; Buckham, J. A. (Chem. Eng. Progr. Symp. Ser. **63** No. 80 [1967] 11/5; C.A. **68** [1968] No. 55712).

[24] Nazare, S.; Ondracek, G.; Thümmler, F. (KFK-1252 [1970] 1/83; C.A. **75** [1971] No. 70285).

[25] Hogg, G. W.; Grimmett, E. S. (AIChE [Am. Inst. Chem. Eng.] Symp. Ser. **67** No. 116 [1971] 190/8; C.A. **77** [1972] No. 22291).

[26] Hogg, G. W. (IN-1422 [1970] 1/29; C.A. **75** [1971] No. 83063).

[27] Fleming, J. D.; Johnson, J. W. (Nucleonics **21** No. 5 [1963] 84/7).

[28] Fleming, J. D.; Johnson, J. W. (TID-7642 [1962] 649/66, C.A. **59** [1963] 9539).

[29] Ondracek, G.; Patrassi, E. (Ber. Deut. Keram. Ges. **45** [1968] 617/21).

[30] Dykstra, L. J. (GA-1479 [1960] 1/21; C.A. **1961** 13975).

[31] Albrecht, W. M.; Koehl, B. G. (Proc. 2nd Intern. Conf. Peaceful Uses At. Energy, Geneva 1958, Vol. 6, pp. 116/21).

[32] Narasimham, A. V.; Roth, S.; Renker, B.; Kafer, K.; Buerkin, J. (Phys. Status Solidi A **57** [1980] K75/K78).

3.1.3.2 Enthalpy, Free Enthalpy, and Entropy of Formation

The enthalpy of formation, $\Delta H_f^\circ(UAl_2)$, was measured by several methods:
From adiabatic calorimetric measurements in aqueous solutions: $\Delta H_f^\circ(298 \text{ K}) = -22.2 \pm 1.4$ kcal/g-$UAl_{1.997}$ [1], also reported in [3], -22.3 ± 2.4 kcal/g-UAl_2 [1], see also [4], also reported in [2, 5 to 8], -7.4 kcal/g-atom [9 from 1].
From high-temperature calorimetric measurements in molten tin: $\Delta H_f(750°C) = -23.8 \pm 1.3$ kcal/mol [5], -7.9 kcal/g-atom [10].
From adiabatic calorimetry: $\Delta H_f^\circ(298 \text{ K}) = -22.1 \pm 2.0$ kcal/mol [6].
From the formation of UAl_2 from α-uranium and solid aluminium at 900 K, emf measurements: $\Delta H_f(900 \text{ K}) = -23.8 \pm 1.3$ kcal/mol, or 7.92 kcal/g-atom [7].

References for 3.1.3.2 on p. 92

The entropy of formation was calculated from emf measurements to be $\Delta S_f(900\ K) = -0.74$ e.u./g-atom [7]. The standard entropy of formation, $S_f^\circ(298\ K) = 8.58$ cal·K^{-1}·g-atom^{-1}, was calculated from Eastman's expression $S_{298}^\circ = R \cdot 3/2 \ln A_{me} + \ln V_{me} - 3/2 \ln T_M) + a$ with A_{me} = molecular weight of the compound, V_{me} = mean atomic volume, T_M = melting point in K, and $a = 12.5 \pm 2$ cal·K^{-1}·g-atom^{-1} [3].

A relation for the standard free energy of formation, based on a combination of an estimate of the free energy at the peritectic temperature of UAl₃ (at about 1350°C) with the value of the enthalpy of formation of [1], is given as $\Delta G_f^\circ = -27.4 + 9.9 \cdot 10^{-3} \cdot T$ (in kcal/mol) [11], also reported in [2]. Emf measurements within the temperature range of 442 to 840°C yield the relation $\Delta G_f^\circ = -20.700 + 0.212 \cdot T$ (442 to 840°C) for U(s) + 2 Al(s) → UAl₂(s) [6]. These data, combined with the enthalpy of formation, yield the relations $\Delta G_f^\circ = -22.790 -2.326 \cdot T \cdot \ln T + 18.37 \cdot T$, $\Delta H_f^\circ = -22.790 + 2.33 \cdot T$, $\Delta S_f^\circ = -16.0 + 2.33 \cdot \ln T$ [6]. A further relation is reported with $\Delta G = -84.520 + 11.3 \cdot T$ [12].

Partial thermodynamic values of the formation of UAl₂ from α uranium and solid aluminium at 900 K obtained from emf measurements are: $\Delta \bar{H}_v = -12.1$ kcal/g-atom, $\Delta \bar{S}_v = 1.8$ e.u.·g-atom·K^{-1} and $\Delta \bar{G}_v = -13.72$ kcal/g-atom [7]. The heat of formation at 900 K is practically the same as at 298 K, indicating that the values are constant over a wide temperature range [7].

References for 3.1.3.2.:

[1] Ivanov, M. I.; Tumbakov, V. A.; Podol'skaya, N. S. (At. Energiya SSSR **5** [1958] 166/70; Soviet J. At. Energy **5** [1958] 1007/11; C.A. **1960** 9481).

[2] Thümmler, F.; Lilienthal, H. E.; Nazare, S. (Powder Met. **12** [1969] 1/22).

[3] Naiborodenko, Yu. S.; Lavrenchuck, G. V.; Filatov, V. M. (Poroshkovaya Metal. **1982** No. 12, pp. 4/8; Soviet Powder Met. Metal Ceram. **21** No. 12 [1982] 909/12; C.A. **98** [1983] No. 76565).

[4] Lebedev, V. A.; Nichkov, I. F.; Raspopin, S. P. (Zh. Fiz. Khim. **48** [1974] 2112/4; Russ. J. Phys. Chem. **48** [1974] 1251/3; C.A. **82** [1975] No. 46843).

[5] Lukas, H. L. (Keram. Z. **20** [1968] 726, 728/9; C.A. **70** [1969] No. 71639).

[6] Chiotti, P., Kateley, J. A. (J. Nucl. Mater. **32** [1969] 135/45).

[7] Lebedev, V. A.; Sal'nikov, V. I.; Nichkov, I. F.; Raspopin, S. P. (At. Energiya SSSR **32** [1972] 115/8; Soviet At. Energy **32** [1972] 129/32; C.A. **76** [1972] No. 145/738).

[8] Rand, M. H.; Kubaschewski, O. (The Thermochemical Properties of Uranium Compounds, Oliver & Boyd, Edinburgh — London 1963).

[9] Katz, S. (J. Nucl. Mater. **6** [1962] 172/81).

[10] Dannöhl, H. D.; Lukas, H. L. (Z. Metallk. **65** [1974] 642/9).

[11] Johnson, I. (Met. Soc. AIME Inst. Metals Div. Spec. Rept. Ser. **10** No. 13 [1964] 171/92).

[12] Kulifeev, V. K.; Stikhin, A. N.; Loskutova, Z. M. (Nauchn. Tr. Mosk. Inst. Stali Splavov' No. 131 [1981] 99/102; C.A. **96** [1982] No. 111206).

3.1.3.3 Crystallographic Properties. Bonding. Lattice Dynamics

Uranium dialuminide crystallizes with laves phase structure, cubic face-centered, isomorphic with MgCu₂, containing eight molecules per unit cell [1 to 3]. The space group is $Fd\bar{3}m - O_h^7$ (No. 227) [1, 2, 4 to 6]. The measured lattice parameters are summarized in Table 3/5. A further value of a = 7.7485 kX is reported in [7 to 9] with reference to [10, 11].

Table 3/5
Measured Lattice Constants of UAl_2.

lattice constant a	preparation	Ref.	also cited in
7.795 kX		[2, 3, 5]	[13]
7.74 Å		[14]	[3, 13, 15, 16]
7.72 Å		[17]	
7.811 Å		[18]	[1, 19 to 22]
7.744 ± 0.001 kX	UAl_2 grains with ~0.3 wt% UAl_3, along the edges diffusion of Al into U	[23]	[24]
7.7475 kX	argon-arc melted, remelted, and homogenized	[10]	[11]
7.766 ± 0.001 Å ⎫ 7.7503 kX ⎬ argon-arc melted		[4, 6, 25]	[26 to 28] [7, 29, 30]
7.711 ± 0.001 Å	precipitates of UAl_2 in „adjusted" uranium	[29, 31]	[30]
7.781 Å	argon-arc melted	[13]	
7.78 Å	single crystals	[5]	
7.766 Å	argon-arc melted	[32 to 34]	
7.75 Å	as cast	[28]	

The X-ray density, calculated from X-ray diffraction measurements, is D (in g/cm^3) = 8.38 [14], 8.28 [4], 8.14 [1 to 3, 35]; measured densities are D (in g/cm^3) = 8.21 [1] or 8.2 [2, 3]. The unit cell volume is 19.52 $Å^3$ (a = 7.766 Å) [27].

The UAl_2 crystal is built up by a uranium sublattice with diamond structure; the site symmetry is $\bar{4}3m$, which is non-centrosymmetric. The near-neighbor environment of uranium consists of four aluminium triangles set in a tetrahedral arrangement and another tetrahedron of four uranium atoms rotated by 90° about a cubic axis with respect to the aluminium tetrahedron [5], see **Fig.** 3-4. The face-centered positions are: 8 U at 0,0,0; 1/4,1/4,1/4. 16 Al at 5/8,5/8,5/8; 5/8,3/8,3/8; 3/8,5/8,3/8; 3/8,3/8,5/8 [2, 3].

The interatomic distances in UAl_2 are: about each U: 4 U at 3.38 kX and 12 Al at 2.58 kX, about each Al: 6 Al at 2.76 kX and 3 U at 2.58 kX [2, 3].

Muffin-tin radii are reported to be R^{MT} (U) = 3.173 bohr and R^{MT} (Al) = 2.550 bohr [36].

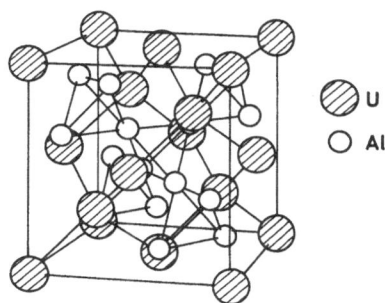

Fig. 3-4. The crystal structure of UAl_2 [38].

References for 3.1.3.3 on pp. 94/5

High-pressure X-ray diffraction, performed at room temperature with an energy dispersive method (fixed Bragg angle) using synchroton radiation, indicated transformation from the cubic face-centered structure to a not-yet-identified new structural UAl$_2$ phase around 10 GPa [33, 37]. The moduli $B_o = 74$ GPa and $B'_o = 7$ were derived from Murnaghan's equation of state, $p = \dfrac{B_o}{B'_o}\left[\left(\dfrac{V_o}{V(p)}\right)^{B'_o} - 1\right]$, from the volume variation with pressure up to 10 GPa [33, 37].

References for 3.1.3.3:

[1] Rough, F. A.; Bauer, A. A. (BMI-1300 [1958] 1/138; N.S.A. **12** [1958] No. 13935).

[2] Rundle, R. E.; Wilson, A. S. (Acta Cryst. **2** [1949] 148/50; AECD-2388 [1948] 1/12).

[3] Rundle, R. E. (CT-2721 [1945] 1/9; N.S.A. **10** [1956] No. 5124).

[4] Katz, G.; Jacobs, A. J. (J. Nucl. Mater. **5** [1962] 338/40).

[5] Rakhecha, V. C.; Lander, G. H.; Arko, A. J.; Moon, R. M. (J. Appl. Phys. **52** [1981] 1636/8).

[6] Katz, G.; Jacobs, A. J. (IA-616 [1961] 1/9; C.A. **61** [1964] 1338).

[7] Petzow, G.; Steeb, S.; Tank, R. (Z. Metallk. **53** [1962] 526/9).

[8] Petzow, G.; Steeb, S.; Kiessler, G. (J. Nucl. Mater. **12** [1964] 271/6).

[9] Steeb, S.; Petzow, G.; Tank, R. (Acta Cryst. **17** [1964] 90/5).

[10] Petzow, G.; Steeb, S.; Ellinghaus, I. (J. Nucl. Mater. **4** [1961] 316/21).

[11] Steeb, S.; Petzow, G. (Naturwissenschaften **48** [1961] 450/1).

[12] Asch, L.; Barth, S.; Gygax, P. N.; Kalvius, G. M.; Kratzer, A.; Litterst, F. J.; Mattenberger, K.; Potzel, W.; Schenck, A.; Spirlet, J. C.; Vogt, O. (J. Magn. Magn. Mater. **63/64** [1987] 160/74).

[13] Badaeva, T. A.; Dashevskaya, L. I. (Fiz. Khim. Splavov Tugoplavkikh Soedin Toriem Uranom **1968** 107/13; C.A. **71** [1969] No. 24297).

[14] Gordon, P.; Kaufmann, A. R. (J. Metals **2** [1950] 182/4).

[15] Kiessling, R. (Proc. 1st Intern. Conf. Peaceful Uses At. Energy, Geneva 1955, Vol. 9, pp. 69/73).

[16] Thümmler, F.; Lilienthal, H. E.; Nazare, S. (Powder Met. **12** [1969] 1/22).

[17] Buzzard, R. W.; Cleaves, H. E. (TID-2501 [1957] 19/47; N.S.A. **12** [1958] No. 17283).

[18] Konobrevskii, S. T.; Zaimovskii, A. S.; Levitsky, B. M.; Sokursky, Y. N.; Chebotarev, N. T.; Bobkov, Y. V.; Egorov, P. P.; Nikolaev, G. N.; Ivanov, A. A. (Proc. 2nd Intern. Conf. Peaceful Uses At. Energy, Geneva 1958, Vol. 6, pp. 194/203).

[19] Seidl, K. (Jad. Energ. [Prague] **8** [1962] 225/30; C.A. **57** [1962] 13381).

[20] Lam, D. J.; Darby, J. B., Jr.; Nevitt, N. V. (in: Freeman, A. J.; Darby, J. B., Jr.; The Actinides: Electronic Structure and Related Properties, Vol. 2, Academic, New York — San Francisco — London 1974, pp. 119/84).

[21] Wright, E. H.; Willey, L. A. (Tech. Paper Alcoa Res. Lab. No. 15 [1960] 1/46; C.A. **61** [1964] 9243).

[22] Hansen, M. (Constitution of Binary Alloys, McGraw-Hill, New York — Toronto — London 1958, pp. 143/4).

[23] Ivanov, M. I.; Tumbakov, V. A.; Podol'skaya, N. S. (At. Energiya SSSR **5** [1958] 166/70; Soviet J. At. Energy **5** [1958] 1007/11; C.A. **1960** 9481).

[24] Elliott, R. P. (Constitution of Binary Alloys, 1st Suppl., McGraw-Hill, New York — Toronto — London 1965, pp. 61/2).

[25] Lam, D. J.; Darby, J. B., Jr.; Downey, J. W.; Norton, L. J. (J. Nucl. Mater. **22** [1967] 22/7).

[26] Shunk, F. A. (Constitution of Binary Alloys, 2nd Suppl., McGraw-Hill, New York — Toronto — London 1969, p. 46).

[27] Dwight, A. E. (in: Giessen, B. C.; Developments in the Structural Chemistry of Alloy Phases, Plenum, New York 1969, pp. 181/226).

[28] Ostberg, G.; Haglund, B. O.; Lehtinen, B.; Storm, L. (J. Microsc. [Paris] 5 [1966] 21/30).

[29] Smith, A. F. (J. Inst. Metals 93 [1964/65] 454/5).

[30] James, P. F.; Fern, F. H. (J. Nucl. Mater. 29 [1969] 191/202).

[31] Smith, A. F. (J. Less-Common Metals 9 [1965] 233/43).

[32] Wire, M. S.; Stewart, G. R.; Johanson, W. R.; Fisk, Z.; Smith, J. L. (Phys. Rev. [3] B 27 [1983] 6518/21).

[33] Itie, J. P.; Olsen, J. S.; Gerward, L.; Benedict, U.; Spirlet, J. C. (Physica B + C 139/140 [1986] 330/2).

[34] Lawrence, J. M.; de Boer, M. L.; Parks, R. D.; Smith, J. L. (Phys. Rev. [3] B 29 [1984] 568/75).

[35] Snyder, M. J.; Duckworth, W. H. (BMI-1223 [1957] 1/35; C.A. 1958 3489).

[36] Boring, A. M.; Albers, R. C.; Stewart, G. R.; Koelling, D. D. (Phys. Rev. [3] B 31 [1985] 3251).

[37] Itie, J. P.; Olsen, J. S.; Gerward, L.; Benedict, U.; Spirlet, J. C. (Rept. Univ. Copenhagen Phys. Lab. 2 No. 85-15 [1985] 1/14; C.A. 103 [1985] No. 204039).

[38] Wire, M. S.; Stewart, G. R.; Roof, R. B. (J. Magn. Magn. Mater. 53 [1985] 283/9).

3.1.3.4 Mechanical Properties

Density

For X-ray densities of UAl_2 see Section 3.1.3.3, p. 93. Experimental densities of 75 to 95% th. d. were obtained from sintered specimens, see Section 3.1.3.1.2, p. 90.

Hardness. Compressibility. Elasticity

The UAl_2 compound is brittle. The brittleness at room temperature increases in accordance with the series: UC, USi_2, UBe_{13}, UAl_2 [1], see also [2].

The microhardness of UAl_2 is given as a function of load in **Fig. 3-5** [3]. Further values for the microhardness are (in kg/mm^2): 820 (25 g, 5 s) [4], 700 [5], 630 (50 g, 15 s) [6], 614 (60 g) [7], see also [8, 9].

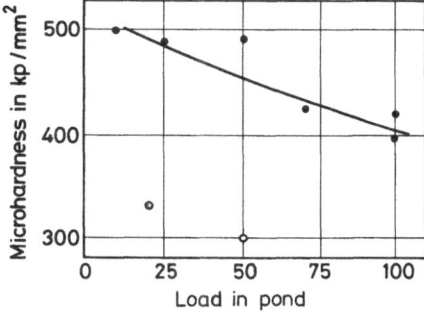

Fig. 3-5. Microhardness of UAl_2 as a function of load [3]. o: values from [6].

References for 3.1.3.4 on p. 97

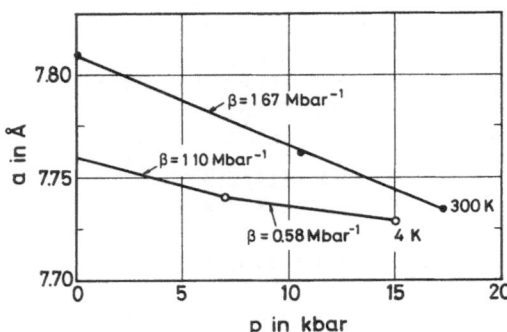

Fig. 3-6. Compressibility of UAl$_2$ versus pressure at 4 K and 300 K [10].

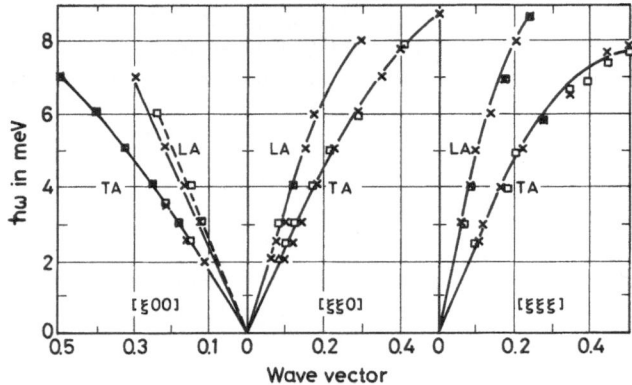

Fig. 3-7. Acoustic phonon branches of UAl$_2$ along the main symmetry directions for 300 K ($\times$) and 5 K ($\square$) [12].

The compressibility, β, of UAl$_2$ was measured at 4 K and 300 K by neutron diffraction experiments [10]. The results are shown in **Fig.** **3-6**. A room temperature value is β (at 300 K) = 1.65 Mbar^{-1} [10].

The elastic constants of UAl$_2$ were measured by ultrasonic techniques using the pulse-echo method at 15 MHz and by inelastic neutron scattering, performed on large single crystals. Room temperature values obtained from sound velocity, measured by ultrasonic technique, are (in 10^{-11} dyn/cm^2): c_{11} = 17.04, c_{12} = 3.92, c_{44} = 5.48 [77], c_{11} = 14.1, c_{12} = 1.9, c_{44} = 112 [11]. At low temperatures (about 10 K) a small peak in the longitudinal mode [12] and a sharp peak in the shear wave [11] were observed, which was interpreted as a lattice transformation from the cubic to the tetragonal state at low temperatures [11, 12]. The elastic modulus in the direction of propagation is a function of the elastic constants of the cubic crystal (c_{11}, c_{12}, c_{44}) [11]. The softening of the lattice with decreasing temperature below 10 K was attributed to a transition of the normal to the superconducting state due to the lattice transformation. These results are in agreement with measurements of the specific resistance and the susceptibility, where sharp changes were also observed at about 12 K [11], see Sections 3.1.3.6 and 3.1.3.8, pp. 103 and 109. A possible magnetic field dependence of the acoustic wave velocity was assumed at 4.2 and 2 K in magnetic fields of up to 10 T [11, 12]. The slopes of a possible T$_2$A branch (measured in magnetic fields of up to 10 Tesla) and of the T_1A branch (zero-field measurement) of 5 K and 300 K along $[\xi,\xi,0]$ are the same within the experimental error [12]. The acoustic phonon dispersion at 300 and 5 K was determined by inelastic neutron scattering. The results are given in **Fig.** 3-7. A temperature effect of about 12% of hardening was observed for the longitudinal branch along [100]. From the slopes of the phonon branches, the following elastic constants were calculated [12] (in 10^{-11} dyn/cm^2):

$c_{11} = 18.6 \pm 1.6$, $c_{12} = 4.6 \pm 1.2$, $c_{44} = 7.8 \pm 0.6$ for 5 K
$c_{11} = 22.2 \pm 1.9$, $c_{12} = 4.2 \pm 1.2$, $c_{44} = 7.1 \pm 0.6$ for 300 K

References for 3.1.3.4:

[1] Dedyurin, A. J.; Gomozov, L. J.; Grigorovich, V. K.; Ivanov, O. S. (Fiz. Khim. Splavov Tugoplavkikh Soedin Toriem Uranom **1968** 165/74; C.A. **71** [1969] No. 52906).
[2] Alekseev, O. A. (At. Tekhn. Rubezhom **1979** No. 8, pp. 18/20; C.A. **92** [1980] No. 66458).
[3] Nazare, S.; Ondracek, G.; Thümmler, F. (J. Nucl. Mater. **56** [1975] 251/9).
[4] Petzow, G.; Steeb, S.; Tank, R. (Z. Metallk. **53** [1962] 526/9).
[5] Stone, H. E. N. (J. Mater. Sci. **10** [1975] 923/34).
[6] Kiessling, R. (Proc. 1st Intern. Conf. Peaceful Uses At. Energy, Geneva 1955, Vol. 9, pp. 69/73).
[7] Badaeva, T. A.; Dashevskaya, L. I. (Fiz. Khim. Splavov Tugoplavkikh Soedin Toriem Uranom **1968** 107/13; C.A. **71** [1969] No. 24297).
[8] Pav, T.; Otruba, J.; Kocik, J.; Saxl, I. (Reaktor. Materialoved. Tr. Konf. Reaktor. Materialoved., Alushta 1978, Vol. 4, pp. 313/22; C.A. **91** [1979] No. 11067).
[9] Gomozov, L. I.; Ivanov, O. S. (Reaktor. Materialoved. Tr. Konf. Reaktor. Materialoved., Alushta 1978, pp. 4/31; C.A. **91** [1979] No. 29285).
[10] Fournier, J. M.; Beille, J. (J. Phys. Colloq. [Paris] **40** [1979] C4-145/C4-146).
[11] Narasimham, A. V. (Indian J. Phys. A **55** [1981] 398/405; C.A. **96** [1982] No. 90207).
[12] Narasimham, A. V.; Roth, S.; Renker, B.; Kafer, K.; Buerkin, J. (Phys. Status Solidi A **57** [1980] K75/K78).

3.1.3.5 **Thermal Properties**

Thermal Expansion

The thermal expansion coefficient, α, of UAl$_2$ increases slightly with temperature. The results of different measurements are summarized in Table 3/6, see also **Fig.** 3-8, p. 98.

Table 3/6
Thermal Expansion Coefficient, α, of UAl$_2$.

temperature range in °C	α in 10^{-6}/K	Ref.	also reported in
0 to 300	14.7	[2]	[4]
0 to 400	15.0	[2]	[4]
0 to 500	15.1	[2]	[4]
0 to 600	15.2	[2]	[4]
20 to 200	13.07	[3]	
20 to 400	13.45	[3]	
20 to 600	13.81	[3]	
20 to 800	14.18	[3]	
20 to 1000	14.55	[3]	

References for 3.1.3.5 on pp. 100/1

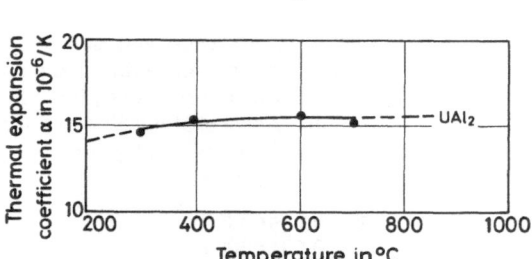

Fig. 3-8. Thermal expansion coefficient of UAl$_2$ (melted samples) [1].

Melting Point

The melting point of UAl$_2$ is 1590°C [5], also reported in [6 to 12] (as 1863 K), [13 to 15] (as 3070 °F), [16 to 19]. A slightly higher value of 1620°C was derived from temperature plots as an average value from measured melting points of 1600 to 1640°C [6], also reported in [4, 10, 20 to 22].

Heat Capacity and Thermodynamic Functions

The heat capacity (in cal · mol^{-1} · K^{-1}) of UAl$_2$ was calculated from the equation

$$C_p(T) = a_0 + a_1 \cdot 10^{-3}\, T + a_{-2} \cdot 10^5\, T^{-2}$$ with $a_0 = 16.71$, $a_1 = 3.84$, and $a_{-2} = -0.27$ [12].

Measurements of the heat capacity on small UAl$_2$ samples were performed by a heat-pulse technique within the temperature range of 1.8 to 400 K using a strong heat-leak calorimeter. The overall accuracy was estimated to be about 1% [23]. The results are presented in **Fig.** 3-9. The electronic contribution to the specific heat becomes large below 50 K, and the values of C_p/T are proportional to $T^2 \cdot \log(T/T_{SF})$ below 6 K, where T_{SF} is identified as the spin-fluctuation temperature. The total specific heat can be written in the low-temperature range

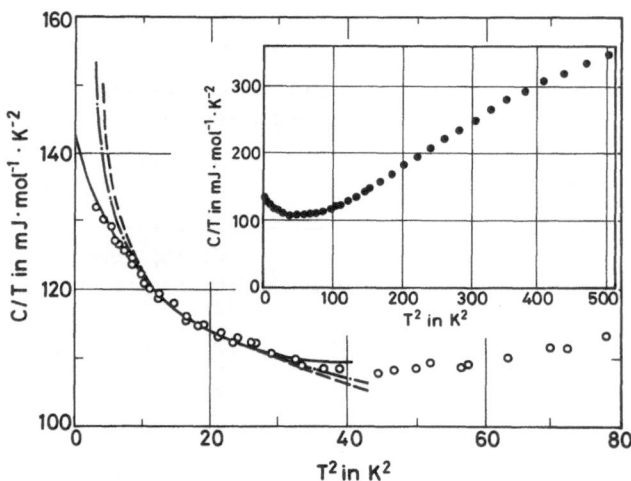

Fig. 3-9. C/T versus T^2 for UAl$_2$ [23]. Fits using the equation $C = A \cdot T + B \cdot T^3 + D \cdot f(T)$, where $f(T) = 1/T^2$ (dashed curve); $f(T) = 1/T$ (dashed-dot curve); $f(T) = T^3 \cdot \log T$ (solid curve).

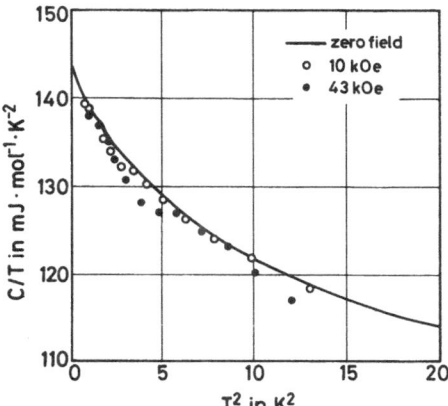

Fig. 3-10. C/T versus T^2 for UAl$_2$ in different applied magnetic fields [23]. The solid line represents the best fit to the zero-field data.

as $C = A \cdot T + B \cdot T^3 + D \cdot f(T)$, where $A = (m^*/m) \cdot \gamma$ (m^*/m represents an electronic mass enhancement for the electrons responsible for spin fluctuations; γ = electronic specific heat coefficient); $B = \beta - (\alpha \cdot \gamma/T^2) \cdot \log T$ (β = coefficient of the lattice term; T_{SF} = spin-fluctuation temperature); $D = \alpha \cdot \gamma/T^2$ and $f(T) = T^3 \cdot \log T$, which characterizes a spin-fluctuation system [23].

The values obtained are $A = 143$, $B = -4.38$, $D = 1.94$ (energy in mJ-units), and $\gamma = 70$ mJ $\cdot$ mol$^{-1} \cdot$ K^{-2} [23], to compare with an approximated value of $\gamma = 88$ mJ $\cdot$ mol$^{-1} \cdot$ K^{-2} [24]. Comparison of B and D yields $T_{SF} = 10.6$ K, assuming a reasonable value for β of 0.2 mJ $\cdot$ mol$^{-1} \cdot$ K^{-4}, corresponding to a Debye temperature of $\Theta_D \approx 300$ K [23]. Values of $\gamma = 74$ to 79 mJ $\cdot$ mol$^{-1} \cdot$ K^{-2} and $T_{SF} = 12$ K are reported, assuming a value for $\Theta_D = 250$ K [25]. Extrapolation to zero temperature yields $m^*/m \approx 2$ [23] or 1.3 to 1.6 [25]. At 300 K, UAl$_2$ exhibits a more typical metallic behavior with $\gamma = 15$ mJ $\cdot$ mol$^{-1} \cdot$ K^{-2} [23].

The specific heat of UAl$_2$ was measured in magnetic fields up to 43 kOe in the range of 0.8 to 25 K [25], see **Fig.** 3-10. The low-temperature region is dominated by a nearly field-independent $T^3 \cdot \ln T$ term [25]. Re-investigation of the former results, using the framework of random-phase approximation, yielded a higher value for the spin-fluctuation temperature of $T_{SF} = 25$ K [26], see also [25], and $\gamma = 142$ mJ $\cdot$ mol$^{-1} \cdot$ K^{-2} [26].

Measurements of the specific heat of a single crystal of UAl$_2$ between 2 and 23 K in magnetic fields up to 12.5 T and from 4 to 16 K in magnetic fields up to 17 T indicate a suppression of about 40% of the spin-fluctuation contribution, which occurs at 2 K in a magnetic field of 12.5 T. The zero-field data can be expressed as $C/T = \gamma_0 (1 + \lambda_{SF} + \lambda_{e-ph}) + \varepsilon \cdot T^2 + \alpha \cdot T^4 + \delta \cdot T^2 \cdot \ln T$ (λ_{e-ph} is the electron$-$photon interaction parameter) [27 to 30]. Resulting data are summarized in Table 3/7 and in Table 3/8, both p. 100. A value of $\Theta_D = 374$ K is reported for the Debye temperature [27].

A value for the standard entropy of UAl$_2$ was estimated to be S° (298 K) = 25.5 $\pm$ 3.0 e.u., assuming the entropy of formation is close to zero [32], also given in [4].

References for 3.1.3.5 on pp. 100/1   7*

Table 3/7
Coefficients of the Low-Temperature Specific Heat of UAl$_2$ Fitted to C/T = $\gamma_o(1 + \lambda_{SF} + \lambda_{e-ph})$ $+ \varepsilon \cdot T^2 + \alpha \cdot T^4 + \delta \cdot T^2 \cdot \ln T$ (C in mJ $\cdot$ mol^{-1} $\cdot$ K^{-1}, T in K).

four-term fit to data over the entire temperature range of 2 to 23 K [29], also cited in [27, 28]:
$\gamma_o(1 + \lambda_{SF} + \lambda_{e-ph})$ = 142.3, ε = $-$3.644, α = 0.00169, δ = 1.566

three-term fit to low-temperature data only (2 to 6.3 K) [29]:
$\gamma_o(1 + \lambda_{SF} + \lambda_{e-ph})$ = 142.1, ε = $-$3.437, δ = 1.413

Table 3/8
Coefficients of the Low-Temperature Specific Heat of UAl$_2$ Fitted to C/T = $\gamma + \beta^* \cdot T^2 +$ $\delta \cdot T^2 \cdot \ln T + \varepsilon \cdot T^4$ (C in mJ $\cdot$ mol U-atoms^{-1} $\cdot$ K^{-1}, T in K).

temperature range	γ	β^*	δ	$\varepsilon \cdot 10^3$	Ref.
1.3 to 23	132.7 $\pm$ 0.2	$-$3.04 $\pm$ 0.03	1.35 $\pm$ 0.01	$-$1.46 $\pm$ 0.02	[30]
1.3 to 8	131.3 $\pm$ 0.1	$-$2.60 $\pm$ 0.03	1.11 $\pm$ 0.02		[30]
1.3 to 10	131.1 $\pm$ 0.1	$-$2.55 $\pm$ 0.02	1.08 $\pm$ 0.01		[30]
0.8 to 6	143	$-$4.38	1.94		[25]
2 to 6.3	142.1	$-$3.44	1.41		[29]
2 to 23	142.3	$-$3.64	1.566	$-$1.7	[29]
0.3 to 2.2	$\sim$126				[30] extracted from [31]

Thermal Conductivity

The thermal conductivity of UAl$_2$ was estimated by extrapolating values of uranium$-$aluminium alloys and uranium to be λ = 0.50 W $\cdot$ cm^{-1} $\cdot$ K^{-1} at 200°C [17], see also [4].

References for 3.1.3.5:

[1] Smith, A. F. (J. Less-Common Metals **9** [1965] 233/43).
[2] Snyder, M. J.; Duckworth, W. H. (BMI-1223 [1957] 1/35; C.A. **1958** 3489).
[3] Mochalov, G. A.; Tagirova, R. Kh.; Terekhov, G. I.; Ivanov, O. S. (Fiz. Khim. Splavov Tugoplavkikh Soedin Toriem Uranom **1968** 53/62; C.A. **71** [1969] No. 24226).
[4] Thümmler, F.; Lilienthal, H. E.; Nazare, S. (Powder Met. **12** [1969] 1/22).
[5] Gordon, P.; Kaufmann, A. R. (J. Metals **2** [1950] 182/4).
[6] Petzow, G.; Steeb, S.; Ellinghaus, I. (J. Nucl. Mater. **4** [1961] 316/21).
[7] Saller, H. A. (in: Finneston, H. M.; Howe, J. P.; Progress in Nuclear Energy, Ser. 5, Metallurgy and Fuels, Vol. 1, Pergamon, London 1956, pp. 535/43).
[8] Kiessling, R. (Proc. 1st Intern. Conf. Peaceful Uses At. Energy, Geneva 1955, Vol. 9, pp. 69/73).
[9] Gibson, G. W.; de Boisblanc, D. R. (Proc. 2nd Intern. Powder Met. Conf., New York 1965 [1966], Vol. 3, pp. 26/35; C.A. **67** [1967] No. 17002).
[10] Petzow, G.; Steeb, S.; Kiessler, G. (J. Nucl. Mater. **12** [1964] 271/6).

[11] Grimmet, E. S.; Ballard, R. K.; Buckham, J. A. (Chem. Eng. Progr. Symp. Ser. **63** No. 80 [1967] 11/5; C.A. **68** [1968] No. 55712).

[12] Naiborodenko, Yu. S.; Lavrenchuk, G. V.; Filatov, V. M. (Poroshkovaya Metal. **1982** No. 12, pp. 4/8; Soviet Powder Met. Metal Ceram. **21** No. 12 [1982] 909/12; C.A. **98** [1983] No. 76565).

[13] Buzzard, R. W.; Cleaves, H. E. (TID-2501 [1957] 19/47; N.S.A. **12** [1958] No. 17283).

[14] Seidl, K. (Jad. Energ. [Prague] **8** [1962] 225/30; C.A. **57** [1962] 13381).

[15] Wright, E. H.; Willey, L. A. (Tech. Paper Alcoa Res. Lab. No. 15 [1960] 1/46; C.A. **61** [1964] 9243).

[16] Cunningham, J. E.; Beaver, R. J.; Thurber, W. C.; Waugh, R. C. (TID-7546 [1958] 269/97; C.A. **1958** 9905).

[17] Rice, W. L. R. (TID-11295 — 3rd Ed. [1964] 105/20; N.S.A. **18** [1964] No. 41933).

[18] Openshaw, P. R.; Shreir, L. L. (Corros. Sci. **3** [1963] 217/37).

[19] Buschow, K. H. J.; van Daal, H. J. (AIP [Am. Inst. Phys.] Conf. Proc. No. 5 [1971/72] 1464/77; C.A. **77** [1972] No. 11324).

[20] Petzow, G.; Steeb, S.; Tank, R. (Z. Metallk. **53** [1962] 526/9).

[21] Ondracek, G.; Petzow, G. (Z. Metallk. **56** [1965] 498/502).

[22] Thümmler, F.; Exner, H. E.; Petzow, G. (J. Nucl. Mater. **24** [1967] 328/39).

[23] Trainor, R. J.; Brodsky, M. B.; Isaacs, L. L. (AIP [Am. Inst. Phys.] Conf. Proc. No. 24 [1974/75] 220/2; C.A. **83** [1975] No. 153591).

[24] Gossard, A. C.; Jaccarino, V.; Wernick, J. H. (Phys. Rev. [2] **128** [1962] 1038/43).

[25] Trainor, R. J.; Brodsky, M. B.; Culbert, H. V. (Phys. Rev. Letters **34** [1975] 1019/22).

[26] Brodsky, M. B.; Trainor, R. J. (Physica B + C **91** [1977] 271/77).

[27] Wire, M. S.; Stewart, G. R.; Roof, R. B. (J. Magn. Magn. Mater. **53** [1985] 283/9).

[28] Boring, A. M.; Albers, R. C.; Stewart, G. R.; Koelling, D. D. (Phys. Rev. [3] B **31** [1985] 3251).

[29] Stewart, G. R.; Giorgi, A. L.; Brandt, B. L.; Foner, S.; Arko, A. J. (Phys. Rev. [3] B **28** [1983] 1524/8).

[30] Frings, P. H.; Franse, J. J. M. (Phys. Rev. [3] B **31** [1985] 4355/60).

[31] Armbruster, H.; Franz, W.; Schlabitz, W.; Steglich, F. (J. Phys. Colloq. [Paris] **40** [1979] C4-150/C4-151).

[32] Rand, M. H.; Kubaschewski, O. (The Thermochemical Properties of Uranium Compounds, Oliver & Boyd, Edinburgh — London 1963).

3.1.3.6 Electrical Properties

Electronic Structure

The electronic structure of UAl_2 comprises a wide 7s-band, a somewhat narrower 6d-band, and a very narrow 5f-band as shown by NMR and photoemission experiments [1, 2]. The uranium 5f-electrons lead to an extensive peak with an abrupt cut-off at the Fermi level in the photoemission spectra [2]. A study of the electronic structure of UAl_2, in terms of a fully relativistic single site scattering, showed that the Fermi energy is situated in the uranium 5f-band [3], in comparison with [2].

The electronic ground states of UAl_2 were determined from self-consistent, semi-re-lativistic, warped muffin-tin linear augmented-plane-wave electronic band structure calcu-lations [4, 5]. The partial density-of-state functions indicate strong d-f hybridization. Spin-fluctuation parameters were obtained from this procedure. The calculations indicate that the

References for 3.1.3.6 on pp. 106/7

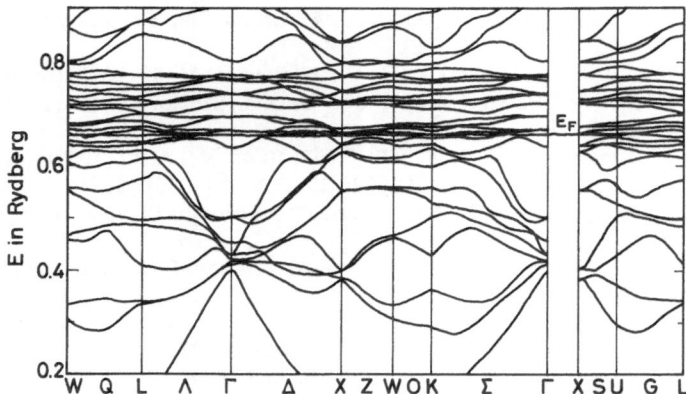

Fig. 3-11. The band structure of UAl₂ along the high symmetry lines of the fcc Brillouin zone [6].

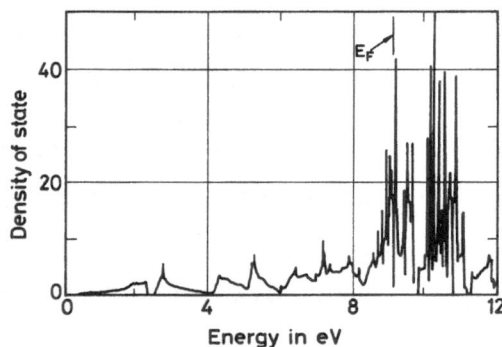

Fig. 3-12. Total density-of-states (DOS) function for UAl₂ [4].

d-f hybridization and spin-orbit coupling are important in the electronic structure [4 to 6], see **Fig. 3-11** and **Fig. 3-12**. The calculation of the band structure gave the following results [4, 5]: 5f-band width = 2.0 eV; occupied 5f-band width = 0.14 eV; total density-of-states at the Fermi energy, $N(O)_{calc} = 7.1$ states per eV per cell; uranium 5f- and 6d-contribution, $N_{5f}(O)_{calc} = 2.2$ states per eV U-atom and $N_{6d}(O)_{calc} = 0.4$ states per eV U-atom (to be multiplied by 2 to give the contribution per unit cell); 5f- and 6d-occupation = 2.25 and 1.95 per U-atom, respectively [4].

The mean structure of the density-of-states is given by the spin-orbit splitting. The fine-structure of the density-of-states around the Fermi energy and the separation into the contributions from the two bands intersecting the Fermi level are given in **Fig. 3-13**. The anomalous behavior of the electric resistivity, specific heat, and magnetic susceptibility can be explained by the fine-structure model without introduction of drastic spin fluctuations or many-body effects [6], see also [7].

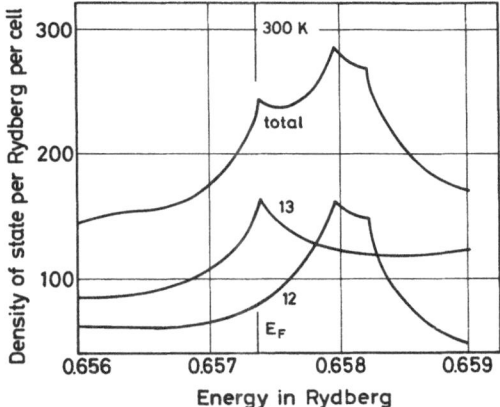

Fig. 3-13. The fine structure of the density-of-states around the Fermi energy and the separation into the contributions from the two bands (12 and 13) intersecting the Fermi level [6].

Specific Resistivity

Large differences were reported for the room temperature electric resistivity, ϱ, of UAl_2 with respect to the densities of the specimen: for a sintered (76% th.d.) specimen $0.5 \times 10^{-2}\,\Omega \cdot cm$ was measured [8], for pressed powders (63 to 90 µm, 90% th.d.) $6 \times 10^{-3}\,\Omega \cdot cm$ with a temperature coefficient of 1.8×10^{-4} [9], also reported in [10], and for cast specimen $0.5 \times 10^{-6}\,\Omega \cdot cm$ [8].

UAl_2 has an anomalous temperature dependence of the resistivity near 100 K. The resistivity increases steeply with temperature below this point and less steeply above it. This effect is discussed using the classical s-d-state model [11].

The specific resistivity of UAl_2 shows, with increasing temperature, a large increase from the residual value of $\varrho_0 = 26\,\mu\Omega \cdot cm$ [12], 7.6 $\mu\Omega \cdot cm$ (at 1.4 K) [13], $\varrho = 8.3\,\mu\Omega \cdot cm$ (at 1.8 K) [14], 26 to 190 $\mu\Omega \cdot cm$ (at 300 K), which cannot be explained by electron-phonon scattering

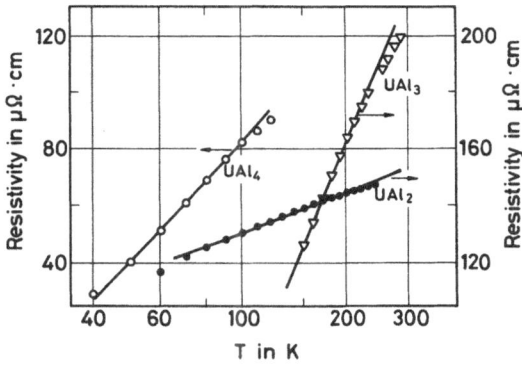

Fig. 3-14. Dependence of the resistivity of UAl_2, UAl_3, and UAl_4 on the logarithm of temperature [16].

References for 3.1.3.6 on pp. 106/7

[12]. Very large values were observed for the Lorentz ratio, $L = \varrho \cdot k/T$, with $L = 2.7 \cdot L_0$ at 300 K and $L = 4.1 \cdot L_0$ at 10 K (L_0 = Sommerfeld constant) [12].

The low-temperature behavior of the electrical resistivity is characterized by the following temperature regions [15 to 17], see **Fig.** 3-**14**, p. 103: at 1.3 to 3.5 K: ϱ proportional to T^2, above 3.5 K: ϱ proportional to $T^{1.5}$, about 40 K: ϱ proportional to T, above 80 K: ϱ proportional to lnT.

The results are discussed in terms of localized spin fluctuations. Spin-fluctuation temperatures, T_{SF}, were calculated for the different temperature regions using the dilute-alloy model (from Rivier und Zlatic [18, 19]) [16], or using the free energy, derived from a Lorentz spectral density function and the relative temperature dependence of the resistivity (see [16, 17]) [20]. The results are summarized in Table 3/9.

Table 3/9
Calculated Spin-Fluctuation Temperatures, T_{SF}, of UAl$_2$ for Different Temperature Regions.

temperature region	T_{SF} in K	Ref.
from slope of T^2 region	57	[16]
	28	[20, 21]
from high-temperature end of T^2 region	16	[16]
from upper limit, T_d, of T^2 region	23	[20]
$T_d = T_{SF}/4$	12 to 16	[20, 22]
$T_d = T_{SF}/2\pi$	19 to 25	[20, 23]
from slope of T region	712	[16]
approach to the high-temperature limit, $-d\varrho/d(1/T) = T_{SF}$	24	[16]
from low-temperature end of lnT region	85	[16]

The pressure dependence of the resistivity of UAl$_2$ was measured at pressures up to 18 kbar at temperatures ranging from 1 K to room temperature [24], see also [25], see **Fig.** 3-**15**. The measured resistance can (empirically) be written as $1/R = 1/R_i + 1/R_s$, where R_i = "ideal" resistance and R_s = "shunt" resistance, whose magnitude is near the saturation value. The values for R_i are proportional to T^2 over an appreciable temperature region, see

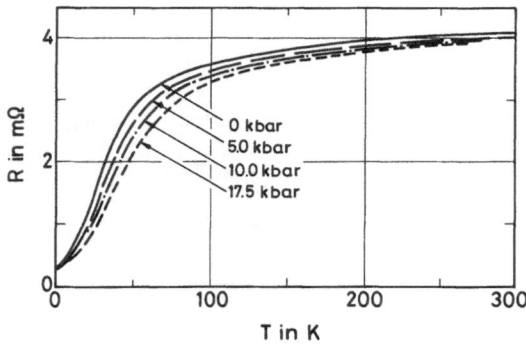

Fig. 3-15. Resistance as a function of temperature for polycrystalline UAl$_2$ at four different clamp pressures [24].

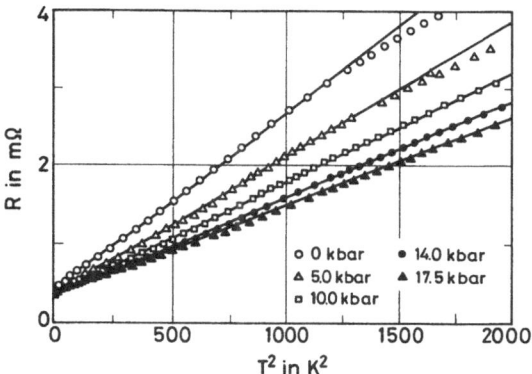

Fig. 3-16. Ideal resistance R_i versus T^2 for UAl_2 at fixed pressures up to 17.5 kbar [27].

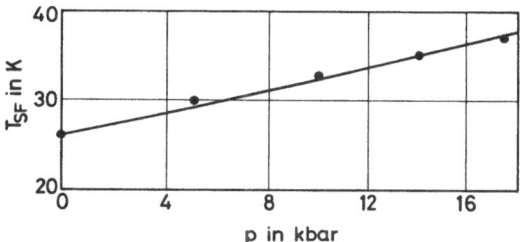

Fig. 3-17. Spin-fluctuation temperature, T_{SF}, versus pressure for UAl_2 [24].

Fig. 3-16. Assuming the ideal resistance to reflect the spin-fluctuation contribution to the resistance, the pressure dependence of the spin-fluctuation temperature can be calculated from $\partial R_i/\partial T^2 = A \cdot T_{SF}$ [24], see also [26]. Values of T_{SF} in dependence on the temperature are given in **Fig. 3-17**.

Thermoelectric Power

The thermoelectric power, S, of UAl_2 was measured from the low-temperature region up to room temperature, see **Fig. 3-18**. The thermoelectric power decreases rapidly below 100 K,

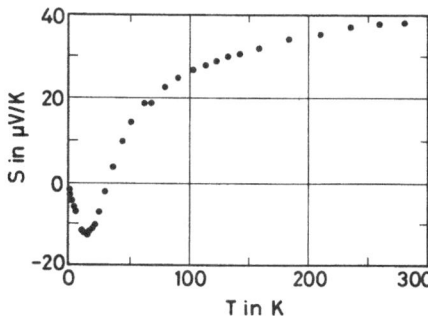

Fig. 3-18. Thermoelectric power of UAl_2 versus temperature [12].

References for 3.1.3.6 on pp. 106/7

changes its sign at 30 K, shows a minimum at 16 K, and finally vanishes proportionally to T with S/T = −1.13 µV/K^2 [12]. Single values are S = 40 µV/K (at room temperature) and −12.5 µV/K (at 16 K) [12].

Hall Coefficient

The Hall coefficient of UAl$_2$ increases rapidly at low temperatures and is associated with the development of skew scattering by fluctuations about the coherent state [28, 29], see **Fig. 3-19**.

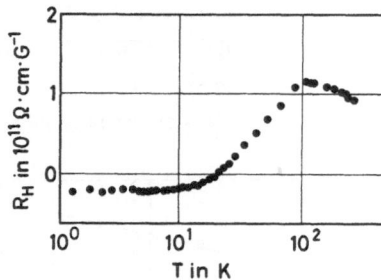

Fig. 3-19. Hall constant versus T for UAl$_2$ [28].

Thermionic Emission

The thermionic emission of UAl$_2$ shows a significantly higher value than the values of other uranium−aluminium alloys [8], see Table 3/10.

Table 3/10
Thermionic Emission of UAl$_2$ and Other Uranium−Aluminium Alloys [8].

at% U	at% Al	thermionic emission in eV
0	100	4.2
25	75	3.2
29	71	3.2
33.33	66.66 (UAl$_2$)	4.2
50	50	3.8
70	30	3.0
100	0	3.5

References for 3.1.3.6:

[1] Gossard, A. C.; Jaccarino, V.; Wernick, J. H. (Phys. Rev. [2] **128** [1962] 1038/43).
[2] Naegele, J. R.; Manes, L.; Spirlet, J. C.; Pellegrini, L.; Fournier, J. M. (Physica B + C **102** [1980] 122/5).
[3] Schadler, G.; Hilscher, G.; Weinberger, P. (J. Magn. Magn. Mater. **29** [1982] 241/6).
[4] Boring, A. M.; Albers, R. C.; Stewart, G. R.; Koelling, D. D. (Phys. Rev. [3] B **31** [1985] 3251).

[5] Boring, A. M.; Albers, R. C.; Koelling, D. D. (J. Magn. Magn. Mater. **54/57** [1986] 543/4).

[6] de Groot, R. A.; Koelling, D. D.; Weger, M. (Phys. Rev. [3] B **32** [1985] 2659/62).

[7] Tyunis, A. V.; Shaburov, V. A.; Savitskii, E. M.; Terekhov, G. I.; Shkatova, T. M. (Splavy Redk. Metall. Osobymi Fiz. Svoistvami Redkozem. Blagorodn. Met. **1983** 111/6; C.A. **99** [1983] No. 200746).

[8] Ondracek, G.; Petzow, G. (Z. Metallk. **56** [1965] 498/502).

[9] Itie, J. P.; Olsen, J. S.; Gerward, L.; Benedict, U.; Spirlet, J. C. (Rept. Univ. Copenhagen Phys. Lab. **2** No. 85-15 [1985] 1/14; C.A. **103** [1985] No. 204039).

[10] Thümmler, F.; Lilienthal, H. E.; Nazare, S. (Powder Met. **12** [1969] 1/22).

[11] Weger, M.; de Groot, R. A.; Mueller, F. M.; Kaveh, M. (J. Phys. F **14** [1984] L207/L213).

[12] Armbruster, H.; Franz, W.; Schlabitz, W.; Steglich, F. (J. Phys. Colloq. [Paris] **40** [1979] C4-150/C4-151).

[13] Wire, M. S.; Stewart, G. R.; Johanson, W. R.; Fisk, Z.; Smith, J. L. (Phys. Rev. [3] B **27** [1983] 6518/21).

[14] Ping, J. Y.; Coles, B. R. (J. Magn. Magn. Mater. **29** [1982] 209/12).

[15] Buschow, K. H. J.; van Daal, H. J. (AIP [Am. Inst. Phys.] Conf. Proc. No. 5 [1971/72] 1464/77; C.A. **77** [1972] No. 11324).

[16] Brodsky, M. B. (Phys. Rev. [3] B **9** [1974] 1381/7).

[17] Arko, A. J.; Brodsky, M. B.; Nellis, W. J. (Phys. Rev. [3] B **5** [1972] 4564/9).

[18] Rivier, N.; Zlatic, V. (J. Phys. F **2** [1972] L87/L92).

[19] Rivier, N.; Zlatic, V. (J. Phys. F **2** [1972] L99/L104).

[20] Trainor, R. J.; Brodsky, M. B.; Knapp, G. S. (Plutonium 1975 Other Actinides, Proc. 5th Intern. Conf., Baden-Baden, FRG, 1975 [1976], pp. 475/85; C.A. **85** [1976] No. 69172).

[21] Beal-Monod, M. T.; Ma, S.-K.; Fredkin, D. R. (Phys. Rev. Letters **20** [1968] 929/32).

[22] Kaiser, A. B.; Doniach, S. (Intern. J. Magn. **1** [1970] 11/22).

[23] Rivier, N.; Zuckermann, M. J. (Phys. Rev. Letters **21** [1968] 904/7).

[24] Wire, M. S.; Thompson, J. D.; Fisk, Z. (Phys. Rev. [3] B **30** [1984] 5591/5).

[25] Wire, M. S.; Giorgi, A. L. (Phys. Rev. [3] B **32** [1985] 1687/90).

[26] Wire, M. S. (Diss. Univ. California, San Diego, La Jolla 1984, pp. 1/123; Diss. Abstr. Intern. B **45** [1985] 2217; C.A. **102** [1985] No. 104701).

[27] Runnalls, O. J. C.; Boucher, R. R. (Trans. AIME **233** [1965] 1726/32).

[28] Hadzic-Leroux, M.; Hamzic, A.; Fert, A.; Haen, P.; Lapierre, F.; Laborde, O. (Europhys. Letters **1** [1986] 579/84).

[29] Lapierre, F.; Haen, P.; Briggs, R.; Hamzic, A.; Fert, A.; Kappler, J. P. (J. Magn. Magn. Mater. **63/64** [1987] 338/40).

3.1.3.7 Optical Properties

The UV photoemission (UPS) spectrum of UAl_2 for He I and He II excitation [1] and its X-ray photoemission (XPS) spectrum [2] have been measured. From these, tentative diagrams of the density-of-states were abstracted, showing an extensive peak with an abrupt cut-off at the Fermi energy level [2 to 5], see also [6] and Section 3.1.3.6 on "Electronic Structure", p. 101.

Photoemission (PES) and bremsstrahlung isochromat (BIS) spectra were taken of UAl_2 to measure the 5f spectral weight, but the one-electron 5f widths are too large to account for the enhanced values of the specific heat coefficient, and too small to account for the measured width [7].

References for 3.1.3.7 on p. 108

Muon spin rotation spectra, performed on UAl$_2$ within the temperature range of 2 to 300 K, show little variation in damping rate and muonic Knight shift. The weak damping was decoupled by a longitudinal field of less than 10 mT [8, 9].

Neutron scattering experiments, using chopper spectrometers [10] or time-of-flight method [11, 12] were performed on UAl$_2$. Experiments with cold neutrons revealed a broad magnetic quasi-elastic line; no inelastic contributions to the magnetic scattering could be resolved at low temperatures using thermal neutrons [12]. The residual width of the quasi-elastic line is given with $\Gamma/2$ ($T \rightarrow 0$) ≈ 25 meV [158], see also [11].

References for 3.1.3.7:

[1] Gossard, A. C.; Jaccarino, V.; Wernick, J. H. (Phys. Rev. [2] **128** [1962] 1038/43).

[2] Naegele, J. R. (Physica B + C **130** [1985] 52/5).

[3] Naegele, J. R.; Manes, L.; Spirlet, J. C.; Pellegrini, L.; Fournier, J. M. (Physica B + C **102** [1980] 122/5).

[4] Naegele, J. R.; Manes, L.; Spirlet, J. C.; Fournier, J. M. (Appl. Surf. Sci. **4** [1980] 510/7).

[5] Sarma, D. D.; Hillebrecht, F. U.; Brooks, M. S. S. (J. Magn. Magn. Mater. **63/64** [1987] 509/11).

[6] Poole, D. M.; Thomas, P. M. (J. Inst. Metals **90** [1962] 228/33).

[7] Allen, J. W.; Oh, S.-J.; Cox, L. E.; Ellis, W. P.; Wire, M. S.; Fisk, Z.; Smith, J. L.; Pate, B. B.; Lindau, I.; Arko, A. J. (Phys. Rev. Letters **54** [1985] 2635/8).

[8] Asch, L.; Barth, S.; Gygax, P. N.; Kalvius, G. M.; Kratzer, A.; Litterst, F. J.; Mattenberger, K.; Potzel, W.; Schenck, A.; Spirlet, J. C.; Vogt, O. (J. Magn. Magn. Mater. **63/64** [1987] 160/74).

[9] Kratzer, A.; Litterst, F. J.; Gygax, F. N.; Asch, L.; Schenck, A.; Kalvius, G. M.; Barth, S.; Potzel, W.; Spirlet, J. C. (Hyperfine Interact. **31** [1986] 309/12).

[10] Loong, C. K.; Loewenhaupt, M.; Vrtis, M. L. (Physica B + C **136** [1986] 413/6).

[11] Horn, S.; Loewenhaupt, M.; Steglich, F.; Just, W. (J. Magn. Magn. Mater. **9** [1978] 54/6).

[12] Loewenhaupt, M.; Horn, S.; Steglich, F.; Holland-Moritz, E.; Lander, G. H. (J. Phys. Colloq. [Paris] **40** [1979] C4-142/C4-144).

3.1.3.8 Magnetic Properties

Magnetization

UAl$_2$ is a simple paramagnet and does not order magnetically at temperatures down to 120 mK [1 to 3]. The induced magnetization density in the unit cell of UAl$_2$ was measured by a polarized-neutron diffractometer on single crystals at 4.2 K with an applied field of 42.5 kOe (induced magnetic moment 0.0344 μ_B per mol). Magnetic scattering amplitudes are given [4].

Magnetization of polycrystalline samples of UAl$_2$ in high magnetic fields (up to 35 T) and low temperatures (1.4 to 77 K) is shown in **Fig. 3-20**. The magnetization curves show a suppression of spin fluctuations near 4.2 K in fields of 15 to 20 T. Field and temperature effects on the paramagnetic contributions were concluded to be closely related to the equation $g \cdot \mu_B \cdot H_{SF} = k_B \cdot T_{SF}$ (g = 2). With the characteristic spin-fluctuation temperature, $T_{SF} = 25$ K, a value of H_{SF} about 19 T was calculated [5 to 8]. The magnetization remains linear up to high fields of 150 kG and down to a temperature of 1 K. These two extreme values correspond to $\mu_B \cdot H/k_B \cdot T = 10$ and an induced moment of about 0.1 μ_B [9].

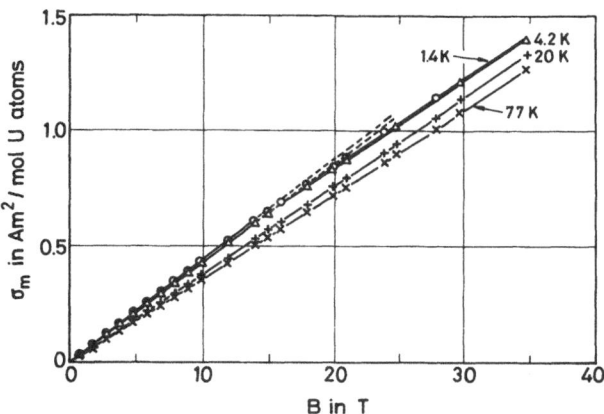

Fig. 3-20. Magnetization σ_m of UAl$_2$ at 1.4, 4.2, 20, and 77 K [7].

Magnetic Susceptibility

The magnetic susceptibility of UAl$_2$ obeys the Curie-Weiss law in the high-temperature region above 100 K [1, 2, 9 to 11] with $\chi = \chi_0 + A/(T + B)$ [12], see **Fig. 3-21**. The temperature dependence of χ is $\chi_g \cdot 10^6 = 3820/(T + 160)$ (at 20 to 800°C) [10]. Room temperature values are (at 20°C): $\chi_g = 8.43 \times 10^{-6}$ emu/g, $\chi_m = 2460$ emu/g-atom, and a magnetic moment of 3.0 μ_B (calculated without correction for diamagnetism in the uranium or aluminium ions), from which the uranium atoms are in the tetravalent state [10], or $\chi_0 \approx 10 \times 10^{-6}$ emu/g; paramagnetic Curie temperature $\Theta_p \approx -230$ K, $\mu_{eff} \approx 2.94 \mu_B$ [9]. A Curie-Weiss behavior of the susceptibility was also observed above 50 K [13]. This temperature region is probably connected with local spin fluctuations from which local moments are induced. The negative value of Θ_p indicates that the fluctuations are antiferromagnetic rather than ferromagnetic [9]. $T_{SF} = 100$ K [11]. Above 50 K the magnetic susceptibility of UAl$_2$ is proportional to $(1 - b \cdot T^2)$, which is probably associated with the narrow band-structure [2].

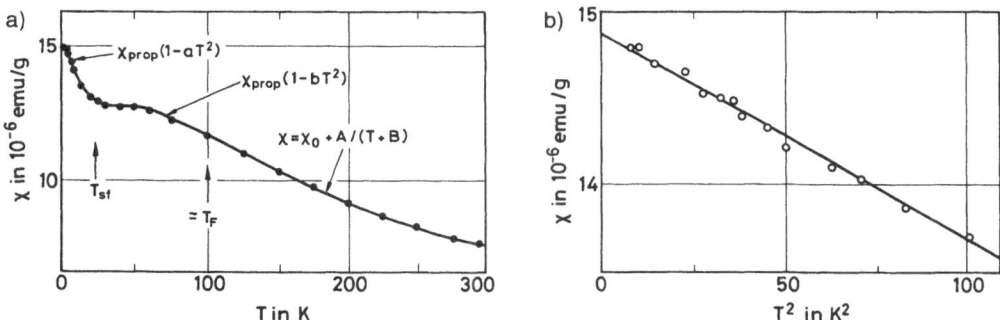

Fig. 3-21. Magnetic susceptibility χ of UAl$_2$ [2]. a) χ as a function of T; b) χ below 10 K as a function of T^2.

            References for 3.1.3.8 on pp. 112/3

In the low-temperature region from 10 to 40 K, a Curie-Weiss law is again obeyed, which is characterized by $\chi_0 \approx 3.2 \times 10^{-6}$ emu/g, $\Theta_p \approx -8$, and $\mu_{eff} \approx 0.4\ \mu_B$ [9]. This temperature region is probably connected with nonlocalized spin fluctuations (paramagnons) [9], see also [14]. In the low-temperature region from 3 to 10 K, the susceptibility of UAl$_2$ is proportional to $(1 - a \cdot T^2)$ [1, 2, 15, 16]. The temperature dependence is represented by the equations: χ (in emu/g) $= 14.86 \times 10^{-6}$ $(1 - 7.94 \times 10^{-4} \cdot T^2)$ [11] or χ (in emu/g) $= 14.15 \times 10^{-6}$ $(1 - 5.6 \times 10^{-4} \cdot T^2)$ [16], also reported in [13]. This low-temperature susceptibility is well described by the ferromagnetic spin-fluctuation theory. The value of "a" corresponds to a spin-fluctuation temperature of T_{SF} about 40 K [1], or $T_{SF} = 36$ K [15] (using the data of [1]), $T_{SF} = 31$ K [15] (using the data of [17]), $T_{SF} = 30$ K [2]. These values are in good agreement with the results from calculations within the Friedel-Anderson model: χ_0 (Pauli-band susceptibility) $= 3.8 \times 10^{-4}$ emu/mol (using the data of [18, 19]), α (Stoner enhancement) $= 3.5$, and $T_{SF} = 29.7$ K [11]. Following the paramagnon model, the susceptibility can be explained in the very low-temperature region, predicting that $T/T_{SF} \ll 1$, as $\chi = \chi_0 [1 - (T/T_{SF})^2]$. From this equation the following values were derived from measurements at 0.12 K to 5 K: $\chi_0 \approx 14.2 \times 10^{-6}$ emu/g and $T_{SF} \approx 13$ K [3]. A re-examination of the susceptibility at 3 to 10 K in magnetic fields of 0.1, 1, 2, and 5 T shows that the dependence of χ versus T is linear rather than quadratic for single crystals and polycrystalline UAl$_2$ [20]. The analysis of the susceptibility at 0.3 to 2.2 K confirms the existence of paramagnons with a spin-fluctuation temperature in the range of 4 to 7 K [14]. Further values are reported: χ_0 (at 4.2 K) $= 57 \times 10^{-9}$ m^3/mol UAl$_2$ [5] and χ (at 1.4 K) $= 14.6 \times 10^{-6}$ emu/g [21].

The differential susceptibilities, derived from field-dependent magnetization experiments (see Fig. 3-20, p. 109) at 1.4 and 4.2 K, decrease with increasing fields, reaching a field independent value above 20 T [6, 7], see also [5, 8]. Differential molar susceptibilities are given in Table 3/11.

Table 3/11
Differential Molar Susceptibilities in 10^{-9} m^3/mol U for UAl$_2$ [6, 7] Compared with Values from [2].

temperature in K	dM/dH at 2 to 5 T	dM/dH at 20 to 35 T	dM/dH from [2]
1.4	56.5	47.3	—
4.2	55.4	47.3	55.0
20	47.7	47.7	—
25	—	—	47.3
50	—	—	46.6
70	45.9	45.9	45.5

The high-pressure dependence of the susceptibility at 4 to 250 K and up to 6.65 kbar is shown in **Fig.** 3-22. The χ versus T curves keep the same shape under pressure. The average pressure dependence is described with $\partial \log \chi / \partial P = -25$ Mbar^{-1} [9]. From $\partial \ln \chi$ (T = 0)/$\partial P = \partial \ln T_{SF} / \partial P$ follows $\partial \ln \chi$ (T = 0)/∂P (at T $\approx$ 0) $= -24$ Mbar^{-1} [22], see also [28, 59].

The volume dependence of the susceptibility is $\partial \log \chi / \partial \log V = +15$ (using compressibility β (at 300 K) $= 1.65$ Mbar^{-1}, see Section 3.1.3.4, p. 95) [9].

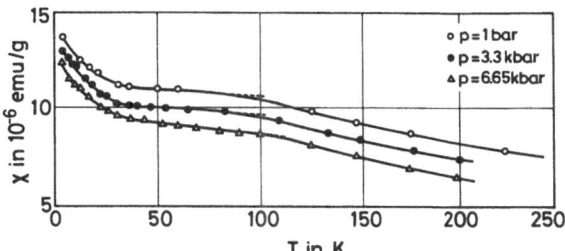

Fig. 3-22. Susceptibility of UAl_2 versus temperature for different pressures [9].

Magnetoresistance

The magnetoresistivity of UAl_2 was observed at single crystals at temperatures of 1.5 K and 5 to 50 K in magnetic fields up to 19.74 T [23], see also [6], see **Fig.** 3-**23** and **Fig.** 3-**24**, p. 112. The magnetoresistivity is consistent with the behavior of a compensated metal; as for a compensated metal, $\Delta\varrho(H)$ increases proportional to H^2 for any direction of H perpendicular to the current. De Haas-van Alpen oscillations were not observed up to 40 T. The small negative magnetoresistivity at temperatures above 30 K may be attributed to spin-fluctuation scattering [23].

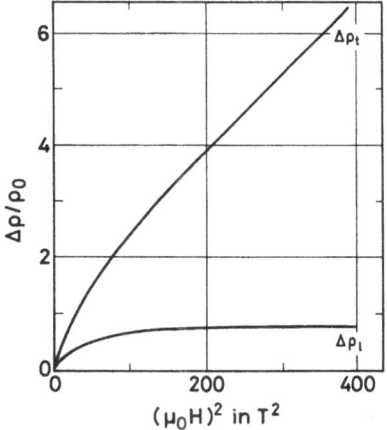

Fig. 3-23. Transverse and longitudinal magnetoresistivity in dependence on the magnetic force μ_0H in Tesla (as $(\mu_0H)^2$ in T^2) for a single crystal of UAl_2 at 1.5 K. The field is parallel to [001] for $\Delta\varrho_t$. The data are recorded in a 5 min sweep from 0 to 19.74 T [23].

Nuclear Magnetic Resonance

NMR measurements of ^{27}Al in UAl_2 at 4 to 300 K showed large Knight shifts, K, with a temperature dependence not of the Curie-Weiss type, indicating that there is no localization of the magnetization of the 5f or 6d electrons of uranium, and that there is no magnetic ordering [24]. The low-temperature relaxation rate $R = (T_1 \cdot T)^{-1}$, with T_1 = nuclear spin-relaxation time, is independent of the applied magnetic field. A value of $\alpha = dK/d\chi = 2.9$ emu/mol is obtained

References for 3.1.3.8 on pp. 112/3

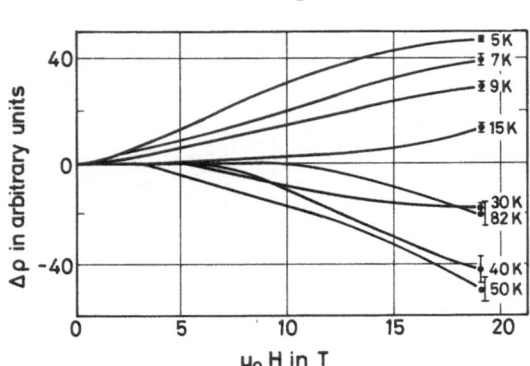

Fig. 3-24. Longitudinal magnetoresistivity of a UAl$_2$ single crystal [23].

from the linear region of the dependence of Knight shift on susceptibility [19], see also [25]. The relaxation rate follows no simple dependence on the measured susceptibility (reduced by the Vleck type orbital contribution $\chi_{orb} = 10 \times 10^{-4}$ emu/mol [25], see also [17, 19]). Within a theoretical model, which considers the spin fluctuations due to the 5f electrons and takes into account the Stoner susceptibility, good agreement was obtained for temperature and resistivity dependence of the relaxation rate [26], see also [19, 27]. For further NMR resonance parameters (at 77 K) and results see [17].

References for 3.1.3.8:

[1] Trainor, R. J.; Brodsky, M. B.; Culbert, H. V. (Phys. Rev. Letters **34** [1975] 1019/22).

[2] Brodsky, M. B.; Trainor, R. J. (Physica B + C **91** [1977] 271/77).

[3] Fournier, J. M. (Solid State Commun. **29** [1979] 111/3).

[4] Rakhecha, V. C.; Lander, G. H.; Arko, A. J.; Moon, R. M. (J. Appl. Phys. **52** [1981] 1636/8).

[5] Franse, J. J. M.; Frings, P. H.; de Boer, F. R.; Menovsky, A. (Phys. Solids High Pressure Proc. Intern. Symp., Bad Honnef, FRG, 1981, pp. 181/91; C.A. **96** [1982] No. 134510).

[6] Franse, J. J. M.; Frings, P. H.; de Boer, F. R.; Menovsky, A.; Beers, C. J.; van Deursen, A. P. J.; Myron, H. W.; Arko, A. J. (Phys. Rev. Letters **48** [1982] 1749/52).

[7] de Boer, F. R.; Franse, J. J. M.; Frings, P. H.; Mattens, W. C. M.; de Chatel, P. F. (High Field Magn. Proc. Intern. Symp., Osaka 1982 [1983], pp. 157/66; C.A. **98** [1983] No. 208787).

[8] Arko, A. J.; Beers, C. J.; van Deursen, A. P. J.; van Kleef, R. P. A. R.; Myron, H. W.; Parker, M. R.; Pepper, M.; Poole, D. A.; Rasing, T. H. M.; Wyder, P. (Lect. Notes Phys. **177** [1983] 479/87; C.A. **99** [1983] No. 15106).

[9] Fournier, J. M.; Beille, J. (J. Phys. Colloq. [Paris] **40** [1979] C4-145/C4-146).

[10] Konobrevskii, S. T.; Zaimovskii, A. S.; Levitsky, B. M.; Sokursky, Y. N.; Chebotarev, N. T.; Bobkov, Y. V.; Egorov, P. P.; Nikolaev, G. N.; Ivanov, A. A. (Proc. 2nd Intern. Conf. Peaceful Uses At. Energy, Geneva 1958, Vol. 6, pp. 194/203).

[11] Brodsky, M. B. (Phys. Rev. [3] B **9** [1974] 1381/7).

[12] Buddery, J. H.; Clark, M. E.; Pearce, R. J.; Stobbs, J. J. (J. Nucl. Mater. **13** [1964] 169/81).

[13] Burzo, E.; Gratz, E.; Lucaci, P. (Solid State Commun. **60** [1986] 241/4).

[14] Armbruster, H.; Franz, W.; Schlabitz, W.; Steglich, F. (J. Phys. Colloq. [Paris] **40** [1979] C4-150/C4-151).

[15] Trainor, R. J.; Brodsky, M. B.; Knapp, G. S. (Plutonium 1975 Other Actinides, Proc. 5th Intern. Conf., Baden-Baden, FRG, 1975 [1976], pp. 475/85; C.A. **85** [1976] No. 69172).

[16] Burzo, E.; Lucaci, P. (Solid State Commun. **56** [1985] 537/9).

[17] Arko, A. J.; Fradin, F. Y.; Brodsky, M. B. (Phys. Rev. [3] B **8** [1973] 4104/18).

[18] Buschow, K. H. J.; van Daal, H. J. (AIP [Am. Inst. Phys.] Conf. Proc. No. 5 [1971/72] 1464/77; C.A. **77** [1972] No. 11324).

[19] Fradin, F. Y.; Brodsky, M. B.; Arko, A. J. (AIP [Am. Inst. Phys.] Conf. Proc. No. 10 [1972/73] 192/6; C.A. **79** [1973] No. 36509).

[20] Foner, S.; Stewart, G. R.; Giorgi, A. L. (Phys. Rev. [3] B **32** [1985] 4768/9).

[21] Wire, M. S.; Stewart, G. R.; Johanson, W. R.; Fisk, Z.; Smith, J. L. (Phys. Rev. [3] B **27** [1983] 6518/21).

[22] Wire, M. S.; Thompson, J. D.; Fisk, Z. (Phys. Rev. [3] B **30** [1984] 5591/5).

[23] van Ruitenbeek, J. M.; van Deursen, A. P. J.; Myron, H. W.; Arko, A. J.; Smith, J. L. (Phys. Rev. [3] B **34** [1986] 8507/11).

[24] Gossard, A. C.; Jaccarino, V.; Wernick, J. H. (Phys. Rev. [2] **128** [1962] 1038/43).

[25] Fradin, F. Y.; Brodsky, M. B.; Arko, A. J. (J. Phys. Colloq. [Paris] **32** [1971] C1-905/C1-906).

[26] Jullien, R.; Coqblin, B. (J. Phys. Lettres Paris **35** [1974] L197/L201).

[27] Fradin, F. Y. (Proc. 2nd Intern. Conf. Electron. Struct. Actinides, Wroclaw 1976 [1977], pp. 247/56; C.A. **87** [1977] No. 93142).

[28] Wire, M. S. (Diss. Univ. California, San Diego, La Jolla 1984, pp. 1/123; Diss. Abstr. Intern. B **45** [1985] 2217; C.A. **102** [1985] No. 104701).

3.1.3.9 Chemical Reactions

UAl_2 melts congruently at 1590°C [1] or 1620°C [2].

For further details of the behavior on heating see Section 3.1.3.5, p. 97.

Reactions With Elements

A burst of acoustic emission was observed when UAl_2 was exposed to hydrogen, but the hydrogen pick-up was only minimal. Further treatment up to 300°C with hydrogen pressures up to 30 MPa under cycling conditions produced no further acoustic emission or hydrogen pick-up [3].

Dense specimens of UAl_2 oxidize rapidly in oxygen or air on heating, forming U_3O_8 as the only oxidation product. UAl_2 powders are even pyrophoric and must be handled under inert gas [4, 5].

The oxidation of UAl_2 in pure oxygen of 1 atm at 200 to 300°C followed a cubic rate law with an activation energy of 26000 cal/mol [6, 7]. The reaction proceeded anisothermally at 300°C; spontaneous heating of the samples occurred with rapidly increasing temperature to above 1000°C (estimated) in less than 1 min [7]. Using an oxygen pressure of 0.1 atm, a linear weight gain was measured within the temperature range of 200 to 300°C. At higher temperatures (325 to 350°C) an initial period of increasing rates of oxidation for 6 to 8 min in 1 atm of oxygen was observed before the linear weight gain proceeded [8], see **Fig.** 3-25, p. 114. The oxidation of UAl_2 occurred rapidly within less than 20 min with an activation energy of 22.1 kcal/mol [8]. Oxidation experiments at higher temperatures (300 to 600°C) resulted in the formation of U_3O_8 as shown only by X-ray diffraction. The formation of both U_3O_8 and Al_2O_3

References for 3.1.3.9 on pp. 115/6

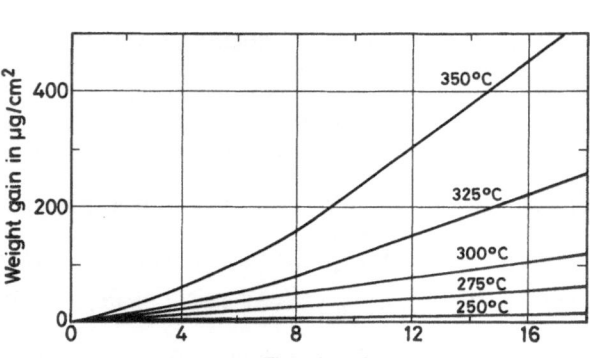

Fig. 3-25. Weight gain versus time for oxidation of UAl$_2$ in the range of 250 to 350°C [8].

should be thermodynamically possible, but, as only U$_3$O$_8$ was observed, the overall reaction of 3 UAl$_2$ + 4 O$_2$ → U$_3$O$_8$ + 6 Al was stated. It is not known whether the aluminium formed by this reaction enters the UAl$_2$ or the U$_3$O$_8$ phase or whether it forms an Al$_2$O$_3$ interface. The oxidation rate in UAl$_2$ is higher than in all other uranium — aluminium compounds [5]. Electron-microscope studies showed characteristic circular eruptions after 5 to 45 min of oxidation at 200 to 500°C in the vicinity of grain boundaries. Those eruptions contained radial cracks and their appearance is suggestive of ruptured blisters. But, these cracks have only little influence on the oxidation rate and it was concluded that they do not penetrate to the substrate [5].

Oxidation in air at 400 to 700°C led to a linear oxidation rate following $\Delta W = 10^7 \cdot \exp(-22000/RT) \cdot A \cdot t$ (with A = surface in cm^2, t = annealing time in min) with an apparent activation energy of 22 kcal/mol [9], see Table 3/12. The free energy of oxidation is ΔG (UAl$_2$) = −218 kcal/mol [10]. A temperature of T_p = 350°C was reported, at which the weight gain reached 1 mg/cm^2 after 4 h of oxidation [11], see also [12]. UO$_2$ was the only oxidation product observed at temperatures of 500 to 550°C [9].

Table 3/12
Oxidation of UAl$_2$ at 420 to 550°C in Air [9].

temperature in °C	annealing time in min	region of linear rate law	oxidation rate in $\mu g \cdot cm^{-2} \cdot min^{-1}$	reaction product	final state of sample
420	150	0 to 150	1.2	—	compact
430	230	0 to 230	1.2	—	compact
450	220	0 to 50	~3.2	—	compact
500	300	10 to 200	7.2	layer	cracks
550	700	50 to 150	14.4	layer	cracks

For reactions of UAl$_2$ with metals other than Al see Section 3.1.7, p. 180, on "Ternary Uranium — Aluminium Compounds".

For reaction of UAl$_2$ with aluminium see Section 3.1.3.10, p. 116, on "UAl$_2$ — Al Dispersions".

Reactions With Compounds

UAl_2 oxidizes in pure CO_2 at similar rates but significantly slower than uranium metal [13, 14], see **Fig. 3-26**.

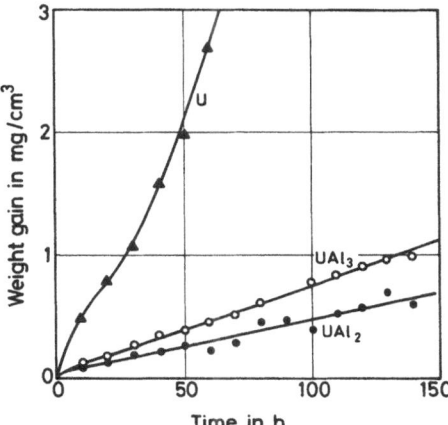

Fig. 3-26. Comparison of oxidation rates of uranium, UAl_2, and UAl_3 in pure CO_2 at 450°C [14].

UAl_2 can be dissolved in HNO_3 with a few drops of HF added [2]. UAl_2 was completely dissolved in a mixture of 600 mL HCl (1.178 g/mL), 400 mL H_3PO_4 (1.39 g/mL), 0.25 g Na_2SiF_6, 0.037 g H_2PtCl_6, 0.12 g $CuSO_4 \cdot 5 H_2O$ for calorimetric experiments [15].

Alcoholic HF (5 to 10% HF) with a few drops of H_2O_2 was used for etching of UAl_2 grain boundaries for metallographic purpose [2].

The following etching treatments are recommended to distinguish metallographically the UAl_2 and UAl_3 phase in the $UAl_2 - UAl_3$ field:

Electrolytically etching with one part of 40% chromic acid and one part of 50% acetic acid maintained for 1 min at 0.05 A/cm^2 at 25°C resulted in darkened and pitted UAl_2 and light yellow UAl_3 [16].

Etching with 1% hydrofluoric acid at 25°C resulted, after 30 s immersion, in light UAl_2 and darkened UAl_3. Excessive etching resulted in pitting of UAl_2 [16].

Etching with 50% nitric acid at 25°C resulted, after 1 min immersion, in light blue UAl_2 and light yellow UAl_3 [16].

References for 3.1.3.9:

[1] Gordon, P.; Kaufmann, A. R. (J. Metals **2** [1950] 182/4).
[2] Petzow, G.; Steeb, S.; Ellinghaus, I. (J. Nucl. Mater. **4** [1961] 316/21).
[3] Northrup, C. J. M.; Kass, W. J.; Baettie, A. G. (Hydrides Energy Storage Proc. Intern. Symp., Geilo, Norway, 1977 [1978], pp. 205/16; C.A. **91** [1979] No. 133205).
[4] Gibson, G. W.; de Boisblanc, D. R. (Proc. 2nd Intern. Powder Met. Conf., New York 1965 [1966], Vol. 3, pp. 26/35; C.A. **67** [1967] No. 17002).

[5] Openshaw, P. R.; Shreir, L. L. (Corros. Sci. **4** [1964] 335/44).

[6] Snyder, M. J.; Duckworth, W. H. (BMI-1223 [1957] 1/35; C.A. **1958** 3489).

[7] Albrecht, W. M.; Koehl, B. G. (Proc. 2nd Intern. Conf. Peaceful Uses At. Energy, Geneva 1958, Vol. 6, pp. 116/21).

[8] Openshaw, P. R.; Shreir, L. L. (Corros. Sci. **3** [1963] 217/37).

[9] Thümmler, F.; Exner, H. E.; Petzow, G. (J. Nucl. Mater. **24** [1967] 328/39).

[10] Thümmler, F.; Lilienthal, H. E.; Nazare, S. (Powder Met. **12** [1969] 1/22).

[11] Stone, H. E. N. (J. Mater. Sci. **10** [1975] 923/34).

[12] Badaeva, T. A.; Dashevskaya, L. I. (Fiz. Khim. Splavov Tugoplavkikh Soedin Toriem Uranom **1968** 107/13; C.A. **71** [1969] No. 24297).

[13] Churchman, A. T.; Buddery, J. H.; Edmondson, B.; Greenwood, G. W.; Hesketh, R. V.; Williamson, G. K. (Proc. 3rd Intern. Conf. Peaceful Uses At. Energy, Geneva 1964 [1965], Vol. 9, pp. 13/22).

[14] Pearce, R. J.; Stobbs, J. J. (Tr. 3rd Mezhdunar. Kongr. Korroz. Metal., Moscow 1966 [1968], Vol. 4, pp. 158/66; Proc. 3rd Intern. Congr. Metal. Corros., Moscow 1966 [1969], Vol. 4, pp. 152/65; C.A. **72** [1970] No. 24061).

[15] Ivanov, M. I.; Tumbakov, V. A.; Podol'skaya, N. S. (At. Energiya SSSR **5** [1958] 166/70; Soviet J. At. Energy **5** [1958] 1007/11; C.A. **1960** 9481).

[16] Hills, R. F. (J. Inst. Metals **86** [1958] 438/41).

3.1.3.10 Uranium Dialuminide – Aluminium Dispersions, UAl$_2$ – Al

UAl$_2$ – Al dispersion fuel elements are of interest in nuclear technology. Research reactors with neutron fluxes of about 10^{15} n·cm^{-2}·s^{-1} utilize plate-type elements having a high uranium content and, in most cases, an aluminium matrix and cladding.

Preparation of UAl$_2$ – Al Dispersions and Fuel Plates

According to nuclear technology aspects, UAl$_2$ (the dispersed phase) is used as a coarse powder (63 to 90 µm in diameter) for dispersions containing 40 wt% uranium mixed with aluminium powder (18.6 µm in diameter, mean value), where dry mixing of the powders achieved sufficient homogeneity and avoided the problems of clustering observed with wet mixing. The compaction behavior of the UAl$_2$ – Al dispersions as a function of pressure and the density of pellets as a function of UAl$_2$ concentration has been determined [1]. Densities of about 94% th.d. were achieved using a solution of stearic acid and petroleum ether as lubricant [1] or mixtures of UAl$_2$ and aluminium powders in desired portions were sintered at 600°C [2, 3]. Fuel plates with aluminium matrix were fabricated with the "picture-frame" technique (see Section 3.1.4.8 on "UAl$_3$ – Al Dispersions", p. 134), by welding and hot rolling (at 550°C) or by forging at 550 to 600°C if the uranium content is less than 30 wt% [4], to obtain a uniform distribution of the fissile material and to prevent possible defects such as tapering or dog-boning [1, 5]. A schematic diagram of the UAl$_2$ – Al fuel plate fabrication is given in Fig. 3-36, p. 135. For more detailed information, see Section 3.1.4.8 on "UAl$_3$ – Al Dispersion", p. 134.

Interaction of UAl$_2$ and Al Matrix

The interaction of UAl$_2$ and the aluminium matrix leads to the formation of UAl$_{4+x}$ (heat of reaction, 9 kcal/mol [6]). The reaction starts at about 400°C with particle sizes of UAl$_2$ of 63 to

90 μm, but the reaction rate is very low at this temperature [1, 7]. A reaction product growth nucleus is formed at this temperature after an incubation period. This nucleation is assumed to be the rate-determining factor for the reaction. The UAl_3 may be present as an intermediate product. The UAl_2-Al reaction leads to a reduction of the aluminium matrix fraction [1]. The UAl_2-Al reaction is accompanied by an increase of about 1.6% in the volume of the dispersion, but cyclic heat treatment at 530°C for 1 h and six cycles did not cause any volume changes of the dispersion-fuel elements [1].

Mechanical Properties

The Young's modulus of elasticity and the rupture strength of UAl_2-Al dispersions are effected by the formation of UAl_4 owing to the reaction of UAl_2 with aluminium. The rupture strength as well as the Young's modulus of elasticity decrease rapidly with increasing amount of the UAl_4 reaction product [7], see also [4, 8].

Neutron Irradiation of UAl_2-Al Dispersion Fuel Elements

The UAl_2 phase (93% enriched in ^{235}U, irradiated at 8.5×10^{13} n·cm^{-2}·s^{-1}) showed a continuous decrease in hardness in the range of burnup at a mean irradiation temperature of 70°C. At higher irradiation temperatures (150°C) the decrease in hardness seemed to occur at lower burnup, followed by a significant increase above 50% burnup. In the recoil zone of the aluminium matrix there was an increase in hardness with burnup [9 to 11]. The swelling of the UAl_2-Al fuel plates under neutron irradiation is shown in **Fig. 3-27**. The increase of

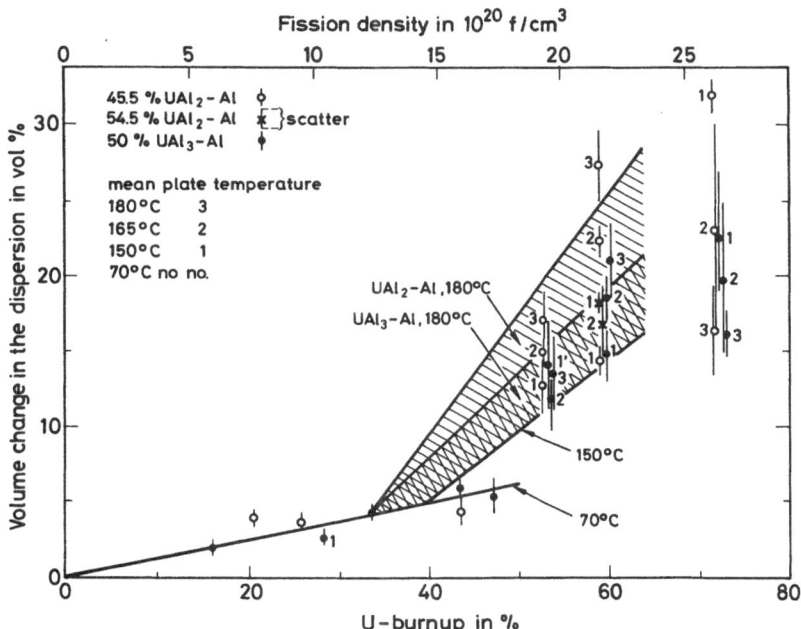

Fig. 3-27. Swelling of UAl_x-Al fuel plates (relative to the fuel dispersion volume) under irradiation as a function of burnup at various irradiation temperatures [9].

References for 3.1.3.10 on p. 118

volume is small with burnup of 33 to 40% uranium, showing an effect of about 0.12 vol%/% burnup. At higher burnup the swelling rate increases with fuel plate temperature [9 to 11].

A two- or three-phase mixture of UAl_2, UAl_3, and UAl_4 (93% enriched in ^{235}U) was dispersed in an aluminium matrix with the fuel composition of 81 wt% U — 19 wt% Al, which is just on the aluminium side of the UAl_2 compound, and fabricated to plate-type elements. The neutron irradiation was performed at about 130°C in a thermal flux of 7×10^{13} n·cm^{-2}·s^{-1} up to 60% burnup. The post-irradiation examination gave no evidence of fission-gas bubble formation in the basic three-phase mixture. However, clusters of small fission-gas bubbles were observed to form on and into small oxide inclusions, which were present as an impurity. It was concluded that the bubble formation can be eliminated by avoiding oxygen exposure during the uranium aluminide preparation [12].

UAl_2 — Al fuel plates show a better operation performance as compared to UAl_3 — Al fuel plates [13], see also [14].

References for 3.1.3.10:

[1] Thümmler, F.; Lilienthal, H. E.; Nazare, S. (Powder Met. **12** [1969] 1/22).
[2] Saller, H. A. (U.S. 2917383 [1959]; C.A. **1960** 18313).
[3] Saller, H. A. (U.S. 2917383 [1959]; N.S.A. **14** [1960] No. 7757).
[4] Seller (Met. Yadern. Energ. Deistvie Obluchen. Mater. Sbornik **1956** 242/57; C.A. **1958** 11707).
[5] Nazare, S.; Ondracek, G.; Thümmler, F.; Bürkin, J. (Ger. Offen. 1809924 [1970]; C.A. **74** [1971] No. 133947).
[6] Rand, M. H.; Kubaschewski, D. (The Thermochemical Properties of Uranium Compounds, Oliver & Boyd, Edinburgh — London 1963).
[7] Hunger, W.; Nazare, S.; Nikolopoulos, P. (Atomkernenergie **31** [1978] 156/9).
[8] Alekseev, O. A. (At. Tekhn. Rubezhom **1979** No. 8, pp. 18/20; C.A. **92** [1980] No. 66458).
[9] Dienst, W.; Nazare, S.; Thümmler, F. (J. Nucl. Mater. **64** [1977] 1/13).
[10] Dienst, W.; Gausmann, G. (KFK-2199 [1975] 1/44; C.A. **85** [1976] No. 38424).

[11] Dienst, W.; Nazare, S.; Thümmler, F. (At. Tekhn. Rubezhom **1977** 32/42; C.A. **88** [1978] No. 199553).
[12] Hofman, G. L. (Nucl. Technol. **77** [1987] 110/5).
[13] Miller, L. G.; Beeston, J. M. (EGG-FT-5273 [1980] 1/74; C.A. **95** [1981] No. 14694).
[14] Miller, L. G.; Brown, K. R.; Beeston, J. M.; McGinty, D. M. (EGG-SE-6464 [1983] 1/15; C.A. **101** [1984] No. 13654).

3.1.4 Uranium Trialuminide, UAl_3

3.1.4.1 Formation and Preparation

Preparation

UAl_3 can be prepared by the peritectic reaction of UAl_2 with a uranium—aluminium alloy containing 60 wt% uranium at 1350°C [1, 2]. However, UAl_3 reacts in the presence of aluminium to form UAl_4, and starting with the metallic components, the amount of uranium increases during vacuum induction melting according to the much higher vapor pressure of aluminium, resulting in a two-phase compound [3 to 7], see also [8, 9]. Yet, metallurgical recoveries of 85 to 90% were obtained with UAl_3 as the only constituent (from X-ray diffraction analysis) using graphite crucibles coated with Mg zirconate. The reaction was carried out at 700°C within 18 h [7, 8, 10, 11]. Monophase (nearly) UAl_3 could be achieved by induction melting of uranium and aluminium with a nonstoichiometric amount of uranium to balance the higher vapor losses of aluminium. The aluminium metal was placed at the bottom of an aluminium oxide crucible (vacuum annealed at 1500°C and 10^{-5} Torr) with the uranium metal on top as a precaution against segregation. The melting procedure was carried out at a maximum temperature of 1600 to 1620°C (heating rate 50°C per min) in helium atmosphere (400 Torr), which plays a role in preventing a possible dissociation of UAl_3. After rapid cooling (30°C per min) to 1250°C the reaction product was homogenized for about 5 h and then cooled to room temperature with an optional cooling rate [3 to 5]. No segregation was observed between top and bottom or between center and periphery of the regulus. Chemical analysis showed only a slightly lower aluminium content near the periphery than that near the center of 25.85 wt% and 26.1 wt% aluminium, respectively [5], see also [12].

UAl_3, containing only small amounts of impurities, was prepared by arc melting of stoichiometric amounts of uranium and aluminium metal in argon atmosphere using tungsten electrodes and water-cooled copper hearths. The compound obtained was remelted several times for homogenization. Metallurgical recoveries yielded 97 to 99% [10, 11, 13], see also [3, 5, 14, 15].

The reaction of uranium hydride, UH_3, with aluminium is somewhat more complicated than melting of the metals. The hydride and aluminium powder were mixed and slowly heated under vacuum to 540°F (= 282.2°C) to decompose the hydride. The temperature was then increased to 1100 to 1130°F (593.3 to 610°C) at which the exothermic reaction started. The UAl_3 is single-phase with high purity. Good metallurgical recoveries resulted [10, 11, 13, 16], see also [3, 5 to 7, 17].

Very pure UAl_3 was prepared by high-temperature interdiffusion of aluminium pieces and dispersed uranium obtained by decomposition of uranium hydride. Uranium and the aluminium metal were placed in stoichiometric amount in a beryllium oxide crucible loaded in a quartz ampoule. The crucible and the quartz tube were degassed before loading by heating in vacuum (10^{-2} Torr) at 900°C for 1.5 h. After loading, the ampoule was filled with hydrogen and heated to 250 to 270°C to convert the uranium to UH_3, which was decomposed at 600°C. The hydrogen evolved was not completely removed so the ampoule was heated in an electric furnace with argon at 1000°C for 3 h, during which the hydrogen pressure reached 400 Torr. After heating to 1100°C for 11 h, the ampoule was cooled to 600°C, the hydrogen pumped off, and then cooled to room temperature. The friable product was unloaded under argon, ground in a jasperite mortar, and again heated under hydrogen to 1300°C for 1.5 h. After cooling to 600°C the hydrogen was completely removed. The chemical analysis of the UAl_3 powder resulted in 25.33 ± 0.03 wt% Al (25.37 wt% theoretical), the stoichiometry yielded $UAl_{2.994}$ [18].

Single-phase UAl_3 was also obtained by interdiffusion between polished aluminium and uranium foils about 0.64 mm and 0.25 mm thick, respectively (to provide an excess of aluminium), clamped together with stainless steel clamps lined with tantalum foil. Complete interdiffusion occurred after heating in a high-purity argon atmosphere at 650°C for 24 h, followed by further annealing at 750°C for 24 h. The resultant foil was removed from the clamps, heated again to 1000°C for 1 h in argon, and then cooled to room temperature. Chemical analysis indicated an aluminium content of 25.4 ± 0.2 wt% [6]. UAl_3 is formed whenever clean surfaces of uranium and aluminium in close contact are heated to 250 to 450°C, yet anodization of the aluminium surface prevents the alloying reaction [19], see also [20, 21]. The activation energy for the diffusion of aluminium into uranium, measured in the temperature range of 200 to 550°C, is 22.0 ± 0.7 kcal/g-atom [22]. The kinetics of the growth

References for 3.1.4.1 on pp. 122/3

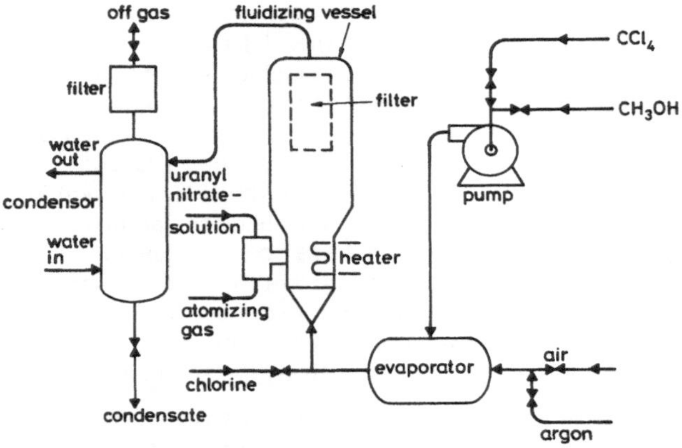

Fig. 3-28. Fluidized-bed production of uranium — aluminium compounds [33].

is characterized by a period of nonparabolic growth, followed by a steady state period of parabolic growth. The annealing temperature, time, and applied pressure are assumed to be correlated with the nature and the distribution of structural defects in the UAl₃ diffusion zone [23, 24], see also [25].

High-purity UAl₃ powder was obtained by vapor phase reaction between uranium and aluminium monochloride (prepared in situ by heating aluminium powder and potassium chloride at 600°C in vacuum). The chemical analysis of the single-phase UAl₃ yielded 25.2 ± 0.2 wt% aluminium [26], also reported in [6].

A four-step fluidized-bed technique was developed for manufacturing uranium — aluminium compounds, especially UAl₃ and UAl₄, starting with a $UO_2(NO_3)_2$ solution and aluminium powder [27 to 30], see also [3, 31, 32]. A flow-sheet for the fluidized-bed production is given in **Fig. 3-28**. The four steps of the process are summarized by the following chemical reactions [27, 28]:

$$UO_2(NO_3)_2 \text{ (l)} \xrightarrow{400°C} UO_3 \text{ (s)} + 2\,NO_2 \text{ (g)} + \tfrac{1}{2}\,O_2 \text{ (g)}$$

$$UO_3 \text{ (s)} + CH_3OH \text{ (g)} \xrightarrow{360°C} UO_2 \text{ (s)} + COH_2 \text{ (g)} + H_2O \text{ (g)}$$

$$UO_2 \text{ (s)} + CCl_4 \text{ (g)} \xrightarrow{Cl_2 \text{ (g), } 360°C} UCl_4 \text{ (s)} + CO_2 \text{ (g)}$$

$$3\,UCl_4 \text{ (s)} + (4 + 3x)\,Al \text{ (l)} \xrightarrow{680°C} 3\,UAl_x \text{ (s)} + 4\,AlCl_3 \text{ (g)}$$

During the first reaction step, uranyl nitrate solution was sprayed into a fluidized bed of aluminium powder (200 to 250 mesh in size, generally spherical) at 400°C to form a layer of UO₃ on the aluminium particles using air as the fluidizing gas. In the second step, the coating of UO₃ was reduced by CH₃OH to UO₂ at 360°C, with argon as the fluidizing gas. The reduction proceeded rapidly when the CH₃OH was injected and vaporized into the fluidizing argon upstream. During the third step, the UO₂ coating was chlorinated by a mixture of 75 mol% CCl₄ and 25 mol% Cl₂ at 330°C injected into the fluidizing argon stream, forming UCl₄ layers on the aluminium particles. In the fourth step, the reaction temperature was raised to 660°C (melting point of aluminium) at which the reaction to UAl_x proceeded very rapidly. In between, most of the UCl₄ was reduced to UCl₃ at about 500°C by the reaction $3\,UCl_4 + Al_x \rightarrow 3\,UCl_3 + AlCl_3$. Any small amount of UO₂ that was not converted to UCl₄ formed Al₂O₃ and UAl_x by the reaction $3 UO_2 + (4 + 3x) Al \rightarrow 2 Al_2O_3 + 3 UAl_x$. Batches of UAl_3 were produced by this process. Chemical analysis, after washing in an ultrasonic water bath, showed less than 0.5 wt% carbon, 0.5 wt% Al_2O_3, and 300 ppm chloride [27, 28, 30].

Violent exothermic reaction with ignition was observed on heating compacts of U_3O_8 and aluminium powder in stoichiometric amount to above the melting point of aluminium. The ignition temperature depends on the U_3O_8 particle size (see Table 3/4, p. 90). X-ray analysis of the products showed the presence of α-Al_2O_3, UO_2, UAl_2, and UAl_3. The reaction is the same whether carried out in vacuum, argon, or air [34, 35], see also [36, 37].

UAl_3 was prepared by electrolytic reduction of UO_2 in the presence of aluminium. UO_2 was suspended in molten $CaCl_2$ or $MgCl_2$ under argon at 850 to 950°C in a covered aluminium oxide crucible. Graphite was used as the anode material and the cathode was aluminium fused in the molten salts. The UAl_3 was formed as a powder at a current of 10 A at 5.5 V [38].

UAl_3 coatings on uranium metal were prepared by electrolytic deposition of aluminium powder onto a freshly electropolished uranium surface to optimize the corrosion resistance of the uranium metal. A coating, which consisted of a primary UAl_3 phase containing UAl_2 at the uranium-coating interfaces, was formed on heating at 650°C for 20 h. UAl_4 was also detected as a thin layer of loose black powder [39]. For aluminium ion-plating to form predominantly UAl_2 or U $-$ 10% Mo alloy to form UAl_2, UAl_3, and UAl_4 coatings see [40] or [41], respectively.

Sintering

UAl_3 powders are extremely friable and easily overground. Hammer-milling led to better results than ball-milling. The powders are pyrophoric; grinding must be carried out in argon atmosphere [13]. For results of hammer-milling under decalin as a protective liquid, see [3 to 5].

Screen analyses of UAl_3 powder after hammer-milling under decalin under various conditions showing particle size distributions, see [5].

A temperature of 900 to 950°C is the lowest sintering temperature for the fine fraction of UAl_3 powder given in [5]. Sintering of UAl_3 powder (36 to 63 μm) under a pressure of 0.19 Mp/cm² in a graphite mold for 1 min sintering time at 960°C yielded 96% th.d. [3].

Single Crystals

Single crystals of UAl_3 of about 1 mm were grown from melts of appropriate mixtures of uranium and aluminium, cooled at a rate of 2°C per h in a gradient of 10°C/cm over a temperature range in which UAl_3 is stable, and then cooled quickly to room temperature. The crystals grew, for the most part, as cubes [42].

Enthalpy, Entropy, and Gibbs Free Energy of Formation

Enthalpy of formation, ΔH_f° (UAl_3), was measured by various methods:

From adiabatic calorimetric measurements in aqueous solutions: ΔH_f° (298 K) $=$ -25.2 ± 1.8 kcal/g-$UAl_{2.994}$ [18]; -25.2 ± 2.2 kcal/g-UAl_2 [18], see also [31, 33, 43]; -6.3 kcal/g-atom [43 from 18]. From adiabatic calorimetry: ΔH_f° (298 K) $= -25.9 \pm 2.0$ kcal/mol [44].

References for 3.1.4.1 on pp. 122/3

From the formation of UAl$_3$ from α uranium and solid aluminium at 900 K, emf measurements: ΔH_f (900 K) $= -7.39$ kcal/g-atom. The entropy of formation, calculated from emf measurements, is ΔS_f (900 K) $= -1.06$ e.u.$\cdot$g-atom$^{-1}\cdot$K^{-1} [45].

Partial thermodynamic values of the formation of UAl$_3$ from α uranium and solid aluminium at 900 K, obtained from emf measurements, are: $\Delta \overline{H}_U = -29.3$ kcal/g-atom, $\Delta \overline{S}_U = -4.1$ e.u.$\cdot$g-atom$^{-1}\cdot$K^{-1}, and $\Delta \overline{G}_U = -25.61$ kcal/g-atom [45].

The free energy of formation of UAl$_3$ is $\Delta G_f^o = -19.81$ kcal/mol at 730°C [46].

For the standard free energy of formation, the equation ΔG_f^o (in kcal/mol) $= -32.9 + 13.2 \times 10^{-3}\cdot$T was estimated from the partial molar free energy of aluminium in the liquid phase, $\overline{G}_{Al}$, and the free energy of formation, ΔG_f^o, ΔG_f^o(UAl$_3$) $= \Delta G_f^o$(UAl$_4$) $- \overline{G}_{Al}$, equilibrium at peritectic temperature [46], see also [31]; emf measurements within the temperature range of 399 to 706°C yield the relation $\Delta G^o = -25.190 \pm 0.487\cdot$T (for 399 to 706°C) [44].

These data, combined with the enthalpy of formation ($\Delta H_f^o = -25.9 \pm 2.0$ kcal/mol), give the relations $\Delta G^o = -22.790 - 2.326\cdot$T$\cdot$lnT $+ 18.37\cdot$T (in cal/mol), $\Delta H^o = -26.290 \pm 1.30\cdot$T (in cal/mol), $\Delta S^o = -9.30 + 1.30\cdot$lnT (in cal$\cdot$mol$^{-1}\cdot$K^{-1}) [48], also reported in [31].

A further relation $\Delta G_f^o = -110.800 + 16.7\cdot$T is reported in [47].

References for 3.1.4.1:

[1] Saller, H. A. (in: Finneston, H. M.; Howe, J. P., Progress in Nuclear Energy, Ser. 5, Metallurgy and Fuels, Vol. 1, Pergamon, London 1956, pp. 535/43).
[2] Openshaw, P. R.; Shreir, L. L. (Corros. Sci. **3** [1963] 217/37).
[3] Nazare, S.; Ondracek, G.; Thümmler, F. (KFK-1252 [1970] 1/83; C.A. **75** [1971] No. 70285).
[4] Nazare, S.; Ondracek, G.; Thümmler, F. (J. Nucl. Mater. **56** [1975] 251/9).
[5] Thümmler, F.; Nazare, S.; Ondracek, G. (Powder Met. **10** [1967] 264/87).
[6] Pearce, R. J. (J. Nucl. Mater. **34** [1970] 151/9).
[7] Gibson, G. W. (IN-1133 [1967] 1/82; C.A. **69** [1968] No. 40509).
[8] Graber, M. J.; Gibson, G. W.; Walker, V. A.; Francis, W. C. (IDO-16958 [1964] 1/53; C.A. **61** [1964] 5161).
[9] Walker, V. A.; Graber, M. J., Jr.; Gibson, G. W. (IDO-17157 [1966] 1/110; C.A. **66** [1967] No. 81543).
[10] Gibson, G. W.; Graber, M. J.; Francis, W. C. (IDO-16934 [1963] 1/7; N.S.A. **18** [1964] No. 8723).

[11] Gibson, G. W.; Zukor, M. (IDO-17154 [1966] 3/6; N.S.A. **20** [1966] No. 19178).
[12] Saller, H. A. (U.S. 2917383 [1959]; C.A. **1960** 18313).
[13] Gibson, G. W.; de Boisblanc, D. R. (Proc. 2nd Intern. Powder Met. Conf., New York 1965 [1966], Vol. 3, pp. 26/35; C.A. **67** [1967] No. 17002).
[14] Badaeva, T. A.; Dashevskaya, L. I. (Fiz. Khim. Splavov Tugoplavkikh Soedin Toriem Uranom **1968** 107/13; C.A. **71** [1969] No. 24297).
[15] van Maaren, M. H.; van Daal, H. J.; Buschow, K. H. J.; Schinkel, C. J. (Solid State Commun. **14** [1974] 145/7).
[16] Eding, H. J.; Carr, E. M. (ANL-6339 [1961] 1/39; C.A. **1961** 16243).
[17] Werner, W. J.; McIlwain, M. C.; Hammond, J. P. (U.S. 3288571 [1966]; C.A. **66** [1967] No. 100903).
[18] Ivanov, M. I.; Tumbakov, V. A.; Podol'skaya, N. S. (At. Energiya SSSR **5** [1958] 166/70; Soviet J. At. Energy **5** [1958] 1007/11; C.A. **1960** 9481).
[19] Bareis, D. W. (AECD-3795 [1949] 1/40; N.S.A. **10** [1956] No. 6215).
[20] Less, C. S. (AERE-G-M-13 [1955] 1/4; C.A. **1956** 13547).

[21] Less, C. S. (AERE-G-M-13 [1948] 1/7; N.S.A. **10** [1956] No. 5652).

[22] Murray, J. R. (AERE-M-R-799 [1951] 1/16; N.S.A. **11** [1957] No. 10497).

[23] Castleman, L. S. (J. Nucl. Mater. **3** [1961] 1/15).

[24] Castleman, L. S.; Seigle, L. (SEP-251 [1958] 1/28; N.S.A. **13** [1959] No. 764).

[25] Subramanyam, D.; Notis, M. R.; Goldstein, J. I. (Met. Trans. A **16** [1985] 589/95).

[26] Pearce, R. J. (J. Nucl. Mater. **17** [1965] 201/2).

[27] Grimmet, E. S.; Ballard, R. K.; Buckham, J. A. (Chem. Eng. Progr. Symp. Ser. **63** No. 80 [1967] 11/5; C.A. **68** [1968] No. 55712).

[28] Hogg, G. W.; Grimmett, E. S. (AIChE [Am. Inst. Chem. Eng.] Symp. Ser. **67** No. 116 [1971] 190/8; C.A. **77** [1972] No. 22291).

[29] Grimmet, E. S.; Ballard, R. K. (IN-1087 [1967] 112/6; N.S.A. **22** [1968] No. 4231).

[30] Grimmet, E. S. (U.S. 3318670 [1967]; C.A. **67** [1967] No. 49755).

[31] Thümmler, F.; Lilienthal, H. E.; Nazare, S. (Powder Met. **12** [1969] 1/22).

[32] Morin, C. (Ger. Offen. 2616828 [1976]; C.A. **87** [1977] No. 119752).

[33] Rand, M. H.; Kubaschewski, D. (The Thermochemical Properties of Uranium Compounds, Oliver & Boyd, Edinburgh — London 1963).

[34] Fleming, J. D.; Johnson, J. W. (Nucleonics **21** No. 5 [1963] 84/7).

[35] Fleming, J. D.; Johnson, J. W. (TID-7642 [1962] 649/66; C.A. **59** [1963] 9539).

[36] Ondracek, G.; Patrassi, E. (Ber. Deut. Keram. Ges. **45** [1968] 617/21).

[37] Dykstra, L. J. (GA-1479 [1960] 1/21; C.A. **1961** 13975).

[38] Gibson, A. R.; Avery, R. G. (Brit. 966807 [1964]; C.A. **61** [1964] 12960).

[39] Pearce, R. J.; Giles, R. D.; Tavender, L. E. (J. Nucl. Mater. **24** [1967] 129/40).

[40] Greenwood, R. C.; Ritchie, A. G.; Randles, S. J.; Peacock, S. (AWRE-02382 [1983] 1/93; C.A. **100** [1984] No. 38177).

[41] Bell, R. T. (Y-1617 [1968] 1/27; C.A. **69** [1968] No. 112578).

[42] McBride, J. J.; Clark, G. W. (ORNL-3470 [1963] 4; N.S.A. **18** [1964] No. 4171).

[43] Katz, S. (J. Nucl. Mater. **6** [1962] 172/81).

[44] Chiotti, P.; Kateley, J. A. (J. Nucl. Mater. **32** [1969] 135/45).

[45] Lebedev, V. A.; Sal'nikov, V. I.; Nichkov, I. F.; Raspopin, S. P. (At. Energiya SSSR **32** [1972] 115/8; Soviet At. Energy **32** [1972] 129/32; C.A. **76** [1972] No. 145738).

[46] Johnson, I. (Met. Soc. AIME Inst. Metals Div. Spec. Rept. Ser. **10** No. 13 [1964] 171/92).

[47] Kulifeev, V. K.; Stikhin, A. N.; Loskutova, Z. M. (Nauchn. Tr. Mosk. Inst. Stali Splavov No. 131 [1981] 99/102; C.A. **96** [1982] No. 111206).

3.1.4.2 Crystallographic Properties

Uranium trialuminide crystallizes simple cubic [1, 2] or face-centered cubic [3], almost certainly isomorphic with $AuCu_3$, containing one molecule per unit cell [1, 2, 4]. The space group is $Pm\bar{3}m - O_h^1$ (No. 221) [1]. The measured lattice parameters are summarized in Table 3/13, p. 124.

The X-ray density is (for one UAl_3 per unit cell) $D_{calc} = 6.8$ g/cm^3 [5], also reported in [1, 3, 6 to 9] or 6.70 g/cm^3 [4]; the measured density is $D_{meas} = 6.4$ g/cm^3 [4, 5], see also [10]. The unit cell volume is $V = 19.70$ Å^3 (for a = 4.287 Å) [11].

The atomic positions for a simple cubic structure are: U at 0,0,0; 3 Al at 0,1/2,1/2; 1/2,0,1/2; 1/2,1/2,0 [2].

References for 3.1.4.2 on pp. 124/5

Table 3/13
Measured Lattice Parameters of UAl$_3$.

lattice constant a	preparation	Ref.	also cited in
4.278 ± 0.001 kX		[2, 4]	[13, 17, 18]
4.287 Å		[11]	[1, 11, 20 to 24]
4.27 Å		[5]	[9, 13, 18, 25]
4.26 Å	induction melting	[12, 15]	
4.827 Å		[16]	
4.254 ± 0.001 Å	single-phase product, interdiffusion of Al into U	[17]	[26, 27]
4.2636 ± 0.0005 Å		[13]	
4.2855 Å	at 300°C		
4.2650 Å	at room temperature, vapor-phase reaction		
4.275 Å	argon-arc melted	[18]	
4.266 ± 0.003 Å	interdiffusion of Al into U	[19]	

The interatomic distances are: U−Al = 3.15 Å [12], 2.88 Å [9]; U−U = 4.26 Å [12].

The temperature dependence of the lattice constant follows the equation $a_T = a_0$ $(1 + 1.6809 \times 10^{-5} \cdot T)$ for temperatures up to 300°C, and within the temperature range of 300 to 750°C: $a_T = a_{300} [1 + 1.35 \times 10^{-5} \cdot \Delta T - 1.6 \times 10^{-8} (\Delta T)^2 + 4.6 \times 10^{-11} (\Delta T)^3]$, where $\Delta T = (T-300)$ and $a_{300} = 4.2855$ Å [13], also reported in [8].

The change of the UAl$_3$ lattice into the UAl$_4$ lattice during the peritectic reaction of UAl$_3$ + liquid → UAl$_4$ can be described by the Lee-Leighly mechanism [14].

References for 3.1.4.2:

[1] Rough, F. A.; Bauer, A. A. (BMI-1300 [1958] 1/138; N.S.A. **12** [1958] No. 13935).

[2] Rundle, R. E.; Wilson, A. S. (Acta Cryst. **2** [1949] 148/50; AECD-2388 [1948] 1/12).

[3] Grimmet, E. S.; Ballard, R. K.; Buckham, J. A. (Chem. Eng. Progr. Symp. Ser. **63** No. 80 [1967] 11/5; C.A. **68** [1968] No. 55712).

[4] Rundle, R. E. (CT-2721 [1945] 1/9; N.S.A. **10** [1956] No. 5124).

[5] Gordon, P.; Kaufmann, A. R. (J. Metals **2** [1950] 182/4).

[6] Gibson, G. W.; de Boisblanc, D. R. (Proc. 2nd Intern. Powder Met. Conf., New York 1965 [1966], Vol. 3, pp. 26/35; C.A. **67** [1967] No. 17002).

[7] Cunningham, J. E.; Beaver, R. J.; Thurber, W. C.; Waugh, R. C. (TID-7546 [1958] 269/97; Fuel Elements Conf., Paris 1957; C.A. **1958** 9905).

[8] Thümmler, F.; Nazare, S.; Ondracek, G. (Powder Met. **10** [1967] 264/87).

[9] Frost, B. R. T.; Maskrey, J. T. (J. Inst. Metals **82** [1954] 171/80).

[10] Kiessling, R. (Proc. 1st Intern. Conf. Peaceful Uses At. Energy, Geneva 1955, Vol. 9, pp. 69/73).

[11] Dwight, A. E. (in: Giessen, B. C., Developments in the Structural Chemistry of Alloy Phases, Plenum, New York 1969, pp. 181/226).

[12] Lee, L. P.; Leighly, H. P., Jr. (Met. Trans. **6** [1975] 135/9).

[13] Pearce, R. J. (J. Nucl. Mater. **17** [1965] 201/2).

[14] Leighly, H. P., Jr.; Edwards, D. R. (Met. Trans. A **12** [1981] 491/5).

[15] Buzzard, R. W.; Cleaves, H. E. (TID-2501 [1957] 19/47; N.S.A. **12** [1958] No. 17283).

[16] Konobrevskii, S. T.; Zaimovskii, A. S.; Levitsky, B. M.; Sokursky, Y. N.; Chebotarev, N. T.; Bobkov, Y. V.; Egorov, P. P.; Nikolaev, G. N.; Ivanov, A. A. (Proc. 2nd Intern. Conf. Peaceful Uses At. Energy, Geneva 1958, Vol. 6, pp. 194/203).

[17] Ivanov, M. I.; Tumbakov, V. A.; Podol'skaya, N. S. (At. Energiya SSSR **5** [1958] 166/70; Soviet J. At. Energy **5** [1958] 1007/11; C.A. **1960** 9481).

[18] Badaeva, T. A.; Dashevskaya, L. I. (Fiz. Khim. Splavov Tugoplavkikh Soedin Toriem Uranom **1968** 107/13; C.A. **71** [1969] No. 24297).

[19] Pearce, R. J. (J. Nucl. Mater. **34** [1970] 151/9).

[20] Lam, D. J.; Darby, J. B., Jr.; Nevitt, N. V. (in: Freeman, A. J.; Darby, J. B., Jr., The Actinides: Electronic Structure and Related Properties, Vol. 2, Academic, New York — San Francisco — London 1974, pp. 119/84).

[21] Wright, E. H.; Willey, L. A. (Tech. Paper Alcoa Res. Lab. No. 15 [1960] 1/46; C.A. **61** [1964] 9243).

[22] Hansen, M. (Constitution of Binary Alloys, McGraw-Hill, New York — Toronto — London 1958, pp. 143/4).

[23] Ambrosio, F. F. (INIS-mf-1356 [1973] 1/64; C.A. **82** [1975] No. 114937).

[24] Gentile, E. F. (Met. ABM [Assoc. Brasil. Metais] **24** No. 124 [1968] 187/92; C.A. **69** [1968] No. 109169).

[25] Sella, C.; Trillat, J. J. (Rev. Met. [Paris] **56** [1959] 105/121).

[26] Elliott, R. P. (Constitution of Binary Alloys, 1st Suppl., McGraw-Hill, New York — Toronto — London 1965, pp. 61/2).

[27] Nazare, S.; Ondracek, G.; Thümmler, F. (KFK-1252 [1970] 1/83; C.A. **75** [1971] No. 70285).

3.1.4.3 Mechanical Properties

The microhardness of UAl_3 is measured as a function of load, of temperature (see **Fig.** 3-**29**), and of grain-size, respectively [1]. Further values for the microhardness are (in

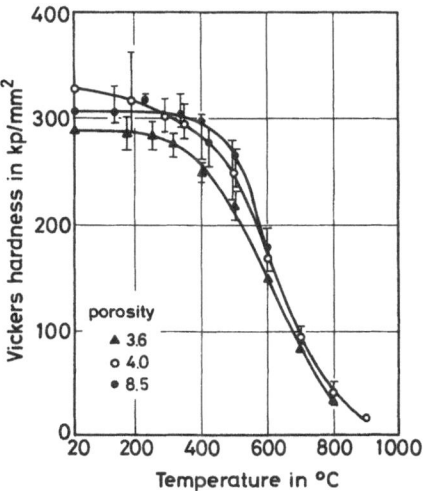

Fig. 3-29. Microhardness of UAl_3 as a function of temperature (1000 g load) [1].
The porosity is given in % of the theoretical density.

References for 3.1.4.3 on p. 126

kg/mm^2): 300 (H$_{4p}$, 50 g, 15 s) [2], 265 (60 g load) [3], see also [4, 5], 249 (average value) [6], 257 (65 g load) [7], 245 (115 g load) [7], 244 (135 g load) [7], and 520 [8]. The Vickers hardness increases with small amounts of silicon in UAl$_3$ [6, 7].

UAl$_3$ is quite brittle at room temperature and maintains considerable hardness at the lower diffusion temperatures. DPH (diamond pyramid hardness) values are: 605 (at 25°C) and 340 (at 450°C) [9].

References for 3.1.4.3:

[1] Nazare, S.; Ondracek, G.; Thümmler, F. (J. Nucl. Mater. **56** [1975] 251/9).

[2] Kiessling, R. (Proc. 1st Intern. Conf. Peaceful Uses At. Energy, Geneva 1955, Vol. 9, pp. 69/73).

[3] Badaeva, T. A.; Dashevskaya, L. I. (Fiz. Khim. Splavov Tugoplavkikh Soedin Toriem Uranom **1968** 107/13; C.A. **71** [1969] No. 24297).

[4] Pav, T.; Otruba, J.; Kocik, J.; Saxl, I. (Reactor. Materialoved. Tr. Konf. po Reactor. Materialoved., Alushta 1978, pp. 313/22; C.A. **91** [1979] No. 11067).

[5] Gomozov, L. I.; Ivanov, O. S. (Reactor. Materialoved. Tr. Konf. po Reactor. Materialoved., Alushta 1978, pp. 4/31; C.A. **91** [1979] No. 29285).

[6] Boucher, R. (Symp. on Solid State Diffusion, Saclay, France, 1958 [1959], pp. 111/7; N.S.A. **15** [1961] No. 26564; AEC-tr-4768 [1960] 1/11; N.S.A. **14** [1960] No. 7826).

[7] Boucher, R. (J. Nucl. Mater. **1** [1959] 13/27).

[8] Stone, H. E. N. (J. Mater. Sci. **10** [1975] 923/34).

[9] Castleman, L. S. (J. Nucl. Mater. **3** [1961] 1/15).

3.1.4.4 Thermal Properties

Thermal Expansion

The thermal expansion coefficient, α, of UAl$_3$ decreases slightly with temperature. Measured values are α (in $10^{-5}/$°C) = 1.6809 (at 0 to 300°C), 1.5898 (at 0 to 400°C), 1.5214 (at 0 to 500°C), and 1.4945 (at 0 to 600°C) [1], also reported in [2, 3].

For measured expansion of the lattice parameter see in Section 3.1.4.2, "Crystallographic Properties", p. 123.

Melting Point

UAl$_3$ decomposes peritectically at 1350°C to UAl$_2$ and aluminium [4], also reported in [2, 5 to 10] (as 2460°F), [11 to 15] (as 1623 K), [3, 16 to 20].

Heat Capacity. Entropy

The specific heat of UAl$_3$ was measured on compacted powders within the temperature range of 1.3 to 13 K using a discontinuously working adiabatic calorimeter. The results are given in **Fig. 3-30**. The electronic and lattice contributions of the specific heat are calculated from the equation $C = \gamma \cdot T + \alpha \cdot T^3$ and resulted in: γ = electronic coefficient: 41.6 mJ $\cdot$ mol$^{-1} \cdot$ K^{-2}, α = lattice coefficient: 0.144 mJ $\cdot$ mol$^{-1} \cdot$ K^{-4}, Θ = Debye temperature: 378 K, whereas the Debye temperature appears to be proportional to $(M(Al_3))^{-1/2}$ with $M(Al_3)$ = atomic mass of aluminium [21].

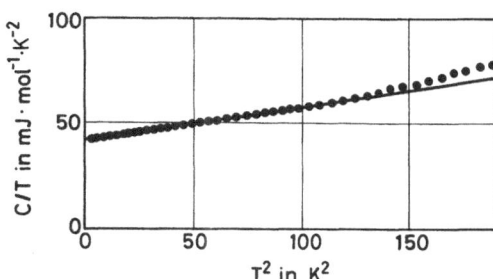

Fig. 3-30. Heat capacity of UAl$_3$ [21]. C/T versus T^2. The straight lines correspond to C = $\gamma \cdot T + \alpha \cdot T^3$ (with γ = 41.6 mJ·mol^{-1}·K^{-2}, α = 0.144 mJ·mol^{-1}·K^{-4}).

Assuming the entropy of formation to be close to zero (see also [22]) a value for the room temperature entropy was estimated to be S (298 K) = 32.5 + 3 e.u. [23], see also [2].

Thermal Conductivity

The thermal conductivity of UAl$_3$, estimated by extrapolating values of uranium−aluminium alloys (up to 58 wt% U) and of pure uranium, is 0.02 ± 0.01 cal·s^{-1}·cm^{-2}·K^{-1} (at 65°C) [19], also reported in [2, 3], or 0.63 W·cm^{-1}·K^{-1} (at 200°C) [17], also reported in [2, 3].

References for 3.1.4.4:

[1] Pearce, R. J. (RD-13-No. 451 [1965] from [13]).
[2] Thümmler, F.; Lilienthal, H. E.; Nazare, S. (Powder Met. **12** [1969] 1/22).
[3] Thümmler, F.; Nazare, S.; Ondracek, G. (Powder Met. **10** [1967] 264/87).
[4] Gordon, P.; Kaufmann, A. R. (J. Metals **2** [1950] 182/4).
[5] Castleman, L. S. (J. Nucl. Mater. **3** [1961] 1/15).
[6] Saller, H. A. (in: Finneston, H. M.; Howe, J. P.; Progress in Nuclear Energy, Ser. 5, Metallurgy and Fuels, Vol. 1, Pergamon, London 1956, pp. 535/43).
[7] Boucher, R. (Symp. on Solid State Diffusion, Saclay, France, 1958 [1959], pp. 111/7; N.S.A. **15** [1961] No. 26564; AEC-tr-4768 [1960] 1/11; N.S.A. **14** [1960] No. 7826).
[8] Kiessling, R. (Proc. 1st Intern. Conf. Peaceful Uses At. Energy, Geneva 1955, Vol. 9, pp. 69/73).
[9] Kraus, V. (Jad. Energ. [Prague] **7** [1961] 48/53; C.A. **1961** 10254).
[10] Gibson, G. W.; de Boisblanc, D. R. (Proc. 2nd Intern. Powder Met. Conf., New York 1965 [1966], Vol. 3, pp. 26/35; C.A. **67** [1967] No. 17002).

[11] Badaeva, T. A.; Dashevskaya, L. I. (Fiz. Khim. Splavov Tugoplavkikh Soedin Toriem Uranom **1968** 107/13; C.A. **71** [1969] No. 24297).
[12] Grimmet, E. S.; Ballard, R. K.; Buckham, J. A. (Chem. Eng. Progr. Symp. Ser. **63** No. 80 [1967] 11/5; C.A. **68** [1968] No. 55712).
[13] Nazare, S.; Ondracek, G.; Thümmler, F. (KFK-1252 [1970] 1/83; C.A. **75** [1971] No. 70285).
[14] Wright, E. H.; Willey, L. A. (Tech. Paper Alcoa Res. Lab. No. 15 [1960] 1/46; C.A. **61** [1964] 9243).
[15] Nazare, S.; Ondracek, G.; Thümmler, F. (J. Nucl. Mater. **56** [1975] 251/9).

[16] Cunningham, J. E.; Beaver, R. J.; Thurber, W. C.; Waugh, R. C. (TID-7546 [1958] 269/97;
     Fuel Elements Conf., Paris 1957; C.A. **1958** 9905).

[17] Rice, W. L. R. (TID-11295, 3rd Ed. [1964] 105/20; N.S.A. **18** [1964] No. 41933).

[18] Openshaw, P. R.; Shreir, L. L. (Corrosion Sci. **3** (No. 4) [1963] 217/37).

[19] Jones, T. I.; Street, K. N.; Scoberg, J. A.; Baird, J. (Can. Met. Quart. **2** No. 1 [1963]
     53/72; C.A. **58** [1963] 12254).

[20] Frost, B. R. T.; Maskrey, J. T. (J. Inst. Metals **82** [1954] 171/80).

[21] van Maaren, M. H.; van Daal, H. J.; Buschow, K. H. J.; Schinkel, C. J. (Solid State
     Commun. **14** [1974] 145/7).

[22] Katz, S. (J. Nucl. Mater. **6** [1962] 172/81).

[23] Rand, M. H.; Kubaschewski, O. (The Thermochemical Properties of Uranium Compounds,
     Oliver & Boyd, Edinburgh — London 1963).

3.1.4.5 Electrical Properties

Specific Resistivity

Room temperature values for the specific resistivity of UAl₃ are $3.0 \times 10^{-3}\,\Omega \cdot cm$ for pressed powders (63 to 90 µm, 90% th. d.) with a temperature coefficient of 2.2×10^{-4} [1] or 2.56×10^{-4} [2], also reported in [3].

The low-temperature dependence of the electrical resistivity is characterized by the following temperature regions: ϱ proportional to T^2 with an estimated upper limit of $T_1 = 100$ K [4], ϱ proportional to T with an estimated upper limit of $T_2 = 200$ K [4], and proportional to $\ln T$ [5], see Fig. 3-14, p. 103 and **Fig. 3-31**. The results are discussed in terms of a localized spin-fluctuation model for nearly metallic intermetallics, whereas the temperature T_1 represents a measure of the characteristic localized spin-fluctuation temperatures, T_{SF}. Further T_{SF} values were calculated using the dilute-alloy model (Rivier and Zlatic [6, 7]) [5]. The results are summarized in Table 3/14.

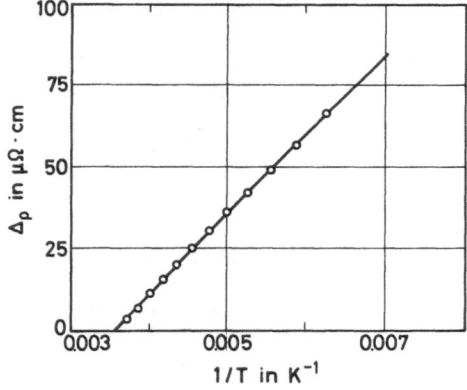

Fig. 3-31. Approach of the resistivity of UAl₃ to the limiting value in dependence on the reciprocal of temperature [5] using data from [4].

Table 3/14
Calculated Spin-Fluctuation Temperatures, T_{SF}, for UAl_3 [5].

region	T_{SF} in K
from slope of T^2 region	480
from high-temperature end of T^2 region	750
from slope of T region	246
approach to the high-temperature limit, $-d\varrho/d\,(1/T) = T_{SF}$	123
from low-temperature end of $\ln T$ region	190

Thermoelectric Power

The thermoelectric power, S, of UAl_3 is characterized by large positive values ranging from 10 to 45 µV/K within the temperature range of 5 to 300 K [8], see **Fig.** 3-32, see also [2]. It is suggested that the thermopower of UAl_3 is dominated by paramagnon scattering [8].

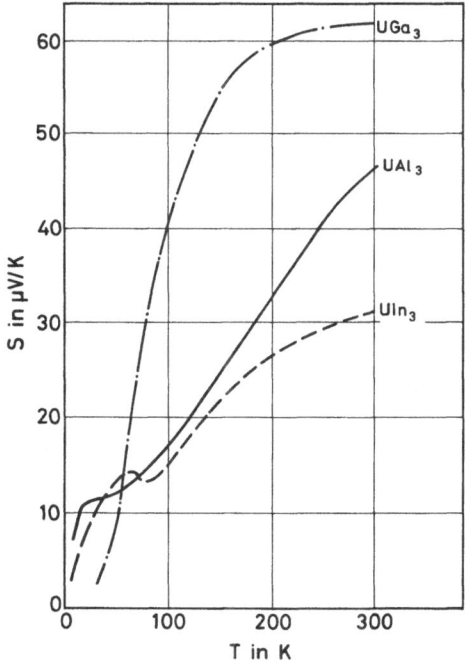

Fig. 3-32. The thermopower, S, of UAl_3 as a function of temperature [8]. For comparison the values for UGa_3 and UIn_3 are shown.

References for 3.1.4.5:

[1] Thümmler, F.; Lilienthal, H. E.; Nazare, S. (Powder Met. **12** [1969] 1/22).
[2] Warren, I. H.; Price, C. E. (Can. Met. Quart. **3** [1964] 245/56; C.A. **62** [1965] 2316).
[3] Nazare, S.; Ondracek, G.; Thümmler, F. (KFK-1252 [1970] 1/83; C.A. **75** [1971] No. 70285).

[4] Buschow, K. H. J.; van Daal, H. J. (AIP [Am. Inst. Phys.] Conf. Proc. No. 5 [1971/72] 1464/77; C.A. **77** [1972] No. 11324).

[5] Brodsky, M. B. (Phys. Rev. [3] B **9** [1974] 1381/7).

[6] Rivier, N.; Zlatic, V. (J. Phys. F **2** [1972] L87/L92).

[7] Rivier, N.; Zlatic, V. (J. Phys. F **2** [1972] L99/L104).

[8] van Daal, H. J.; Buschow, K. H. J.; van Aken, P. B. (Thermoelectr. Met. Conduct. Proc. 1st Intern. Conf., East Lansing, Mich., 1977 [1978], pp. 107/15; C.A. **92** [1980] No. 32756).

3.1.4.6   Magnetic Properties

Paramagnetic susceptibilities were measured for UAl$_3$ within the temperature range 20 to 800°C, from which room-temperature values (20°C) are $\chi_g = 4.85 \times 10^{-6}$ emu/g or $\chi_a = 1540 \times 10^{-6}$ emu/g-atom [1]. Within this temperature range (approximately [2]) a Curie-Weiss law is obeyed with $\chi_g \cdot 10^6 = 4580/T + 700$ [1], from which an effective magnetic moment of 3.5 μ_B (without corrections for the diamagnetism of the uranium and aluminium ions) was derived [1]. At lower temperatures the susceptibility becomes temperature-independent, and in the low-temperature range χ is proportional to $(1 - b \cdot T^2)$ [2]. The results are discussed in terms of a localized spin-fluctuation model, whereas the constant b and the Curie temperature may be seen as a measure of the lifetime of the localized spin fluctuations [2]. Calculations within the Friedel-Anderson model led to following results: χ (Pauli-band susceptibility) = 15.2×10^{-4} emu/mol (using the data of [2]), α (Stoner enhancement) = 2.6, and $T_{SF} = 79.7$ K [3].

References for 3.1.4.6:

[1] Konobrevskii, S. T.; Zaimovskii, A. S.; Levitsky, B. M.; Sokursky, Y. N.; Chebotarev, N. T.; Bobkov, Y. V.; Egorov, P. P.; Nikolaev, G. N.; Ivanov, A. A. (Proc. 2nd Intern. Conf. Peaceful Uses At. Energy, Geneva 1958, Vol. 6, pp. 194/203).

[2] Buschow, K. H. J.; van Daal, H. J. (AIP [Am. Inst. Phys.] Conf. Proc. No. 5 [1971/72] 1464/77; C.A. **77** [1972] No. 11324).

[3] Brodsky, M. B. (Phys. Rev. [3] B **9** [1974] 1381/7).

3.1.4.7   Chemical Reactions

Behavior on Heating

UAl$_3$ decomposes peritectically at 1350°C to UAl$_2$ and aluminium [1]. For further details see Section 3.1.4.4, "Thermal Properties", p. 126.

Reactions with Elements

Dense specimens of UAl$_3$ oxidize rapidly in oxygen or air on heating, forming U$_3$O$_8$ as the only oxidation product. UAl$_3$ powders are pyrophoric and must be handled under inert gas conditions [2, 3].

When the oxidation of UAl$_3$ was carried out in pure oxygen of 1 atm, a linear weight gain was observed after an initial period of decreasing rates of oxidation for about 100 min at oxidation temperatures of 450 to 550°C and 5 to 10 min at 550 to 600°C [4], see also [5], see **Fig.** 3-**33**. An activation energy of 23.6 kcal/mol and a weight gain of 1.7×10^5 µg $\cdot$ cm$^{-2} \cdot$ s^{-1}

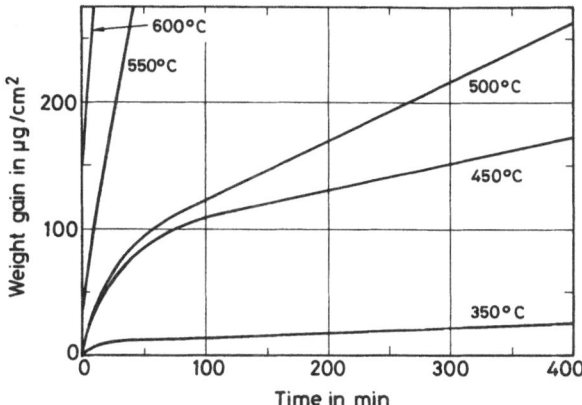

Fig. 3-33. Weight gain versus time for the oxidation of UAl$_3$ in pure oxygen at 350, 450, 500, 550, and 600°C [4].

was derived from the linear oxidation region [4], see Table 3/15. The surface of the specimen showed a slight roughness after an oxidation time of 80 min at 400 to 600°C, which became more pronounced with time. The surface roughness disappeared after 108 h and cracks in the surface became evident. The formation of both U$_3$O$_8$ and Al$_2$O$_3$ should be thermodynamically possible, but as only U$_3$O$_8$ was observed, the overall reaction of 3 UAl$_3$ + 4 O$_2$ → U$_3$O$_8$ + 9 Al was stated. It was not possible to state if the aluminium formed by this reaction enters the UAl$_3$ or the U$_3$O$_8$ phase, or forms an Al$_2$O$_3$ interface [3]. A temperature of T$_p$ = 390°C is reported at which the weight gain reaches 1 mg/cm^2 after 4 h of oxidation in air (1 atm) [22], see also [5].

Diffusion couples of UAl$_3$-coated uranium and magnox ("Al$_3$Mg$_4$" or "Al$_3$Mg") showed no evidence of interaction after heating in argon (1 atm) at 600°C for 750 h, and no tendency of the UAl$_3$ to diffuse into the uranium even after 21 days at 600°C [6, 7].

For reaction of UAl$_3$ with aluminium see Section 3.1.4.8, UAl$_3$ – Al Dispersions, p. 134.

Table 3/15
Comparison of Linear Oxidation Rates for Oxidation of UAl$_3$ with CO$_2$ and O$_2$.

temperature in °C	linear oxidation rates in mg · cm^{-2} · h^{-1}		
	UAl$_3$ + CO$_2$ [8]	UAl$_x$-coated U + CO$_2$ [7]	UAl$_3$ + O$_2$ [4]
350	0.0022	—	0.0025
400	0.0044	—	—
	0.0052		
450	0.0054	0.0022	0.013
	0.0073		
500	0.021	—	0.03
550	0.039	0.02	0.34
	0.045		
600	0.108	—	0.6

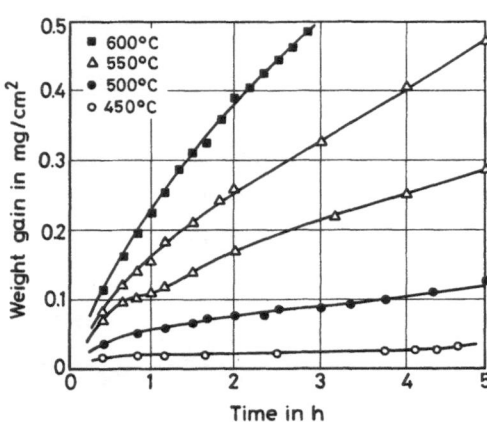

Fig. 3-34. Early stages of oxidation of UAl₃ in flowing CO_2 [8].

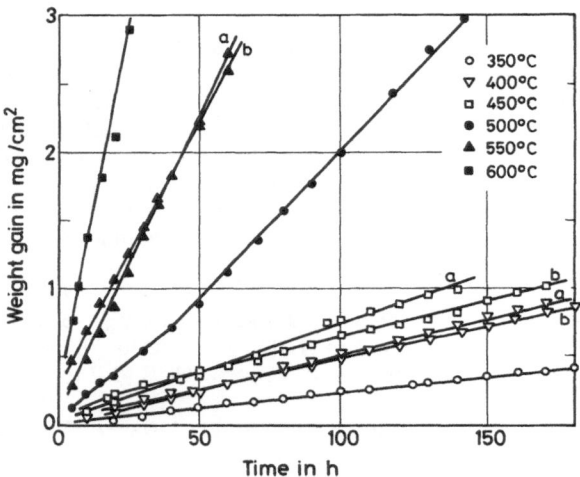

Fig. 3-35. Long term oxidation of UAl₃ in flowing CO_2 [8].

Reactions with Compounds

The oxidation of UAl₃ in flowing CO_2 at 450 to 600°C first follows an initial parabolic rate law with an activation energy of 47 ± 3 kcal/mol, then a linear rate law with an activation energy of 19 ± 2 kcal/mol; see Table 3/15, p. 131, and **Fig.** 3-**34** for short time, and **Fig.** 3-**35** for long time oxidation. This transition is thought to be associated with cracking of the oxide layer by diffusion of oxygen through a layer of UO_2 modified by inclusions of Al^{3+} ions. UO_2 was observed by X-ray diffraction analysis and Al_2O_3 from gravimetric analysis of oxidized UAl₃ powder as the reaction products. The overall chemical reaction $2\,UAl_3 + 13\,CO_2 \rightarrow 2\,UO_2 + 3\,Al_2O_3 + 13\,CO$, followed in the microcracks by the reaction $2\,UAl_3 + 13\,CO \rightarrow 2\,UO_2 + 3\,Al_2O_3 + 13\,C$ was stated for the oxidation process [8].

UAl₃-coated uranium shows, after an initial period of linear oxidation, an accelerating oxidation rate when oxidized in flowing (50 cm³/min) or static (1 atm) CO_2 at 450, 550, and

650°C [6, 7, 9, 10], compare Fig. 3-26, p. 115. The accelerating oxidation is associated with localized attack through the UAl_3 layer on uranium. A weight gain of 0.002 $\mu g \cdot cm^{-2} \cdot h^{-1}$ in flowing CO_2 and 0.0005 $\mu g \cdot cm^{-2} \cdot h^{-1}$ in static CO_2 at 450°C was observed in the linear oxidation region before the localized attack occurred [7]. Only UO_2 was found as the oxidation product by X-ray analysis [10].

UAl_3 can be dissolved in HNO_3 [11]. It was completely dissolved in a mixture of 600 mL HCl (1.178 g/mL), 400 mL H_3PO_4 (1.39 g/mL), 0.25 g Na_2SiF_6, 0.037 g H_2PtCl_6, 0.12 g $CuSO_4 \cdot$ 5 H_2O for calorimetric experiments [12].

Metallographic Etching

Etching of metallographic samples was carried out with 50% HNO_3 [10, 13] with HF added [14], dilute HF (1%) [15], or with a solution of ammonium persulfate (5%) [14].

Etching treatments which are recommended to distinguish the UAl_2 and UAl_3 phase in the $UAl_2 - UAl_3$ field and the UAl_3 and UAl_4 phase in the $UAl_3 - UAl_4$ field metallographically are described in [16], see also [17 to 21].

References for 3.1.4.7:

[1] Gordon, P.; Kaufmann, A. R. (J. Metals **2** [1950] 182/4).
[2] Gibson, G. W.; de Boisblanc, D. R. (Proc. 2nd Intern. Powder Met. Conf., New York 1965 [1966], Vol. 3, pp. 26/35; C.A. **67** [1967] No. 17002).
[3] Openshaw, P. R.; Shreir, L. L. (Corros. Sci. **4** [1964] 335/44).
[4] Openshaw, P. R.; Shreir, L. L. (Corros. Sci. **3** (No. 4) [1963] 217/37).
[5] Badaeva, T. A.; Dashevskaya, L. I. (Fiz. Khim. Splavov Tugoplavkikh Soedin Toriem Uranom **1968** 107/13; C.A. **71** [1969] No. 24297).
[6] Churchman, A. T.; Buddery, J. H.; Edmondson, B.; Greenwood, G. W.; Hesketh, R. V.; Williamson, G. K. (Proc. 3rd Intern. Conf. Peaceful Uses At. Energy, Geneva 1964 [1965], Vol. 9, pp. 13/22).
[7] Buddery, J. H.; Clark, M. E.; Pearce, R. J.; Stobbs, J. J. (J. Nucl. Mater. **13** [1964] 169/81).
[8] Pearce, R. J. (J. Nucl. Mater. **34** [1970] 151/9).
[9] Pearce, R. J.; Stobbs, J. J. (Tr. 3rd Mezhdunar. Kongr. Korroz. Metal., Moscow 1966 [1968], Vol. 4, pp. 158/66; Proc. 3rd Intern. Congr. Metal. Corros., Moscow 1966 [1969], Vol. 4, pp. 152/65; C.A. **72** [1970] No. 24061).
[10] Pearce, R. J.; Giles, R. D.; Tavender, L. E. (J. Nucl. Mater. **24** [1967] 129/40).

[11] Rough, F. A.; Bauer, A. A. (BMI-1300 [1958] 1/138; N.S.A. **12** [1958] No. 13935).
[12] Ivanov, M. I.; Tumbakov, V. A.; Podol'skaya, N. S. (At. Energiya SSSR **5** [1958] 166/70; Soviet J. At. Energy **5** [1958] 1007/11; C.A. **1960** 9481).
[13] Nazare, S.; Ondracek, G.; Thümmler, F. (KFK-1252 [1970] 1/83; C.A. **75** [1971] No. 70285).
[14] Castleman, L. S. (J. Nucl. Mater. **3** [1961] 1/15).
[15] Gentile, E. F. (Met. ABM [Assoc. Brasil. Metais] **24** No. 124 [1968] 187/92; C.A. **69** [1968] No. 109169).
[16] Hills, R. F. (J. Inst. Metals **86** [1958] 438/41).
[17] Saller, H. A. (in: Finneston, H. M.; Howe, J. P.; Progress in Nuclear Energy, Ser. 5, Metallurgy and Fuels, Vol. 1, Pergamon, London 1956, pp. 535/43).
[18] Boucher, R. (J. Nucl. Mater. **1** [1959] 13/27).
[19] Boucher, R. (Symp. Solid State Diffusion, Saclay, France, 1958, pp. 111/7; N.S.A. **15** [1961] No. 26564; AEC-tr-4768 [1960] 1/11; N.S.A. **14** [1960] No. 7826).
[20] Zelezny, W. F. (IDO-17154 [1966] 56/73; N.S.A. **20** [1966] No. 19178).

[21] Nazare, S.; Ondracek, G. (Prakt. Metallog. **6** [1969] 742/6; C.A. **72** [1970] No. 50 161).

[22] Stone, H. E. N. (J. Mater. Sci. **10** [1975] 923/34).

3.1.4.8 Uranium Trialuminide — Aluminium Dispersions, UAl$_3$—Al

UAl$_3$—Al dispersion-fuel elements are of interest in nuclear technology. Research reactors with neutron fluxes of about 10^{15} n·cm^{-2}·s^{-1} utilize plate-type elements having a high uranium content and, in most cases, an aluminium matrix and cladding.

Preparation of UAl$_3$—Al Dispersions and Fuel Plates

The UAl$_3$—Al dispersions were prepared by mixing, pressing, and compaction of UAl$_3$ and Al powders. According to nuclear aspects, the UAl$_3$ powder was used as a coarse powder, with an optimum grain size of 67 µm, to avoid recoil-zone interactions [1, 2], see also [3]. In practice, grain sizes of UAl$_3$ of 44 to 88 µm [4] or 63 to 90 µm [1, 2] were blended in air or argon with aluminium powder of about 44 µm [4, 5] or 18.6 µm [1, 2] in size by dry mixing within 4 h. Sometimes small amounts of boron or B$_4$C (0.19 wt%) were added to UAl$_3$ [4, 6 to 8], and instead of pure aluminium (ALCOA 101 [1]), X8001 or 6061 Al alloy were also used [4, 6, 7]. These powder mixtures were compacted by pressing (about 30 t.s.i.) using stearic acid in methanol as a lubricant and were sintered at 450 to 600°C for 4 to 96 h to densities of 75 to 78% th.d., based on the theoretical density of the UAl$_3$—Al compact [6, 7, 9 to 12]. The UAl$_3$—Al dispersion is affected by the reaction of UAl$_3$ with aluminium to form UAl$_4$ during sintering, at least to a small degree, starting at 450°C, and total reaction at 600°C within 1 h with a volume increase of 1.5% [13 to 16]. Core materials cold-pressed to high density showed longitudinal dimensional changes of about 8% as a result of the UAl$_3$ → UAl$_4$ phase transformation after vacuum degassing at 590°C. This effect was thought to take place owing to changes of particle shape during the transformation. Only small volume changes were observed when UAl$_3$ transformed to UAl$_4$ in an aluminium matrix [17]. The UAl$_3$ phase can be stabilized by addition of Si, Ge, Pd, Mg, Zr, Zn, Hf, or Ti in small amounts (0.3 to 3 wt%) [6, 13 to 16, 18]. Tin did not appear to be an effective stabilizer to prevent the UAl$_3$ → UAl$_4$ reaction [18].

In order to fabricate UAl$_3$—Al fuel plates for nuclear application the mixed UAl$_3$—Al powder was compacted to a specified rectangular shape of high density, whereas the "green" density depends not only on the pressure but also on the total amount of aluminium or UAl$_3$, respectively [1, 2, 4, 6]. This "core" was assembled by the "picture-frame technique" using a picture frame and cover plates of aluminium or aluminium-alloy materials similar to those used for the blending of the UAl$_3$ powder. If aluminium alloys instead of pure aluminium were used as frame and cover plates, the alloy materials were plated with pure aluminium to obtain sufficient bonding between core and cladding [1, 4, 9, 19]. This assembly was kept together (by welding as an example [1]), hot-rolled at 500°C for bonding and cold-rolled to specified dimensions [1, 4, 7 to 9, 19]. Finally, some tests were undertaken for quality control, such as an X-ray structure test to confirm the absence of core failures ("dog-boning") and a "blister-test" at 500°C for 1 h to point out bonding defects [1, 2, 4, 7, 20 to 22]. Further data for the fabrication of UAl$_3$—Al irradiation test plates are given in [2]. A flow sheet for the fabrication process is given in **Fig. 3-36**.

Interaction of UAl$_3$ and Al Matrix

The interaction of UAl$_3$ and the aluminium matrix leads to the formation of UAl$_{4+x}$. The reaction starts at about 400°C with particle sizes of UAl$_3$ of 63 to 90 µm, but the reaction rate is very low at this temperature. A reaction product growth nucleus is formed at this temperature

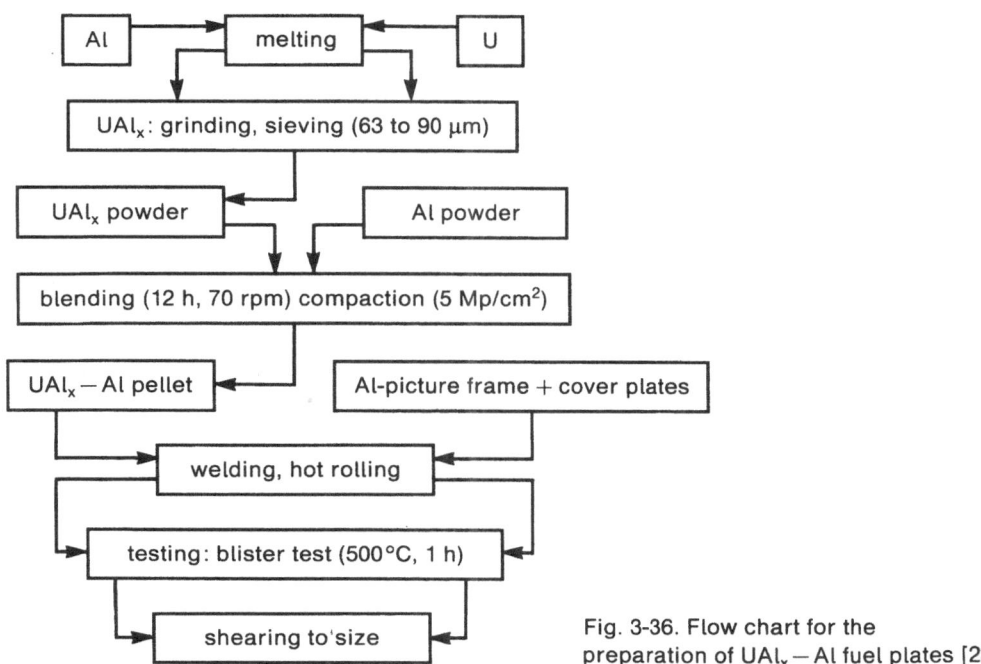

Fig. 3-36. Flow chart for the preparation of $UAl_x - Al$ fuel plates [23].

after an incubation period. It seems that this nucleation is the rate-determing factor for the peritectic reaction. The reaction is probably accelerated by the heat released [1, 3]. The heat of reaction is 6 kcal/mol [24]. The reaction is accompanied by an increase in the volume of about 4.6% of the dispersion, but, cyclic heat treatment at 530°C for 1 h with 6 cycles caused only small volume changes in the range of about 0.2% on hot-rolled dispersion fuel elements [1, 3], see also [23].

From experiments with $U-Al$ diffusion couples, the solubility of uranium in the UAl_3 reaction zone must be at least 2.6%. Otherwise, it was shown that the growth of UAl_4 and UAl_2 phases was slower in special UAl_3-Al and UAl_3-U couples, where the UAl_3 was prepared as a layer on uranium metal by a separate diffusion reaction in contact with Al, as compared to the reaction in $U-Al$ couples. Heat treatment at 600°C for 4 and 24 h at a pressure of 2.5 t.s.i. led to a UAl_4 layer growth of 0.087 ± 0.006 mm and 0.146 ± 0.026 mm, respectively [25], see also the section "Interdiffusion of Uranium and Aluminium", Chapter 3.1.2, p. 85. UAl_4, formed by peritectic reaction in molten samples, shows a long needlelike shape with a characteristic fishtail morphology when cut along the long axis [26]. The stabilization of UAl_3 by addition of small amounts of ternary metals was suggested to result from simple reduction of the transformation rate [13, 14, 27, 28]. Investigations of the ternary systems $U-Al-Ni$ and $U-Al-Zr$ showed that the alloys formed are in thermodynamic equilibrium with the aluminium matrix [29], see Section 3.1.7.13, p. 215, and Section 3.1.7.6, p. 198.

Normally, etching with HNO_3 is used for metallographic examination. Polarized light and dark-field illumination [30] or covering of the specimen by vapor deposition of zinc selenide to increase the contrast of the phases [31] are of interest in the quantitative metallurgy.

                    References for 3.1.4.8 on pp. 137/8

Mechanical Properties

The rupture strength and Young's modulus of UAl_3 — Al dispersions decrease with increasing amount of the dispersed phase. Furthermore, the reaction between UAl_3 and aluminium leads to a further decrease of the rupture strength and Young's modulus [1], see also [2, 23]. Tensile strength [2, 23], yield strength [2], and Young's modulus [1] are shown in graphs.

Thermal Expansion

The linear thermal expansion coefficient of UAl_3 — Al dispersions measured at 20 to 500°C decreases with increasing amount of UAl_3 [1, 23]. The coefficient as a function of UAl_3 concentration is shown in [23].

Thermal Conductivity

The thermal conductivity of UAl_3 — Al dispersions was measured on pressed powders of high density (about 92% th.d.) at 94°C. The thermal conductivity decreases with increasing amount of UAl_3. The measured values are significantly smaller than those of the molten components ($\lambda_{Al} = 0.53$ cal·K^{-1}·cm^{-1}·s^{-1}, $\lambda_{UAl_3} = 0.02$ cal·K^{-1}·cm^{-1}·s^{-1}), which is assumed to be an effect of oxide layers surrounding the powder particles [1, 23].

Electrical Resistivity

The specific resistivity of UAl_3 — Al dispersions was measured on pressed powders of high density (about 92% th.d.) at room temperature and the temperature coefficient in the range of 20 to 120°C. The specific resistivity increases only slightly with increasing amount of UAl_3. The measured values of the components are much higher than those of the molten ones ($\varrho_{UAl_3} = 2.56 \times 10^{-4}$ Ω·cm, $\varrho_{Al} = 2.9 \times 10^{-6}$ Ω·cm) [1, 23].

Neutron Irradiation of UAl₃ — Al Dispersion Fuel Elements

The UAl_3 phase (93% enriched in ^{235}U, irradiated at 8.5×10^{13} n·cm^{-2}·s^{-1}) showed a significant increase in hardness of about 30% up to 6.5% burnup at a mean irradiation temperature of 70°C. At higher burnup, in the range of 6.5 to 35%, no further change in hardness was observed before the hardness began to decrease [32 to 34], see also [19, 35]. At higher irradiation temperatures (150°C), the decrease in hardness seemed to occur at lower burnup, followed by a significant increase above 50% burnup. In the recoil zone of the aluminium matrix there was an increase in hardness with burnup [32 to 34], see also [4]. The stress-strain diagram, taken from a UAl_3 — Al fuel plate section at 180°C after 16% burnup, showed the earlier fracture of the less ductile dispersion, followed by the fracture of the cladding. The fracture of the irradiated dispersion occurred at a load of 45 ± 4 kp with a plastic strain of $0.05 \pm 0.03\%$, as compared to 36 ± 0.5 kp with a strain of $0.35 \pm 0.2\%$ for the unirradiated sample [32 to 34].

The reaction of the UAl_3 phase with the aluminium matrix was induced by neutron irradiation. After 10 to 15% burnup 50 vol% of the particles reacted, and the reaction was essentially complete after 50% burnup at temperatures of about 150°C [32 to 34]. The swelling of the UAl_3 — Al fuel plates under neutron irradiation is shown in Fig. 3-27, p. 117. The increase in volume is small with burnup of 33 to 40%, showing an effect of about 0.12 vol%/% burnup. At higher burnup, the swelling rate increases with fuel plate temperature [32 to 34].

50 wt% UAl_3 — Al (fully enriched in uranium) yield a swelling rate of $\Delta V/V$ in the range of 0.8 to 5.1% after radiation exposure of 10.7×10^{20} to 16.3×10^{20} fissions/cm^3 [4, 36]. The volume of the fuel particles increased from 21 vol% UAl_3 to 28 vol% (totally reacted to) UAl_4 after 40% burnup in a flux of 2.8×10^{14} n·cm^{-2}·s^{-1} at 175°C [4, 19, 36 to 38].

Neutron irradiation was performed on UAl_3 — Al fuel plate compositions, slightly to the uranium side of UAl_3, of 74 wt% U — 26 wt% Al (93% enriched in ^{235}U) in a thermal flux of 7×10^{-13} n·cm^{-2}·s^{-1} up to 60% burnup at about 130°C and of 75 wt% U — 25 wt% Al (40% enriched in ^{235}U) in a thermal flux of 1.5×10^{14} n·cm^{-2}·s^{-1} up to 60% burnup at about 120°C. The post-irradiation examination gave no evidence of fission gas (He, Xe, Kr) bubble formation in the basic two-phase mixture of UAl_3 and UAl_4. However, clusters of small fission gas bubbles were observed to form on and into small uranium oxide inclusions, which were present as an impurity. It was concluded that the bubble formation could be eliminated by avoiding oxygen exposure during the uranium aluminide preparation [36, 39], see also [40]. From post-irradiation annealing experiments it was observed that the principal release of fission gas (He, Xe, Kr) occurred with UAl_3 pellets at about 1548 K. The major part of the fission gas was released with UAl_3 — Al clad fuel plates at about 858 K [41, 42], see also [43 to 45].

UAl_3 — Al fuel plates show a lower operation performance compared to UAl_2 — Al fuel plates [46], see also [6, 47], and the swelling or blistering behavior seems to be better than that of uranium oxide — aluminium dispersions [4, 17, 48].

References for 3.1.4.8:

[1] Nazare, S.; Ondracek, G.; Thümmler, F. (KFK-1252 [1970] 1/83; C.A. **75** [1971] No. 70285).
[2] Thümmler, F.; Nazare, S.; Ondracek, G. (Powder Met. **10** [1967] 264/87).
[3] Thümmler, F.; Lilienthal, H. E.; Nazare, S. (Powder Met. **12** [1969] 1/22).
[4] Walker, V. A.; Graber, M. J., Jr.; Gibson, G. W. (IDO-17157 [1966] 1/110; C.A. **66** [1967] No. 81543).
[5] Gibson, G. W.; Zukor, M. (IDO-17154 [1966] 3/6; N.S.A. **20** [1966] No. 19178).
[6] Rice, W. L. R. (TID-11295, 3rd Ed. [1964] 105/20; N.S.A. **18** [1964] No. 41933).
[7] Graber, M. J.; Gibson, G. W.; Walker, V. A.; Francis, W. C. (IDO-16958 [1964] 1/53; C.A. **61** [1964] 5161).
[8] Sumpter, K. C.; Crandall, J. K; Gibson, G. W. (IN-1131 [1968] 8/18; N.S.A. **22** [1968] No. 23761).
[9] Saller, H. A. (in: Finneston, H. M.; Howe, J. P., Progress in Nuclear Energy, Ser. 5, Metallurgy and Fuels, Vol. 1, Pergamon, London 1956, pp. 535/43).
[10] Saller, H. A. (U.S. 2917383 [1959]; C.A. **1960** 18313).

[11] Snyder, M. J.; Duckworth, W. H. (BMI-1223 [1957] 1/35; C.A. **1958** 3489).
[12] Saller, H. A. (U.S. 2917383 [1959]; N.S.A. **14** [1960] No. 7757).
[13] Boucher, R. (J. Nucl. Mater. **1** [1959] 13/27).
[14] Thurber, W. C.; Erwin, J. H.; Beaver, R. J. (ORNL-2351 [1958] 1/35; N.S.A. **12** [1958] No. 6613).
[15] Gualandi, D.; Schileo, G. (Alluminio **28** [1959] 449/55).
[16] Petzow, G.; Exner, H. E. (BMwF-FBK-67-88 [1967] 1/69; C.A. **69** [1968] No. 38273).
[17] Gregg, J. L.; Crouse, R. S.; Werner, W. J. (ORNL-4056 [1967] 1/14; C.A. **67** [1967] No. 49445).
[18] Zelezny, W. F. (IDO-17154 [1966] 56/73; N.S.A. **20** [1966] No. 19178).
[19] Gibson, G. W. (IN-1133 [1967] 1/82; C.A. **69** [1968] No. 40509).

[20] Gibson, G. W.; Graber, M. J.; Francis, W. C. (IDO-16934 [1963] 1/7; N.S.A. **18** [1964] No. 8723).

[21] Pearce, R. J. (J. Nucl. Mater. **34** [1970] 151/9).
[22] Graber, M. J.; Marchbanks, M. F. (IN-1131 [1968] 43/5; N.S.A. **22** [1968] No. 23843).
[23] Nazare, S.; Ondracek, G.; Thümmler, F. (J. Nucl. Mater. **56** [1975] 251/9).
[24] Rand, M. H.; Kubaschewski, O. (The Thermochemical Properties of Uranium Compounds, Oliver & Boyd, Edinburgh — London 1963).
[25] Castleman, L. S. (J. Nucl. Mater. **3** [1961] 1/15).
[26] Lee, L. P.; Leighly, H. P., Jr. (Met. Trans. **6** [1975] 135/9).
[27] Thurber, W. C.; Beaver, R. J. (ORNL-2602 [1959] 1/53; C.A. **1959** 15797).
[28] Gualandi, D.; Paganelli, M.; Schileo, G. (Alluminio **28** [1959] 507/15).
[29] Exner, H. E.; Petzow, G. (5th Intern. Leichtmetalltagung, Leoben 1968 [1969], pp. 77/83; C.A. **74** [1971] No. 37351).
[30] Nazare, S.; Ondracek, G. (Prakt. Metallog. **6** [1969] 742/6; C.A. **72** [1970] No. 50161).

[31] Hünlich, A.; Schreibmaier, J.; Spieler, K. (Prakt. Metallog. **9** [1972] 227/9; C.A. **77** [1972] No. 12844).
[32] Dienst, W.; Nazare, S.; Thümmler, F. (J. Nucl. Mater. **64** [1977] 1/13).
[33] Dienst, W.; Gausmann, G. (KFK-2199 [1975] 1/44; C.A. **85** [1976] No. 38424).
[34] Dienst, W.; Nazare, S.; Thümmler, F. (At. Tekhn. Rubezhom **1977** 32/42; C.A. **88** [1978] No. 199553).
[35] Gibson, G. W.; Graber, M. J.; Francis, W. C. (IDO-16934 [1963] 21/6; N.S.A. **18** [1964] No. 8723).
[36] Graber, M. J.; Zelezny, W. F.; Moen, R. A. (IDO-17154 [1966] 7/28; N.S.A. **20** [1966] No. 19178).
[37] Gibson, G. W.; de Boisblanc, D. R. (Proc. 2nd Intern. Powder Met. Conf., New York 1965 [1966], Vol. 3, pp. 26/35; C.A. **67** [1967] No. 17002).
[38] Gibson, G. W.; Graber, M. J.; Francis, W. C. (IDO-16934 [1963] 8/10; N.S.A. **18** [1964] No. 8723).
[39] Hofman, G. L. (Nucl. Technol. **77** [1987] 110/5).
[40] Leitten, C. F., Jr.; Richt, A. E.; Beaver, R. J. (ORNL-3470 [1963] 87/8; N.S.A. **18** [1964] No. 4173).

[41] Francis, W. C. (IN-1437 [1970] 1/75; N.S.A. **25** [1971] No. 11458).
[42] Beeston, J. M.; Hobbins, R. R.; Gibson, G. W.; Francis, W. C. (Nucl. Technol. **49** [1980] 136/49).
[43] Shibata, T.; Kanda, K.; Mishima, K.; Tamai, T.; Hayashi, M.; Snelgrove, J. L.; Stahl, D.; Matos, J. E.; Travelli, A.; et al. (Proc. Intern. Meeting Res. Test React. Core Convers. HEU LEU Fuels, Argonne, Ill., 1982 [1983], pp. 99/116; C.A. **101** [1984] No. 80371, Nucl. Sci. Eng. **87** [1984] 405/17; C.A. **101** [1984] 118746).
[44] Hofman, G. L. (J. Nucl. Mater. **140** [1986] 156/63).
[45] Zelezny, W. F.; Gibson, G. W.; Graber, M. J. (3rd Natl. Symp. Develop. Irradiat. Test. Technol. Papers, Sandusky, Ohio, 1969, pp. 516/32; C.A. **74** [1971] No. 149748).
[46] Miller, L. G.; Beeston, J. M. (EGG-FT-5273 [1980] 1/74; C.A. **95** [1981] No. 14694).
[47] Miller, L. G.; Brown, K. R.; Beeston, J. M.; McGinty, D. M. (EGG-SE-6464 [1983] 1/15; C.A. **101** [1984] No. 13654).
[48] Graber, M. J.; Gibson, G. W.; Walker, V. A.; Francis, W. C. (IDO-16958 [1964] 1/53; C.A. **61** [1964] 5161).

3.1.5   Uranium Tetraaluminide, UAl$_4$

3.1.5.1   Formation and Preparation

Preparation

UAl$_4$ was produced by a eutectic reaction of UAl$_3$ with a uranium – aluminium alloy containing 18 wt% uranium [1, 2]. Pure UAl$_4$ was obtained by separation from melts of 25 wt% uranium – 75 wt% aluminium alloys. The melt was cooled from 950 to 650°C at a rate of 500°C/h and then air-cooled to room temperature. Filings from the ingot were treated with a solution of NaOH, from which a black powder was separated containing 64.2 to 66.34 wt% uranium, corresponding to a stoichiometry ranging from UAl$_{4.5}$ to UAl$_{4.9}$. From crystallographic considerations, this UAl$_{4+x}$ compound is interpreted as UAl$_4$ with some uranium sites unoccupied or some occupied by aluminium atoms [3]. But, the homogeneity range of UAl$_4$ is possibly much smaller, as observed with alloys ranging from 63.5 to 68.8 wt% uranium after annealing at 700°C for 700 h in the region, liquid + UAl$_4$ [4].

The interdiffusion of uranium and aluminium powders proceeds rapidly at 620°C under vacuum with a very obvious exothermic reaction. The pyrophoric compounds are brittle, thus they can easily be crushed [5]. This process can be better controlled with compacted powders using extrusion techniques at elevated temperatures [5].

Melting of uranium and aluminium at 100°C above the alloy liquidus, stirring with a graphite rod and casting into graphite molds, gave good results only with uranium contents up to 58 wt% [6], see also [7]. Arc melting of stoichiometric amounts of uranium and aluminium resulted in alloys with different phases [8], otherwise UAl$_4$ was successfully prepared by arc melting under argon atmosphere [9]. The separation of UAl$_4$ with selective dissolution by NaOH was unsuccessful [2], see also [10]. Arc melting was predominantly used for the preparation of UAl$_2$ and UAl$_3$ (see pp. 88 and 118) [11]. Induction melting and subsequent annealing at 720°C, or less than 800°C, led to products of restricted UAl$_4$ content with 81 wt% UAl$_4$ [10], or 70% by area of primary crystals of UAl$_4$ in a matrix of UAl$_4$ – Al eutectic [2], with stoichiometries ranging from UAl$_{4.0}$ to UAl$_{4.8}$ [10]. For a schematic diagram of preparation and chemical analysis see **Fig. 3-37**, p. 140.

Vacuum induction melting, using graphite crucibles coated with magnesium zirconate, was successful but it was complicated by loss of uranium soaking into the crucible and loss of aluminium by evaporation. The pick-up of carbon ranges from 0.1 to 0.4 wt%, metallurgical recoveries yielded 85 to 90% [8], see also [12]. Segregation occurred during melting due to the large differences in density of aluminium (2.7 g/cm^3) and UAl$_4$ (5.7 g/cm^3). Alloys containing larger amounts of UAl$_4$ were achieved with a centrifugal furnace at temperatures at which the eutectic was molten while the UAl$_4$ was solid. The principals of the centrifugal furnace are shown in **Fig. 3-38**, p. 140. Optimum melts were obtained with a speed of rotation of about 3000 rev/min (maximum value) at temperatures of 700 to 900°C and times of up to 20 min, yielding 96 wt% or 90% UAl$_4$ by area [2, 13].

The reaction of uranium hydride and aluminium led to compounds with high UAl$_4$ content. A sharp jump in temperature indicated the occurrence of an exothermic reaction at 600 to 615°C. A monophase compound with UAl$_4$ as the only constituent (X-ray diffraction determination) was obtained only for the composition of UAl$_{4.9}$ [14 to 16], see also [10, 17].

Very pure UAl$_4$ was prepared by high-temperature interdiffusion of aluminium pieces and dispersed uranium obtained by decomposition of uranium hydride. Uranium and the aluminium metal were placed in stoichiometric amount in a beryllium oxide crucible loaded in a quartz ampoule. The crucible and the quartz tube were degassed before loading by heating in vacuum (10^{-2} Torr) at 900°C for 1.5 h. After loading, the ampoule was filled with hydrogen

   References for 3.1.5.1 on pp. 142/3

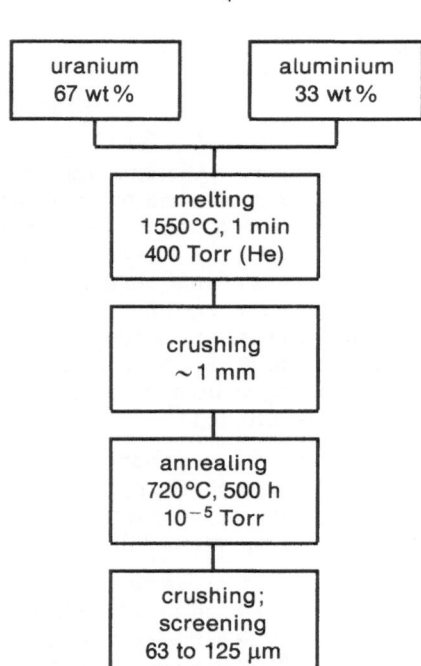

Fig. 3-37. Schematic diagram of the preparation of UAl$_4$ powder [10].

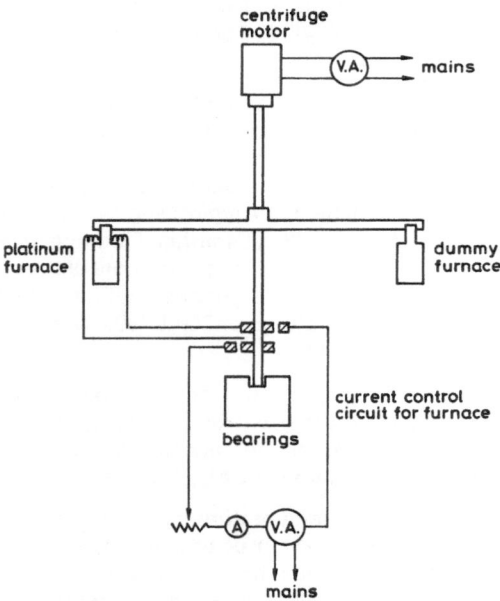

Fig. 3-38. Centrifugal furnace for the preparation of uranium aluminides [2].

and heated to 250 to 270°C to convert the uranium to UH_3, which was decomposed at 600°C. The hydrogen evolved was not completely removed at this stage, therefore the ampoule was heated in an electric furnace with argon at 1000°C for 3 h. The hydrogen pressure reached 400 Torr. After further heating to 1100°C for 11 h, the ampoule was cooled to 600°C, the hydrogen pumped off, and then the sample was cooled to room temperature. The friable product was unloaded under argon, ground in a jasperite mortar, and again heated under hydrogen at 1300°C for 1.5 h. After cooling to 600°C the hydrogen was completely removed. The chemical analysis of the UAl_4 powder resulted in 31.02 ± 0.04 wt% aluminium (31.18 wt% theoretical), the stoichiometry yielded $UAl_{3.970}$ [18].

A four-step fluidized-bed technique was developed for manufacturing uranium – aluminium compounds, especially UAl_3 and UAl_4, starting with a $UO_2(NO_3)_2$ solution, and aluminium powder [19 to 22], see also [17, 23, 24]. A flow sheet for the fluidized-bed production is given in Fig. 3-28, p. 120. The four steps of the process are summarized by the following chemical reactions [19, 20, 25]:

$$UO_2(NO_3)_2 \text{ (l)} \xrightarrow{400°C} UO_3 \text{ (s)} + 2\,NO_2 \text{ (g)} + \tfrac{1}{2}\,O_2 \text{ (g)}$$

$$UO_3 \text{ (s)} + CH_3OH \text{ (g)} \xrightarrow{360°C} UO_2 \text{ (s)} + COH_2 \text{ (g)} + H_2O \text{ (g)}$$

$$UO_2 \text{ (s)} + CCl_4 \text{ (g)} \xrightarrow{Cl_2 \text{ (g)},\ 360°C} UCl_4 \text{ (s)} + CO_2 \text{ (g)}$$

$$3\,UCl_4 \text{ (s)} + (4 + 3x)\,Al \text{ (l)} \xrightarrow{680°C} 3\,UAl_x \text{ (s)} + 4\,AlCl_3 \text{ (g)}$$

During the first reaction step, uranyl nitrate solution was sprayed into a fluidized bed of aluminium powder (200 to 250 mesh in size, generally spherical) at a temperature of 400°C to form a layer of UO_3 on the aluminium particles using air as the fluidizing gas. In the second step, the coating of UO_3 was reduced by methanol to UO_2 at 360°C, with argon as the fluidizing gas. The reaction proceeded rapidly when the methanol was injected and vaporized into the fluidizing argon upstream. During the third step, the UO_2 coating was chlorinated by a mixture of 75 mol% CCl_4 and 25 mol% Cl_2 at 330°C injected into the fluidizing argon stream to form UCl_4 layers on the aluminium particles. In the fourth step, the reaction temperature was raised to 660°C (melting point of aluminium) at which the reaction to UAl_x proceeded very rapidly. In between, most of the UCl_4 was reduced to UCl_3 at about 500°C by the reaction: $3\,UCl_4 + Al \rightarrow 3\,UCl_3 + AlCl_3$. Any small amounts of UO_2 that was not converted to UCl_4 formed Al_2O_3 and UAl_x by the reaction: $3\,UO_2 + (4 + 3x)\,Al \rightarrow 2\,Al_2O_3 + 3\,UAl_x$. Batches of UAl_4 were produced by this process. After washing in an ultrasonic water bath, chemical analysis resulted in less than 0.5 wt% C, 1.0 wt% Al_2O_3, and 15 ppm chloride [19, 20, 22].

Violent exothermic reaction with ignition was observed when heating compacts of U_3O_8 and aluminium powder in stoichiometric amount above the melting point of aluminium. The ignition temperature depends on the U_3O_8 particle size (see Table 3/4, p. 90). No indication was found of UAl_4 or of residual U_3O_8 or aluminium [26, 27], see also [28, 29].

Single Crystals

Single crystals of UAl_4 were obtained from a melt of 25 wt% uranium – aluminium after slowly cooling to 650°C at 50°C per h and then cooling in air, and separating the UAl_4 crystals from filings by leaching with NaOH. The single crystals were isolated as small black needles [3]. Single crystals of about 1 mm were grown from melts of appropriate mixtures of uranium and aluminium, cooled at a rate of 2°C per h in a gradient of 10°C/cm. The orthorhombic UAl_4 crystals were prismatic in shape with (110) faces dominant [30].

Enthalpy, Entropy, and Gibbs Free Energy of Formation

Enthalpy of formation, ΔH_f° (UAl$_4$), measured by various methods: From adiabatic calorimetric measurements in aqueous solutions: ΔH_f° (298 K) = -31.1 ± 2.6 kcal/g-UAl$_{3.970}$ [18]; or 31.2 ± 3.1 kcal/g-UAl$_4$ [18], see also [31, 32]; 6.2 kcal/g-atom [33 from 18]. From adiabatic calorimetry: ΔH_f° (298 K) = -29.8 ± 2.0 kcal/mol [34]. The formation of UAl$_4$ from α uranium and solid aluminium at 900 K, emf measurements: ΔH_f (900 K) = -5.94 kcal/g-atom [35]. The entropy of formation was calculated from emf measurements to be ΔS_f (900 K) = -0.86 e. u./g-atom [35].

Emf measurements within the temperature range of 401 to 639°C yield the relation: ΔG° = $-26040 + 1.121 \cdot T$ (401 to 639°C). These data, combined with the enthalpy of formation (ΔH_f° (298 K) = -29.8 ± 2.0 kcal/mol), yield the relations ΔG° = $-32040 - 7.52 \cdot T \cdot \ln T + 59.07$ (in cal/mol); ΔH° = $-32040 + 7.52 \cdot T$ (in cal/mol); ΔS° = $-51.5 + 7.52 \cdot \ln T$ (in cal $\cdot$ mol^{-1} $\cdot$ K^{-1}) [34].

Partial thermodynamic values of the formation of UAl$_4$ from α uranium and solid aluminium at 900 K obtained from emf measurements are: $\Delta \overline{H}_U$ = -29.7 kcal/g-atom, $\Delta \overline{S}_U$ = -4.3 e. u./g-atom, and $\Delta \overline{G}_U$ = -25.83 kcal/g-atom [35].

The activity coefficient of uranium in liquid aluminium is 9.7×10^{-4} at 686°C [240], see also [32]. Together with the solubility of uranium in aluminium of about 2.0 at% [36] the free energy of formation at 686°C is approximately -20.8 kcal/mol [32]. Combined with the calorimetric data for the enthalpy, the following equation was obtained: ΔG_f° = $-41.5 \pm 21.6 \times 10^{-3} \cdot T$ (in kcal/mol) [32].

References for 3.1.5.1:

[1] Saller, H. A. (in: Finneston, H. M.; Howe, J. P.; Progress in Nuclear Energy, Ser. 5, Metallurgy and Fuels, Vol. 1, Pergamon, London 1956, pp. 535/43).

[2] Openshaw, P. R.; Shreir, L. L. (Corros. Sci. **3** [1963] 217/37).

[3] Borie, B. S., Jr. (J. Metals **3** [1951] 800/2; ORNL-810 [1950] 1/14).

[4] Boucher, R. (J. Nucl. Mater. **1** [1959] 13/27).

[5] Montagne, R.; Meny, L. (TID-7546 [1958] 142/56; C.A. **1958** 9904).

[6] Jones, T. I.; Street, K. N.; Scoberg, J. A.; Baird, J. (Can. Met. Quart. **2** No. 1 [1963] 53/72; C.A. **58** [1963] 12254).

[7] Colombi, G.; Gabaglio, M.; Liscia, A.; Trittoni, E. (Met. Ital. **22** [1967] 359/63).

[8] Gibson, G. W.; de Boisblanc, D. R. (Proc. 2nd Intern. Powder Met. Conf., New York 1965 [1966], Vol. 3, pp. 26/35; C.A. **67** [1967] No. 17002).

[9] Badaeva, T. A.; Dashevskaya, L. I. (Fiz. Khim. Splavov Tugoplavkikh Soedin Toriem Uranom **1968** 107/13; C.A. **71** [1969] No. 24297).

[10] Jesse, A.; Ondracek, G.; Thümmler, F. (Powder Met. **14** [1971] 289/97).

[11] Petzow, G.; Steeb, S., Kiessler, G. (J. Nucl. Mater. **12** [1964] 271/6).

[12] Gibson, G. W.; Graber, M. J.; Francis, W. C. (IDO-16934 [1963] 1/7; N.S.A. **18** [1964] No. 8723).

[13] Openshaw, P. R.; Shreir, L. L. (Corros. Sci. **4** [1964] 335/44).

[14] Gibson, G. W.; Zukor, M. (IDO-17154 [1966] 3/6; N.S.A. **20** [1966] No. 19178).

[15] Eding, H. J.; Carr, E. M. (ANL-6339 [1961] 1/39; C.A. **1961** 16243).

[16] Werner, W. J.; McIlwain, M. C.; Hammond, J. P. (U.S. 3288571 [1966]; C.A. **66** [1967] No. 100903).

[17] Thümmler, F.; Lilienthal, H. E.; Nazare, S. (Powder Met. **12** [1969] 1/22).

[18] Ivanov, M. I.; Tumbakov, V. A.; Podol'skaya, N. S. (At. Energiya SSSR **5** [1958] 166/70; Soviet J. At. Energy **5** [1958] 1007/11; C.A. **1960** 9481).

[19] Grimmet, E. S.; Ballard, R. K.; Buckham, J. A. (Chem. Eng. Progr. Symp. Ser. **63** No. 80 [1967] 11/5; C.A. **68** [1968] No. 55712).

[20] Hogg, G. W.; Grimmet, E. S. (AIChE [Am. Inst. Chem. Eng.] Symp. Ser. **67** No. 116 [1971] 190/8; C.A. **77** [1972] No. 22291).

[21] Grimmet, E. S.; Ballard, R. K. (in: Bower, J. R., IN-087 [1967] 112/6; N.S.A. **22** [1968] No. 4231).

[22] Grimmet, E. S. (U.S. 3318670 [1967]; C.A. **67** [1967] No. 49755).

[23] Nazare, S.; Ondracek, G.; Thümmler, F. (KFK-1252 [1970] 1/83; C.A. **75** [1971] No. 70285).

[24] Morin, C. (Ger. Offen. 2616828 [1976]; C.A. **87** [1977] No. 119752).

[25] Hogg, G. W. (IN-1422 [1970] 1/29; C.A. **75** [1971] No. 83063).

[26] Fleming, J. D.; Johnson, J. W. (Nucleonics **21** No. 5 [1963] 84/7).

[27] Fleming, J. D.; Johnson, J. W. (TID-7642 [1962] 649/66; C.A. **59** [1963] 9539).

[28] Ondracek, G.; Patrassi, E. (Ber. Deut. Keram. Ges. **45** [1968] 617/21).

[29] Dykstra, L. J. (GA-1479 [1960] 1/21; C.A. **1961** 13975).

[30] McBride, J. J.; Clark, G. W. (ORNL-3470 [1963] 4; N.S.A. **18** [1964] No. 4171).

[31] Rand, M. H.; Kubaschewski, O. (The Thermochemical Properties of Uranium Compounds, Oliver & Boyd, Edinburgh — London 1963).

[32] Johnson, I. (Met. Soc. AIME Inst. Metals Div. Spec. Rept. Ser. **10** No. 13 [1964] 171/92).

[33] Katz, S. (J. Nucl. Mater. **6** [1962] 172/81).

[34] Chiotti, P.; Kateley, J. A. (J. Nucl. Mater. **32** [1969] 135/45).

[35] Lebedev, V. A.; Sal'nikov, V. I.; Nichkov, I. F.; Raspopin, S. P. (At. Energiya SSSR **32** [1972] 115/8; Soviet At. Energy **32** [1972] 129/32; C.A. **76** [1972] No. 145738).

[36] Gordon, P.; Kaufmann, A. R. (J. Metals **2** [1950] 182/4).

3.1.5.2 Crystallographic Properties

UAl_4 exists in a wide homogeneity range, from $UAl_{4.0}$ to $UAl_{4.8}$ [1], see also [2, 3], owing to a lattice defect structure with unoccupied uranium positions. It was originally referred to as UAl_5 [2].

UAl_4 crystallizes in a body-centered orthorhombic structure with four molecules per unit cell. The space group is $C_{2v}^{22} - I2ma$ (No. 46) or $D_{2h}^{28} - Imma$ (No. 74) [3], see also [4, 5]. The measured lattice parameters are given in Table 3/16. The calculated X-ray density is D_{calc} (in g/cm^3) = 6.5 (based on UAl_5) [3], 6.06 [6], also reported in [7 to 10], or 6.10 (based on UAl_4) [7]. The measured densities are significantly lower: D_{meas} (in g/cm^3) = 5.7 $\pm$ 0.3 [3], see also [4, 7, 11, 12] or 5.6 $\pm$ 0.1 [7, 10].

Table 3/16
Measured Lattice Constants of UAl_4.

lattice parameter in Å	preparation	Ref.	also reported in
a = 4.41 ± 0.02 b = 6.27 ± 0.02 c = 13.71 ± 0.03	separation from melt	[3]	[4, 5, 7, 10 to 12, 14, 16 to 23]

References for 3.1.5.2 on pp. 145/6

Table 3/16 (continued)

lattice parameter in Å	preparation	Ref.	also reported in
a = 4.39 b = 6.25 c = 13.72	solid state reaction	[13]	[5]
a = 4.384 ± 0.001 b = 6.241 ± 0.001 c = 13.670 ± 0.001	UAl$_{3.970}$, solid state reaction, with traces of eutectics between UAl$_4$ and Al	[14]	[24]
a = 4.397 ± 0.001 b = 6.251 ± 0.001 c = 13.714 ± 0.002	dissolution of U in molten Al	[7]	[1, 10]
a = 4.42 b = 6.30 c = 13.63	argon arc melting	[15]	

The atomic positions of UAl$_4$ concerning the space group Imma are [3] (origin at center $2/m2_1 1$ [17]):

4 U in (e): $0, 1/4, z$; $0, 3/4, \bar{z}$; with $z/c = 0.111$
4 Al in (e): $0, 1/4, z$; $0, 3/4, z$; with $z/c = -0.111$
4 Al in (b): $0, 0, 1/2$; $0, 1/2, 1/2$
8 Al in (h): $0, y, z$; $0, \bar{y}, \bar{z}$; $0, 1/2 + y, \bar{z}$; $0, 1/2 - y, z$; with $y/b = -0.033$, $z/c = 0.314$

The atomic distances are summarized in Table 3/17.

Table 3/17
Interatomic Distances of UAl$_4$ [3].

atom	number of neighbors	kind of neighbors	distance in Å	atom	number of neighbors	kind of neighbors	distance in Å
U (4e)	3	Al (4e)	3.14	Al (8h)	2	U (4e)	3.02
	4	Al (4b)	3.10		1	U (4e)	3.30
	4	Al (8h)	3.02		2	Al (4e)	2.79
	2	Al (8h)	3.30		1	Al (4e)	3.09
Al (4e)	3	U (4e)	3.14		1	Al (4b)	2.57
	4	Al (4b)	3.10		2	Al (8h)	2.81
	4	Al (8h)	2.79		1	Al (8h)	3.56
	2	Al (8h)	3.09		1	Al (8h)	2.72
Al (4b)	4	U (4e)	3.10				
	4	Al (4e)	3.10				
	2	Al (4b)	3.14				
	2	Al (8h)	2.57				

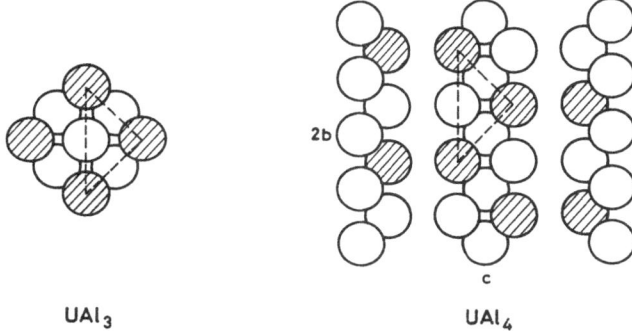

UAl$_3$ UAl$_4$

Fig. 3-39. Comparison of two unit cells of UAl$_4$, with the eight-fold Al atoms omitted, and a unit cell of UAl$_3$. Analogous units in the two structures are shown with dotted lines. Uranium atoms are shown in shadowed circles [3].

The structure of UAl$_4$ may be thought of as UAl$_3$ plates held together by extra aluminium atoms [10, 17], see **Fig. 3-39**. The morphology of the UAl$_4$ crystals results from the peritectic reaction of liquid + UAl$_3 \rightarrow$ UAl$_4$, to show a "fish-tail" form, which is the result of a shear and diffusion controlled transformation, whereas (001) (UAl$_3$) plates undergo the shear mechanism (001) [11$\bar{1}$] or (001) [1$\bar{1}$1] for a distance of 1/2 [111] [25]. A theoretical investigation led to a homogeneous matrix operator combining rotation, coordinate change, slip, and distortion, which describes the relationship between the crystallographic directions and plane normals for the UAl$_3$ and UAl$_4$ crystal lattices [25].

References for 3.1.5.2:

[1] Jesse, A. (J. Nucl. Mater. **37** [1970] 340/2).

[2] Gordon, P.; Kaufmann, A. R. (J. Metals **2** [1950] 182/4).

[3] Borie, B. S., Jr. (J. Metals **3** [1951] 800/2; ORNL-810 [1950] 1/14).

[4] Kiessling, R. (Proc. 1st Intern. Conf. Peaceful Uses At. Energy, Geneva 1955, Vol. 9, pp. 69/73).

[5] Cesoni, G.; Gualandi, D.; Paganelli, M.; Schileo, G. (Proc. 2nd Intern. Conf. Peaceful Uses At. Energy, Geneva 1958, Vol. 6, pp. 451/62).

[6] Cunningham, J. E.; Beaver, R. J.; Thurber, W. C.; Waugh, R. C. (TID-7546 [1958] 269/97; C.A. **1958** 9905).

[7] Runnalls, O. J. C.; Boucher, R. R. (Trans. AIME **233** [1965] 1726/32).

[8] Gibson, G. W.; de Boisblanc, D. R. (Proc. 2nd Intern. Powder Met. Conf., New York 1965 [1966], Vol. 3, pp. 26/35; C.A. **67** [1967] No. 17002).

[9] Grimmet, E. S.; Ballard, R. K.; Buckham, J. A. (Chem. Eng. Progr. Symp. Ser. **63** No. 80 [1967] 11/5; C.A. **68** [1968] No. 55712).

[10] Rice, W. L. R. (TID-11295, 3rd Ed. [1964] 105/20; N.S.A. **18** [1964] No. 41933).

[11] Rough, F. A.; Bauer, A. A. (BMI-1300 [1958] 1/138; N.S.A. **12** [1958] No. 13935).

[12] Ambrosio, F. F. (INIS-mfn-1356 [1973] 1/64; C.A. **82** [1975] No. 114937).

[13] Gualandi, D.; Paganelli, M.; Schileo, G. (Alluminio **28** [1959] 507/15).

[14] Ivanov, M. I.; Tumbakov, V. A.; Podol'skaya, N. S. (At. Energiya SSSR **5** [1958] 166/70; Soviet J. At. Energy **5** [1958] 1007/11; C.A. **1960** 9481).

[15] Badaeva, T. A.; Dashevskaya, L. I. (Fiz. Khim. Splavov Tugoplavkikh Soedin Toriem Uranom **1968** 107/13; C.A. **71** [1969] No. 24297).

[16] Konobrevskii, S. T.; Zaimovskii, A. S.; Levitsky, B. M.; Sokursky, Y. N.; Chebotarev, N. T.; Bobkov, Y. V.; Egorov, P. P.; Nikolaev, G. N.; Ivanov, A. A. (Proc. 2nd Intern. Conf. Peaceful Uses At. Energy, Geneva 1958, Vol. 6, pp. 194/203).

[17] Lam, D. J.; Darby, J. B., Jr.; Nevitt, N. V. (in: Freeman, A. J.; Darby, J. B., Jr., The Actinides: Electronic Structure and Related Properties, Vol. 2, Academic, New York — San Francisco — London 1974, pp. 119/84).

[18] Wright, E. H.; Willey, L. A. (Tech. Paper Alcoa Res. Lab. No. 15 [1960] 1/46; C.A. **61** [1964] 9243).

[19] Hansen, M. (Constitution of Binary Alloys, McGraw-Hill, New York — Toronto — London 1958, pp. 143/4).

[20] Lee, L. P.; Leighly, H. P., Jr. (Met. Trans. **6** [1975] 135/9).

[21] Gentile, E. F. (Met. ABM [Assoc. Brasil. Metais] **24** No. 124 [1968] 187/92; C.A. **69** [1968] No. 109169).

[22] Casteels, F.; Diels, P.; Cools, A. (J. Nucl. Mater. **24** [1967] 87/94).

[23] Bramfitt, B. L.; Leighly, H. P., Jr. (Metallography **1** [1968] 165/93).

[24] Elliott, R. P. (Constitution of Binary Alloys, 1st Suppl., McGraw-Hill, New York — Toronto — London 1965, pp. 61/2).

[25] Leighly, H. P., Jr.; Edwards, D. R. (Met. Trans. A **12** [1981] 491/5).

3.1.5.3 Mechanical Properties

Microhardness

The microhardness of UAl$_4$ is given in **Fig. 3-40** as a function of load. Further values for the microhardness are (in kg/mm^2): 260 (50 g load, 15 s) [1], 364 (60 g load) [2], see also [3, 4], 391 (average value) [5], 400 (65 g load) [6], 388 (115 g load) [6], 382 (135 g load) [6], 380 (50 p load) for UAl$_4$ with 67 wt% uranium [7], 340 [8], and 322 (60 p load) for UAl$_4$ as a product from reaction of UO$_2$ with aluminium [9].

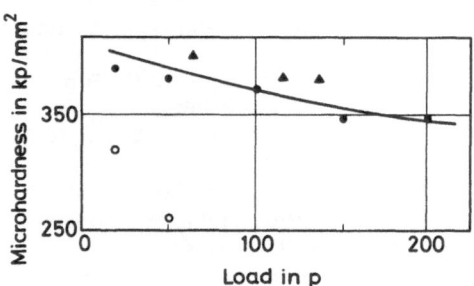

Fig. 3-40. Microhardness as a function of load [10]. ▲: values from [6]; ○: values from [1].

References for 3.1.5.3:

[1] Kiessling, R. (Proc. 1st Intern. Conf. Peaceful Uses At. Energy, Geneva 1955, Vol. 9, pp. 69/73).

[2] Badaeva, T. A.; Dashevskaya, L. I. (Fiz. Khim. Splavov Tugoplavkikh Soedin Toriem Uranom **1968** 107/13; C.A. **71** [1969] No. 24297).

[3] Pav, T.; Otruba, J.; Kocik, J.; Saxl, I. (Reaktor. Materialoved. Tr. Konf. Reaktor. Materialoved., Alushta 1978, Vol. 4, pp. 313/22; C.A. **91** [1979] No. 11067).

[4] Gomozov, L. I.; Ivanov, O. S. (Reaktor. Materialoved. Tr. Konf. Reaktor. Materialoved., Alushta 1978, pp. 4/31; C.A. **91** [1979] No. 29285).

[5] Boucher, R. (Symp. Solid State Diffusion, Saclay, France, 1958, pp. 111/7; N.S.A. **15** [1961] No. 26564; AEC-tr-4768 [1960] 1/11; N.S.A. **14** [1960] No. 7826).

[6] Boucher, R. (J. Nucl. Mater. **1** [1959] 13/27).

[7] Jesse, A. (J. Nucl. Mater. **37** [1970] 340/2).

[8] Stone, H. E. N. (J. Mater. Sci. **10** [1975] 923/34).

[9] Ondracek, G.; Patrassi, E. (Ber. Deut. Keram. Ges. **45** [1968] 617/21).

[10] Nazare, S.; Ondracek, G.; Thümmler, F. (J. Nucl. Mater. **56** [1975] 251/9).

3.1.5.4 Thermal Properties

Thermal Expansion

The thermal expansion coefficient, α, of UAl_4, within the temperature range of 20 to 500°C, is $\alpha = 1.65 \times 10^{-5}\,K^{-1}$ [1].

Melting Point

UAl_4 decomposes peritectically to UAl_3 and aluminium at 730°C (stoichiometry of $UAl_{4.9}$) [2, 3], see also [4 to 6]; 732°C [7], see also [8], or at 750°C [9], see also [10, 11 (as 1350°F)]. Different values were observed for the peritectic decomposition and formation at a sample containing 40 wt% uranium: 745°C (on heating) [12], see also [1, 13] and 712°C (on cooling) [12].

Thermal Conductivity

The thermal conductivity of UAl_4 was estimated by extrapolating values of uranium − aluminium alloys up to 58 wt% uranium and pure uranium to be $0.02 \pm 0.01\,cal \cdot s^{-1} \cdot cm^{-2} \cdot K^{-1}$ (at 65°C) [14], see also [1], $0.72\,W \cdot cm^{-1} \cdot K^{-1}$ (at 200°C) [5], see also [1], or $0.16\,W \cdot cm^{-1} \cdot K^{-1}$ (UAl_4 with 67 wt% U) at 93°C [1].

References for 3.1.5.4:

[1] Jesse, A. (J. Nucl. Mater. **37** [1970] 340/2).

[2] Gordon, P.; Kaufmann, A. R. (J. Metals **2** [1950] 182/4).

[3] Kiessling, R. (Proc. 1st Intern. Conf. Peaceful Uses At. Energy, Geneva 1955, Vol. 9, pp. 69/73).

[4] Grimmet, E. S.; Ballard, R. K.; Buckham, J. A. (Chem. Eng. Progr. Symp. Ser. **63** No. 80 [1967] 11/5; C.A. **68** [1968] No. 55712).

[5] Rice, W. L. R. (TID-11295, 3rd Ed. [1964] 105/20; N.S.A. **18** [1964] No. 41933).

[6] Openshaw, P. R.; Shreir, L. L. (Corros. Sci. **3** [1963] 217/37).

[7] Borie, B. S., Jr. (J. Metals **3** [1951] 800/2; ORNL-810 [1950] 1/14).

[8] Wright, E. H.; Willey, L. A. (Tech. Paper Alcoa Res. Lab. No. 15 [1960] 1/46; C.A. **61** [1964] 9243).

[9] Cunningham, J. E.; Beaver, R. J.; Thurber, W. C.; Waugh, R. C. (TID-7546 [1958] 269/97; C.A. **1958** 9905).

[10] Boucher, R. (Symp. Solid State Diffusion, Saclay, France, 1958, pp. 111/7; N.S.A. **15** [1961] No. 26564; AEC-tr-4768 [1960] 1/11; N.S.A. **14** [1960] No. 7826).

[11] Gibson, G. W.; de Boisblanc, D. R. (Proc. 2nd Intern. Powder Met. Conf., New York 1965 [1966], Vol. 3, pp. 26/35; C.A. **67** [1967] No. 17002).

[12] Abramson, R.; Boucher, R.; Fabre, R.; Monti, H. (Proc. 2nd Intern. Conf. Peaceful Uses At. Energy, Geneva 1958, Vol. 6, pp. 174/83).

[13] Jesse, A.; Ondracek, G.; Thümmler, F. (Powder Met. **14** [1971] 289/97).

[14] Jones, T. I.; Street, K. N.; Scoberg, J. A.; Baird, J. (Can. Met. Quart. **2** No. 1 [1963] 53/72; C.A. **58** [1963] 12254).

3.1.5.5 Electrical Properties

Electronic Structure

The electronic structure of UAl$_4$ is characterized by a mixture of a narrow 5f band, a broad 6d band from uranium, and an aluminium 3p band, to form a band similar to a metallic band structure. This feature results from molecular orbital calculations based on the Hartree-Fock-Slater one-electron theory. From Mulliken population analysis, the bonding in the UAl$_4$ clusters is given by interactions between the uranium 6d band and the aluminium 3s and 3p orbitals, whereas the 5f electrons hardly contribute to bonding. The calculated bond length is 3.238 Å [1, 2].

The local density-of-states (DOS) is shown in **Fig.** 3-**41** with the Fermi energy level, E$_F$, located within the high density band [1, 2].

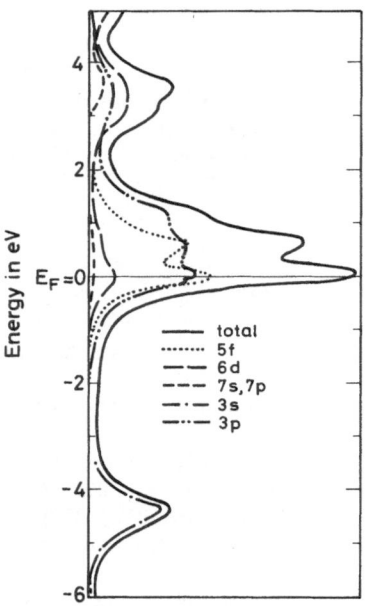

Fig. 3-41. Local density-of-states for UAl$_4$ [2].

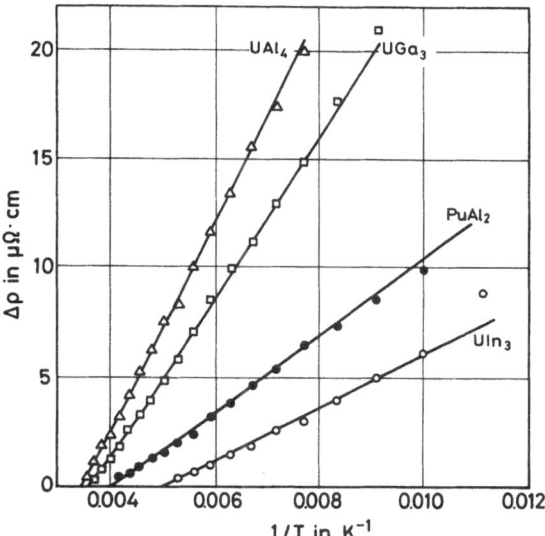

Fig. 3-42. Resistivity of UAl$_4$ versus 1/T of UAl$_4$, compared to UGa$_3$, UIn$_3$, and PuAl$_2$ (data from [4]) [5].

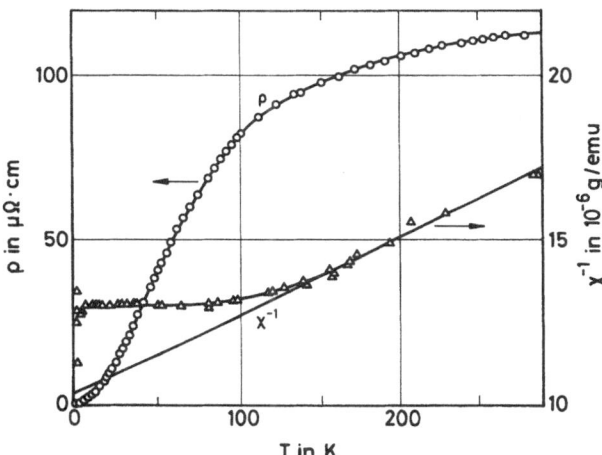

Fig. 3-43. Temperature dependence of the electrical resistivity ϱ and the reciprocal suscepti-bility 1/χ of UAl$_4$ [4].

Specific Resistivity

The specific resistivity of UAl$_4$ (67 wt% U) is 2.0 $\Omega \cdot$ cm at 20°C with a temperature coef-ficient of 4.0×10^{-4} K^{-1} measured at 20 to 120°C [3].

                      References for 3.1.5.5 on p. 150

The low temperature dependence of the electrical resistivity is characterized by the following temperature regions [4, 5], see Fig. 3-14, p. 103, **Fig. 3-42**, p. 149, and **Fig. 3-43**, p. 149:

proportional to T^2 with an estimated upper limit of T_1 about 30 K,
proportional to T with an estimated upper limit of T_2 about 100 K,
proportional to ln T.

The results are discussed in terms of a localized spin-fluctuation model for nearly magnetic intermetallics, whereas the temperature T_1 represents a measure of the characteristic localized spin-fluctuation temperature, T_{SF} [4, 5]. Further T_{SF} values were calculated using the dilute-alloy model (Rivier and Zlatic [6, 7]) [5], see Table 3/18.

Table 3/18
Calculated Spin-Fluctuation Temperatures, T_{SF}, for UAl$_4$ [5].

part of the curve	T_{SF} in K
from slope of T^2 region	550
from high-temperature end of T^2 region	220
from slope of T region	130
approach to the high-temperature limit, $-d\varrho/d \, (1/T) = T_{SF}$	43
from low-temperature end of ln T region	45

References for 3.1.5.5:

[1] Adachi, H.; Shiokawa, S.; Imoto, S. (J. Phys. Soc. Japan **45** [1978] 1423/4).
[2] Shiokawa, S.; Adachi, H.; Imoto, S. (Technol. Rept. Osaka Univ. **29** [1979] 45/50; C.A. **91** [1979] No. 9733).
[3] Jesse, A. (J. Nucl. Mater. **37** [1970] 340/2).
[4] Buschow, K. H. J.; van Daal, H. J. (AIP [Am. Inst. Phys.] Conf. Proc. No. 5 [1971/72] 1464/77; C.A. **77** [1972] No. 11324).
[5] Brodsky, M. B. (Phys. Rev. [3] B **9** [1974] 1381/7).
[6] Rivier, N.; Zlatic, V. (J. Phys. F **2** [1972] L87/L92).
[7] Rivier, N.; Zlatic, V. (J. Phys. F **2** [1972] L99/L104).

3.1.5.6 Magnetic Properties

Paramagnetic susceptibilities were measured for UAl$_4$ within the temperature range 20 to 650°C, from which room temperature values (at 20°C) are $\chi_g = 6.14 \times 10^{-6}$ emu/g or $\chi_a = 2120 \times 10^{-6}$ emu/g-atom [1]. Within this temperature range a Curie-Weiss law is obeyed with $\chi_g \cdot 10^6 = 3760/(T + 320)$, from which an effective paramagnetic moment of 3.2 μ_B (without corrections for the diamagnetism of the uranium and aluminium ions) was derived [1]. Up to about 100 K the susceptibility of UAl$_4$ is almost temperature independent and above this temperature the Curie-Weiss behavior sets in with $\Theta = -450$ K [2], see Fig. 3-43, p. 149. The results are discussed in terms of a localized spin-fluctuation model [2, 3]. Calculations within the Friedel-Anderson model led to the following results: χ (Pauli band susceptibility) = 16.6×10^{-4} emu/mol (using data of [2]), α (Stoner enhancement) about 3.5, and $T_{SF} = 68$ K [3].

References for 3.1.5.6:

[1] Konobrevskii, S. T.; Zaimovskii, A. S.; Levitsky, B. M.; Sokursky, Y. N.; Chebotarev, N. T.; Bobkov, Y. V.; Egorov, P. P.; Nikolaev, G. N.; Ivanov, A. A. (Proc. 2nd Intern. Conf. Peaceful Uses At. Energy, Geneva 1958, Vol. 6, pp. 194/203).

[2] Buschow, K. H. J.; van Daal, H. J. (AIP [Am. Inst. Phys.] Conf. Proc. No. 5 [1971/72] 1464/77; C.A. **77** [1972] No. 11324).

[3] Brodsky, M. B. (Phys. Rev. [3] B **9** [1974] 1381/7).

3.1.5.7 Chemical Reactions

Behavior on Heating

UAl_4 decomposes peritectically at 730°C to UAl_3 and aluminium [1]. For further details see Section 3.1.5.4, "Thermal Properties", p. 147.

Reaction in Oxygen or Air

UAl_4 is only slightly pyrophoric as a coarse powder and can be handled by conventional metallurgical techniques [2].

The oxidation of UAl_4 was investigated on thin sheets of two-phase x% UAl_4/Al' alloys, where x refers to the percentage surface area of primary UAl_4, and Al' refers to the UAl_4/Al eutectic with 13 wt% uranium. U_3O_8 was observed as the only oxidation product after heating in pure oxygen of 1 atm at 350 to 600°C [3, 4].

The x% UAl_4/Al' alloys (with x = 30, 65, and 90%) show linear oxidation rates after an initial period of logarithmic weight gain. This initial period becomes shorter with increasing amounts of UAl_4. As an interesting feature of these alloys, a temperature anomaly was observed at oxidation temperatures of 600°C, i.e., the rate of oxidation was slower than that at 550 or 500°C in long-term tests as well as in short-term tests (see **Fig.** 3-**44** [3]). As an

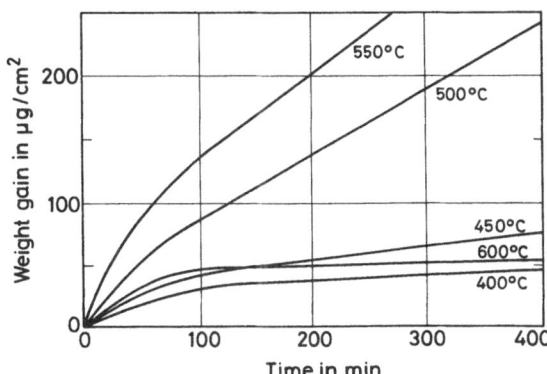

Fig. 3-44. Weight gain versus temperature for oxidation of 65% UAl_4/Al' in the range of 400 to 600°C [3]. Weight gain/time curves at these temperatures for 30 and 90% UAl_4/Al' are also found in [3].

References for 3.1.5.7 on p. 153

exception, the oxidation at 600 °C, performed on 90% UAl$_4$/Al' alloys, followed a cubic rate law. Pre-oxidation of the 65 and 90% UAl$_4$/Al' alloys at 600 °C markedly reduced the rate of oxidation at 500 °C. The oxide layers on the surface showed concentric or random cracking as confirmed by electron microscopy. The less rapid oxidation at 600 °C was postulated to be an effect of sintering of the oxide [3, 4]. Different activation energies obtained are summarized in Table 3/19.

Table 3/19
Activation Energies Observed After Oxidation of x% UAl$_4$/Al' Alloys [3].

x in %	initial logarithmic region	linear region final stage, short-term test	linear region final stage, long-term test
30	7.5	12.9	15.4
65	—	21.4	—
90	—	33.4	—

The formation of both U$_3$O$_8$ and Al$_2$O$_3$ should be thermodynamically possible, but as only U$_3$O$_8$ was observed, the overall reaction of 3 UAl$_4$ + 4 O$_2$ → U$_3$O$_8$ + 12 Al was stated. It was not possible to state if the aluminium formed by this reaction enters the UAl$_4$ or the U$_3$O$_8$ phase, or forms an Al$_2$O$_3$ interphase [4]. It was observed that the rate of diffusion of uranium ions in UAl$_4$ is more rapid than that of aluminium ions [7]. Thus, the rapid diffusion of the uranium ions to the metal/oxide interface may result in U$_3$O$_8$ being the predominant oxide in the film [4]. A temperature of T$_p$ = 510 °C is reported at which the weight-gain reaches 1 mg/cm^2 after 4 h of oxidation in air (1 atm) [8], see also [9].

For reaction of UAl$_4$ with aluminium see Section 3.1.5.8 on "UAl$_4$ — Al Dispersions", p. 153.

Reactions with Compounds

UAl$_4$ and Al$_2$O$_3$ are assumed to undergo reaction as revealed by measurements of the hardness. The measured microhardness of the unknown phase was 385 kp/mm^2 as compared to 322 kp/mm^2 at 60 p load observed for pure UAl$_4$ [10].

UAl$_4$ is dissolved by acids more rapidly than pure aluminium [11]. UAl$_4$ was completely dissolved in a mixture of 600 mL HCl (1.178 g/mL), 400 mL H$_3$PO$_4$ (1.39 g/mL), 0.25 g Na$_2$SiF$_6$, 0.037 g H$_2$PtCl$_6$, and 0.12 g CuSO$_4 \cdot$ 5 H$_2$O for calorimetric experiments [12].

The following etching treatments are recommended to distinguish the UAl$_3$ and UAl$_4$ phase metallographically in the UAl$_3$ — UAl$_4$ field:

Electrolytic etching with 1 part of 40% chromic acid and 1 part of 50% acetic acid maintained for 3 min at 0.05 A/cm^2 at 25 °C resulted in orange-yellow UAl$_3$ and blue-grey UAl$_4$ [13 to 15]. The same results were observed with a solution of 170 g Cr$_2$O$_3$ in 200 mL H$_2$O and 800 mL CH$_3$COOH with 0.2 A/cm^2 at 25 V [16, 17].

Etching with 1% HF at 25 °C resulted, after 40 s immersion, in light UAl$_3$ and dark, heavily attacked UAl$_4$. Prolonged etching leads to pitting of UAl$_3$ [14].

Etching with 50% HNO$_3$ at 250 °C resulted, after 3 min immersion, in yellow, sometimes bluish, and definitely roughened UAl$_3$ and light grey UAl$_4$, attacked to a lesser degree [14].

Polarized light- and dark-field illumination was successfully used to separate the UAl_3 and UAl_4 phases for quantitative image analysis [18].

And to distinguish the UAl_3 and UAl_4 phase metallographically in the UAl_4-Al field:

Electrolytic etching with 1 part of 40% chromic acid and 1 part of 50% acetic acid maintained for 3 min at 0.05 A/cm^2 at 25°C resulted in yellow UAl_3 and blue-grey UAl_4 [14].

Etching with 50% HNO_3 at 25°C resulted, after 1 min immersion, in light straw UAl_3 and distinct grey UAl_4 [14].

Etching with 1% HF at 25°C gave no clear distinction [14].

References for 3.1.5.7:

[1] Gordon, P.; Kaufmann, A. R. (J. Metals **2** [1950] 182/4).
[2] Rice, W. L. R. (TID-11295-3rd Ed. [1964] 105/20; N.S.A. **18** [1964] No. 41933).
[3] Openshaw, P. R.; Shreir, L. L. (Corros. Sci. **3** [1963] 217/37).
[4] Openshaw, P. R.; Shreir, L. L. (Corros. Sci. **4** [1964] 335/44).
[5] Green, D. R. (HW-53486 [1957] 1/16; N.S.A. **12** [1958] No. 6607).
[6] Northrup, C. J. M.; Kass, W. J.; Baettie, A. G. (Hydrides Energy Storage Proc. Intern. Symp., Geilo, Norway, 1977 [1978], pp. 205/16; C.A. **91** [1979] No. 133205).
[7] LeClaire, A. D.; Bear, I. J. (AERE-M-R-879 [1952] from [4]).
[8] Stone, H. E. N. (J. Mater. Sci. **10** [1975] 923/34).
[9] Badaeva, T. A.; Dashevskaya, L. I. (Fiz. Khim. Splavov Tugoplavkikh Soedin Toriem Uranom **1968** 107/13; C.A. **71** [1969] No. 24297).
[10] Ondracek, G.; Patrassi, E. (Ber. Deut. Keram. Ges. **45** [1968] 617/21).

[11] Wire, M. S. (Diss. Univ. California, San Diego, La Jolla 1984, pp. 1/123; Diss. Abstr. Intern. B **45** [1985] 2217; C.A. **102** [1985] No. 104701).
[12] Ivanov, M. I.; Tumbakov, V. A.; Podol'skaya, N. S. (At. Energiya SSSR **5** [1958] 166/70; Soviet J. At. Energy **5** [1958] 1007/11; C.A. **1960** 9481).
[13] Saller, H. A. (in: Finneston, H. M.; Howe, J. P., Progress in Nuclear Energy Ser. 5 Metallurgy and Fuels, Vol. 1, Pergamon, London 1956, pp. 535/43).
[14] Hills, R. F. (J. Inst. Metals **86** [1958] 438/41).
[15] Zelezny, W. F. (IDO-17154 [1966] 56/73; N.S.A. **20** [1966] No. 19178).
[16] Boucher, R. (J. Nucl. Mater. **1** [1959] 13/27).
[17] Boucher, R. (Symp. Solid State Diffusion, Saclay, France, 1958, pp. 111/7; N.S.A. **15** [1961] No. 26564; AEC-tr-4768 [1960] 1/11; N.S.A. **14** [1960] No. 7826).
[18] Nazare, S.; Ondracek, G. (Prakt. Metallog. **6** [1969] 742/6; C.A. **72** [1970] No. 50161).

3.1.5.8 Uranium Tetraaluminide — Aluminium Dispersions, UAl_4-Al

UAl_4-Al dispersion fuel elements are of interest in nuclear technology. Research reactors with neutron fluxes of about 10^{15} $n \cdot cm^{-2} \cdot s^{-1}$ utilize plate-type elements with a high uranium content and, in most cases, an aluminium matrix as cladding.

Preparation of UAl_4-Al Dispersions and Fuel Plates

The UAl_4-Al dispersions are prepared by mixing, pressing, and compaction of UAl_4 and aluminium powder. According to nuclear aspects the UAl_4 is used as a coarse powder, 43 µm [1], 63 to 125 µm [2, 3] in diameter. Typical fabrication data are given in [2].

References for 3.1.5.8 on pp. 155/6

154 UAl₄ — Al Dispersions →

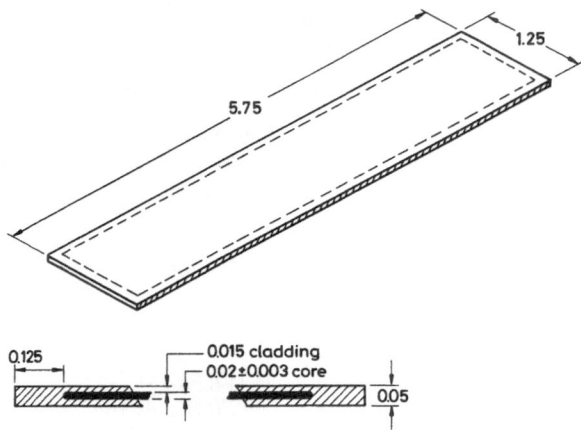

Fig. 3-45. Sample fuel plate as used in neutron irradiations [15]. Dimensions in inches.

In order to fabricate UAl$_4$ — Al fuel plates for nuclear application, the mixed UAl$_4$ and aluminium powder was compacted to a specified rectangular shape of high density [2 to 4], or the mixed powders were sintered at 600°C [5, 6]. This core was assembled by the "picture-frame technique", using a picture frame and cover plates of aluminium or aluminium-alloy materials similar to those used for the blending of the UAl$_4$ powder [2 to 4, 7, 8]. A typical fuel plate is shown in **Fig. 3-45**. If aluminium alloys instead of pure aluminium were used as frame and cover plates, the alloy material was plated with pure aluminium to obtain sufficient bonding between core and cladding [9 to 12]. This assembly was hot rolled at 550°C [2, 4, 13] or 600°C [14] for bonding and then cold rolled to specified dimensions. Finally, tests were undertaken for quality control, such as an X-ray structure test to point out core-end failures ("dog-boning" or "fish-tailing"), which were observed to a certain degree [2, 7], and a "blister test" at 500°C (for 4 × 1 h) to point out bonding defects [2, 3].

A special fabrication method was developed to produce UAl$_4$ — Al fuel plates of 300 mm in length, 30 mm exterior width, and 3 mm total thickness by fritting extrusion of mixed UAl$_4$ (43 μm) and aluminium (8 μm, less than 4% Al$_2$O$_3$) powders inside a box of aluminium. The fabrication steps are described in [1], see also [16]: 1) preparation of an aluminium box with a characteristic geometry to obtain a clad with regular thickness and a tubular core; 2) cold pressing of the mixed powders into the box; 3) closure of the box by a cast aluminium disk; 4) extrusion, at which the closed and preheated (500°C) box was drawn at the same temperature between a drawplate and a punch integral with the presser. The fabrication was carried out with a 150-ton vertical press.

Interaction of UAl$_4$ and Al Matrix

UAl$_4$ is compatible in contact with aluminium up to the point of liquefaction [7]. Annealing and tempering at 550°C decreased the strength of cold-rolled UAl$_4$ — Al dispersions [13].

Physical Properties

The linear thermal expansion coefficient of UAl$_4$ — Al dispersions, measured at 20 to 200°C, decreases with increasing amount of UAl$_4$ [2], see also [3].

The thermal conductivity of UAl_4 — Al dispersions was measured with cast alloys and after heat treatment of pressed specimens at 893 K for 5 h. The thermal conductivity decreases with increasing amount of UAl_4. The results are in good agreement with theoretical calculations [2, 3].

The electrical resistivity of pressed UAl_4 — Al dispersions, measured at room temperature, increases with increasing amount of UAl_4 [3].

Neutron Irradiation of UAl_4 — Al Dispersion Fuel Elements

UAl_4 — Al dispersions show a markedly good irradiation behavior concerning the mechanical stability of the fuel plates [3]. The post-irradiation examination gave no evidence of fission-gas (He, Xe, Kr) bubble formation in a basic two-phase mixture of UAl_3 and UAl_4. However, clusters of small fission-gas bubbles were observed to form on and into small uranium oxide inclusions which were present as an impurity [17, 18], see also [19]. The fission-gas bubble formation appears to be suppressed by accomodation of the gaseous atoms in the UAl_4 fuel particles, i.e., in solution [8, 18], see also [20 to 23]. The retention behavior is attributed to the UAl_4 defect structure [8, 18, 24, 25], see also [20 to 23]. From post-irradiation annealing experiments, the principal release of fission gas from pellets of UAl_4 was observed to occur at about 1003 K. The major part of the fission gas was released from fuel plates above the solidus temperature of the cladding material, i.e., about 858 K [8, 22, 24], see also [23].

References for 3.1.5.8:

[1] Montagne, R.; Meny, L. (TID-7546 [1958] 142/56; C.A. **1958** 9904).
[2] Jesse, A.; Ondracek, G.; Thümmler, F. (Powder Met. **14** [1971] 289/97).
[3] Nazare, S.; Ondracek, G.; Thümmler, F. (J. Nucl. Mater. **56** [1975] 251/9).
[4] Ondracek, G. (Ger. Offen. 3042424 [1980]; C.A. **97** [1982] No. 135543).
[5] Saller, H. A. (U.S. 2917383 [1959]; C.A. **1960** 18313).
[6] Saller, H. A. (U.S. 2917383 [1959]; N.S.A. **14** [1960] No. 7757).
[7] Cunningham, J. E.; Beaver, R. J.; Thurber, W. C.; Waugh, R. C. (TID-7546 [1958] 269/97; C.A. **1958** 9905).
[8] Beeston, J. M.; Hobbins, R. R.; Gibson, G. W.; Francis, W. C. (Nucl. Technol. **49** [1980] 136/49).
[9] Saller, H. A. (in: Finneston, H. M.; Howe, J. P., Progress in Nuclear Energy Ser. 5 Metallurgy and Fuels, Vol. 2, Pergamon, London 1956, pp. 535/43).
[10] Nazare, S.; Ondracek, G.; Thümmler, F. (KFK-1252 [1970] 1/83; C.A. **75** [1971] No. 70285).

[11] Gibson, G. W. (IN-1133 [1967] 1/82; C.A. **69** [1968] No. 40509).
[12] Walker, V. A.; Graber, M. J., Jr.; Gibson, G. W. (IDO-17157 [1966] 1/110; C.A. **66** [1967] No. 81543).
[13] Seller (Met. Yadern. Energ. Deistvie Obluchen. Mater. Sbornik **1956** 242/57; C.A. **1958** 11707).
[14] Colombi, G.; Gabaglio, M.; Liscia, A.; Trittoni, E. (Met. Ital. **22** [1967] 359/63).
[15] Gibson, G. W.; Zukor, M. (IDO-17154 [1966] 3/6; N.S.A. **20** [1966] No. 19178).
[16] Montagne, R.; Meny, L. (CEA-784 [1957] 1/33; N.S.A. **13** [1959] No. 5118).
[17] Hofman, G. L. (Nucl. Technol. **77** [1987] 101/5).
[18] Graber, M. J.; Zelezny, W. F.; Moen, R. A. (IDO-17154 [1966] 7/28; N.S.A. **20** [1966] No. 19178).
[19] Leitten, C. F., Jr.; Richt, A. E.; Beaver, R. J. (ORNL-3470 [1963] 87/8; N.S.A. **18** [1964] No. 4173).

[20] Shibata, T.; Kanda, K.; Mishima, K.; Tamai, T.; Hayashi, M.; Snelgrove, J. L.; Stahl, D.; Matos, J. E.; Travelli, A. et al. (Proc. Intern. Meeting Res. Test React. Core Convers. HEU LEU Fuels, Argonne, Ill., 1982 [1983], pp. 99/116; C.A. **101** [1984] No. 80371; Nucl. Sci. Eng. **87** [1984] 405/17; C.A. **101** [1984] No. 118746).

[21] Hofman, G. L. (J. Nucl. Mater. **140** [1986] 156/63).

[22] Zelezny, W. F.; Gibson, G. W.; Graber, M. J. (3rd Natl. Symp. Develop. Irradiat. Test. Technol. Papers, Sandusky, Ohio, 1969, pp. 516/32; C.A. **74** [1971] No. 149748).

[23] Woodley, R. E. (HEDL-7598 [1986] 1/105; C.A. **106** [1987] No. 127612).

[24] Francis, W. C. (IN-1437 [1970] 1/75; N.S.A. **25** [1971] No. 11458).

[25] Francis, W. C.; Moen, R. A. (IDO-17218 [1966] 1/160; N.S.A. **21** [1967] No. 18203).

3.1.6   Uranium – Aluminium Alloys

3.1.6.1   Uranium-Rich Alloys

Small additions of aluminium increase the hot and cold mechanical characteristics of pure uranium. With additions in the range of up to 0.4 wt% aluminium, the mechanical properties in the α-uranium phase depend markedly on the methods of fabrication as a result of its intrinsic anisotropy and allotropic α, β, and γ modifications [1 to 7].

The transformation temperatures of γ to β and β to α uranium decreased with increasing addition of aluminium in the range of 650 to 1650 ppm aluminium. The grain size of the alloys decreased with increasing cooling rates, and feathery or acicular structures were obtained when severely cooled from the γ-uranium range. UAl_2 precipitates appeared within the grains or on the grain boundaries. The microhardness increased when the aluminium content and the cooling rate was increased as a (supposed) result from the size of the precipitates and lattice hardening due to martensitic transformation [8, 9], see also [10 to 12] and Section 3.1.1, p. 82.

The addition of aluminium did not alter the general appearances of the various network substructures, produced in uranium by the eutectoid decompositions of the γ and β phases, but resulted in the formation of UAl_2 placed in the networks of black lines during α annealing after quenching, or by slower cooling rates [13], see also [14]. In the range of about 0.1% addition of aluminium globular precipitates of UAl_2 were observed, the mean distances of which depended on the heat treatment of the alloys. There is a linear relation observed between hardness (H_{V30}) at room temperature and the reciprocal value of the mean distance [15]. The UAl_2 precipitation kinetics were determined, in the temperature range of 550 to 650°C, with alloys containing 0.7 to 1.4 at% aluminium by a continuous measurement of the electrical resistivity during isothermal aging. The precipitation mechanism yields the Johnson-Mehl relation, $y = 1 - \exp(-k \cdot t^n)$, with n = 0.8 to 1.0. The apparent activation energy was 1.65 to 1.75 eV [16].

Uranium alloys with 0.15 to 0.5% aluminium showing finely dispersed UAl_2, free from plates, network, or an unstable needle-like structure, and with an optimum in mechanical properties, were obtained by the following procedure: aluminium was added in the form of UAl_2 to the melt of uranium and, after annealing and homogenization at 1040 to 1100°C for 4 h, the alloys were quenched in oil and subsequently drawn to a diffusion heating at 850°C for 2 h [17], see also [18].

A significant increase of the Vickers hardness from 210 (ordinary uranium) to 420 kp/mm^2 (U $-$ 1.5% Al) was observed after heating the alloy at 1050°C in vacuum for 200 to 300 h and subsequent cooling and re-heating to 600°C in vacuum for stabilization [19].

Small additions of aluminium (0.9 at% Al) affected the hot-rolling textures of pure uranium by increasing the [010] component and reducing the [110] component. Thus, for the same temperature of rolling, the uranium alloys possess a much higher axial growth index than unalloyed uranium [20], see also [21, 22].

There is an increase in creep strength with uranium $-$ aluminium alloys up to 2 at% aluminium, as observed from heat-treated alloys containing 0.5 to 8.0 at% aluminium in tensile creep tests at 500°C at a stress of 4480 psi [23], see also [24].

Uranium, alloyed with 1 at% aluminium, shows a significantly lower ignition temperature as compared to pure uranium: 575°C (BMI-uranium) to 595°C (ANL-uranium), 355°C (U-1 at% Al). Self-accelerating reaction rates were observed [25], see also [26]. Small amounts of aluminium (500 to 1000 ppm) in uranium lower the oxidation rate in CO_2 at 450 to 650°C. The oxidation is controlled by the diffusion of oxygen ions through a thin layer of uncracked UO_2 next to the metal [27]. The oxidation of uranium, alloyed with 1 at% aluminium, follows a parabolic rate law between 600 and 1200°C. The observed reaction rates do not differ significantly from those found with pure uranium [28].

References for 3.1.6.1:

[1] Boudouresques, B.; Englander, M. (in: Finneston, H. M.; Howe, J. P., Progress in Nuclear Energy Metallurgy and Fuels, Ser. 5, Vol. 2, Pergamon, London $-$ New York $-$ Paris $-$ Los Angeles 1959, pp. 621/31).

[2] Lehmann, J.; Aubert, H. (Symp. Special Metallurgy, Saclay 1957, pp. 41/6; N.S.A. **13** [1959] No. 2244).

[3] Badaeva, T. A.; Alekseenko, G. K.; Aleksandrova, L. N. (Str. Svoistva Splavov At. Energ. **1973** 45/52; C.A. **81** [1974] No. 174511).

[4] Englander, M. (CEA-776 [1958] 1/79; C.A. **1959** 16728).

[5] Allen, N. P.; Grogan, J. D. (Brit. 811841 [1959]; C.A. **1959** 13034).

[6] Laniesse, J.; Aubert, H. (CEA-R-2584 [1964] 1/24; C.A. **62** [1965] 15710).

[7] Althaus, W. A.; Cook, M. M. (NLCO-978 [1966] 1/13; C.A. **66** [1967] No. 68276).

[8] Jung, S. H.; Park, W. K.; Hong, J. H. (Kumsok Hakhoe Chi **14** [1976] 519/27; C.A. **87** [1977] No. 155735).

[9] Taplin, D. M. R.; Martin, J. W. (J. Nucl. Mater. **12** [1964] 50/5).

[10] Kaderabek, E.; Pochob, P. (Czech. 141031 [1971] 1/3; C.A. **77** [1972] No. 9063).

[11] Gomozov, L. I. (Izv. Akad. Nauk SSSR Metally **1982** No. 2, pp. 165/6; C.A. **97** [1982] No. 149208).

[12] Shober, F. R. (N-64-17298 [1964] 1/79; C.A. **61** [1964] 14293).

[13] Angerman, C. L.; Huntoon, R. T. (J. Less-Common Metals **9** [1965] 338/53).

[14] Marsh, D. J.; Slattery, G. F.; Dewey, M. A. P.; Brammar, I. S. (J. Inst. Metals **93** [1964/65] 471/5).

[15] Truchly, J.; Novak, J. (Kovove Mater. **5** [1967] 135/49; C.A. **68** [1968] No. 5562).

[16] Stelly, M. (J. Nucl. Mater. **40** [1971] 84/92).

[17] Cabane, G.; Englander, M.; Lehmann, J. (Comm. Energie At. [France] Rappt. No. 444 [1955] 1/25; C.A. **1956** 2401).

[18] Macherey, R. E.; Dunworth, R. J. (ANL-5341 [1956] 1/26; N.S.A. **11** [1957] No. 13766).

[19] Englander, M.; Stohr, J. (Fr. 1129082 [1957]; C.A. **1959** 21577).

[20] Slattery, G. F.; Conolly, D. E. (J. Nucl. Energy A/B **18** No. 6 [1964] 347/59).

[21] James, P. F.; Fern, F. H. (J. Nucl. Mater. **29** [1969] 191/202).

[22] Huet, J. J.; Massaux, H. (ATB Met. [Acta Tech. Belg.] **11** [1961] 167/76; C.A. **1962** 11144).

[23] Chamberlain, J.; Johnson, M. P.; Kench, J. R.; Young, A. G. (J. Nucl. Mater. **19** [1966] 121/32).

[24] Foster, E. L., Jr.; Shober, F. R.; Daniel, N. E.; Bauer, A. A.; Thompson, C. R.; Melton, C. W. (BMI-1664 [1964] 1/76; C.A. **61** [1964] 4029).

[25] Baker, L., Jr.; Bingle, J. D. (J. Nucl. Mater. **20** [1966] 11/21).

[26] Schnizlein, J. G.; Pizzolato, P. J.; Porte, H. A.; Bingle, J. D.; Fischer, D. F.; Mishler, L. W.; Vogel, R. C. (ANL-5974 [1959] 1/207; N.S.A. **13** [1959] No. 18645).

[27] Stobbs, J. J.; Whittle, I. (J. Nucl. Mater. **19** [1966] 160/8).

[28] Wilson, R. E.; Barnes, C., Jr.; Koonz, R.; Baker, L., Jr. (Nucl. Sci. Eng. **25** [1966] 109/15).

3.1.6.2 Aluminium-Rich Alloys

The aluminium-rich alloys were usually prepared from the pure metals by melting and casting techniques. Uranium — aluminium compounds, UAl_3 and UAl_4, are formed during solidification, resulting in alloys consisting of particles of UAl_x compounds surrounded by primary aluminium owing to the uranium — aluminium phase diagram, compare Fig. 3-1, p. 83.

An investigation of the structure of the alloys with contents of up to 30 wt% uranium showed that hypoeutectic alloys consist of primary dendritic aluminium surrounded by a continuous network of the eutectic, which appeared to be of a "chevron" type of morphology. On the other side, the hypereutectic alloys consist of primary crystals of UAl_4 surrounded by halos of aluminium and a eutectic matrix. The presence of UAl_3 was noted with higher percentages of uranium. The eutectic exhibited "chevron" and "rhombic spiral" type of morphology ("chevron type" means a directional dendritic growth at fast cooling rates). This unusual type of eutectic structure is believed to be a result of the growth rate anisotropy of the two phases [1], see also [2 to 6]. The formation of UAl_3 was also found with eutectic alloys (13 wt% U) as confirmed by X-ray analysis. The amount of UAl_3 decreased from bottom to top of the castings because the formation of UAl_3 decreases with decreasing cooling rate. Thus, UAl_3 nucleates before the UAl_4 is formed [7]. With alloys of up to 20 wt% uranium, a pronounced thermal effect was observed on heating or cooling near the UAl_4 — Al eutectic at 646°C. On cooling, (a type of) β-UAl_4 precipitates initially during eutectic solidification, and after an induction period of 5 to 20 min, the nucleation and growth of the low temperature (type of) α-UAl_4 produced an evolution of energy from which the temperature increases by 1 to 2°C [8].

The particle size of the UAl_x compounds decreases with increasing pouring temperature from 1040 to 1290°C with alloys containing 40 wt% uranium. The thickness of the castings also had a marked effect and the particle size increased with increasing mold temperature (25 to 230°C), but that effect was smaller than the influence of the pouring temperature [9, 10].

3.1.6.2.1 Preparation

The uranium — aluminium alloys were usually prepared by melting the pure metals. The uranium was used in form of chunks (from shearing of vacuum cast slabs [11]) or uranium powder prepared from decomposition of uranium hydride [12]. The melts were performed in crucibles of graphite (or clay-graphite) [10, 14, 15 to 17], beryllium oxide (or beryllium oxide lined) to prevent pick-up of carbon [12], or zirconium dioxide which was employed to obtain larger inductive stirring [14, 15]. Different heating systems and atmospheres were used to prepare the melt: resistance furnace under vacuum [12, 15], HF-induction furnace under vacuum [12, 14, 18 to 20], flowing argon [12, 16], or air [10, 13, 14, 21], arc melting [20], or conventional open heating by gas-air mixtures [10, 13], see also [22]. Also a special method, the levitation melting under argon, was used to avoid problems of segregation [23].

The preparation of the melt starts with the melting of the aluminium part, which was superheated to 800 to 900°C [10, 11, 13, 15, 17]. The uranium part then was added slowly and the melt kept for a considerable time [10, 11, 13, 15] to reach equilibrium in the liquid state, as particles of UAl$_2$ persist in the melt for a considerable time [10, 13]. The melt was stirred with a graphite rod [15, 19], which was especially necessary with a resistance furnace [10, 12, 13], at increased temperature of 50 to 100°F above the liquidus under 0.5 atm of helium [14, 18] (1175 to 1225°C), [11, 19] (1175°C), [15]. Two main problems, uranium segregation and gas porosity, occur in the preparation, especially of alloys containing more than 25 wt% uranium [14, 15, 24]. Segregation may be avoided by eddy current stirring, which was especially successful using the levitation technique [23]. Remelting is often desirable to improve the homogeneity [12, 14], see also [25, 26]. The alloys possess a marked tendency to absorb gas, particularly hydrogen, in the molten state and blow-holes occur as the temperature drops. Melts containing appreciable amounts of gas bleed as the eutectic temperature of 640°C is approached [14, 15]. The melt may be degassed by bubbling helium through the melt for 10 min [15] or by high-frequency melting in vacuum [14]. Sound ingots with only small deviation in composition containing 25% uranium were also obtained by melting aluminium in an air furnace under circulating nitrogen or a layer of cryolite and then adding the uranium. The pouring temperature was kept at 950°C [22].

Castings of the alloys were performed by pouring the melt into copper or copper-inserted steel molds [22] under helium at room temperature [15, 18, 19], or into preheated graphite molds [11, 18, 19], see also [27 to 29].

Uranium fluorides or oxides were used as starting materials if the preparation of uranium—aluminium alloys were performed with enriched uranium. The process involves the melting of aluminium under cryolite (Na$_3$AlF$_6$) flux, then the cryolite plus uranium oxide or fluoride was added with constant stirring for 1 to 5 min. The flux was then removed from the low-melting alloys by allowing it to freeze around a carbon rod. The metal and the flux then were separated mechanically and the metal subsequently remelted and cast into shape at pouring temperatures of 700 to 1100°C, depending on the uranium content. The temperature of the mold was kept below 150°C [10, 13], see also [30 to 35].

Uranium—aluminium alloys with 2.0 ± 0.5 and 4.4 ± 0.5 wt% uranium [36, 37] or 20 to 30 wt% uranium [38] were also prepared from uranium compounds dissolved in molten alkali chlorides (NaCl—KCl) by a reduction process using excess amounts of liquid aluminium: $4 \, Al + 3 \, UCl_4 \rightarrow 4 \, AlCl_3 + 3 \, U$ (rapid reaction) or $4 \, Al + 3 \, UF_4 \rightarrow 4 \, AlF_3 + 3 \, U$ (rapid reaction) [38], see also [30, 39 to 41], by electrolytic deposition on the molten aluminium cathode at 973 K [36, 37], see also [42, 43].

The kinetics of the UAl$_3 \rightarrow$ UAl$_4$ transformation, observed at 40 wt% uranium—aluminium alloys quenched from the liquid + UAl$_3$ region, is strongly influenced by small additions of silicon in the range of up to 3 wt% [14, 21, 44]. Normally this transformation needs about 1 h if annealed at 600°C. With additions of 0.1 wt% Si, the transformation time is about ten times longer and there is practically no transformation observed with 0.6 wt% Si added [44 to 46], see also [47 to 49]. This effect was also observed with other elements as ternary additions, resulting in different effectiveness. The formation of U(Al, Si)$_3$ in those alloys was confirmed

References for 3.1.6.2 on pp. 175/9

by X-ray analysis [21]. Larger amounts of silicon, i.e., more than one Si atom per U atom, resulted in the precipitation of pure Si [50].

Castings of such Si-doped alloys (up to 50 wt% uranium and about 3 wt% Si) were prepared by conventional open-air melting techniques using graphite crucibles and molds. A gassy melt often resulted and several presolidification steps were necessary to degas the alloys. These presolidification steps could be avoided by using a vacuum-melted 50 wt% Si—U master alloy as ternary addition. In practice, two main effects were observed with silicon alloying: 1) elimination of hot tearing at the juncture of the head and the body of the castings; 2) alteration of the macrostructure from completely columnar to completely equiaxed morphology. Alloys with a U:Si ratio of nearly 1 showed a maximum in softness as pointed out by a maximum in tensile strength [50]. In general, the quality of the ingots was improved by addition of about 3 wt% Si [14, 21], see also [49, 51 to 63].

3.1.6.2.2 Physical Properties

Density

Measured densities of uranium—aluminium alloys with 0.5 to 60 wt% uranium are given in **Fig. 3-46**. The measured values are in good agreement with calculations based on the volumes of α aluminium and UAl_4 present in the alloys under equilibrium conditions, showing a significant deviation from a linear relationship [19], see also [64]. Otherwise, a linear relationship is reported for the same region of alloys [57].

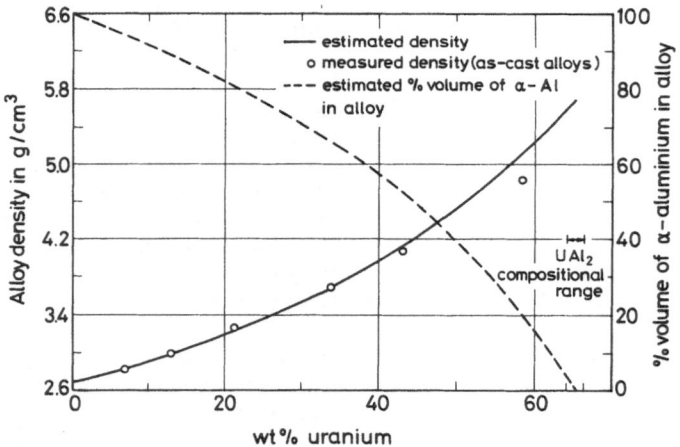

Fig. 3-46. Relationships between composition, density, and % volume in uranium—aluminium alloys [19].

Thermal Expansion

The coefficient of linear thermal expansion of uranium—aluminium alloys containing up to 30.5 wt% uranium decreases with increasing uranium content as summarized in Table 3/20 [10, 13]. In **Fig. 3-47** the coefficient of linear thermal expansion of an 18 wt% uranium—aluminium alloy is compared to pure aluminium and a 48 wt% uranium—aluminium alloy doped with 3 wt% silicon [21].

Table 3/20
Coefficient of Linear Expansion of U — Al Alloys [10].

temperature range in °C	coefficient of linear expansion (in 10^{-6}/K) for different uranium contents			
	0 wt % U	12.5 wt % U	22.7 wt % U	30.5 wt % U
20 to 100	23.9	20.0	20.0	19.4
20 to 200	24.6	21.1	21.2	20.8
20 to 300	25.5	22.1	21.9	21.3
20 to 400	26.5	23.1	22.5	21.6
20 to 500	27.7	23.5	22.7	22.1
100 to 500	—	24.4	23.2	22.6

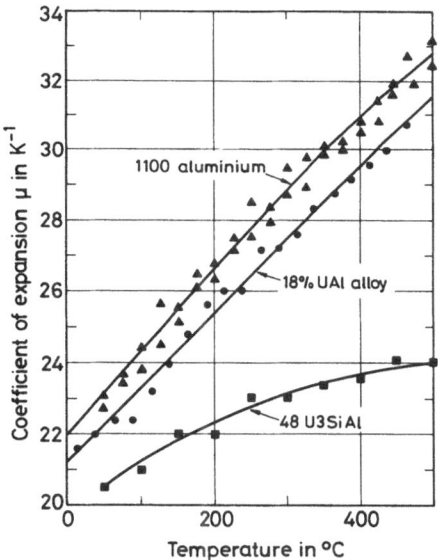

Fig. 3-47. Coefficient of linear thermal expansion for pure aluminium and uranium — aluminium alloys [21].

Hardness

The hardness of uranium — aluminium alloys containing up to 16.8 wt% uranium was measured on sheets of 0.080 inch thickness, which were solution-heated at 550°C and then aged at 175°C for 2 to 8 h. The results gave an indication of some solubility of uranium in aluminium (see Table 3/21, p. 162) [10, 13]. A decrease in hardness was observed with increasing temperature from 70 kg/mm^2 (diamond pyramid hardness) at room temperature to 4 to 8 kg/mm^2 (DPH) at 550°C [15], see also [27, 65, 66].

References for 3.1.6.2 on pp. 175/9

Table 3/21
Hardness Data of Uranium — Aluminium Sheets [10].

| wt% U | Rockwell hardness H | | | | |
| | as quenched[*] | time aged at 175°C in h | | | |
		2	4	6	8
2.43	25	27 to 29	29 to 31	35 to 36	24
3.10	29 to 31	32 to 33	36	37 to 38	33
4.95	41 to 42	43 to 44	45 to 47	43	—
7.00	53 to 54	57	55 to 56	56	—
9.00	57 to 59	60 to 61	61	59 to 61	—
11.3	61 to 62	64 to 65	64 to 65	65	—
14.3	56	58	58 to 59	58 to 59	—
16.3	58 to 59	60 to 62	61 to 62	62 to 63	—
16.8	58 to 59	60 to 62	61 to 62	60	—

[*] Quenched from 550°C after holding for 3 h.

Tensile Properties and Creep Behavior

Tensile properties of uranium — aluminium alloys containing 2.43 to 16.8 wt% uranium were measured on 0.080 in. thick sheets and 1 in. forged bars. The alloy sheets were tested as cold rolled, annealed at 550°C for 3 h, water quenched after annealing at 550°C for 3 h, and water quenched and aged at 175°C for 6 h after quenching. After annealing or quenching, all alloys showed a decrease in yield strength and ultimate tensile strength. The strengths increased with ultimate tensile strength and with increasing uranium content with alloys of up to 9 wt% uranium. A further increase in uranium showed no appreciable effect on strength. Only little variation was observed with alloys quenched or slowly cooled from 550°C. Aging of the quenched alloys gave no significant change in the tensile properties [10, 13], see also [66]. The tensile properties of forged specimens with 12.5 to 30.5 wt% uranium were measured on standard samples with 0.505 in. after annealing at 370°C for 0.5 h. The results are summarized in Table 3/22 [10, 13].

Table 3/22
Tensile Properties of Forged Uranium — Aluminium Alloys [10].

wt% U	average modulus of elasticity[1] in 10^6 psi	tensile strength in psi	0.2% offset yield strength in psi	elongation in %	reduction of area in %
0	10.0	13000	5000	45	—
12.5[2]	10.4	22500	10800	20	34
22.7	10.9	18600	11600	4	7
22.7	11.3	23500	14500	13	14
30.5	11.3	26100	14850	10.5	11.5

[1] Estimated to be correct within ±3%. — [2] Single specimens were used except in the 22.7% uranium alloy.

High-temperature properties for specimens with 11.3 and 17.3 wt% uranium are given in Table 3/23, see also [13, 67], or with 18.4 wt% uranium, see [68].

Table 3/23
High-Temperature Properties of Uranium—Aluminium Alloys (tests on metal annealed at 370°C for 1 h) [10].

alloy	room temperature		150°C		300°C
	tensile strength in psi	0.2% offset yield strength in psi	tensile strength in psi	0.2% offset yield strength in psi	tensile strength in psi
Al (type 2 SO)	13000	5000	7500	3500	2500
11.3 wt% U	19700	11320	14125	9200	8630
17.3 wt% U	20080	11480	14700	10000	8780
	19030	8550	14000	7230	8130
	—	—	13650	7020	8210
	—	—	13300	6480	—

Different values of yield strengths of specimens with 24.7 to 44.8 wt% uranium are given in [14] (room temperature) and [11] (450°C). In general, the uranium—aluminium alloys are stronger but less ductile as compared to pure aluminium.

The lack of ductility in the uranium—aluminium alloys with 40 to 50 wt% uranium can be partly ameliorated by doping the alloys with ternary additions such as silicon (see **Fig. 3-48**) [14, 21], see also [50, 52].

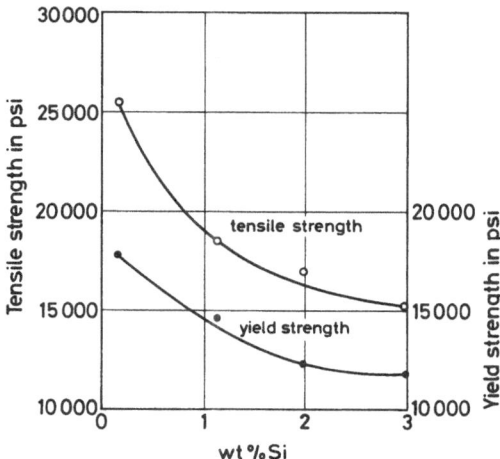

Fig. 3-48. Effect of Si on the room temperature tensile properties of 48 wt% U—Al alloys [14].

Creep rates for a U — Al alloy with 17.3 wt % uranium, annealed at 370°C for 1 h and tested at 150°C for 2000 h, are given in Table 3/24 [10, 13].

Table 3/24
Creep Rate Data for a 17.3 wt % U — Al Alloy at 150°C [10].

load in psi	creep rate in % elongation/h	total deformation in % elongation
3000	0.00004	0.134
4000	0.00035	0.420
5000	0.0027	4.21

The adiabatic modulus of elasticity was measured by electro-acoustic methods at U — Al alloys containing 16 wt % U, resulting in a value for the velocity of sound of 4972.4 m/s and an elastic modulus of 8004.9 kg/mm^2 [69].

Activity Factors and Partial Thermodynamic Characteristics

Activity factors of uranium in liquid aluminium were derived from emf measurements [70] or calculated from the uranium — aluminium phase diagram [71]. The results from emf measurements are summarized in Table 3/25.

Table 3/25
Activity Factor of Uranium in Aluminium Alloys [70] (standard state: γ uranium).

temperature in K	$x_U \cdot 10^4$	$\gamma_U \cdot 10^4$	temperature in K	$x_U \cdot 10^4$	$\gamma_U \cdot 10^4$
934	145	0.66	980	155	1.9
934	156	0.60	1077	83	8.8
958	76	1.1	1080	140	9.7
959	157	1.2	1085	199	10
961	205	1.3	1144	200	18
980	100	1.9	1147	275	20

Partial thermodynamic characteristics of γ uranium in uranium — aluminium alloys containing 10 wt % uranium [72] and 25 to 40 wt % uranium [70] were derived from emf measurements. The results are summarized in Table 3/26.

Diffusion Coefficients

Diffusion coefficients of uranium in molten aluminium were derived from diffusivity measurements using the capillary-bath method [73]. The results are summarized in Table 3/27.

From these measurements, a value for the heat of activation for diffusion was calculated to be $\Delta H_D = 33400$ cal/mol [73]. The measurements of the diffusion coefficients of uranium in molten aluminium gave the relation: log D (in cm^2/s) = $-3450 + (-1.03 \times 10^4/T)$ [57].

Table 3/26
Partial Thermodynamic Characteristics of γ Uranium in Uranium – Aluminium Alloys.

wt% U	phase composition	temperature in °C	$-\Delta\overline{H}_U$ in kg/g-atom	$-\Delta\overline{S}_U$ in e.u./g-atom	$-\Delta\overline{G}_U$ in kg/g-atom (at 1000 K)	Ref.
25 to 40	liq. + UAl$_3$	870 to 730	38.6 ± 0.8	14.9 ± 0.8	23.6 ± 0.1	[70]
25 to 40	liq. + UAl$_4$	730 to 640	41.5 ± 0.8	17.8 ± 0.8	23.6 ± 0.1	[70]
25 to 40	Al + UAl$_4$	640 to 537	29.1 ± 2.3	4.1 ± 2.8	(25.0 ± 0.3)	[70]
77 to 80	UAl$_3$ + UAl$_2$	880 to 660	13.9 ± 2.8	1.0 ± 2.8	(12.9 ± 0.4)	[70]
10	Al + UAl$_4$	401 to 639	28.1 ± 0.6	3.2 ± 0.8	(24.9 ± 0.1)	[70, 72]
71.4	UAl$_4$ + UAl$_3$	399 to 706	24.3 ± 0.6	0.2 ± 0.6	24.1 ± 0.1	[70, 72]
77	UAl$_3$ + UAl$_2$	442 to 840	13.6 ± 0.6	1.5 ± 0.6	12.1 ± 0.1	[70, 72]

Table 3/27
Diffusion Coefficients of Uranium in Molten Aluminium.

temperature in °C	wt% U	mean diffusion coefficient in 10^{-5} cm/s	Ref.	temperature in °C	wt% U	mean diffusion coefficient in 10^{-5} cm/s	Ref.
700	0.580	0.45	[73]	1000	4.25	5.7	[57]
700	0.527	0.65	[73]	1000	8.15	6.0	[57]
750	0.990	1.56	[73]	1000	11.8	6.5	[57]
800	0.273	2.42	[73]	1000	15.25	7.5	[57]
800	0.731	2.40	[73]	1000	18.40	7.3	[57]
850	0.182	4.48	[73]				

Thermal Conductivity

The thermal conductivity of uranium – aluminium alloys decreases with increasing uranium content [10, 13, 19], see **Fig. 3-49**. Concerning the 13 wt% uranium – aluminium alloys,

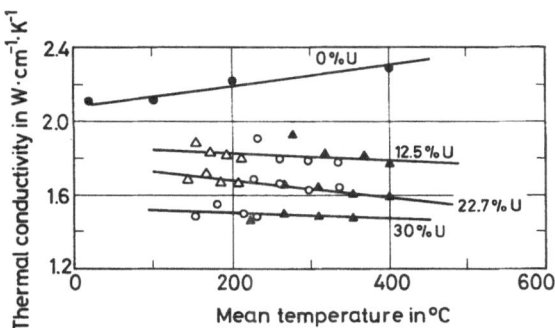

Fig. 3-49. Thermal conductivity of uranium – aluminium alloys [13]. △, ○, ▲, and ● denote results obtained with different runs.

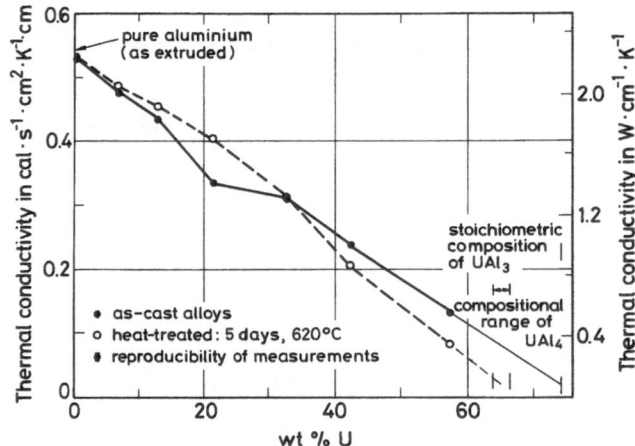

Fig. 3-50. Effect of composition on thermal conductivity at 65°C on cast and heat-treated U — Al alloys [19].

no significant influence on the thermal conductivity was found with sample conditions: quenched in water, cooled in air, or casting condition (100°C or 700°C preheated graphite mold). Improved thermal conductivity was observed after heat treatment at 620°C for 5 d, which was most pronounced for samples containing 21 wt% uranium. Those alloys consist entirely of an α aluminium — UAl₄ eutectic in the as-cast condition. After heat treatment the eutectic changed from a lamellar to a spheroidal morphology. With the as-cast alloys the conductivity decreased fairly uniformly for alloys with up to 21 wt% uranium. A marked change occurred in the gradient of the thermal conductivity with alloys containing 21 to 33 wt% uranium. This effect is seen with a considerable quantity of free α aluminium, confirmed with alloys containing 33 to 42 wt% uranium, whereas no free α aluminium was found with 21 wt% uranium — aluminium alloys [19], see **Fig.** 3-**50**. The measured values are summarized in Table 3/28.

Table 3/28
Thermal Conductivities of Uranium — Aluminium Alloys.

wt% U	temperature in °C	thermal conductivity			Ref.
		as cast	heat treated 620°C, 5 d	forged bars 370°C, 0.5 h	
		in cal·s^{-1}·cm^{-2}·K^{-1}·cm		in W·cm^{-1}·K^{-1}	
0.15	65	0.533 ± 0.002	0.537, 0.527		[19]
0.5	65	0.523, 0.525	0.538, 0.535, 0.532		[19]
6.97	65	0.476 ± 0.005	0.488 ± 0.006		[19]
12.91	65	0.437 ± 0.006	0.454 ± 0.006		[19]
21.43	65	0.336 ± 0.006	0.408 ± 0.003		[19]
32.94	65	0.313 ± 0.005	0.314 ± 0.003		[19]
42.49	65	0.239 ± 0.002	0.207 ± 0.008		[19]
58.40	65	0.134, 0.133, 0.135	0.082, 0.080		[19]

Table 3/28 (continued)

wt % U	temperature in °C	thermal conductivity			Ref.
		as cast	heat treated 620°C, 5 d	forged bars 370°C, 0.5 h	
		in $cal \cdot s^{-1} \cdot cm^{-2} \cdot K^{-1} \cdot cm$		in $W \cdot cm^{-1} \cdot K^{-1}$	
12.5	200			1.83	[10, 13]
	300			1.81	[10, 13]
	400			1.79	[10, 13]
22.7	200			1.68	[10, 13]
	300			1.64	[10, 13]
	400			1.59	[10, 13]
30.5	200			1.51	[10, 13]
	300			1.49	[10, 13]
	400			1.48	[10, 13]

3.1.6.2.3 Chemical Reactions

Corrosion Behavior

The corrosion behavior of uranium — aluminium alloys with 30 to 45 wt % uranium in steam was found to be essentially independent of the uranium content. The corrosion experiments

Table 3/29
Corrosion Test Results of Uranium — Aluminium Alloys in 150°C Water [14].

wt % U	melting procedure	heat treatment	total weight gain in mg/cm², in indicated time, in days				
			1	5	10	20	30
30	melted in ZrO_2	none; as cast	0.36	0.23	0.38	0.46	0.39
35	crucibles under		0.36	0.38	0.40	0.52	0.44
40	vacuum and cast		0.34	0.34	0.41	0.58	0.53
45	into copper molds		0.34	0.40	0.45	0.31	0.63
30	melted in air and	none; as cast	0.30	0.33	0.36	0.49	0.49
35	poured in graphite;		0.30	0.48	0.54	0.66	0.61
40	remelted in vacuum		0.38	0.41	0.48	0.62	0.63
45 *)	and poured in copper		0.45	0.65	0.92	1.42	1.82
30	melted in ZrO_2	sealed in evacuated	0.35	0.27	0.38	0.49	0.41
35	crucibles under	vycor and heat-	0.34	0.29	0.39	0.34	0.49
40	vacuum and cast	treated at 1022°F	0.43	0.48	0.52	0.72	0.67
45	into copper molds	for 24 h	0.40	0.38	0.42	0.65	0.44
25 aluminium				0.20	0.32	0.32	

*) Sample very porous; therefore, not a good indication of the corrosion resistance of alloy.

References for 3.1.6.2 on pp. 175/9

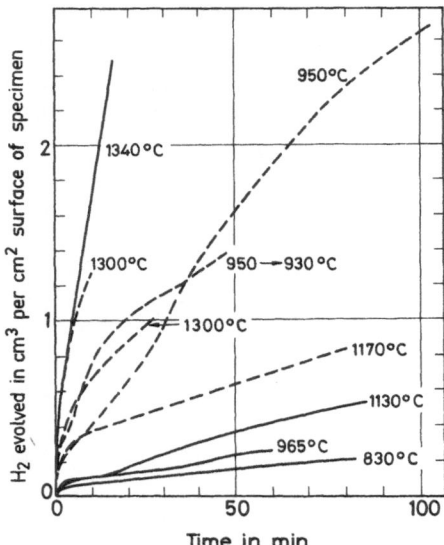

Fig. 3-51. Influence of temperature on the reactivity of uranium — aluminium alloys (-----) and of aluminium (———) in 100% H_2O flowing at 70 cm^3/min [74].

at 150 °C for 30 days showed that the effect was about two times as great as for pure aluminium. A tenacious oxide film inhibits further attack, see Table 3/29, p. 167 [14]. High-temperature experiments, performed on 20 wt% uranium — aluminium alloys heated by electromagnetic levitation technique under 1 atm of steam, gave the following results: provided the oxide layer does not break down, a parabolic rate law is followed, as analyzed by the volume of hydrogen liberated. At 950 to 990 °C the reaction was several times faster than that at 950 to 1 150 °C. The corrosion rate curve ceased to fluctuate between 1 000 °C and 1 060 °C. Between 1 060 °C and 1 400 °C the uranium — aluminium alloys are somewhat less reactive. The results are discussed in terms of hydrogen dissolving into the metal, the formation of an oxide film, cracking of the oxide layer, formation of suboxides, and disruption of the oxide layer by hydrogen pockets, see **Fig.** 3-51 [74], see also [75 to 77]. Cast and wrought uranium — aluminium alloys with 6 wt% uranium or more showed a good corrosion resistance against water at 350 °C in short term tests for 1 d. The corrosion resistance of the cast alloys was comparable to alloys with ternary additions of Si [78], see also [79]. Only little surface attack was observed with 48 wt% U — Al alloys containing 3 wt% Si at 60 °C and 90 °C within 3, 6, and 12 weeks [52]. The corrosion resistance of alloys with uranium contents in the range of 15 to 87% U and with additional Ni contents of up to 2% and Fe to 0.5% were tested in distilled water at 290 °C and 350 °C for up to 60 d. In the range of 15 to 53% U, the corrosion rate was found to be 4 mils/year or less. With U contents of more than 80% there was a breakdown of the corrosion resistance. Further addition of 0.1% Ti or 2% Nb showed no influence in the materials behavior [80].

Explosive reactions may occur if molten particles of U — Al alloys are contacted with water, resulting in the formation of oxides and hydrogen [81 to 83].

Dissolution of Uranium — Aluminium Alloys

Continuous dissolution of U—Al alloys for reprocessing of the materials is carried out by dissolving the alloys in HNO_3 (4 to 8 M [84], 6.6 M [85], or 3 M [86]) catalyzed with mercury (or fluorides [87]), $Hg(NO_3)_2$ 0.002 M [86], 0.005 M [84, 88], or 0.007 M [85], using 4.8 mol acid per mol of Al [85] under boiling conditions (90 to 95°C [84], 105°C [89, 90]), see also [91 to 97]. For rates of dissolution see **Fig. 3-52**. The dissolution rate decreases with increasing U and Si content (see Table 3/30, p. 170 [85]). Continuous dissolution was also carried out in 6 M HCl with $HgCl_2$ as catalyst [98].

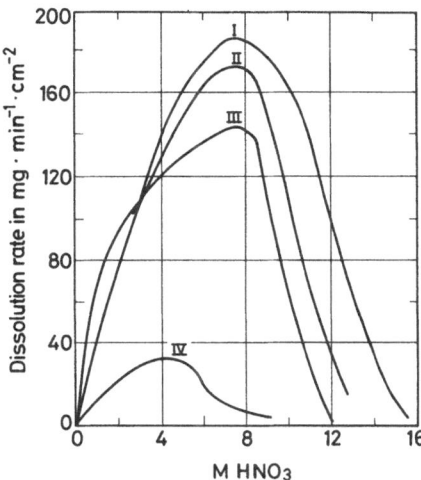

Fig. 3-52. Dissolution rates of U—Al alloys in HNO_3. I) 15% U, extruded, 0.005 M $Hg(NO_3)_2$; II) same, 0.002 M $HgNO_3$; III) 5% U, cast, 0.005 M $Hg(NO_3)_2$; IV) 7.5% U, cast, 0.002 M $Hg(NO_3)_2$ [88].

There is a marked difference in the dissolving rates of cast slugs and extruded slugs for alloys with 7.5 wt% U owing to the different morphology of the alloys. Depending on the mode of solidification and fabrication, the UAl_4 particles are more or less uniformly distributed in a matrix of aluminium, whereas in the cast slugs, the UAl_4 forms a network around the boundaries of pure aluminium. Thus, the extruded alloys are more rapidly dissolved because the acid can more readily attack the material if the UAl_4, which is easily dissolved by acids, is uniformly dispersed [99], see also [100, 101]. Flat plates and flattened tubes are much less dependent on catalyst concentration (1.5×10^{-6} to 7.5×10^{-4} M Hg), with round rods or tubes over twice of the catalyst concentrations are required [102]. The reprocessing of alloys with ternary additions of Si (0.4 to 0.7 wt% or up to 11% [50]) in HNO_3, using Hg catalysis, resulted in insoluble residues with Al and Si as the major constituents [50, 103].

Al-rich alloys were also successfully dissolved in NaOH solutions [88], see also [104 to 106]. Electrolytic dissolution was carried out in $NaOH-NaNO_3-Ba(NO_3)_2$ solutions at 80 to 90°C at 5 to 6 V [107]. For analytical purposes, dissolution in concentrated HCl or HNO_3 [108, 109], or decomposition in HCl/H_2O_2 (especially for small amounts of Al with 0.01% upwards) was used [110].

References for 3.1.6.2 on pp. 175/9

Table 3/30
Instantaneous Dissolution Rates for Uranium—Aluminium—
Silicon Alloys (in mg·cm^{-2}·min^{-1}), Determined at 103°C at
a Level of 30 g Al/L in Solution [85].

wt% Si	no U	22 wt% U	31 wt% U
0.0	26	9	4
	26	9	4
0.6	23	—	8
	22	—	8
1.5	26	—	10
	23	—	8
3.0	—	26	12
	—	25	11
	—	—	11
	—	—	13
	—	—	12
5.0	27	16	10
	28	13	11
11.0	9	—	—
	7	—	—

Metallographic Etching

Metallographic etching of U—Al alloys usually was carried out with NaOH or other aluminium etchants [10, 13]: with 1% NaOH (at 16 wt% U—Al) [48, 53], 2% NaOH (at 33 wt%
U—Al) [19], 10% NaOH for 10 to 15 s (at 30 wt% U—Al) [1]. Etching was also carried out with
0.5% HF (at 30 wt% U—Al) [48, 53], or by anodic oxidation at 25 V in 3% ammonium tartrate
(pH 5.5) [69]. The following etching treatment is recommended to distinguish parts of UAl$_3$
and UAl$_4$ phases metallographically in the UAl$_4$—Al field:

Electrolytic etching with one part of 40% chromic acid and one part of 50% acetic acid
maintained for 3 min at 0.05 A/cm^2 at 25°C resulted in yellow UAl$_3$ and blue-grey UAl$_4$ [111].

Etching with 50% HNO$_3$ at 25°C resulted, after 1 min immersion, in light straw-colored UAl$_3$
and distinct grey UAl$_4$ [111].

Etching with 1% HF at 25°C gave no clear distinction [111].

3.1.6.2.4   Fabrication of Alloys and Fuel Elements

Uranium—aluminium alloys were fabricated by hot-forging, hot rolling, or in some cases,
cold-rolling. Forging was carried out at 550 to 600°C on alloys containing up to 20 wt% U.
Alloys in the range of 20 to 30 wt% U can only be forged with great difficulty. Alloys with more
than 30 wt% U cannot be forged. Hot-rolling, performed at 550 to 600°C, shows similar
characteristics as hot-forging. After the cast structure is broken down, cold-rolling is easily
performed [10, 13], see also [11, 17, 27, 112 to 116].

Special fabrication methods were used for alloys containing more than about 35 wt% U.
Such alloys at first show a large amount of UAl$_3$ after casting. As solidification continues, this

UAl_3 reacts with Al to form the brittle compound UAl_4. Suppression of the peritectic reaction by rapid cooling leads to a structure with a smaller amount of hard and brittle compound and more soft matrix material. This improves the fabricability. A combination of rapid cooling and jacketing in aluminium prior to fabrication allows the alloys with higher uranium content to be hot-rolled also. After initial breakdown by hot-rolling in an aluminium jacket and subsequent cold-rolling, reduction of 50 to 60% can be achieved without cracking [10, 13], see also [11, 14, 117]. Typical defects of hot-rolled sandwich-plates by the "picture-frame" technique (see p. 154) are "dog-boning", tapering, "fish-tailing" (double tapering), variation in core thickness and heterogeneity in composition owing to multiphase alloys, or rotating and bending of the core, wedge shape, or various irregularities [118].

Uranium—aluminium alloys containing up to 24.5 wt% U were extruded without difficulty [27, 112]. Extrusion was also successfully carried out with alloys containing 48 wt% U. Billets of 3 in. diameter were successfully extruded into rods of 3/4 in. diameter through a 45° cone-shaped die [14, 52], see also [119].

Si-modified alloys were hot-rolled without jacketing in aluminium [14, 52]. Addition of 3 wt% Si led to a slight gain in ductility and a decrease of about 25% in yield strength, which significantly relieved the fabrication problems [55].

In the cold-rolled textures of alloys containing 5 wt% and 13 wt% U and of pure Al there is an increase in spread and a decrease in intensity with increasing U content. The results are analyzed in terms of a change from a $\{5, 6, 16\}\langle\overline{10}, \overline{13}, 8\rangle$ (near $-\{113\}\langle\overline{3}\overline{4}3\rangle$) texture in pure Al to a near $-\langle111\rangle$ fiber texture in the alloy containing 13 wt% U [120, 121].

The U—Al alloys are fabricated to nuclear fuel elements in the form of Al-covered plates or tubular rods to be used in MTR-type research reactors with neutron fluxes of about $10^{15}\,n\cdot cm^{-2}\cdot s^{-1}$. The fuel plates are normally fabricated by hot-rolling using the "picture frame" technique similar to the UAl_2—Al (see p. 116), or UAl_4—Al (see p. 153 and Fig. 3-45, p. 154) dispersion fuel elements [10, 14, 55, 112, 113, 122, 123], see also [11, 52, 114, 116, 124]. The tubular elements are fabricated using extrusion techniques [119, 123, 125, 126], see also [127 to 131], or centrifugal casting techniques [132].

Table 3/31
Design Data on Assembly Containing 48 wt% Uranium—Aluminium Alloy for Irradiation Testing in Material Test Reactors (MTR) [11].

number of fuel plates per element	18
nominal water gap spacing (in in.)	0.117
width and length of active portion of plate (in in.)	$2.5 \times 23^5/_8$
overall width and length of fuel plates (in in.)	(2) $2.8 \times 28^5/_8$
	(16) $2.8 \times 24^5/_8$
total thickness of fuel plates (in in.)	0.060
core thickness of fuel plates (in in.)	0.028
nominal core composition (in wt%)	uranium: 48
	aluminium: 52
nominal ^{235}U enrichment (in %)	20.00
^{235}U content per plate (in g)	10.56
^{235}U content per assembly (in g)	190
^{235}U fuel distribution (in g/cm^2)	0.027
overall dimensions of side plates (in in.)	$3.169 \times 0.188 \times 28^5/_8$

References for 3.1.6.2 on pp. 175/9

A typical design specification of an MTR fuel plate element is given in Table 3/31, p. 171. The design specifications stipulate a curvature with 5½ in. radius. The complete element assembles 18 of the fuel plates (for an example) [11], see also [55]. Different designs of tubular elements are given in **Fig.** 3-**53** and **Fig.** 3-**54**.

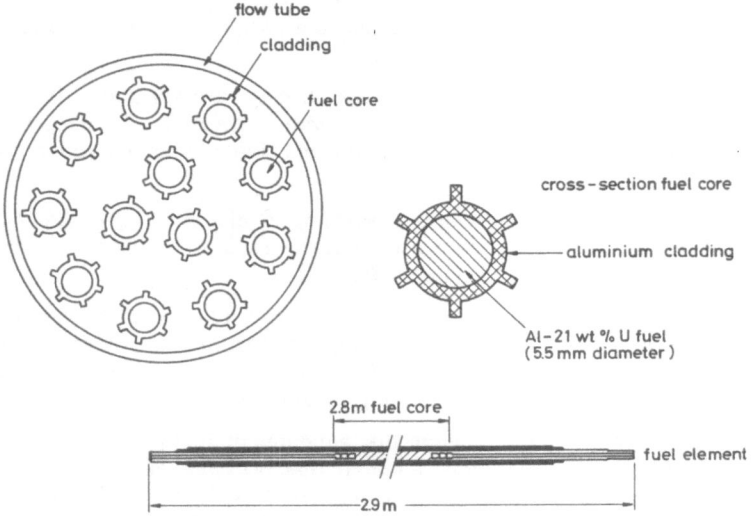

Fig. 3-53. Design of tubular aluminium-clad U — Al fuel elements [126].

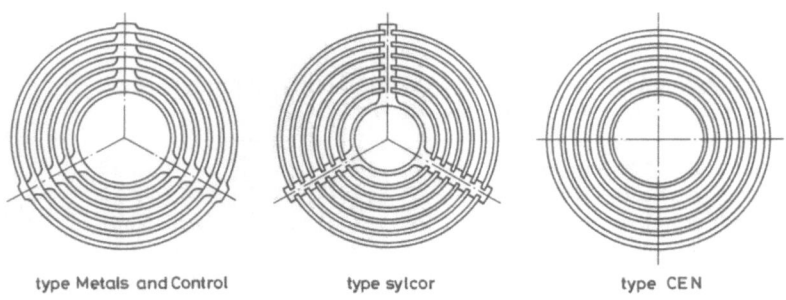

Fig. 3-54. Special design of a tubular aluminium-clad U — Al fuel element [119].

Mechanical Properties

The tensile strength of the aluminium-clad U — Al alloys showed the expected trends of increasing strength and decreasing ductility with increasing U content [14], see also [66].

Corrosion Properties

Corrosion tests were performed on aluminium-clad U−Al alloys in wrought and cast condition with defects in the cladding of 1/16 in. in diameter in deionized water at 350°C for 24 h. Samples containing less than 6 wt% U were destroyed in all cases, the claddings were either split or badly swollen. Samples containing 6 wt% U in the as-cast condition showed only very slight swelling at one slot. Samples containing additional 12 wt% Si showed no dimensional changes [75]. In general, the corrosion resistance increases with increasing U content [75], see also [52].

Uranium−Aluminium Alloys Containing Boron and/or Lithium

The introduction of "burnable poisons" such as boron or lithium into a reactor core improves the reactor performance. Therefore, three types of U−Al alloys were developed containing additionally up to 2.09 wt% Li and up to 0.168 wt% B: U−B−Al, U−Li−Al, and U−B−Li−Al [53, 69, 128, 133 to 136]. The addition of about 0.15 wt% B leads structurally to the formation of a complex boride; Li removes the position of the U−Al eutectic and leads to the formation of the eutectic Al−UAl$_4$−AlLi [53]. The solubility of Li in the matrix is about 0.6 wt% in the annealed alloys [69]. The addition of small amounts of boron leads to no substantial modifications in the mechanical or electrochemical properties of the basic alloys containing 16 wt% U [69]. The addition of about 2 wt% Li produced a marked rise of the tensile properties, in the elastic modulus, and the electrochemical potential. The density is reduced by about 7.5%. The presence of Li involves serious technological difficulties in casting and plastic working, therefore the Li content should be reduced to about 0.1 wt% [69], see also the section on ternary uranium−aluminium compounds, Section 3.1.7.2, p. 180.

3.1.6.2.5 Irradiation Behavior of Uranium−Aluminium Alloys

The behavior of uranium-rich U−Al alloys (0.78 wt% Al) under neutron irradiation is characterized by two effects: 1) secondary phase precipitation occurred, enhanced by a non-equilibrium concentration of vacancies, and 2) redissolution occurred owing to the action of fission products. This dynamic equilibrium is characterized by an enhanced solid solubility which decreases with increasing temperature, measured at 200, 450, and 550°C. Neutron radiation enhances the diffusion of solute atoms, but this effect vanishes at 550°C [137 to 139].

No defects or dimensional changes were observed with Al-rich alloys containing 48 wt% U (20% enriched in ^{235}U) after 25% burnup in a neutron flux of 1.9×10^{14} n·cm^{-2}·s^{-1} [11], see also [140 to 143]. After a maximum burnup of 58%, local corrosion and swelling occurred, caused by combination of high burnup and high fuel temperature [144]. The ranges of different fission fragments in U−Al alloys were measured at U−Al targets by radiochemical methods. The measured results followed the empirical relation $R/E^{2/3} = 0.03723 + 0.02011 \cdot A/Z^{1/2}$, with R = range of fission fragments (in mg/cm^2), E = energy (in MeV), A and Z = effective mass and atomic number [145].

The aluminium in U−Al alloys (20 wt% U−Al irradiated in a flux of 8.42×10^{16} n·cm^{-2}·s^{-1}) plays an important role in retarding the ejection of uranium atoms from the surface, compared to the irradiation of pure uranium [146].

References for 3.1.6.2 on pp. 175/9

The fission gas behavior of irradiated U — Al alloys (20 wt% U) was observed at temperatures of above 640°C, the eutectic temperature. No measurable loss of radiokrypton was found at temperatures below 640°C after annealing of up to three weeks. Above 640°C evolution occurred with a time dependence, which was in rough agreement with the theoretical prediction for diffusion of spherical particles [147], see also [148]. Further investigations showed that mostly ^{133}Xe was released. Only small amounts of ^{131}I and ^{137}Ce were observed [149].

3.1.6.2.6   Reprocessing of Irradiated Uranium — Aluminium Alloys

Dissolution of Uranium — Aluminium Fuel Elements

Irradiated U — Al fuel plates (containing 17.6 wt% U coated with Al) were dissolved in HNO$_3$ catalyzed with Hg ions. The dissolution rate of the U — Al alloy is very small for acid-deficient conditions (< 1.2 M) and a partial reduction of the U^{6+} ions to the tetravalent state occurred. For acid concentrations of > 1.2 M the rate of dissolution is very slightly less than that of pure aluminium. The dissolution generally decreases with increasing catalyst concentration [150, 151]. Precipitation of MnO$_2$ effected a maximum removal of the fission products when the solutions were acid-deficient [152].

Concentration of Uranium in Irradiated Fuels

The concentration of U and/or Pu after irradiation in the fuels is of special interest because Al is the major component in the fuel elements. Following this, a volatile halide process was developed in which ZnCl$_2$ or AlF$_3$ was used in quantities less than stoichiometric for conversion of all of the aluminium. Zinc chloride, e.g., which melts at 450 to 500°C, reacts to form AlCl$_3$, which sublimes off. At higher temperatures (i.e., 650°C) a Zn — Al alloy was formed and at about 1000°C unreacted ZnCl$_2$ and Zn volatilize, preferably at reduced pressure. With AlF$_3$, the U — Al alloy was melted and a subfluoride was formed. The AlF$_3$ was maintained in the solid state at 800 to 1000°C. This process is preferably carried out under vacuum conditions. A large amount of the fission products is removed by these methods, only about 5% of the γ activity is retained [153]. The decontamination factor is higher than 10^3 for ^{137}Ba, ^{137}Cs, and ^{90}Sr; for ^{106}Ru, ^{106}Rh, ^{95}Zr, and ^{95}Nb this factor is about 10 [154]. The advantage of this process is in the handling of small volumes [153, 154].

Concentration of U and/or Pu in the irradiated U — Al alloys was also achieved by an electrolytic process, using the alloy as the anode of an electrolytic cell in contact with molten cryolite (Na$_3$AlF$_6$) as the electrolyte. The cell was operated with 22 to 26 A · h at a temperature of the bath of 970°C. After removal of the Al, an alloy was found with increased U and/or Pu content [155, 156].

A further process for concentration of the fissionable material was developed by melting the U — Al fuel element and precipitating the enriched U by addition of magnesium in which the uranium is insoluble. The uranium was concentrated at the bottom of the ingot as UAl$_3$. After cooling, the U-containing part was cut off. In laboratory experiments a concentration of the U-containing volume of 20:1 was achieved [157].

Other Reprocessing Methods

Reprocessing of irradiated U — Al alloys was carried out by fractional sublimation of the chlorides. The alloys were treated with dry HCl at increasing temperatures and the separation of aluminium yielded 99%. The separation of uranium was found to be somewhat smaller owing to partial oxidation during the process. A second sublimation step of the uranium part was useful in removing the fission products ^{141}Ce, ^{140}La, and ^{103}Ru [158]. In a similar process, chlorination and pyrolysis of the volatile chlorides was used with subsequent fluorination of the uranium part to UF_6 and adsorption on NaF. Excellent decontamination factors were observed for the uranium part with less than 5×10^{-8} Ci/g U [159], see also [160].

Separation of the fission products and Pu from irradiated U−Al fuels was also carried out by dissolving the alloys in $KAlCl_4$ with additional aluminium at temperatures of 260 to 800°C. During the process, the formation of Al_2O_3 was prevented by bubbling CCl_4 through the molten mass. A separation into two phases occurred when the mass was cooled to 400 to 500°C. The metallic phase consisted of a U−Al alloy and the second phase (salt phase) contained Pu and the major part of the fission products in form of the chlorides [161, 162].

References for 3.1.6.2:

[1] Bramfitt, B. L.; Leighly, H. P., Jr. (Metallography **1** [1968] 165/93).

[2] Ambrozio, F. F.; Vieira, R. R. (Met. ABM [Assoc. Brasil. Metais] **31** No. 207 [1975] 73/80; C.A. **83** [1975] No. 31816).

[3] Marques de Lima, R.; Ambrozio, F. F. (Congr. Anu. ABM **39** [1984] 247/60; C.A. **104** [1986] No. 232499).

[4] Marques de Lima, R.; Filho, F. A. (Metalurgia [Sao Paulo] **40** [1984] 599/603; C.A. **102** [1985] No. 224333).

[5] Filho, F. A.; Vieira, R. R. (IEA-433 [1976] 1/14; C.A. **87** [1977] No. 71936; C.A. **89** [1978] No. 115926).

[6] Filho, F. A.; Vieira, R. R. (IEA-436 [1976] 1/17; C.A. **87** [1977] No. 71935).

[7] Peckner, D. (5th Nucl. Eng. Sci. Conf., Cleveland, Ohio, 1959, Preprint V-105; C.A. **1959** 17849).

[8] Runnalls, O. J. C.; Boucher, R. R. (Trans. AIME **233** [1965] 1726/32).

[9] Saller, H. A.; Keeler, J. R.; Eddy, N. S. (BMI-T-16 [1949] 1/19; N.S.A. [1949] No. 1977).

[10] Saller, H. A. (Proc. 1st Intern. Conf. Peaceful Uses At. Energy, Geneva 1956, Vol. 9, pp. 214/20).

[11] Thurber, W. C.; Erwin, J. H.; Beaver, R. J. (ORNL-2351 [1958] 1/35; N.S.A. **12** [1958] No. 6613).

[12] Gordon, P.; Kaufmann, A. R. (J. Metals **2** [1950] 182/4).

[13] Saller, H. A. (in: Finneston, H. M.; Howe, J. P., Progress in Nuclear Energy, Ser. 5, Metallurgy and Fuels, Vol. 1, Pergamon, London 1956, pp. 535/43).

[14] Cunningham, J. E.; Beaver, R. J.; Thurber, W. C.; Waugh, R. C. (TID-7546 [1958] 269/97; C.A. **1958** 9905).

[15] Daniel, N. E.; Foster, E. L.; de Mastry, J. A.; Bauer, A. A.; Dickerson, R. F. (BMI-1183 [1957] 1/40; N.S.A. **11** [1957] No. 11196).

[16] Allen, B. C.; Isserow, S. (Acta Met. **5** [1957] 465/72).

[17] Manly, W. D. (ORNL-766 [1950] 1/20; N.S.A. **11** [1957] No. 8530).

[18] Storhok, V. W.; Bauer, A. A.; Dickerson, R. F. (Trans. Am. Soc. Metals **53** [1961] 837/42).

[19] Jones, T. I.; Street, K. N.; Scoberg, J. A.; Baird, J. (Can. Met. Quart. **2** No. 1 [1963] 53/72; C.A. **58** [1963] 12254).

[20] Haynes, W. B.; Lorenz, F. R. (WAPD-PWR-FEM-106 [1956] 1/20; C.A. **1959** 13017).

[21] Thurber, W. C.; Beaver, R. J. (Met. Soc. AIME Inst. Metals Div. Spec. Rept. Ser. **5** No. 7 [1958] 57/62).

[22] Huel, J. J. (NP-7144 [1959] 1/21; N.S.A. **13** [1959] No. 5568).

[23] van Audenhove, J.; Joyeux, J. (J. Nucl. Mater. **19** [1966] 97/102).

[24] Bomar, E. S. (CF-53-5-190 [1953] 1/9; N.S.A. **11** [1957] No. 13897).

[25] Larson, W. L.; Klein, J. L. (NMI-1168 [1956] 1/65; C.A. **1957** 11965).

[26] Kneppel, D. S. (NMI-1206 [1958] 1/40; C.A. **57** [1962] 5669).

[27] Dürrschnabel, W.; Lange, E. (Z. Metallk. **57** [1966] 451/4).

[28] Dürrschnabel, W. (Z. Metallk. **60** [1969] 521/5).

[29] Titmuss, R. J. (NLCO-1052 [1970] 1/15; C.A. **74** [1971] No. 44766).

[30] United Kingdom Atomic Energy Authority (Brit. 798687 [1958]; C.A. **1959** 3026; Nucl. Eng. **4** [1959] 146; N.S.A. **13** [1959] No. 10099).

[31] Bridges, W. H. (ORNL-1727 [1954] 1/120; N.S.A. **14** [1960] No. 8673).

[32] Baird, J.; Runnalls, O. C. J. (AECL-1418 [1961] 1/14; C.A. **1962** 12623).

[33] Cacciari, A., de Leone, R.; Fizzotti, C.; Gabaglio, M. (Can. 614071 [1961]; N.S.A. **15** [1961] No. 20781).

[34] Huet, J. J. (ATB Met. [Acta Tech. Belg.] **1** No. 8 [1959] 217/20; C.A. **1960** 16344).

[35] Runnalls, O. J. C. (Can. 642140 [1962]; C.A. **58** [1963] 318).

[36] Gol'dshtein, S. L.; Gudkov, S. V.; Raspopin, S. P.; Fedorov, V. L. (Elektrokhimiya **17** [1981] 1267/72; Soviet Electrochem. **17** [1981] 1043/7; C.A. **95** [1981] No. 177575).

[37] Gol'dshtein, S. L.; Raspopin, S. P.; Fedorov, V. A. (Elektrokhimiya **17** [1981] 1350/3; Soviet Electrochem. **17** [1981] 1110/3; C.A. **95** [1981] No. 211856).

[38] Trzebiatowski, W.; Bogacz, A.; Barycka, I. (Nucleonika **4** [1959] 591/8; C.A. **1960** 18242).

[39] Seki, Y.; Mitamura, N. (Genshiryoku Kogyo, **11** [1965] 37/40; C.A. **67** [1967] No. 49365).

[40] Moore, R. H. (Belg. 614861 [1962]; U.S. 3052536 [1962]; C.A. **57** [1962] 10902, **57** [1962] 12243).

[41] Zappi, D. A. (Ger. 1212304 [1966]; C.A. **64** [1966] 15342).

[42] Ogawa, Y.; Hisamatsu, Y.; Kawamura, K. (Nippon Kogyo Kaishi **75** [1959] 223/8; C.A. **1961** 9224).

[43] Gol'dshtein, S. L.; Raspopin, S. P.; Sergeev, V. L.; Fedorov, V. A. (Fiz. Khim. Elektrokhim. Rasplavl. Tverd. Elektrolitov Tezisy Dokl. Vses. 7th Konf. Fiz. Khim. Ionnyk Rasplavov Tverd. Elektrolitov 1979, Sverdlovsk, Vol. 2, pp. 73/5; C.A. **93** [1980] No. 122537).

[44] Boucher, R. (CEA-1298 [1959] 111/7; C.A. **1960** 2224).

[45] Boucher, R. (J. Nucl. Mater. **1** [1959] 13/27).

[46] Boucher, R. (Symp. Solid State Diffusion, Saclay, France, 1958, pp. 111/7; N.S.A. **15** [1961] No. 26564; AEC-tr-4768 [1960] 1/11; N.S.A. **14** [1960] No. 7826).

[47] Gualandi, D.; Schileo, G. (Alluminio **28** [1959] 449/55).

[48] Gualandi, D.; Paganelli, M.; Schileo, G. (Alluminio **28** [1959] 331/6; C.A. **1960** 8547).

[49] Exner, H. E.; Petzow, G. (Metall **23** [1969] 220/5).

[50] Paige, B. E.; Gibson, G. W.; Rohde, K. L. (IN-1194 [1968] 1/89; N.S.A. **23** [1969] No. 9508).

[51] Less, C. S. (AERE-GM-13 [1948] 1/7; N.S.A. **10** [1956] No. 5652).

[52] Thurber, W. C.; Beaver, R. J. (ORNL-2602 [1959] 1/53; C.A. **1959** 15797).

[53] Gualandi, D.; Paganelli, M.; Schileo, G. (Alluminio **28** [1959] 507/15).

[54] Exner, H. E.; Petzow, G. (5th Intern. Leichtmetalltagung, Leoben 1968 [1969], pp. 77/83; C.A. **74** [1971] No. 37351).

[55] Beaver, R. J.; Cunningham, J. E. (TID-7559-Pt. 1 [1959] Paper 4; N.S.A. **13** [1959] No. 23049).

[56] Gibson, G. W.; Pollard, R. J.; Carlson, B. G.; Hatch, M.; Angilella, A. G. (IN-1131 [1968] 2/8; N.S.A. **22** [1968] No. 23761).

[57] Mitamura, N.; Maruya, K.; Kimura, J. (Nippon Genshiryoku Gakkaishi **5** [1963] 467/75; C.A. **61** [1964] 14136).

[58] Daniel, N. E.; Foster, E. L.; Dickerson, R. F. (BMI-1388 [1959] 1/31; C.A. **57** [1962] 10890).

[59] Waugh, R. C. (ORNL-2701 [1959] 1/67; C.A. **1959** 157969).

[60] Miller, E. C.; Bridges, W. H. (ORNL-910 [1951] 1/70; N.S.A. **14** [1960] No. 14012).

[61] Williams, R. O. (CF-50-7-160 [1950] 1/90; N.S.A. **11** [1957] No. 7749).

[62] Epremian, E. (WASH-703 [1957] 1/40; N.S.A. **14** [1960] No. 15027).

[63] Depoitier, J.; Timmermans, W.; Huberlant, M.; Mathieu, F.; Marchal, F.; Detavernier, W.; Michiels, H. (EUR-2522.f [1966] 1/94; C.A. **66** [1967] No. 42869).

[64] Aronin, L. R.; Klein, J. L. (NMI-1118 [1954] 1/44; N.S.A. **11** [1957] No. 12489).

[65] Cintra, S. H. L.; Gentile, E. F.; Haydt, H. M.; Capocchi, J. D. T. (Met. ABM [Assoc. Brasil. Metais] **24** No. 131 [1968] 781/7; C.A. **71** [1969] No. 26810).

[66] Lillie, D. W. (WASH-155 [1954] 1/66; N.S.A. **14** [1960] No. 9725).

[67] Saller, H. A.; Rough, F. H.; Chubb, W. (BMI-1113 [1957] 1/27; C.A. **1960** 20782).

[68] Freitas de Quadros, N. (IEA-451 [1976] 1/18; C.A. **87** [1977] No. 89103; C.A. **88** [1978] No. 93513).

[69] Cesoni, G.; Gualandi, D.; Paganelli, M.; Schileo, G. (Proc. 2nd Intern. Conf. Peaceful Uses At. Energy, Geneva 1958, Vol. 6, pp. 451/62).

[70] Lebedev, V. A.; Sal'nikov, V. I.; Nichkov, I. F.; Raspopin, S. P. (At. Energiya SSSR **32** [1972] 115/8; Soviet At. Energy **32** [1972] 129/32; C.A. **76** [1972] No. 145738).

[71] Jacobs, A. J. (IA-627 [1961] 1/12; C.A. **61** [1964] 1313).

[72] Chiotti, P.; Kateley, J. A. (J. Nucl. Mater. **32** [1969] 135/45).

[73] Hesson, J. C.; Hootman, H. E.; Burris, L., Jr. (Electrochem. Technol. **3** [1965] 240/4).

[74] Robertson, A., Rostron, A. J. (Corros. Sci. **5** [1965] 425/47).

[75] Bowen, H. C.; Dillon, R. L. (Corrosion [Houston] **17** [1961] 9/11).

[76] Coriou, H.; Grall, L.; Kurka, G. (Mem. Sci. Rev. Met. **58** [1961] 209/14; C.A. **1961** 25702).

[77] Zelezny, W. F. (IDO-16629 [1960] 1/55; C.A. **1961** 12077).

[78] Bowen, H. C.; Dillon, R. L. (HW-55352 [1958] 1/10; C.A. **1958** 16168).

[79] Jones, T. I. (CRFD-810 [1958] 1/24; AECL-735 [1959] 1/22; C.A. **1959** 12133).

[80] Ruther, W. E.; Draley, J. E. (ANL-6053 [1959] 1/9; C.A. **1960** 8546).

[81] Ruebsamen, W. C.; Chrisney, J. B. (NAA-SR-Memo-75 [1951] 1/8; N.S.A. **11** [1957] No. 2511).

[82] Ruebsamen, W. C.; Shon, F. J.; Chrisney, J. B. (NAA-SR-197 [1952] 1/30; N.S.A. **11** [1957] No. 2253).

[83] Pearlman, H. (NAA-SR-Memo-858 [1954] 1/22; C.A. **1960** 20746).

[84] Bresee, J. C.; Foster, D. L.; Nurmi, E. O. (Chem. Eng. Progr. Symp. Ser. **55** No. 22 [1959] 25/32; N.S.A. **13** [1959] No. 16816; N.S.A. **14** [1960] No. 3576; C.A. **1961** 7081).

[85] Bower, J. R. (IN-1087 [1967] 7/10; N.S.A. **22** [1968] No. 4231).

[86] Rodrigues, G. C. (JAERI-M-84-073 [1984] 409/15; C.A. **102** [1985] No. 193411).

[87] Andelin, R. L.; Tingey, F. H.; Slansky, C. M. (IDO-14407 [1957] 1/45; N.S.A. **11** [1957] No. 10470).

[88] Wymer, R. E.; Blanco, R. E. (Ind. Eng. Chem. **49** No. 1 [1957] 59/61).

[89] Michel, P. (Compt. Rend. 1st Colloq. Fr. Esp. Trait. Combust. Irradies, Fontenay-aux-Roses, Fr., 1967, pp. 11/27; C.A. **72** [1970] No. 27397).

[90] Michel, P. (Bull. Inform. Sci. Tech. Commis. Energ. At. [Paris] No. 127 [1968] 61/73; C.A. **69** [1968] No. 102195).

[91] Faugeras, P.; LeRoy, P.; Lheureux, C. (Ind. Chim. Belge Suppl. **1** [1959] 672/9; C.A. **1960** 10564).

[92] Burns, R. E.; Holm, C. H. (HW-18414 [1952] 1/19; C.A. **57** [1962] 6826).

[93] Wymer, R. G. (CF-54-5-74 [1954] 1/10; N.S.A. **12** [1958] No. 893).

[94] Boeglin, A. F.; Buckham, J. A.; Chajson, L.; Lemon, R. B.; Paige, D. M.; Stoops, C. E. (IDO-14321 [1954] 1/93; N.S.A. **15** [1961] No. 23566).

[95] Boeglin, A. F.; Buckham, J. A. (IDO-14359 [1955] 1/26; N.S.A. **15** [1961] No. 26110).

[96] Martin, M. D.; Buckham, J. A. (IDO-14361 [1955] 1/31; C.A. **57** [1962] 6939).

[97] Faugeras, P.; LeRoy, P.; Lheureux, C. (Ind. Chim. Belge Suppl. **1** [1959] 1/8; N.S.A. **15** [1961] No. 5391).

[98] Crocker, I. H. (Anal. Chim. Acta **18** [1958] 231, 523; AECL-744 [1958] 1/20).

[99] Hamby, D. E. (CF-50-12-23 [1950] 1/11; N.S.A. **11** [1957] No. 7658).

[100] Culler, F. L., Jr. (CF-51-10-91 [1951] 1/7; C.A. **1961** 17444).

[101] Paige, B. E.; Jacobson, M. E.; LuDell Evans, T.; Vernon, H. (IN-1364 [1970] 1/47; C.A. **74** [1971] No. 37388).

[102] Boeglin, A. F.; Buckham, J. A. (IDO-14425 [1957] 1/26; N.S.A. **12** [1958] No. 7122).

[103] Parrett, O. W.; Rhode, K. L. (IDO-14441 [1958] 1/25; C.A. **1959** 4079).

[104] Foster, D. L. (CF-55-11-123 [1957] 1/14; N.S.A. **12** [1958] No. 704).

[105] Hibbs, R. (NYO-2700 [1959] 3/8; C.A. **1959** 21224).

[106] Delpech, A. (Fr. 1539924 [1968]; C.A. **72** [1970] No. 35094).

[107] Jasny, G. R.; Barkman, J. R.; Sprague, T. P.; Smith, R. P. (Ind. Eng. Chem. **50** [1958] 1777/80).

[108] Bulyanista, L. S.; Ivanova, K. S.; Ryzhinskii, M. V.; Alekseeva, N. A.; Solutseva, L. F.; Shereshevskaya, I. I. (Radiokhimiya **20** [1978] 859/65; Soviet Radiochem. **20** [1978] 733/8; C.A. **90** [1979] No. 161625).

[109] Krtil, J.; Sus, F.; Kuvik, V.; Klosova, E. (J. Radioanal. Nucl. Chem. **106** [1986] 319/26).

[110] Vinogradov, A. V.; Shpinel, V. S. (Zavodsk. Lab. **29** [1963] 804; Ind. Lab. [USSR] **29** [1963] 860; C.A. **59** [1963] 9304).

[111] Hills, R. F. (J. Inst. Metals **86** [1958] 438/41).

[112] Eschnauer, H.; Wolf, A. (Atomwirtsch. Atomtech. **11** [1966] 82/5).

[113] Haydt, H. M.; Cintra, S. H. (Met. ABM [Assoc. Brasil. Metais] **23** No. 121 [1967] 955/60; C.A. **68** [1968] No. 55714).

[114] Gray, R. J. (TID-7523 [1956] 67/73; C.A. **1957** 5580).

[115] Jones, T. I.; McGee, I. J.; Norlock, L. R. (FD-43 [1960] 1/29; AECL-1215 [1961] 1/29; C.A. **1961** 19697).

[116] Ambrozio Filho, F.; Freitas de Quadros, N. (IEA-443 [1976] 1/14; C.A. **87** [1976] No. 108037).

[117] Thurber, W. C. (CF-57-3-160 [1957] 1/11; N.S.A. **12** [1958] No. 276).

[118] de Meester, P. J. A.; Deruyttere, A.; Brabers, M. J. (J. Inst. Metals **98** [1970] 86/94).

[119] Feraday, M. A.; Foo, M. T.; Davidson, R. D.; Winegar, J. E. (Nucl. Technol. **58** [1982] 233/41).

[120] Thurber, W. C.; McHargue, C. J. (Trans. AIME **218** [1960] 141/4).

[121] Thurber, W. C. (ORNL-2635 [1958] 1/64; C.A. **1959** 7937).

[122] Jonckheere, E.; Defreyn, B.; van Asbroeck, Ph.; Flipot, J.; van Geel, J. (Fuel Elem. Fabr. Proc. Symp. **1** [1960] 343/71; C.A. **1961** 26732).

[123] Yamamoto, T. (Genshiryoku Kogyo **13** [1967] 33/7; C.A. **67** [1967] No. 16996).

[124] Walker, D. E.; Noland, R. A. (ANL-5559 [1958] 1/35; N.S.A. **12** [1958] No. 5370).

[125] Strasser, A. (5th Nucl. Eng. Sci. Conf., Cleveland, Ohio, 1959, Preprint V-85, pp. 1/15; C.A. **1959** 17696).

[126] Flipot, A. J.; Massaux, H. (J. Nucl. Mater. **22** [1967] 177/91).

[127] Smith, C. D. (ORNL-216 [1948] 1/8; N.S.A. **11** [1957] No. 12494).

[128] Lillie, D. W. (Ger. 1027809 [1958]; C.A. **1961** 8107).

[129] Garrett, E. E.; Alm, G. V.; Binstock, M. H. (NAA-SR-5120 [1960] 1/47; N.S.A. **14** [1960] No. 21954).

[130] United Kingdom Atomic Energy Authority (Fr. 1434757 [1966]; C.A. **66** [1967] No. 15851).

[131] Peacock, H. B. (DP-MS-78-89 [1978]1/26; C.A. **91** [1979] No. 165 147).

[132] Daniel, N. E.; Foster, E. L.; Dickerson, R. F. (BMI-1363 [1959] 1/24; N.S.A. **14** [1960] No. 10 764).

[133] Lillie, D. W. (U.S. 2 951 801 [1960]; C.A. **1961** 2 304).

[134] Thurber, W. C. (U.S. 2 967 812 [1961]; C.A. **1961** 7 099).

[135] Thurber, W. C.; Milko, J. A.; Beaver, R. J. (TID-7 526 [1957] 116/28; C.A. **1957** 11 965; ORNL-2 149 [1957]; C.A. **1958** 1 027).

[136] Tamura, S.; Toida, Y.; Yonezawa, C.; Tamura, K. (JAERI-M-82-070 [1982] 1/17; C.A. **99** [1983] No. 147 991).

[137] Lazarevic, Dj. (Phys. Metall React. Fuel Elem. Proc. Intern. Conf., Berkeley, Engl., 1973 [1975], pp. 282/5; C.A. **84** [1976] No. 127 812).

[138] Lazarevic, Dj. (Radiat. Damage React. Mater. Proc. 2nd Symp., Vienna 1969, Vol. 2, pp. 439/46; C.A. **72** [1970] No. 61 943).

[139] Bobkov, Yu. V.; Naskidashivili, I. A.; Petrosyem, V. V.; Sokurskii, Yu. N. (At. Energia **38** [1957] 20/2; Soviet At. Energy **38** [1975] 22/4; C.A. **83** [1975] No. 14 413).

[140] Perez, E.; Kohut, C.; Giorsetti, D.; Copeland, G.; Suelgrove, J. (JAERI-M-84-073 [1984] 67/76; C.A. **102** [1985] No. 193 407).

[141] Naskidashvili, I. A.; Maile, Kh. E.; Dolize, V. M. (V sb. Vopr. Atom. Nauki i Tekhn. Ser. Fiz. Radiatsion. Povrezhdenii i Radiatsion. Materialoved. **1** [1975] 81/3; C.A. **85** [1976] No. 196 560).

[142] Savage, H. C.; Drosten, F. (ORNL-473 [1949] 1/10; N.S.A. **10** [1956] No. 6 302).

[143] Francis, W. C.; Gibson, G. W.; Scarrah, W. P. (TID-7 642-2 [1962] 444/68; C.A. **59** [1963] 7 127).

[144] Gavin, A. P.; Crothers, C. C. (ANL-6 180 [1960] 1/46; C.A. **1961** 144).

[145] Saeki, M.; Ishimori, T. (Nippon Genshiryoku Gakkaishi **14** [1972] 278/82; C.A. **77** [1972] No. 82 721).

[146] Hashimoto, T. (Radioisotopes [Tokyo] **21** [1972] 635/40; C.A. **78** [1973] No. 91 583).

[147] Reynolds, M. B. (Nucl. Sci. Eng. **3** [1958] 428/34).

[148] Tachimori, S.; Amano, H. (J. Nucl. Sci. Technol. [Tokyo] **11** [1974] 488/94).

[149] Tamai, T.; Hayashi, M.; Mishima, K.; Sagane, T.; Yoshida, H.; Kanda, K.; Shibata, T.; Posey, J. C.; Rimshaw, S. J. (Ann. Rept. Res. Reactor Inst. Kyoto Univ. **15** [1982] 15/26; C.A. **98** [1983] No. 115 375).

[150] Beone, G.; Mazzoleni, G. P. (RT-CHI-23 [1965] 1/31; C.A. **64** [1966] 9 184).

[151] Beone, G.; Lojacono, R.; Mazzoleni, G. P.; Moccia, A. (RT-CHI-8 [1967] 1/14; C.A. **68** [1968] 64 958).

[152] Henry, H. E. (DP-347 [1959] 1/12; C.A. **1960** 23 979).

[153] U.S. Atomic Energy Commission (Brit. 855 490 [1960]; N.S.A. **15** [1961] No. 11 092).

[154] Roake, W. E.; Lyon, W. L. (U.S. 2 931 721 [1960]; C.A. **1960** 19 212).

[155] Roake, W. E.; Lyon, W. L. (U.S. 2 930 738 [1960]; C.A. **1960** 19 212).

[156] U.S. Atomic Energy Commission (Brit. 861 700 [1961]; N.S.A. **15** [1961] No. 13 012).

[157] David, C.; Junger, J. M.; Lorenz, R.; Montellanico, P.; Wurm, J. G. (EUR-4 243 [1968] 1/20; C.A. **72** [1970] No. 27 408).

[158] Speeckaert, P. (AEC-tr-4 553 [1960]; N.S.A. **15** [1961] No. 17 007).

[159] Bourgeois, M.; Perrot, M.; Rochedreux, Y. (CEA-CONF-1 537 [1969] 1/16; C.A. **74** [1971] No. 48 753).

[160] Ramaswami, D.; Levitz, N. M.; Jonke, A. A. (ANL-6 830 [1965] 1/48; C.A. **67** [1967] No. 69 763).

[161] U.S. Atomic Energy Commission (Fr. 1 229 237 [1960]; N.S.A. **15** [1961] No. 32 254).

[162] Moore, R. H. (U.S. 2 948 586 [1960]; C.A. **1961** 1 234).

3.1.7   Ternary Uranium − Aluminium Compounds

3.1.7.1   The Quasi-Binary System UAl₂ − UC

UAl$_2$ is used in solid-state sintering of uranium monocarbide as a removable sintering aid. From this, the quasi-binary system UAl$_2$−UC was investigated [1, 2].

The quasi-binary UAl$_2$−UC phase diagram (see **Fig.** 3-**55**) shows a simple eutectic system with a eutectic temperature of 1540°C at 12 wt% UC; solid solubilities between the two phases are negligibly small. There are no ternary compounds known in the system UAl$_2$−UC [1, 2].

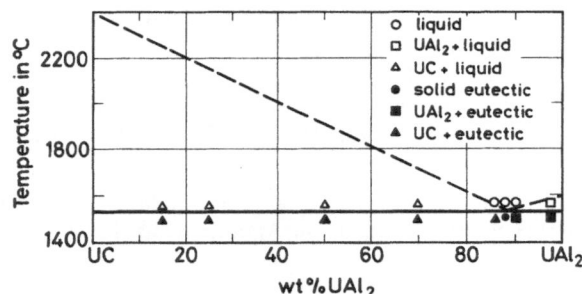

Fig. 3-55. Phase diagram of the quasi-binary system UAl$_2$−UC [1].

References for 3.1.7.1:

[1] Hammond, J. P.; Adamson, G. M., Jr. (Carbides Nucl. Energy Proc. Symp., Harwell, Engl., 1963 [1964], Vol. 2, pp. 648/67; C.A. **63** [1965] 1433).

[2] Hammond, J. P.; Adamson, G. M. (Proc. 2nd Intern. Powder Met. Conf., New York 1965 [1966], Vol. 3, pp. 3/25; C.A. **67** [1967] No. 16999).

3.1.7.2   U − Al − B;   U − Al − Li;   U − Al − B − Li

Boron and lithium are used as ternary (and also quarternary) additions in uranium−aluminium alloys as burnable poisons in plate-like fuel elements for nuclear application, to permit a correction of the neutron flux [1].

The addition of 0.05 to 0.15% B leads to the formation of an intermetallic compound, probably a complex boride [1, 2]. The solubility of lithium in the U−Al matrix is about 0.6% in the annealed alloys. With contents of Li of about 2% a eutectic Al−UAl$_4$−AlLi composition is formed [1, 2].

The alloys were prepared in two steps: 1) preparation of the U−Al alloy at 1100°C in vacuum (10^{-2} Torr) by reaction of the elements in a graphite crucible, and preparation of AlLi and AlB master alloys; 2) after melting of the U−Al alloy and the master alloy under argon, the resulting product was cast at 780 to 790°C in an iron-mould covered with colloidal graphite. The ingots were then homogenized at 480°C for 11 h [1, 2], see also [3, 4]. The presence of 2% Li involves serious technological difficulties both in casting and in plastic working [1, 2].

The Brinell hardness, elastic moduli, and potentials of dissolution versus saturated calomel electrode of different alloys are summarized in Tables 3/32 and 3/33.

Table 3/32
Physical Properties of Boron- and/or Lithium-Containing Uranium — Aluminium Alloys [1].

alloy	Brinell hardness in kg/mm^2	elastic modulus in kg/mm^2
$Al-16\%$ U	40	8004.9
$Al-16\%$ U-0.15% B	42	8121.2
$Al-16\%$ U-2% Li	55	8384.8
$Al-16\%$ U-2% Li-0.05% B	55	8165.6
$Al-16\%$ U-0.1% Li-0.05% B	41	7974.4

Table 3/33
Potentials of Dissolution of Boron- and/or Lithium-Containing Uranium — Aluminium Alloys in mV [1]. H_2O stands for demineralized water.

alloy	in H_2O		in H_2O + 60 ppm Na_2CrO_4	
	after 1 min	after 24 h	after 1 min	after 24 h
$Al-16\%$ U	740	500	840	400
$Al-16\%$ U-0.15% B	740	364	680	360
$Al-16\%$ U-2% Li	1244	480	880	480
$Al-16\%$ U-2% Li-0.05% B	1210	464	840	460
$Al-16\%$ U-0.1% Li-0.05% B	830	364	680	420

References for 3.1.7.2:

[1] Cesoni, G.; Gualandi, D.; Paganelli, M.; Schileo, G. (Proc. 2nd Intern. Conf. Peaceful Uses At. Energy, Geneva 1958, Vol. 6, pp. 451/62).

[2] Gualandi, D.; Paganelli, M.; Schileo, G. (Alluminio **28** [1959] 507/15).

[3] Thurber, W. C.; Milko, J. A.; Beaver, R. J. (TID-7526 [1957] 116/28; C.A. **1957** 11965; ORNL-2149 [1957] 1/23; C.A. **1958** 1027).

[4] Tamura, S.; Toida, Y.; Yonezawa, C.; Tamura, K. (JAERI-M-82-070 [1982] 1/17; C.A. **99** [1983] No. 147991).

3.1.7.3   U−Al−Si

3.1.7.3.1   Phase Relationships

Silicon can replace aluminium in UAl_2 to at least 12 at% at 900°C, and in the as-cast condition even higher silicon substitutions are possible under non-equilibrium conditions, as observed from X-ray diffraction analysis [1]. Otherwise, it was stated that the solubility of silicon in UAl_2 extends only to about 1 at% at 1300°C [2, 3], see also [4 to 6]. UAl_3 is isostructural with USi_3 (Cu_3Au type). They form a complete series of solid solutions at 900°C.

    References for 3.1.7.3 on pp. 189/90

There are some indications for a miscibility gap at some lower temperatures [1], see also [7, 8]. UAl_4, which is formed by a peritectic reaction at 730°C (liquid + $UAl_3 \rightarrow UAl_4$), is not present in ternary U—Al—Si alloys containing more than 3 at% Si [9 to 13].

Only small amounts of Al can be dissolved in U_3Si: 0.5 wt% [14] or about 1 at% at 650°C [2, 3]. There was no measurable solution of Al in U_3Si_2 [1], or about 1 at% at 1360°C [2, 3]. USi also shows a very low solubility for Al [1]. Solubility data of Al in U_3Si_5 are not available, but the defect-type crystal structure should permit a ternary element, such as Al, to either occupy an empty lattice site or displace an Si atom [1]. The solubility of Al in USi_{2-x} (defect structure with tetragonal $ThSi_2$ type) is at least 16 at%. The compounds show a distinctive banded structure in the micrographs [1]. USi_2 (with stable AlB_2-type structure) seems to disappear when entering the three-phase region [1].

A ternary intermetallic compound is reported with the composition of U_2AlSi_2 or possibly $U_2Al_2Si_3$ [1]. Two further compounds, UAl_2Si_2 with a cubic [15, 16], and UAlSi with a hexagonal crystal symmetry [15], are reported. UAlSi could not be confirmed by [1].

A series of alloys in the range of U_3Si to Al are of interest in nuclear metallurgy: $U_{60}Al_{20}Si_{20}$ to $U_{40}Al_{20}Si_{40}$ [1] and an alloy with the composition U + 3.5 wt% Si + 1.5 wt% Al ("UAlSi"), which is of interest as "UAlSi"—Al dispersion in nuclear fuel elements [17, 18], see also [1, 14, 19].

The ternary system U—Al—Si was investigated particularly in the U-rich region extending to the quasi-binary system U_3Si_2—UAl_2 [2, 3, 5]. The U-rich corner at the 900°C isothermal section is shown in **Fig. 3-56**. A special investigation of the quasi-binary system U_3Si_2—UAl_2 was based on thermal and microscopic measurements as well as X-ray determinations. Melting surfaces (see **Fig. 3-57**) and melting equilibria, isothermal sections at 950, 850, 650, and 400°C (see **Fig. 3-58**), and four temperature-concentration sections were determined. Especially in the quasi-ternary system U—UAl_2—U_3Si_2 four non-variant 4-phase equilibria were found. The following transformation reactions have been observed [3] (see **Fig. 3-59**, p. 184):

melt $\rightarrow$ γ-U + U_3Si_2 + UAl_2, at 960°C; γ-U + U_3Si_2 $\rightarrow$ U_3Si + UAl_2, at 890°C; γ-U $\rightarrow$ β-U + U_3Si + UAl_2, at 730°C; β-U $\rightarrow$ α-U + U_3Si + UAl_2, at 580°C.

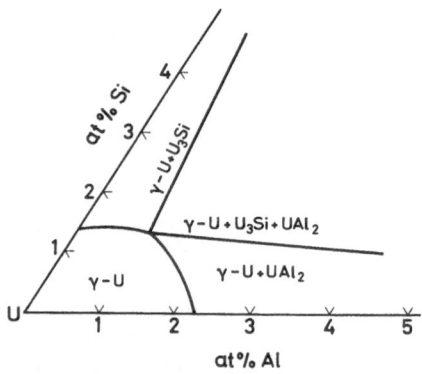

Fig. 3-56. Isothermal section at 900°C in the U-rich corner of the U—Al—Si phase diagram [1], reproduced from [5].

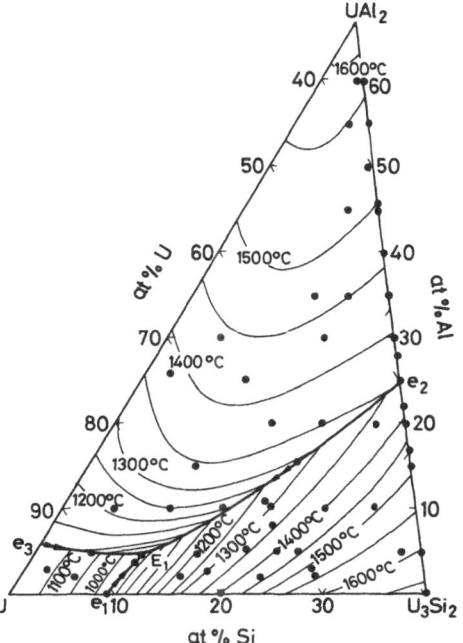

Fig. 3-57. Alloys and melting surfaces in the ternary system $U - U_3Si_2 - UAl_2$ [3].

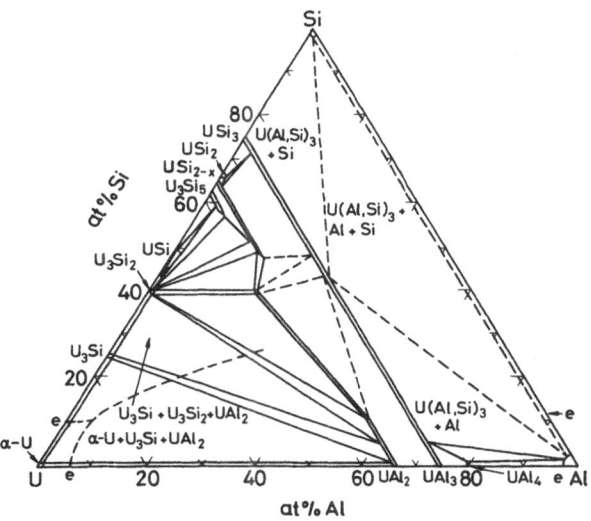

Fig. 3-58. Isothermal section at 400°C of the ternary system U — Al — Si [1].

References for 3.1.7.3 on pp. 189/90

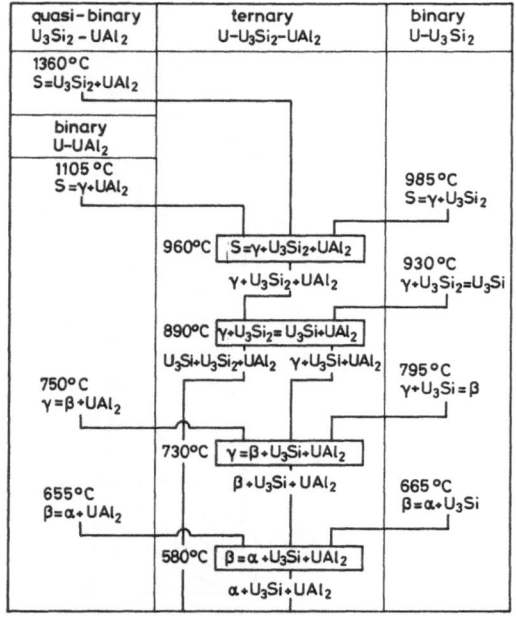

Fig. 3-59. Schematic diagram of the reactions in the system $U - U_3Si_2 - UAl_2$ [3].

The uranium solid solution fields extend only slightly into the ternary region, whereas the intermetallic phases U_3Si, U_3Si_2, and UAl_2 showed no extension into this region [3].

The intermetallic phases of the quasi-binary system $UAl_2 - U_3Si_2$ form a simple eutectic system with a eutectic temperature of 1360°C and a eutectic composition of 51 at% U, 23 at% Al, 26 at% Si. A complete miscibility gap was observed at 850°C.

The solubility of Si in UAl_2 is similar to that of Al in U_3Si_2, extending to 1 at% at the eutectic temperature [2, 3], see also [4, 5]. A phase diagram is given in [3]. The X-ray density is 10.4 g/cm³. The oxidation velocity in air was determined at 400°C for 100 min to be $8.0 \ \mu g \cdot cm^{-2} \cdot min^{-1}$ [4].

3.1.7.3.2 Preparation and Properties

U(Al, Si)₂

Compounds of the composition $U(Al, Si)_2$ in form of buttons were prepared by arc-melting. The $U(Al, Si)_2$ compounds crystallize cubically with $MgCu_2$-type structure [1]. The lattice parameters decrease with increasing Si content as shown in **Fig. 3-60**.

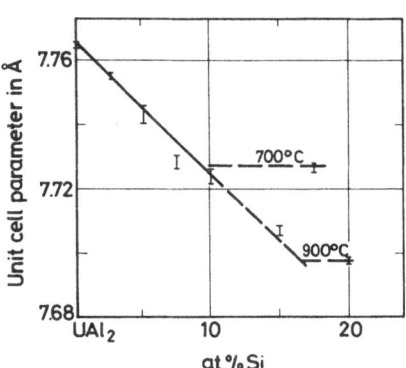

Fig. 3-60. Unit-cell parameters of U(Al, Si)$_2$ compounds [1].

U(Al, Si)$_3$

Formation of U(Al, Si)$_3$ compounds was observed in castings of different ternary alloys by X-ray analysis [20, 21], see Table 3/34.

From the reaction of U$_3$Si–Al and U$_3$Si$_2$–Al dispersions with the aluminium matrix, the following reactions are stated by the volume change of aluminium-covered fuel plates: 1) U$_3$Si + 8 Al $\rightarrow$ 3 U(Al, Si)$_3$, whereas the stoichiometric reaction would occur at about 38 vol% U$_3$Si and 62 vol% Al; 2) U$_3$Si$_2$ + 7 Al $\rightarrow$ 3 U(Al, Si)$_3$, whereas the stoichiometric reaction would occur at 48 vol% U$_3$Si$_2$ and 52 vol% Al [306].

Table 3/34
Ternary Alloys Containing U(Al, Si)$_3$ [20].

alloy composition in wt%			U:Si ratio	X-ray analysis	rate of dissolution in HNO$_3$ in mg·cm^{-2}·min^{-1}
U	Si	Al			
22	3	75	1.2	U(Al, Si)$_3$, Al, Si	26
22	5	73	1.9	U(Al, Si)$_3$, Al, Si	16
32	3	65	0.8	U(Al, Si)$_3$, Al	12
32	5	63	1.3	U(Al, Si)$_3$, Al, Si	11

UAl$_3$ and USi$_3$ are isomorphous with ordered structures of the AuCu$_3$ type [22]. A continuous series of solid solutions is formed [1]. The lattice constants of the mixed crystals decrease with increasing Si content [1, 21] as given in **Fig. 3-61**, p. 186.

The dissolution rate for U(Al, Si)$_3$ compounds in HNO$_3$ decreases with increasing Si content (see Table 3/34) [20].

References for 3.1.7.3 on pp. 189/90

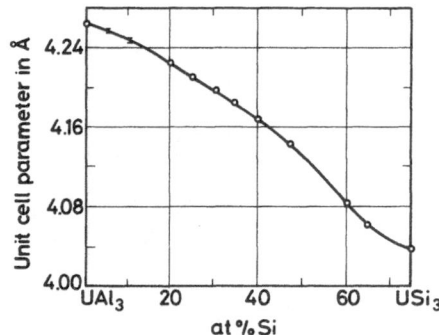

Fig. 3-61. Unit cell parameters of U(Al, Si)$_3$ compounds [1].

U$_2$AlSi$_2$

A compound with the composition U$_2$AlSi$_2$ (or possibly U$_2$Al$_2$Si$_3$) was prepared by melting of the elements. Samples of the as-cast condition showed a sharply formed X-ray diagram, but the crystal structure was not identified. From microstructure analysis, the samples showed the ternary U$_2$AlSi$_2$ phase and a eutectic of the ternary phase + UAl$_2$ embedded in an alloy of composition U$_{37.5}$Al$_{27.5}$Si$_{35}$. After heat treatment at 900°C and lower, a reaction occurred which was completed after 3 days. The nature of this reaction is not clear, but a completely different X-ray diagram was observed. The observed d spaces and intensities of the X-ray diffraction pattern are given in the original paper. The Si atoms can be substituted by Ge atoms to form U$_2$AlGe$_2$, which is isostructural with the heat-treated U$_2$AlSi$_2$ [1].

UAlSi

The ternary compound UAlSi was prepared by melting stoichiometric amounts of the elements plus Al in 0.2 to 0.5% excess to compensate for losses by evaporation. The melts were carried out in an arc furnace under argon atmosphere. The samples were remelted, placed in Ta containers, sealed in evacuated quartz ampoules, and heated at 850 to 900°C for 7 days. UAlSi was obtained as a fragile and well recrystallized material [15]. This compound was not confirmed by [1].

UAlSi crystallizes with a hexagonal structure. The lattice constants, obtained from X-ray diffraction pattern, are a = 10.778 ± 0.008 Å, c = 8.433 ± 0.013 Å [15].

UAl$_2$Si$_2$

The ternary compound UAl$_2$Si$_2$ was prepared by melting stoichiometric amounts of the elements with Al in 0.2 to 0.5% excess using exactly the same method as that for UAlSi (see above). UAl$_2$Si$_2$ was obtained as a fragile and well recrystallized material [15, 16]. This compound could not be confirmed by [1]. Probably this compound has to be regarded as U(Al, Si)$_3$ with AuCu$_3$-type structure [1].

UAl$_2$Si$_2$ crystallizes cubically in a partly disordered Perovskite type or a AuCu$_3$-type structure with Z = 1.

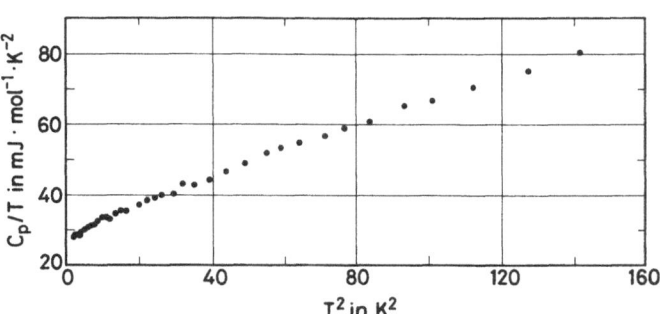

Fig. 3-62. Low-temperature specific heat of UAl$_2$Si$_2$ between 1.5 and 12 K [16].

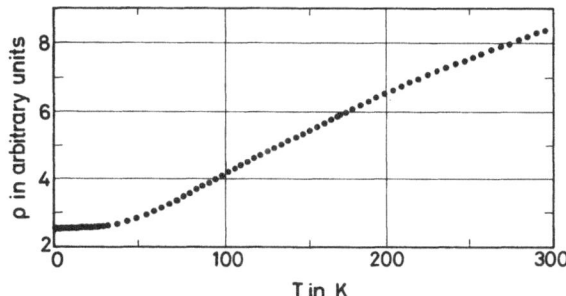

Fig. 3-63. Temperature dependence of the electrical resistivity of UAl$_2$Si$_2$ between 1.6 and 300 K [16].

The space group is probably Pm$\overline{3}$m $-$ O$_h^1$ (No. 221) [16]. Lattice constant from X-ray diffraction pattern: a $=$ 4.145 $\pm$ 0.002 Å [15, 16]. A slight variation of the lattice constant was observed with different samples: a $=$ 4.163 $\pm$ 0.004 Å, 4.176 $\pm$ 0.006 Å [16].

The specific heat of UAl$_2$Si$_2$ was measured in the range of 1.5 to 12 K (see **Fig.** 3-62). Only at the lowest temperatures the specific heat may be represented by the equation $C_p = \gamma \cdot T + \beta \cdot T^3$, where $\gamma = 27.9$ mJ $\cdot$ mol$^{-1} \cdot$ K^{-2} (parameter of the electronic specific heat), $\beta = 0.35$ mJ $\cdot$ mol$^{-1} \cdot$ K^{-4} (lattice contribution term) [16].

The temperature dependence of the electrical resistivity of UAl$_2$Si$_2$ was measured in the range of 1.6 to 300 K (see **Fig.** 3-**63**). In the low-temperature region up to 40 K the resistivity follows a T^3 law rather than a T^2 law [16].

The magnetic susceptibility was measured in the range of 4 to 280 K. As shown in **Fig.** 3-**64**, p. 188, the magnetic susceptibility of UAl$_2$Si$_2$ is nearly temperature-independent [16]. From measurements of the magnetic susceptibility UAl$_2$Si$_2$ was shown to be a superconductor with a transition temperature of 1.34 $\pm$ 0.05 K, indicating the absence of well-defined magnetic moments on the uranium atoms [16].

                    References for 3.1.7.3 on pp. 189/90

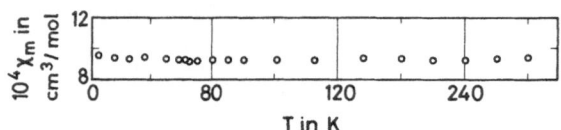

T in K

Fig. 3-64. Temperature dependence of the magnetic susceptibility of UAl_2Si_2 between 4 and 300 K [16].

U + 3.5 wt% Si + 1.5 wt% Al Alloy ("UAlSi")

Molten "UAlSi" (containing U $+3.5$ wt% Si $+1.5$ wt% Al) cast into molds contains the uranium-rich phases U_3Si and UAl_2 which are transformed by a heat-treatment at 1073 K within 72 h to δ-U_3Si containing 0.5 wt% Al and particles of U_3Si_2 and UAl_2 [7, 17, 18], see also [1, 14, 19, 23]. This material was used in aluminium-covered fuel test elements for application in nuclear technology experiments.

"UAlSi" — Al Dispersions and Fuel Elements

"UAlSi"—Al dispersions are of interest in nuclear technology to be used as "Reduced Enrichment Fuels for Research Reactors". Al dispersions of "UAlSi" with a liquidus temperature of 1648 K were prepared by blending "UAlSi" powder (<150 µm) with aluminium powder (<44 µm), pressing the mixtures into billets, which are then hot-extruded into rods [17].

Tensile properties of two dispersions are given in Table 3/35.

Table 3/35
Tensile Properties of "UAlSi" — Al Dispersions at 423 K [17].

dispersion	volume fraction of Al	0.2% yield stress in MPa	total elongation in %
Al — 63 wt% "UAlSi"	0.76	67	2.2
Al — 75 wt% "UAlSi"	0.64	61	0.3

The thermal conductivity of the hot extruded dispersions are $180 \text{ W} \cdot \text{m}^{-1} \cdot \text{K}^{-1}$ and $121 \text{ W} \cdot \text{m}^{-1} \cdot \text{K}^{-1}$ for compositions with 55 wt% and 75 wt% "UAlSi", respectively [17].

"UAlSi" cast at about 1500 °C in vacuum and annealed at 800 °C for 72 h showed an average corrosion rate of $<0.03 \text{ kg} \cdot \text{m}^{-2} \cdot \text{h}^{-1}$ at 300 °C under a pressure of 7 MN/m² after 5 h in water and 0.02 to 0.20 $\text{kg} \cdot \text{m}^{-2} \cdot \text{h}^{-1}$ at 550 °C under a pressure of 0.1 MN/m² after 2 h in streaming water vapor [14], see also [1, 23].

Tubular "UAlSi"—Al fuel test elements were fabricated by cladding the dispersion with aluminium using extrusion techniques (test rods: 184 mm in length, 5.5 mm core diameter, 0.76 mm clad wall thickness) [17, 18]. Vacuum annealing of those fuel elements at temperatures of up to 523 K for 2160 h showed no dimensional changes for Al—55 wt% "UAlSi", whereas about 5% elongation was observed with Al—75 wt% "UAlSi". Elongations of the fuels were found to be 6 to 10% after vacuum annealing at 673 K for 2232 h. This elongation is associated with diffusion of Al into the "UAlSi" particles. The Al penetrates along grain boundaries to form UAl_3 containing dissolved Si [17, 18].

The aqueous corrosion resistance of the dispersion was tested at small fuel rods with a 1.0 mm diameter hole drilled through the Al cladding in the mid-section of the rods. The corrosion tests were carried out in autoclaves at 423 K and 453 K for 528 and 336 h, resulting in weight increases by 0.5 to 1.0% [17].

Neutron irradiation tests were carried out in a flux of $1.1 \times 10^{18} \, n \cdot m^{-2} \cdot s^{-1}$ at 76.6 to 131.0 kW/m to burnups of up to 80%. The fuel core swelling at 18.8 and 42.4 at% burnup was 1.82 and 3.35, 3.39% volume change, respectively, for dispersions of Al−61.5 wt% "UAlSi", which was less than the volume change in U−Al alloys irradiated in the same assembly [17, 18], see also [23 to 26].

Irradiation in a high-voltage electron microscope (HVEM) at 500 to 1 180 keV and at 300 to 660 K showed that the atomic displacement rates produced in the HVEM apparatus are about two orders of magnitude higher than for neutron irradiation at typical power reactor ratings [27].

References for 3.1.7.3:

[1] Dwight, A. E. (ANL-82-14 [1982] 1/44; C.A. **98** [1983] No. 186482).

[2] Petzow, G.; Kvernes, I. (Z. Metallk. **52** [1961] 693/5).

[3] Petzow, G.; Kvernes, I. (Z. Metallk. **53** [1962] 248/56).

[4] Thümmler, F.; Exner, H. E.; Petzow, G. (J. Nucl. Mater. **24** [1967] 328/39).

[5] Svistunova, Z. V.; Ivanov, O. S. (Stroenie Svoistva Splavov Urana Toriya Tsirkoniya Sb. Statei **1963** 9/15; C.A. **59** [1963] 7211).

[6] Ludwig, R. L. (Y-2213 [1980] 1/38; C.A. **95** [1981] No. 101417).

[7] Wiencek, T. C.; Domagala, R. F.; Thresh, H. R. (Nucl. Technol. **71** [1985] 608/16).

[8] Rough, F. A.; Bauer, A. A. (BMI-1300 [1958] 1/138; N.S.A. **11** [1958] No. 13935).

[9] Boucher, R. (CEA-1298 [1959] 111/7; C.A. **1960** 22247).

[10] Cunningham, J. E.; Beaver, R. J.; Thurber, W. C.; Waugh, R. C. (TID-7546 [1958] 269/97; Fuel Elements Conf., Paris 1957; C.A. **1958** 9905).

[11] Thurber, W. C.; Beaver, R. J. (Met. Soc. AIME Inst. Metals Div. Spec. Rept. Ser. **5** No. 7 [1958] 57/62).

[12] Boucher, R. (Symp. Solid State Diffusion, Saclay 1958 [1959], pp. 111/7; N.S.A. **15** [1961] No. 26564; AEC-tr-4768 [1960] 1/11; N.S.A. **14** [1960] No. 7826).

[13] Boucher, R. (J. Nucl. Mater. **1** [1959] 13/27).

[14] Ross, A. M. (Can. 987135 [1976]; GER. Offen. 2427265 [1975]; U.S. 4023992 [1977]).

[15] Zygmunt, A. (in: Mulak, J.; Suski, W.; Troc, R., Proc. 2nd Intern. Conf. Electron. Struct. Actinides, Wroclaw 1976 [1977], pp. 335/41; C.A. **87** [1977] No. 76741).

[16] Ott, H. R.; Hulliger, F.; Rudigier, H.; Fisk, Z. (Phys. Rev. B [3] **31** [1985] 1329/33).

[17] Wood, J. C.; Foo, M. T.; Berthiaume, L. C. (3rd Proc. Ann. Conf. Can. Nucl. Soc., Toronto 1982, pp. F1/F7; C.A. **98** [1983] No. 97370).

[18] Feraday, M. A.; Foo, M. T.; Davidson, R. D.; Winegar, J. E. (Nucl. Technol. **58** [1982] 233/41).

[19] Wyatt, B. S. (Ger. Offen. 2029789 [1971]; U.S. 3717454 [1973]).

[20] Paige, B. E.; Rohde, K. L. (Nucl. Appl. **5** [1968] 218/23).

[21] Thurber, W. C.; Beaver, R. J. (ORNL-2602 [1959] 1/53; C.A. **1959** 15797).

[22] Lam, D. J.; Darby, J. B., Jr.; Nevitt, N. V. (in: Freeman, A. J.; Darby, J. B., Jr., The Actinides: Electronic Structure and Related Properties, Academic, New York − San Francisco − London 1974, Vol. 2, pp. 119/84).

[23] Feraday, M. A.; Fehrenbach, P. J.; Cotuam, K. D.; Morel, P. A. (AECL-5028 [1975] 1/9; C.A. **84** [1976] No. 10148).

[24] Caillibot, P. F.; Hastings, I. J. (J. Nucl. Mater. **59** [1976] 257/62).
[25] Domagala, R. F.; Wiencek, T. C.; Thresh, H. R. (Nucl. Technol. **62** [1983] 353/60).
[26] Hofman, G. L. (J. Nucl. Mater. **140** [1986] 256/63).
[27] Hastings, I. J. (J. Nucl. Mater. **56** [1975] 76/80).

3.1.7.4 U−Al−Mg

Ternary additions of 0.5 and 0.54 wt% Mg to binary U−Al alloys cause no change in the liquidus temperature: 20 wt% U−0.5 wt% Mg (balance Al): 835°C, 49.6 wt% U−0.54 wt% Mg (balance Al): 1254°C [1].

Reference for 3.1.7.4:

[1] Storhok, V. W.; Bauer, A. A.; Dickerson, R. F. (Trans. Am. Soc. Metals **53** [1961] 837/42).

3.1.7.5 U−Al−RE (Sc, Y, La; Ce, Sm, Gd, Dy, Lu)

3.1.7.5.1 Phase Relationships

Additions of Sm or Dy to U−Al alloys showed, at most, only a slight influence on the melting points of the U−Al alloys [1, 2]. UAl$_2$ and SmAl$_2$ form a complete series of solid solutions with the composition of U$_{1-x}$Sm$_x$Al$_2$ compounds [1]. With La [3] and Dy [2] miscibility gaps were observed which are also assumed for other rare earth elements. This gap is placed at $0.25 < x < 0.75$ for the U$_{1-x}$La$_x$Al$_2$ compounds [3]. Only 500 ppm of U can be dissolved in LaAl$_2$ [4]. The solubility of U in hexagonal SmAl$_3$ and β-DyAl$_3$ was found, by microscopic and structural analysis, to be 35 wt% [1] and 14 wt% [2], respectively. Sm is soluble in cubic UAl$_3$ up to 14 wt% [1]. UAl$_4$ and SmAl$_4$ form a complete series of solid solutions, U$_{1-x}$Sm$_x$Al$_4$ [1]. (U, Dy)Al$_4$ compounds are stable up to 20 wt% addition of Dy to UAl$_4$ [2]. A compound with

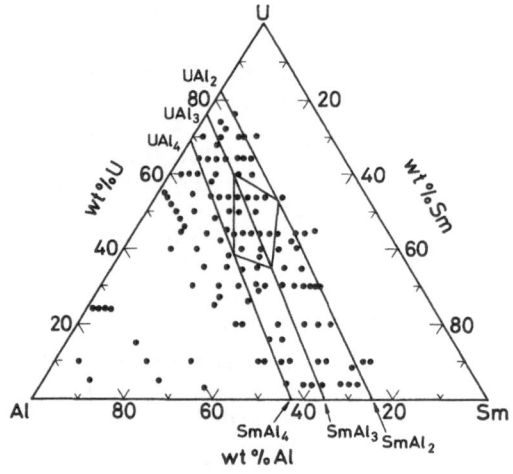

Fig. 3-65. Isothermal section of the phase diagram of Al−UAl$_2$−SmAl$_2$ at 600°C [1].

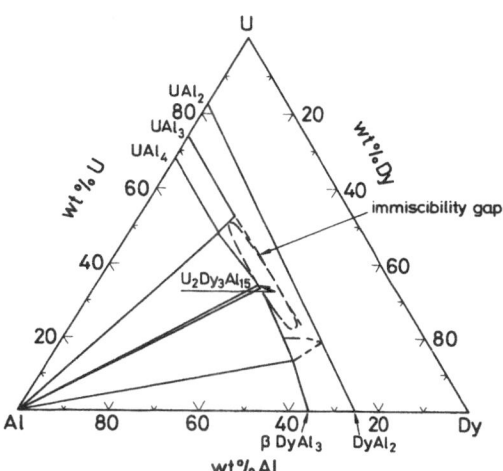

Fig. 3-66. Isothermal section of the phase diagram of Al−UAl$_2$−DyAl$_2$ at 600°C [2].

hexagonal structure, U$_2$Al$_{15}$Dy$_3$, was found in the Al-rich corner of the Al−UAl$_2$−DyAl$_2$ system (section at 600°C) [2]. Phase diagrams of the isothermal section at 600°C of the Al−UAl$_2$−SmAl$_2$ and Al−UAl$_2$−DyAl$_2$ systems are given in **Fig. 3-65** and **Fig. 3-66**.

3.1.7.5.2   U$_{1-x}$Ln$_x$Al$_2$ (Ln = Sc, Y, La; Ce, Sm, Gd, Dy, Lu)

Preparation

The samples were prepared by melting of the elements in an argon-arc furnace followed by an annealing step for homogenization with the samples sealed in capsules, which were filled with argon. Different homogenization methods are reported: at 600°C in Vycor capsules under argon for samples containing Sm or Dy [5, 6]; at 750°C for 4 to 5 d in vacuum for samples containing Gd [7]; at 900°C in vacuum after casting into chilled molds for samples containing Gd [5, 6]; at 1000 to 1200°C in tantalum or quartz tubes under 500 Torr argon; above 1200°C in the vacuum chamber of an rf-induction furnace for samples containing any of the rare earth elements [3, 8, 9]; at 1450°C for 9 h under argon for any of the rare earth elements [8, 9].

Samples of U$_{1-x}$La$_x$Al$_2$ were also prepared by melting UAl$_2$ and LaAl$_2$ in an induction furnace under argon with the samples placed in water-cooled copper crucibles. Homogenization of the samples was carried out by annealing in vacuum at 600°C [4]. Also quaternary compounds (La, Y, U)Al$_2$ were prepared by the same method [4].

Crystallographic Properties

The U$_{1-x}$Ln$_x$Al$_2$ compounds crystallize cubically with MgCu$_2$-type structure (C 15) as reported for the Y [10], La [3, 4, 9], Sm [1], and Gd [5] containing mixed crystals. Small positive deviations from Végard's law are reported for the Sm-containing compounds in the system UAl$_2$−SmAl$_2$ [1] and for the Gd-containing compounds [5]. The lattice constants for the quasi-binary UAl$_2$−ULa$_2$ system for samples of different annealing steps are shown in [3].

References for 3.1.7.5 on p. 198

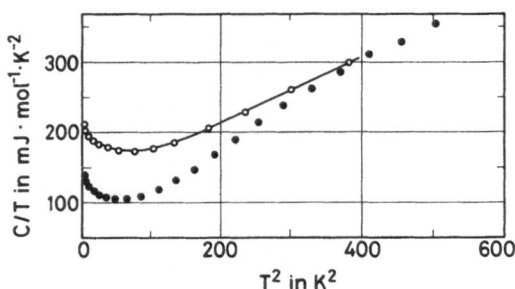

Fig. 3-67. Specific heat versus T^2 for $U_{0.80}La_{0.20}Al_2$ and, for comparison, for UAl_2 [3] from [9].
○ $U_{0.80}La_{0.20}Al_2$, ● UAl_2.

Specific Heat

The specific heat of $U_{1-x}La_xAl_2$ compounds (see **Fig.** 3-**67**) is in agreement with the spin-fluctuation theories [3]. The low-temperature specific heat follows the equation $C = \gamma \cdot T + B \cdot T^3 + \delta \cdot T^3 \cdot \ln T + \alpha \cdot T^5$, where $\gamma = \gamma_0(1 + \lambda_{e\text{-ph}} + \lambda_{SF})$ with γ_0, $\lambda_{e\text{-ph}}$, λ_{SF} = electron-photon and spin-fluctuation interaction parameters; B is a constant containing the Debye temperature Θ_D, the spin-fluctuation temperature T_{SF}, the coefficient of the spin-fluctuation term δ, and the Stoner enhancement factor S; α is the coefficient of the next order term in lattice specific heat [3, 9], see also [4]. The calculated values are summarized in Table 3/36.

Table 3/36
Coefficients in the Temperature-Dependence Equation $C = \gamma \cdot T + B \cdot T^3 + \delta \cdot T^3 \cdot \ln T + \alpha \cdot T^5$ of the Low-Temperature Specific Heat (C in $mJ \cdot mol^{-1} \cdot K^{-1}$) of $U_{1-x}La_xAl_2$ Compounds. Coefficients for UAl_2 are shown for comparison [3].

parameter	UAl_2	Ref.	$U_{0.85}La_{0.15}Al_2$	Ref.	$U_{0.80}La_{0.20}Al_2$	Ref.
γ	142.3	[11]	238.3	[9]	214.7	[3]
B	−3.64	[11]	−5.21	[9]	−3.52	[3]
δ	1.57	[11]	2.04	[9]	1.42	[3]
α	−0.0017	[11]	−0.0019	[9]	−0.0012	[3]
Θ_D in K	374	[3]	384	[3]	387	[3]
S	7.5	[3]	12	[3]	11	[3]
γ_0	40	[3]	54	[3]	72	[3]
T_{SF} in K	34	[3]	49	[3]	64	[3]

The $La_{1-x}U_xAl_2$ compounds ($100 \ll x \ll 1200$ ppm) show a second order (normal → superconducting) phase transition in the temperature dependence of the zero-field specific heat. In **Fig.** 3-**68** the reduced transition temperatures $T_c/T_{c,0}$ are given as a function of reduced impurity concentration x/x_{cr} [4].

Thermoelectric Power

The thermoelectric power of $La_{1-x}U_xAl_2$ (x = 0.07 at%) was measured at 4.2 to 300 K (see **Fig.** 3-**69**). A pronounced shoulder was observed in the thermoelectric power at 20 K, which is related to an upper boundary for the Kondo temperature, i.e., $T_K < 20$ K [4].

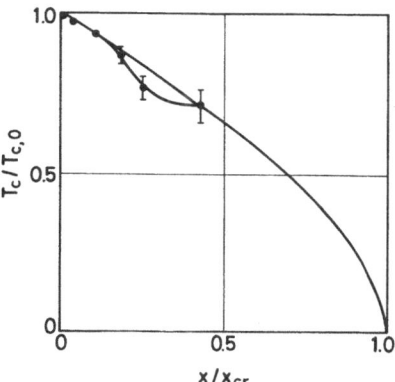

Fig. 3-68. Reduced transition temperature $T_c/T_{c,0}$ of $La_{1-x}U_xAl_2$ compounds versus x/x_{cr}. $T_{c,0} = 3.28$ K, $x_{cr} = 0.28$ at% [4].

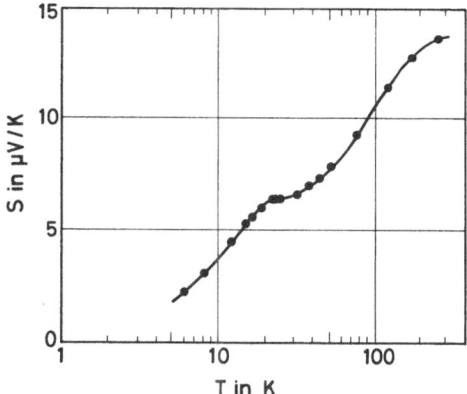

Fig. 3-69. Thermoelectric power of $La_{1-x}U_xAl_2$ (x = 0.07 at%) as a function of temperature [4].

Electrical Resistivity

The temperature dependence of the electrical resistivity of $U_{1-x}Y_xAl_2$ compounds is given in **Fig. 3-70**, p. 194. With increasing substitution of Y for U in UAl_2 the behavior can be described by a transition from a compound model with local spin-fluctuations (where in the low-temperature region ϱ is proportional to T^2) to a dilute alloy model (where in the low-temperature region ϱ is proportional to $(1-c \cdot T^2)$). This transition occurs at about x = 0.67, which is also apparent from the behavior of the residual resistivities ϱ_0 (at T = 0 K) as a function of x [10].

The electrical resistivity of UAl_2 as well as of $U_{0.97}La_{0.03}Al_2$ follows a T^2 law only in the very low-temperature region (see **Fig. 3-71**, p. 194). Using a "parallel resistor model" with $1/\varrho = 1/\varrho_i + 1/\varrho_s$ (ϱ_i = "ideal resistivity", ϱ_s = "shunt resistivity") it was shown that the ideal resistivity of UAl_2 follows a T^2 law up to higher temperatures, which was expected for a clean spin

References for 3.1.7.5 on p. 198

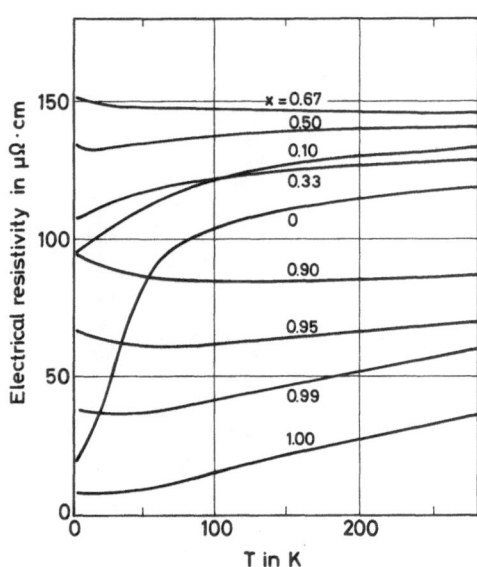

Fig. 3-70. Temperature dependence of the electrical resistivity for $U_{1-x}Y_xAl_2$ compounds [10].

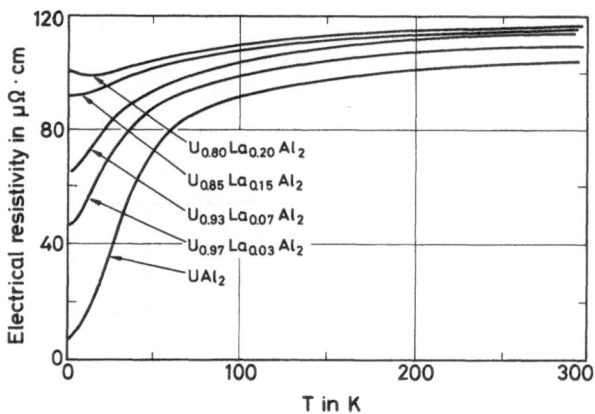

Fig. 3-71. Resistivity versus temperature for $U_{1-x}La_xAl_2$ compounds [8].

fluctuator, whereas ϱ_i for $U_{0.97}La_{0.03}Al_2$ follows a $T^{3/2}$ law. A large increase of the residual resistivity ϱ_0 was observed with increasing content of La [8, 9]. Residual resistivities at 1.4 K for $U_{0.9}Y_{0.1}Al_2$, $U_{0.93}La_{0.07}Al_2$, and $U_{0.93}Sc_{0.07}Al_2$ are given in [9]. The same effect of a large increase of the residual resistivity with increasing substitution of the U atoms was shown for $U_{1-x}Gd_xAl_2$ compounds. The most interesting feature of these compounds is the low-temperature resistivity behavior down to 1.8 K. Well-defined minima at 11, 18, 26.5, and 40 K were observed in compounds where more than 10% of the U atoms were substituted by Gd atoms [5]. The results are summarized in Table 3/37.

Table 3/37
Residual Resistivities of $U_{1-x}Gd_xAl_2$ Compounds with x between 0 and 0.30 at 1.8 K [5].

x	$\varrho_{1.8K}$ in $\mu\Omega \cdot$ cm	x	$\varrho_{1.8K}$ in $\mu\Omega \cdot$ cm
0	8.3	0.10	98
0.01	39	0.15	118
0.03	59	0.20	127
0.05	80	0.30	148

Magnetic Susceptibility

The magnetic susceptibility of $U_{1-x}Ln_xAl_2$ compounds generally increases with increasing values for x. This feature is most pronounced for Y- and La-containing compounds (see also [9]) and is much smaller for Sc-containing compounds. Calculated derivates of the susceptibility $d\chi(T = 0)/dx$ are (in 10^{-3} cm$^3 \cdot$ mol^{-1} U): 30 for Y, 28 for La, 9.5 for Lu, 8.8 for Ce, and 0.7 for Sc [3]. In $U_{1-x}Y_xAl_2$ compounds the boundary between Curie-Weiss behavior and temperature independence is close to 70 K, which is almost the same temperature at which the temperature dependence of the electrical resistivity starts to flatten off (see Fig. 3-70 [10]).

The magnetic susceptibility of lanthanum-rich $La_{1-x}U_xAl_2$ compounds is shown in **Fig. 3-72**. A Curie-Weiss behavior was observed for compounds with $x \ll 6100$ ppm below

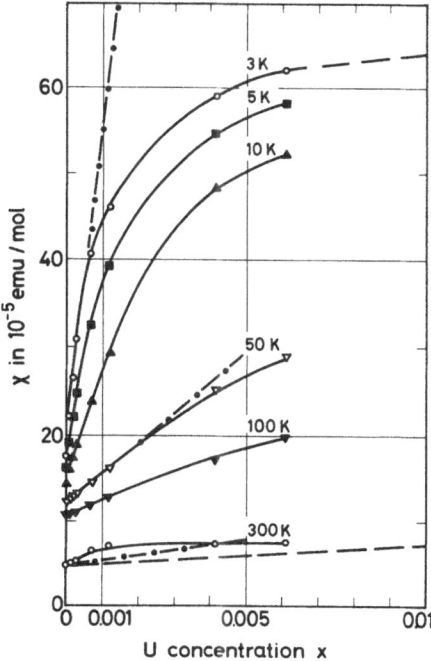

Fig. 3-72. Magnetic susceptibility of $La_{1-x}U_xAl_2$ compounds [4]. $-\cdot-\cdot-\cdot$ susceptibility in the case of non-interacting ions with $\mu_{eff} = 3.60$ μ_B. $----$ slope of $\chi(x) = (1-x)\chi_{LaAl_2} + x \cdot \chi_{UAl_2}$.

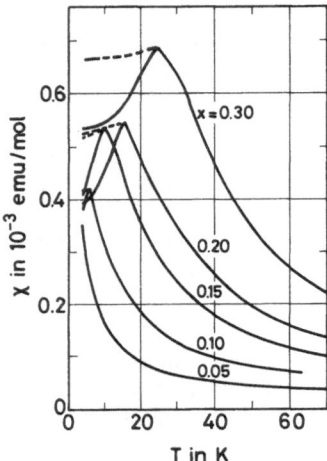

Fig. 3-73. Low-temperature susceptibility for $U_{1-x}Gd_xAl_2$ compounds [5]. ········ values found on warming the samples after cooling in a field of 150 Oe.

200 K in each case. The impurity part of the susceptibility, $\Delta\chi(T) = \chi(T) - \chi(T)_{LaAl_2}$ is proportional to the U concentration only for compounds with $x \ll 500$ ppm. The Curie-Weiss temperature is $\Theta = -2 \pm 0.3$ K and the effective moment is $\mu_{eff} = 3.6 \pm 0.2$ μ_B. With U concentrations above 500 ppm the susceptibility remains below its single ion value, whereas the negative Curie-Weiss temperature increases. This was interpreted as the formation of nonmagnetic clusters of UAl_2 ($\Theta \approx 300$ K [12]) [4].

The temperature dependence of the susceptibility of $U_{1-x}Gd_xAl_2$ compounds shows pronounced maxima in the low-temperature region for a value of $x \leqq 0.1$, from which spin-glass formation is suggested, see **Fig.** 3-73 [5]. Those cusps were confirmed by a very-low-field (150 Oe) ac susceptibility measurement [5, 6]. Curie-Weiss temperatures and effective magnetic moments of Gd^{3+} were derived from the higher temperature region as summarized in Table 3/38. The observed magnetic transition temperatures are summarized in **Fig.** 3-**74**.

Table 3/38
Susceptibility Cusp Temperatures T_f, Curie-Weiss Temperatures Θ_p, and Effective Magnetic Moments of Gd^{3+}, μ_{eff}, for $U_{1-x}Gd_xAl_2$ Compounds [5].

composition	cusp temperature T_f in K	Curie-Weiss temperature Θ_p in K	magnetic moment of Gd^{3+}, μ_{eff} in μ_B
0.03	—	0	8.9
0.05	—	0	8.5
0.10	6	6	8.3
0.15	10	18	8.35
0.20	16	20	8.32
0.30	25	27	8.32

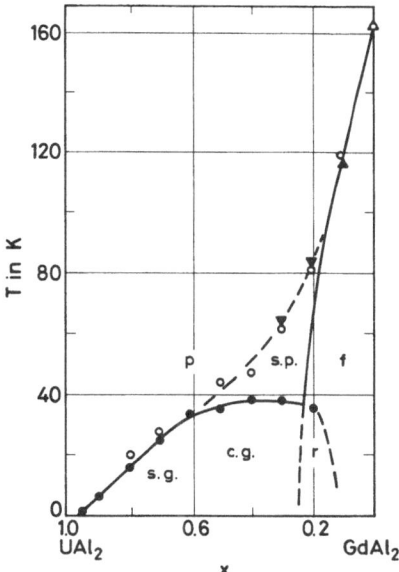

Fig. 3-74. Magnetic transition temperatures for $Gd_{1-x}U_xAl_2$ compounds [6]. ▲ Curie temperatures; ▼ quasi-Curie temperatures; ○ paramagnetic Curie temperatures; ● spin-glass (s.g.), cluster-glass (c.g.), or re-entrant glass (r) freezing temperatures T; p paramagnetic; s.p. super-paramagnetic; f ferromagnetic.

3.1.7.5.3 $U_{1-x}Ln_xAl_3$ and $U_{1-x}Ln_xAl_4$ (Ln = Sm, Dy)

The formation of a $(U,Sm)Al_3$ compound with hexagonal structure was observed in the microstructure of an Al−66 wt% Sm−4 wt% U alloy, together with the formation of $(U,Sm)Al_2$, and an Al−32 wt% Sm−35 wt% U alloy and the orthorhombic $(Sm,U)Al_4$. After heat treatment at 600°C for 30 d, the intermetallic compounds $SmAl_3$, $(Sm,U)Al_3$, and $(Sm,U)Al_4$ were identified at Al−56 wt% Sm−64 wt% U ternary diffusion couples. A study of the diffusion of Al into $(U,Sm)Al_2$ showed a large contribution of grain-boundary diffusion of Al in $(U,Sm)Al_2$ with the formation of $(U,Sm)Al_3$. The different phases can be distinguished metallographically by etching with a solution of 2.5 parts of HF and 12.5 parts of HNO_3 in 85 parts of water. After etching, the $(U,Sm)Al_2$ phase remains white, the $(Sm,U)Al_3$ phase is brownish colored, and the $(U,Sm)Al_4$ phase has a light blue-green color [1]. It was concluded from microstructural and microprobe analysis that the Al-rich solid solution in the Al−Dy−U system is in equilibrium with three intermetallic compounds, $(Dy,U)Al_4$, $U_2Dy_3Al_{15}$ (Al−34.8 wt% U−35.5 wt% Dy), β-$(Dy,U)Al_3$, and with two 2-phase mixtures: $(Dy,U)Al_3$, $U_2Dy_3Al_{15}$ and $U_2Dy_3Al_{15}$, β-$(Dy,U)Al_3$. The different phases can be distinguished metallographically by etching similar to the $(U,Sm)Al_x$ compounds [2].

3.1.7.5.4 $U_2Al_{15}Dy_3$

$U_2Al_{15}Dy_3$ was identified by microstructural and microprobe analysis to crystallize in a hexagonal lattice with the parameters a = 6.05 Å and c = 14.35 Å [2].

References for 3.1.7.5 on p. 198

References for 3.1.7.5:

[1] Casteels, F.; Diels, P.; Cools, A. (J. Nucl. Mater. **24** [1967] 87/94).
[2] Casteels, F.; Diels, P.; Cools, A. (J. Nucl. Mater. **24** [1967] 95/100).
[3] Wire, M. S.; Stewart, G. R.; Roof, R. B. (J. Magn. Magn. Mater. **53** [1985] 283/9).
[4] Schlabitz, W.; Steglich, F.; Bredl, C. D.; Franz, W. (Physica B + C **102** [1980] 321/5).
[5] Ping, J. Y.; Coles, B. R. (J. Magn. Magn. Mater. **29** [1982] 209/12).
[6] Coles, B. R.; Ping, J. Y.; Bennett, M. H. (Phil. Mag. [8] B **50** [1984] 1/9).
[7] Larica, C.; Coles, B. R. (Phil. Mag. [8] B **52** [1985] 1097/1105).
[8] Wire, M. S.; Giorgi, A. L. (Phys. Rev. B [3] **32** [1985] 1687/90).
[9] Wire, M. S.; Stewart, G. R.; Johanson, W. R.; Fisk, Z. (Phys. Rev. B [3] **27** [1983] 6518/21).
[10] Buschow, K. H. J.; van Daal, H. J. (AIP [Am. Inst. Phys.] Conf. Proc. No. 5 [1972] 1464/77).
[11] Stewart, G. R.; Giorgi, A. L.; Brandt, B. L.; Foner, S.; Arko, A. J. (Phys. Rev. B [3] **28** [1983] 1524/8).
[12] Brodsky, M. B.; Trainor, R. J. (Physica B + C **91** [1977] 271/7).

3.1.7.6   U−Al−Zr

3.1.7.6.1   Phase Relationships

Only the Al−UAl$_2$−ZrAl$_2$ section of the ternary system U−Al−Zr was studied in detail, based on thermal analysis, micrography, and X-ray determination. This section of the ternary system is dominated by extended heterogeneous equilibria involving, in most cases, the

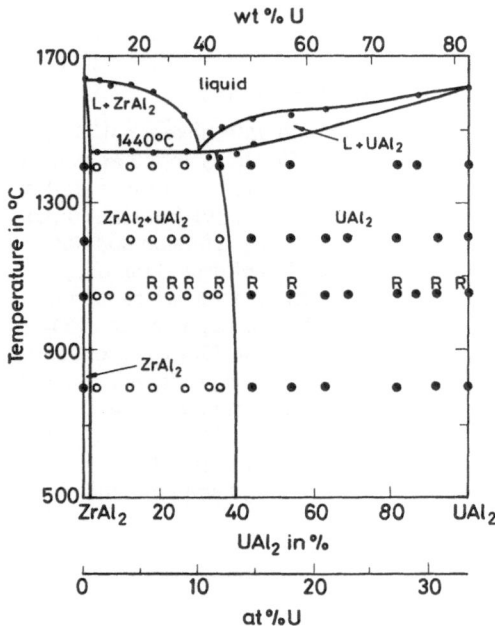

Fig. 3-75. Phase diagram of the quasi-binary system UAl$_2$−ZrAl$_2$ [2]. ● monophase, ○ 2-phase, • thermal analysis, R X-ray determination.

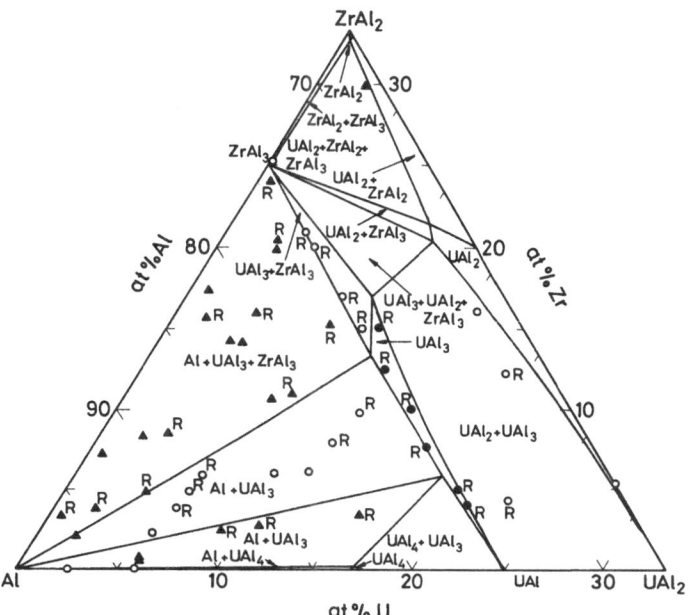

Fig. 3-76. Isothermal section of the system Al−UAl$_2$−ZrAl$_2$ at 600°C [1]. ● monophase,
○ 2-phase, ▲ 3-phase.

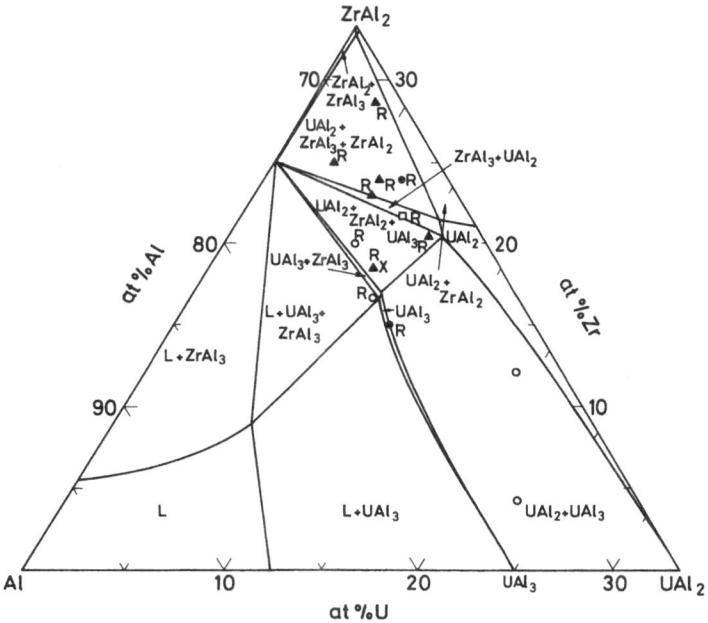

Fig. 3-77. Isothermal section of the system Al−UAl$_2$−ZrAl$_2$ at 1300°C [1].

References for 3.1.7.6 on pp. 202/3

phase UAl_3. The process of the reactions implies an extended two-phase range marked by UAl_3 inclusions in an Al matrix [1].

The quasi-binary system UAl_2-ZrAl_2 was found to form a simple eutectic system with the eutectic point placed at about 10 at% U and a eutectic temperature of $1440 \pm 10°C$ (see **Fig.** 3-75, p. 198) [2, 3], see also [1]. No solubility of U in $ZrAl_2$ was observed, but 22.5 at% Zr can be dissolved in UAl_2. The monophase $(U,Zr)Al_2$ mixed crystals crystallize with cubic structure ($MgCu_2$ type, C 15) [2, 3].

The solubility of Zr in UAl_2 was determined to be 16.9 at%, whereas the solubility of Zr in UAl_4 was very small [1]. The observed melting surfaces with six areas of primary crystallization, the melting equilibria and the position of the 4-phase areas and their extension are demonstrated in figures and tables in [1]. The isothermal sections at 600°C and 1300°C are shown in **Fig.** 3-76 and **Fig.** 3-77, p. 199, demonstrating the dominant role of the UAl_3 phase (especially at the 600°C section), which forms equilibria with all phases excluded $ZrAl_2$. The determined temperature-concentration sections are shown in [1].

3.1.7.6.2 Preparation and Properties

Preparation

Ternary U—Al—Zr samples were prepared by melting the elementary metals in an arc furnace with a tungsten electrode in water-cooled copper hearths under argon atmosphere. Losses of Zr at Zr-rich and of Al at Al-rich alloys in the range of 1 to 2 wt%, respectively, were minimized by calculated excess of Zr or Al, respectively. Samples were remelted to obtain homogeneous alloys [1, 2], see also [4].

Crystallographic Properties

Monophase solid solutions of $(U,Zr)Al_2$ compositions are formed in the range of $0 \le x \le 22.5$ at% Zr. The mixed crystals crystallize with cubic $MgCu_2$-type (C 15) structure. The lattice constants decrease nearly linearly from 7.7475 kX (UAl_2) to 7.575 kX ($(U,Zr)Al_2$ with 13 at% U) [2], see also [3]. Further measured values are summarized in Table 3/39.

Table 3/39
Crystallographic Properties of Monophase $(U,Zr)Al_2$ Solid Solutions [4].

composition in at%	lattice constant in Å	X-ray density in g/cm^3
Al—33.3 U (UAl_2)	7.76	8.3
Al—32 U—1.4 Zr	7.76	8.2
Al—30 U—3.4 Zr	7.75	8.0
Al—27.5 U—5.9 Zr	7.74	7.7
Al—22 U—11.4 Zr	7.68	7.1
Al—19 U—14.4 Zr	7.66	6.8

Monophase solid solutions of $(U,Zr)Al_3$ are formed in the range of $0 \le x \le 16.9$ at% Zr. The lattice constant of UAl_3 (cubic $AuCu_3$-type structure [5 to 7]) decreases linearly with increasing content of Zr [1].

Thermal Properties

The solidus and liquidus points of some monophase $(U, Zr)Al_2$ alloys are summarized in Table 3/40.

Table 3/40
Solidus and Liquidus Points of Monophase $(U, Zr)Al_2$ Alloys [4]. See also [2, 3].

composition in at%	solidus temperature in °C	liquidus temperature in °C
Al − 33.3 U (UAl_2)	1620	−
Al − 32 U − 1.4 Zr	1610	1615
Al − 30 U − 3.4 Zr	1590	1610
Al − 27.5 U − 5.9 Zr	1570	1600
Al − 22 U − 11.4 Zr	1520	1560
Al − 19 U − 14.4 Zr	1480	1550

Electrical Resistivity

Using a "parallel resistor model" with $1/\varrho = 1/\varrho_i + 1/\varrho_s$ (ϱ_i = "ideal" resistivity, ϱ_s = "shunt" resistivity) it was shown that the ideal resistivity of $U_{0.93}Zr_{0.07}Al_2$ follows a $T^{3/2}$ law (see **Fig. 3-78**) as also observed for $U_{1-x}Ln_xAl_2$ compounds (see p. 193) [8].

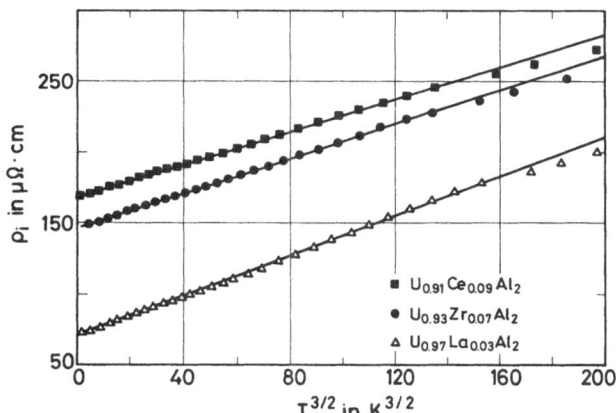

Fig. 3-78. Ideal resistivity ϱ_i versus $T^{3/2}$ for $U_{0.93}Zr_{0.07}Al_2$ [8]. The shunt resistivity used was 130 $\mu\Omega \cdot cm$.

Magnetic Properties

The substitution of Zr for U in UAl_2 decreases the susceptibility (at low temperatures), although the change is not significant. A derivate of the susceptibility with respect to $U_{1-x}Zr_xAl_2$ is $d\chi \, (T = 0)/dx = -3.5 \times 10^{-3} \, cm^3 \cdot mol^{-1} \, U$ [9].

References for 3.1.7.6 on pp. 202/3

Oxidation Behavior

The oxidation behavior of different $(U,Zr)Al_2$ alloys was measured manometrically in the temperature range of 500 to 700°C in air. Minor amounts of substitution of Zr for U in UAl_2 (1.4 at%) increase the oxidation velocity, whereas higher Zr additions (>3.4 at% Zr) retard the oxidation velocity. Homogeneous $U_{1-x}Zr_xAl_2$ solid solutions with x > 10 at% are resistant in air up to 700°C. The apparent activation energy for UAl_2 and $(U,Zr)Al_2$ alloys is about 22 kcal/mol [4], see also [10].

Stabilizing of UAl_3 in Aluminium-Rich Nuclear Fuels

The formation of a brittle UAl_x phase in U—Al alloys used as nuclear fuels leads to difficulties in the fabrication of fuel elements. Small additions of certain elements, such as Si, prevent the peritectic transformation of UAl_3 to UAl_4. Ternary additions of Zr stabilize the UAl_3 phase [11, 12], see also [13]. The investigation of the ternary system U—Al—Zr has shown that the UAl_3 phase in the stabilized alloys is in thermodynamic equilibrium with the Al matrix. The amount of the Al—UAl_3 phase is determined only by the solubility of Zr in the UAl_3 phase [11, 12], see also [1, 14]. The section of U—Al—Zr alloys, which are of interest in nuclear technology, is shown in **Fig. 3-79**.

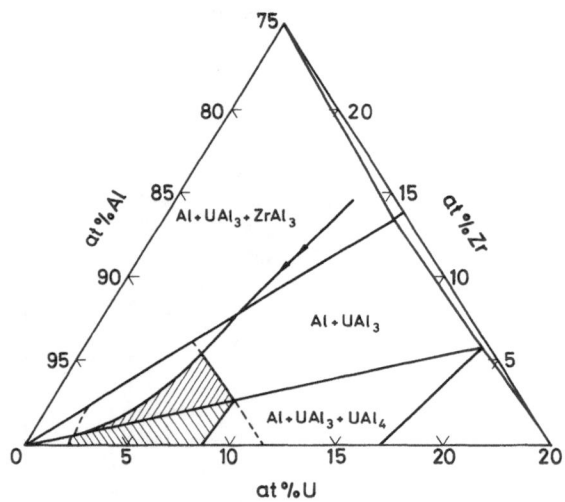

Fig. 3-79. Section of $(U,Zr)Al_2$ alloys which are of interest as nuclear fuels [11].

References for 3.1.7.6:

[1] Petzow, G.; Exner, H. E.; Chakraborty, A. K. (J. Nucl. Mater. **25** [1968] 1/15).
[2] Petzow, G.; Steeb, S.; Ellinghaus, I. (J. Nucl. Mater. **4** [1961] 316/21).
[3] Steeb, S.; Petzow, G. (Naturwissenschaften **48** [1961] 450/1).
[4] Thümmler, F.; Exner, H. E.; Petzow, G. (J. Nucl. Mater. **24** [1967] 328/39).
[5] Rundle, R. E.; Wilson, A. S. (Acta Cryst. **2** [1949] 148/50).
[6] Rundle, R. E. (CT-2721 [1945] 1/9; N.S.A. **10** [1956] No. 5124).

[7] Rough, F. A.; Bauer, A. A. (BMI-1300 [1958] 1/138; N.S.A. **12** [1958] No. 13935).

[8] Wire, M. S.; Giorgi, A. L. (Phys. Rev. B [3] **32** [1985] 1687/90).

[9] Wire, M. S.; Stewart, G. R.; Roof, R. B. (J. Magn. Magn. Mater. **53** [1985] 283/9).

[10] Reynolds, J. E.; Berry, W. E.; Ogden, H. R.; Peoples, R. S.; Jaffee, R. I. (BMI-1087 [1956] 1/36; C.A. **1961** 17445).

[11] Exner, H. E.; Petzow, G. (5th Intern. Leichtmetalltagung, Leoben 1968 [1969], pp. 77/83; C.A. **74** [1971] No. 37351).

[12] Exner, H. E.; Petzow, G. (Metall **23** [1969] 220/5).

[13] Daniel, N. E.; Foster, E. L., Jr.; Dickerson, R. F. (BMI-1454 [1960] 1/30; C.A. **1961** 1370).

[14] Petzow, G.; Exner, H. E. (BMwF-FBK-67-88 [1967] 1/69; C.A. **69** [1968] No. 38273).

3.1.7.7   U−Al−Th

3.1.7.7.1   Phase Relationships

The Al corner of the ternary system U−Al−Th was investigated using thermal and microscopic analysis. A wide range of compositions was found that can be maintained in monophase liquid solutions at temperatures lower than 640°C (or 630°C [1]), ranging from 13 wt% U at 0 wt% Th to 0 wt% U at 25 wt% Th [2]. The data indicate a ternary eutectic near 76 wt% Al, 18 wt% Th, and 6 wt% U. The products of the eutectic decomposition are thought to be Al, UAl_3, and $ThAl_3$ [3]. No solid solutions were found for the substitution of U for Th in Th_2Al [4]. $U_{1-x}Th_xAl_2$ alloys with $0 < x < 0.10$ were found to be Laves phases with cubic structure. Further alloys with hexagonal structure of the composition $U_{1-x}Th_xAl_3$ (with x = 0.5 and x = 0.8) [5], and $U_{1-x}Th_xAl_2$ with x = 0.95 [6, 7] were prepared.

3.1.7.7.2   Preparation and Properties

Preparation

$U_{1-x}Th_xAl_2$ samples with x in the range of 0 to 0.10 were prepared by arc melting of the elements [6, 7]. $U_{0.5}Th_{0.5}Al_3$ and $U_{0.2}Th_{0.8}Al_3$ were prepared by arc melting under argon in two steps: 1) preparation of a U−Th alloy, 2) adding stoichiometric amounts of Al to the U−Th alloy, and preparing the quasi-binary alloy (U, Th)Al_3 by repeated arc melting. The weight loss was in all cases below 0.2% [5]. A small amount of UAl_3 (about 5%) was analyzed by X-ray diffraction as a second phase in the $U_{0.5}Th_{0.5}Al_3$ sample. Annealing this sample inductively in vacuum at 1050°C for 90 h resulted in an increase of the second phase (UAl_3) and in an increase of the lattice parameters. From this, it was suggested that there is an upper limit of the U concentration at x = 0.5 for $U_{1-x}Th_xAl_3$ solid solutions [5]. U−Al−Th alloys were also prepared by direct reduction of ThO_2 and U_3O_8 with Al at 1200°C within 5 min [1].

Crystallographic Properties

The $U_{1-x}Th_xAl_2$ alloys with x ranging from 0 to 0.10 are Laves phases with cubic structure (C 15). For samples with x = 0.10, a breakdown of the long-range order was observed. A sample with x = 0.95 showed a hexagonal structure (C 32) [6, 7]. $U_{1-x}Th_xAl_3$ alloys (with a suggested limit of x = 0.5) crystallize with a hexagonal structure. Measured lattice parameters are reported for two compositions: $U_{0.2}Th_{0.8}Al_3$: a = 6.450 Å, c = 4.609 Å, c/a = 0.72; $U_{0.8}Th_{0.2}Al_3$: a = 6.399 Å, c = 4.606 Å, c/a = 0.72. For $U_{0.5}Th_{0.5}Al_3$ the distance U−U is 4.354 Å [5].

                     References for 3.1.7.7 on p. 207

Electronic Structure

The substitution of U for Th in diamagnetic ThAl$_2$ results in the formation of localized 5f-like magnetic states. Extremely large electronic specific heats were measured in the low-temperature region, which are associated with the formation of impurity states at the Fermi level [8], see also [6, 7, 9]. The change in the density of states per added U atom is shown in a figure in [8].

Specific Heat

The low-temperature specific heat of U$_{1-x}$Th$_x$Al$_2$ solid solutions was measured in the range of 1.7 to 20 K (see **Fig.** 3-**80**) [8, 9]. The results are discussed in terms of the spin-fluctuation theory. The electronic contribution, γ, to the specific heat increases with increasing Th content [9]. For the total specific heat the equation

$$C = \gamma \cdot T \left[m^*/m + \alpha \cdot (T/T_{SF})^2 \cdot \ln (T/T_{SF}) \right] + \beta \cdot T^3, \text{ is used,}$$

with m^*/m = zero-temperature many-body mass enhancement, γ = coefficient of the electronic specific heat, α is proportional to $S(1-S^{-1})^2$ (S = Stoner enhancement factor), $\beta \cdot T^3$ = lattice contribution to the specific heat. The following values were calculated for U$_{1-x}$Th$_x$Al$_2$ alloys: γ (in mJ$\cdot$mol$^{-1}\cdot$K^{-2}) = 142 (x = 0), 147 (x = 0.02), 157 (x = 0.05), 171 (x = 0.10) [6, 7]. The difference $\Delta\gamma = \gamma(\text{Th}_{1-x}\text{U}_x\text{Al}_2) - \gamma(\text{ThAl}_2) = (C_{T,x} - C_{T,0})/T$ of the electronic specific coefficients from measurements between 1.7 and 20 K increases markedly with increasing U content [8], see Table 3/41.

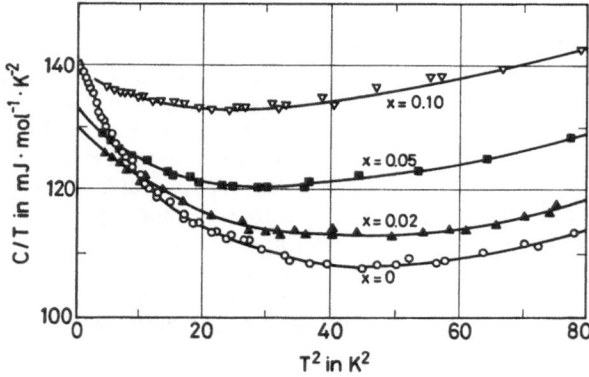

Fig. 3-80. Specific heat, given as C/T, versus T^2 for U$_{1-x}$Th$_x$Al$_2$ alloys above 9 K [9].

Measurements of the low-temperature specific heat of U$_{0.5}$Th$_{0.5}$Al$_3$ showed a T$^3 \cdot \ln (T/T_{SF})$ term, from which long-range ferromagnetic spin fluctuations are considered (see **Fig.** 3-**81**). The coefficient of the electronic contribution to the specific heat is $\gamma = 360$ mJ$\cdot$mol$^{-1}\cdot$K^{-2} [5].

Table 3/41
Electronic Specific Heat Coefficient, Expressed as $\Delta\gamma = \gamma(Th_{1-x}U_xAl_2) - \gamma(ThAl_2)$ for $Th_{1-x}U_xAl_2$ Alloys [8].

uranium content, x	$\Delta\gamma$ in 10^{-4} cal·mol^{-1}·K^{-2}	$\Delta N_{E(F)}$, in states·eV^{-1}·(U-atom)$^{-1}$
0	—	—
0.005	5.62	199.5
0.015	9.5	112.6
0.02	15.5	137.5
0.035	25.6	129.5
0.05	32.5	115.3
0.079	40.5	91
0.10	33.2	59
0.12	53.1	78.6
0.14	80	101.5
0.18	81	79.7

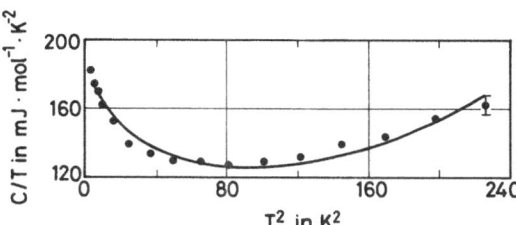

Fig. 3-81. Specific heat, given as C/T, versus T^2 for $U_{0.5}Th_{0.5}Al_3$ [5]. The upturn in C/T and its agreement with a $T^3 \cdot \ln(T/T_{SF})$ dependence is a strong indication of ferromagnetic spin fluctuations.

Electrical Resistivity

The electrical resistivity of $U_{0.5}Th_{0.5}Al_3$, which was measured between 1.4 and 300 K, increases linearly up to about 50 K. Above 50 K the value of R (300 K)/R (50 K) is 90, which is the same value as for UPt$_3$ [5].

Magnetic Properties

The magnetic susceptibility of $U_{1-x}Th_xAl_2$ solid solutions with x < 0.10 was measured at temperatures of up to 300 K (see **Fig. 3-82**, p. 206). Above 60 K a Curie-Weiss behavior was observed with only little variation in the parameters (see Table 3/42, p. 206). In the low-temperature region a T^2 dependence of the susceptibility and an apparent Fermi liquid contribution were observed [6, 7].

References for 3.1.7.7 on p. 207

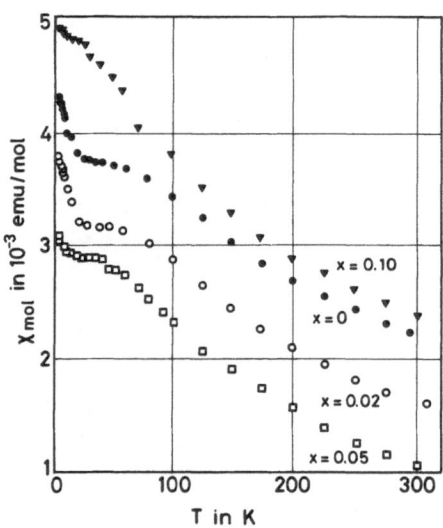

Fig. 3-82. Molar susceptibility of $U_{1-x}Th_xAl_2$ alloys [7].

Table 3/42
Magnetic Properties of $U_{1-x}Th_xAl_2$ Alloys [7].

| | $U_{1-x}Th_xAl_2$ | | | |
	x = 0	x = 0.02	x = 0.05	x = 0.10
χ_0 in 10^{-3} emu/mole	4.38	4.33	4.18	4.65
Θ (Curie-Weiss) in K	245	237	234	193
μ_{eff} in μ_B	3.1	3.0	3.0	2.8

The magnetic susceptibility of $U_{0.5}Th_{0.5}Al_3$, measured at 1.4 to 180 K, is shown in **Fig. 3-83**. Above 100 K a Curie-Weiss behavior is obeyed from which an effective moment of $\mu_{eff} = 3.0\ \mu_B$ was derived [5].

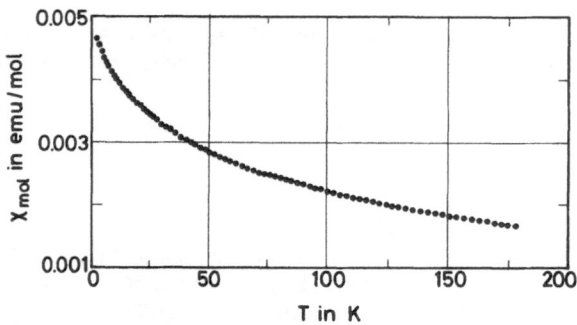

Fig. 3-83. Molar magnetic susceptibility of $U_{0.5}Th_{0.5}Al_3$ [5] (Formula unit of $U_{0.5}Th_{0.5}Al_2$ = 0.5 mol U).

References for 3.1.7.7:

[1] Capocchi, J. D. T.; Gentile, E. F.; Tracanella, R. B. (Met. ABM [Assoc. Brasil. Metais] **25** No. 144 [1969] 831/42; C.A. **72** [1970] No. 92457).
[2] Bobeck, G. E.; Wilhelm, H. A. (ISC-832 [1956] 1/26; C.A. **1958** 3639).
[3] Rough, F. A.; Bauer, A. A. (BMI-1300 [1958] 1/138; N.S.A. **12** [1958] No. 13935).
[4] Farkas, M. S.; Bauer, A. A.; Dickerson, R. F. (BMI-1568 [1962] 1/20; C.A. **1962** 13732).
[5] Giorgi, A. L.; Stewart, G. R.; Wire, M. S.; Willis, J. O. (Phys. Rev. B [3] **32** [1985] 3010/3).
[6] Brodsky, M. B. (Proc. 3rd Intern. Conf. Electron. Struct. Actinides, Grenoble 1978, pp. 1/8; C.A. **90** [1979] No. 196776).
[7] Brodsky, M. B. (J. Phys. Colloq. [Paris] **40** [1979] C4-147/C4-149).
[8] Scott, W. R.; Jaccarino, V.; Wernick, J. H.; Maita, J. P. (J. Appl. Phys. **35** [1964] 1092/3).
[9] Brodsky, M. B.; Trainor, R. J. (Physica B + C **91** [1977] 271/7).

3.1.7.8 U – Al – Ge

The two alloys UAlGe and UAl_2Ge_2 were found to exist in the ternary system U – Al – Ge, but no detailed information on the ternary system has been reported [1, 2].

UAlGe and UAl_2Ge_2 were prepared by melting stoichiometric amounts of the elements in an arc furnace under argon with a slight excess of Al or Ge to prevent losses by evaporation. The samples were remelted for homogenization. The final weight loss was less than 0.1% of Al or Ge. Well-recrystallized alloys were obtained after annealing the samples at 850 to 900°C for 7 d with the samples placed in tantalum containers, which were sealed in evacuated quartz ampoules [1], see also [2].

UAlGe crystallizes with a hexagonal structure as determined by X-ray diffraction, and UAl_2Ge_2 crystallizes with a cubic structure. The lattice parameters are a = 10.05 $\pm$ 0.02 Å, c = 8.54 $\pm$ 0.02 Å for UAlGe; a = 4.217 $\pm$ 0.005 Å for UAl_2Ge_2 [1].

The space group for UAl_2Ge_2 is probably $Pm\bar{3}m - O_h^1$ (No. 221), from which the structure is assumed to be a partly disordered $AuCu_3$ type (or a perovskite type) [2]. The density of UAl_2Ge_2 was measured pycnometrically to be D = 7.73 $\pm$ 0.06 g/cm^3 [2].

UAl_2Ge_2 was found to be a superconductor by measurements of the susceptibility with a critical temperature of T_c = 0.87 $\pm$ 0.08 K [2].

3.1.7.9 U – Al – Sn

A complete series of solid solutions $U(Al, Sn)_3$ was observed in the ternary system U – Al – Sn [3], see also [4]. A phase diagram is given in Fig. 4-8, p. 307. Five three-phase regions were observed [3]:

T_1 = U   + UAl_2 + U_5Sn_4,
T_2 = UAl_2 + U_5Sn_4 + ~ $U_{25}Al_{55}Sn_{20}$ (saturated solid solution),

References for 3.1.7.8 and 3.1.7.9 on p. 208

$T_3 = U_5Sn_4 + U_3Sn_5 + \sim U_{25}Al_{34}Sn_{41}$ (saturated solid solution),
$T_4 = Al \quad + UAl_4 \quad + \sim U_{25}Al_{70}Sn_5$ (saturated solid solution),
$T_5 = Al \quad + Sn \quad + \sim U_{25}Al_{13}Sn_{62}$ (saturated solid solution).

More details of the system U−Al−Sn are given in Section 4.1.10.1, p. 307.

References for 3.1.7.8 and 3.1.7.9:

[1] Zygmunt, A. (Proc. 2nd Intern. Conf. Electron. Struct. Actinides, Wroclaw 1976 [1977], pp. 335/41; C.A. **87** [1977] No. 76741).
[2] Ott, H. R.; Hulliger, F.; Rüdiger, H.; Fisk, Z. (Phys. Rev. B [3] **31** [1985] 1329/33).
[3] Marazza, R.; Ferro, R.; Rambaldi, G.; Mazzone, D. (J. Less-Common Metals **51** [1977] 51/4).
[4] Dwight, A. E. (ANL-82-14 [1982] 1/44; C.A. **98** [1983] No. 186482).

3.1.7.10   U−Al−V. U−Al−Ta

Ternary compositions of UM_2Al_{21} (with M = V, Ta) were prepared by melting of the elements in graphite crucibles under argon and stirring the melt for some hours. The crystallized alloys were separated from the bottom of the melts [1, 2].

The UM_2Al_{21} alloys (with M = V, Ta) crystallize with four molecules per unit cell. Measured lattice parameters are a = 10.24 Å, c = 14.482 Å for UV_2Al_{21} [1]; a = 10.33 Å, c = 14.609 Å for UTa_2Al_{21} [2]. The calculated X-ray densities (in g/cm^3) are 3.963 for UV_2Al_{21} [1], 4.967 for UTa_2Al_{21} [2].

References for 3.1.7.10:

[1] Layne, G. S.; Huml, J. O. (U.S. 3198629 [1965]; C.A. **63** [1965] 9626).
[2] Layne, G. S.; Huml, J. O. (U.S. 3198628 [1965]; C.A. **63** [1965] 9626).

3.1.7.11   U−Al−Cr. U−Al−Mo. U−Al−W

Phase Relationships

The phase diagrams of the ternary systems U−Al−Cr and U−Al−W are not yet investigated in detail. Concerning the ternary system U−Al−Mo, the melting and phase equilibria were investigated for the system $U-UAl_2-Al_8Mo_3-Mo$ in detail by microscopic, X-ray, and thermal analysis [1, 2]. The melting surfaces of primary solidification and the composition of the investigated alloys are graphically shown in [1].

A ternary Laves phase, $U(Al, Mo)_2$, with hexagonal structure (C 14 type) was found. The system is characterized by extensive heterogeneous phase fields, whereas the phases UAl_2 and $U(Al, Mo)_2$ play a dominant role. The reaction of the $U(Al, Mo)_2$ phase with the binary phases UAl_2 and $AlMo_3$ results in two nonvariant three-phase equilibria (see **Fig.** 3-**84**). The two high-temperature phases in the binary system $Mo-Al_8Mo_3$ cannot be stabilized by addition of uranium. Above 1450°C these two phases undergo four-phase reactions. In total, eight four-phase melting reactions have been established, seven of which are transition equilibria, and one is of eutectic nature. The abbreviations used in the phase diagram are summarized in Table 3/43 [1].

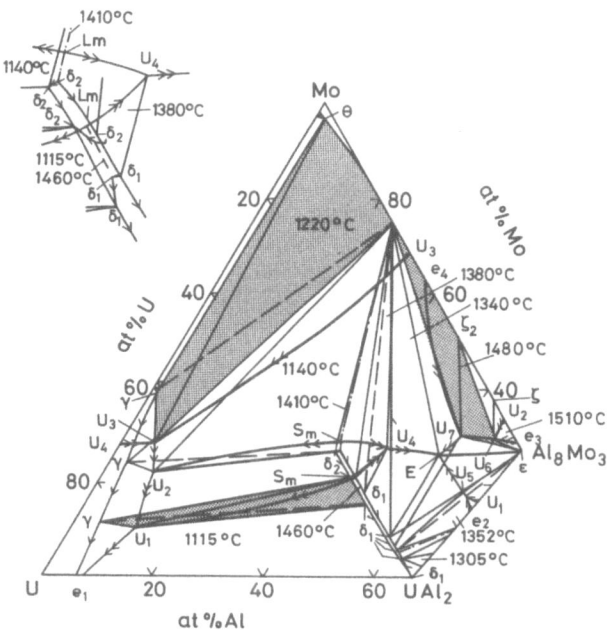

Fig. 3-84. Melting equilibria in the system U − UAl$_2$ − Al$_8$Mo$_3$ − Mo [1].

Table 3/43

Crystal Structure, Melting Surfaces of Primary Solidification, and Abbreviations of the Compositions in the System U − UAl$_2$ − Al$_8$Mo$_3$ − Mo [1].

crystal	crystal structure	abbreviation	melting surface
α uranium	orthorhombic, A 20 type	α	—
β uranium	tetragonal, β-U type	β	—
γ uranium	bcc, A 2 type	γ	U-e$_1$-U$_1$-U$_2$-U$_3$-u$_4$-U
MoU$_2$	tetragonal body-centered C 11b type	γ′	—
UAl$_2$	fcc, C 15 type	δ$_1$	e$_1$-UAl$_2$-e$_2$-U$_5$-E-U$_4$-U$_1$-e$_1$
U(Al, Mo)$_2$	hexagonal, C 14 type	δ$_2$	U$_1$-U$_4$-U$_2$-U$_1$
Al$_8$Mo$_3$	monoclinic	ε	u$_1$-Al$_8$Mo$_3$-e$_3$-U$_6$-U$_7$-E-U$_5$-u$_1$
Al$_{62}$Mo$_{38}$	unknown	ζ$_1$	e$_3$-u$_2$-U$_6$-e$_3$
Al$_{48}$Mo$_{52}$	unknown	ζ$_2$	u$_2$-e$_4$-U$_7$-U$_6$-u$_2$
AlMo$_3$	cubic, A 15 type	η	e$_4$-u$_3$-U$_3$-U$_2$-U$_4$-E-U$_7$-e$_4$
molybdenum	bcc, A 2 type	Θ	u$_4$-U$_3$-u$_3$-Mo-u$_4$
UAl$_3$	fcc, AuCu$_3$ type	—	—
U$_9$Al$_{75}$Mo$_{16}$	unknown	T	e$_2$-u$_1$-U$_5$-e$_2$

A schematic flow sheet for the reactions of the melting equilibria is given in [1]. The areas of the nonvariant melting equilibria are summarized in [1]. The transformation equilibria based on the polymorphism of uranium are shown in [1]. Several temperature-concentration sections are shown in the original paper [1].

Preparation and Properties

The preparation of the U — Al — Mo alloys is carried out in small quantities by melting the samples in an arc furnace [1]; larger quantities, especially for uranium-base alloys with small amounts of Mo and Al, are produced by induction melting under vacuum and casting of the melts in graphite molds [3 to 6].

Uranium-base alloys with small quantities of Mo and Al, as low-solubility additions, could provide nucleation centers in the β phase and prevent the growth of the α grains. This is of interest in nuclear technology with respect to the swelling behavior of the metallic fuel [3], see also [4]. Such uranium-base alloys are characterized by very fine grains, resulting in increased mechanical properties [4 to 7], see also [8, 9].

$UAl_{21}M_2$ Alloys (M = Cr, Mo, W)

Ternary alloys of $UAl_{21}M_2$ (with M = Cr, Mo, W) have been prepared by melting of the elements in graphite crucibles under argon and stirring the melt for some hours. The crystallized alloys were separated from the bottom of the melts [10 to 12]. The $UAl_{21}M_2$ alloys (with M = Cr, W) crystallize with four molecules per unit cell. Measured lattice parameters are a = 10.197 Å, c = 14.412 Å for $UAl_{21}Cr_2$ [10]; a = 10.247 Å, c = 14.491 Å for $UAl_{21}W_2$ [12]. The calculated X-ray densities (in g/cm³) are: 4.03 for $UAl_{21}Cr_2$ [10], 5.115 for $UAl_{21}W_2$ [12]. $UAl_{21}Mo_2$ alloys crystallize with tetragonal structure, space group I4/amd (probable correct space group $I4_1/amd - D_{4h}^{19}$, No. 141), with four molecules per unit cell. Measured lattice parameters are a = 10.266 Å, c = 14.543 Å. The density is D = 4 to 4.37 g/cm³ [11].

UAl_8Cr_4

A further ternary alloy with the composition of UAl_8Cr_4 was prepared by arc melting of stoichiometric amounts of the elements. UAl_8Cr_4 crystallizes with body-centered tetragonal structure ($ThMn_{12}$ type), the space group is $I4/mmm - D_{4h}^{17}$ (No. 139). The lattice parameters, based on X-ray diffraction, are a = 8.907 Å, c = 5.122 Å with c/a = 0.575 and a unit cell volume of 406.35 Å³. The calculated X-ray density is D = 5.41 g/cm³ as compared to a measured value of D = 5.43 g/cm³. The magnetic susceptibility of UAl_8Cr_4 was measured in the range of 4.2 to 300 K (see **Fig. 3-85**). A distinct anomaly was found close to 180 K, related to antiferromagnetic

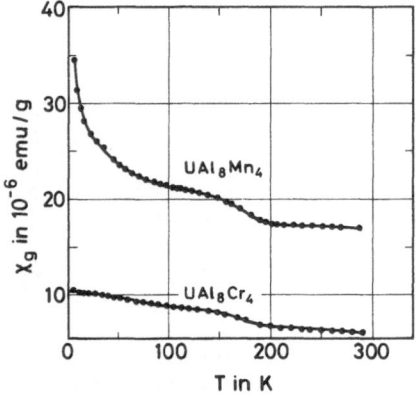

Fig. 3-85. Temperature dependence of the magnetic susceptibility for UAl_8Cr_4 and UAl_8Mn_4 [13].

ordering. The low-temperature magnetization at 4.2 K shows a linear dependence on applied external fields, which was measured up to 100 kOe [13].

References for 3.1.7.11:

[1] Petzow, G.; Rexer, J. (Z. Metallk. **60** [1969] 449/53).
[2] Petzow, G.; Rexer, J. (Z. Metallk. **62** [1971] 34/8).
[3] Djuric, B.; Drobujak, Dj.; Lazarevic, Dj.; Malcic, S.; Mihailovic, A.; Milosevic, S. (Proc. 3rd Intern. Conf. Peaceful Uses At. Energy, Geneva 1964 [1965], Vol. 11, pp. 183/9).
[4] Huet, J. J.; Massaux, H. (ATB [Acta Tech. Belg.] Met. **11** [1961] 167/76; C.A. **56** [1962] 11144).
[5] USAEC (Brit. 912960 [1962]; C.A. **58** [1963] 9850).
[6] UKAEA (Belg. 631107 [1963]; Belg. 631108 [1963]; C.A. **61** [1964] 4033).
[7] Saxl, I. (UJV-2802-M [1972] 1/46; C.A. **78** [1973] No. 61359).
[8] Khan, T.; Brun, G.; Decours, J. (Compt. Rend. C **269** [1969] 1490/3).
[9] Faussat, A.; Millet, P.; Ollier, H.; Peray, R.; Blanchard, P.; Pelce, J.; Guichard, C. (Proc. 4th Intern. Conf. Peaceful Uses At. Energy, Geneva 1971 [1972], Vol. 4, pp. 459/73).
[10] Layne, G. S.; Huml, J. O. (U.S. 3198626 [1965]; C.A. **63** [1965] 9626).
[11] Layne, G. S.; Teitel, R. J. (U.S. 3098742 [1963]; C.A. **59** [1963] 8443).
[12] Layne, G. S.; Huml, J. O. (U.S. 3198627 [1965]; C.A. **63** [1965] 9626).
[13] Baran, A.; Suski, W.; Mydlarz, T. (J. Magn. Magn. Mater. **63/64** [1987] 196/8).

3.1.7.12   U−Al−Mn

3.1.7.12.1   Phase Relationships

The quasi-binary system $UAl_2 - UMn_2$ and the uranium-rich section $U - UAl_2 - UMn_2$ of the ternary system $U - Al - Mn$ have been investigated in detail based on microscopic, X-ray, and thermal analysis [1, 2], see also [3].

Concerning the quasi-binary system $UAl_2 - UMn_2$, there is almost no solubility of Al in UMn_2 observed, whereas up to 16 at% Mn can be dissolved in UAl_2 at 1190°C with contraction of the unit cell [1], see also [4]. A ternary Laves phase was found with hexagonal structure ($MgZn_2$ type, C 14), ranging from 18 at% Al at 1090°C to 37 at% Al at 1190°C, or ranging from the composition of $U_4Al_3Mn_5$ to $U_4Al_4Mn_4$, from which this phase is named as UAlMn. A eutectic three-phase equilibrium (L = UMn_2 + UAlMn) is formed, extending from 1 to 18 at% Al at 1090°C with the eutectic point at 16 at% Al. A peritectic three-phase equilibrium (L = UAl_2 + UAlMn) extends from 31 to 51 at% Al at 1190°C [1]. A phase diagram is given in **Fig. 3-86**, p. 212.

The system $U - UAl_2 - UMn_2$ is dominated by extended heterogeneous areas involving the ternary Laves phase $U(Al, Mn)_2$, crystallizing with hexagonal structure ($MgZn_2$ type, C 14). The melting surfaces of primary solidification are given in [2], showing the eutectic point E_1 at 705°C to be the lowest melting point in the investigated system for an alloy with 74 at% U and 6 at% Al. Four non-variant melting equilibria at 850, 730, 710, and 705°C have been observed, whereas the equilibrium at 705°C is of eutectic nature. Lastly, three four-phase solid-state transformations are caused by the γ and β transformations of uranium at 720, 640, and 610°C. The compositions of the corners of the four-phase equilibria are summarized in [2]. The solubility of the homogeneous phases is reported to be low, and for U_6Mn practically zero [2]. Phase diagrams at different temperatures and temperature-concentration sections are graphically shown in [2].

References for 3.1.7.12 on p. 215   14*

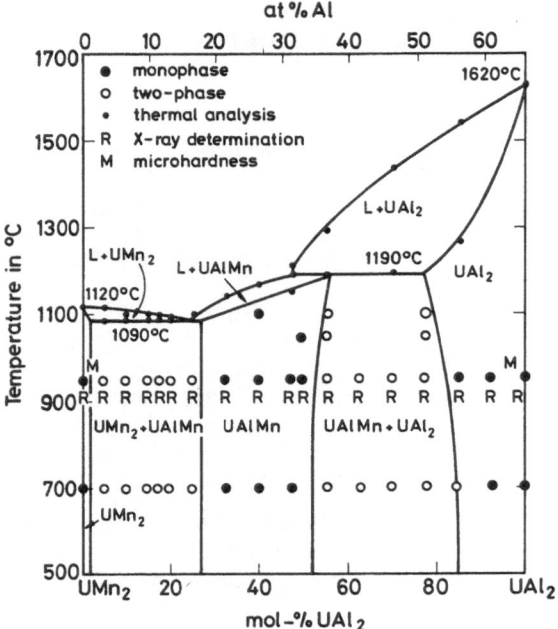

Fig. 3-86. Phase diagram of the quasi-binary system UAl_2-UMn_2 [1].

3.1.7.12.2 Preparation

Ternary U—Al—Mn alloys were prepared by arc melting of the elements under argon with tungsten electrodes in water-cooled copper hearths. To minimize losses by evaporation (up to 6 at% Al, and 4 at% Mn) an excess of Mn and Al was used. The samples were remelted for 3 to 5 times to obtain homogeneous products [1, 2, 5, 6], see also [4, 7], or annealed in the range of 700 to 800°C for one week [8]. Samples with small amounts of Mn, such as $U(Al_{1-x}Mn_x)_2$ with $0.2 \le x \le 1.0$ were homogenized by annealing at 850°C for one week [4].

The brittle alloys can easily be crushed to powders. Sintered specimen can be obtained by cold pressing (without binding agents) at 8 to 10 t/cm^2 and sintering in vacuum (10^{-4} to 10^{-5} Torr). Sintering starts at about 1293 K, which is $(0.88 \pm 0.02) \cdot T_M$ (with melting temperature of $U(Al, Mn)_2 = 1453$ K). This high sintering temperature is related to the Laves phase structure. The sintering process is suggested to be based on solid state diffusion [6].

3.1.7.12.3 UAlMn, U(AlMn)$_2$

Crystallographic Properties

UAlMn and $U(Al, Mn)_2$ solid solutions are Laves phases, crystallizing with hexagonal MgZn$_2$-type structure (C 14). UAlMn has a large homogeneity range, extending from 18 at% Al at 1090°C to 37 at% at 1190°C [1]. The space group is $P6_3/mmc-D_{6h}^4$ (No. 194) [5, 9], with four molecules per unit cell [9].

UAl_2 and UMn_2 are Laves phases crystallizing with cubic $MgCu_2$-type structure (C 15) [1, 9]. The measured lattice parameters are summarized in Table 3/44. The calculated X-ray density for stoichiometric UAlMn is 10.58 g/cm³ [5], see also [7], the measured density is 10.1 g/cm³ [5].

Table 3/44
Measured Lattice Parameters for Ternary U−Al−Mn Phases.

phase	composition in at %			lattice parameters		Ref.	also given in
	U	Al	Mn	a	c		
UAlMn	33.3	33.3	33.3	5.240 kX	8.391 kX	[5]	[1]
UAlMn				5.25 Å	8.408 Å	[10]	[9]
$U(Al, Mn)_2$	33.3	33.3	33.3	5.21 Å	8.27 Å	[11]	[7]
$U(Al, Mn)_2$	33.3	21.7	45	5.181 kX	8.175 kX	[5]	

The atomic positions of the $MgZn_2$-type structure are (with origin at center $\bar{3}$m1) [9]:

4 U in f (3m): 1/3, 1/3, z; 2/3, 1/3, $\bar{z}$; 2/3, 1/3, 1/2 + z; 1/3, 2/3, 1/2 − z.
2 B in a ($\bar{3}$m): 0, 0, 0; 0, 0, 1/2.
6 B in h (mm): x, 2x, 1/4; 2$\bar{x}$, $\bar{x}$, 1/4; x, $\bar{x}$, 1/4; $\bar{x}$, 2$\bar{x}$, 3/4; 2x, x, 3/4; $\bar{x}$, x, 3/4.

The Al and Mn atoms (B atoms) are statistically distributed in the B positions [5]. The atomic distances are U−U: 3.17 kX (UAlMn), 3.16 kX ($U(Al, Mn)_2$), B−B: 2.54 kX (UAlMn), 2.52 kX ($U(Al, Mn)_2$) [5].

$U(Al_{1-x}Mn_x)_2$ alloys with $0 \leq x \leq 1.0$ crystallize with cubic $MgCu_2$-type structure. The nearly linear concentration dependence of the cubic lattice parameters agrees with Végard's law [4].

Physical Properties

Measured values of microhardness (50 g load, 5 s) for samples of UMn_2, UAlMn, and UAl_2 annealed at 950°C for 30 h are 700 to 750 kp/mm² [1].

The thermal expansion coefficient of $U(Al, Mn)_2$ is given in **Fig. 3-87**, p. 214, as compared to UAl_2 and UMn_2 [6].

The specific electrical resistance of $U(Al, Mn)_2$ (molten samples) is 2.7 $\mu\Omega \cdot$ cm [6].

The magnetic properties of UAl_2 were explained within the framework of spin-fluctuation theory [12]. Concerning the $U(Al_{1-x}Mn_x)_2$ alloys, the magnetic spin-fluctuation behavior is gradually changed to Pauli-type paramagnetism with increasing content of Mn. The effective U moments decrease nearly linearly, suggesting that the simple dilution model may describe the magnetic behavior above 200 K (see **Fig. 3-88**, p. 214) [4].

                          References for 3.1.7.12 on p. 215

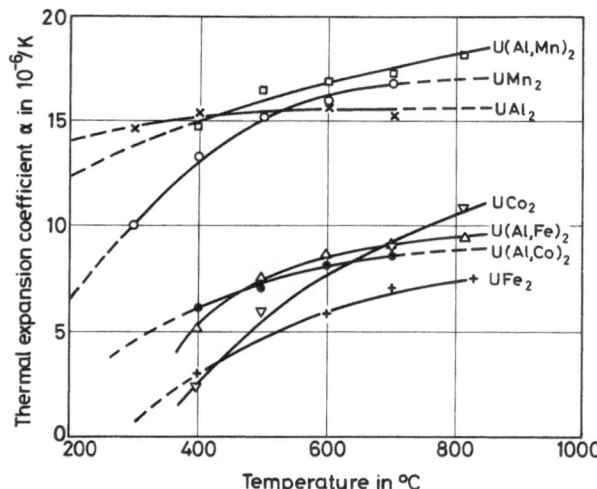

Fig. 3-87. Temperature dependence of the thermal expansion coefficients of molten Laves phase alloys [6].

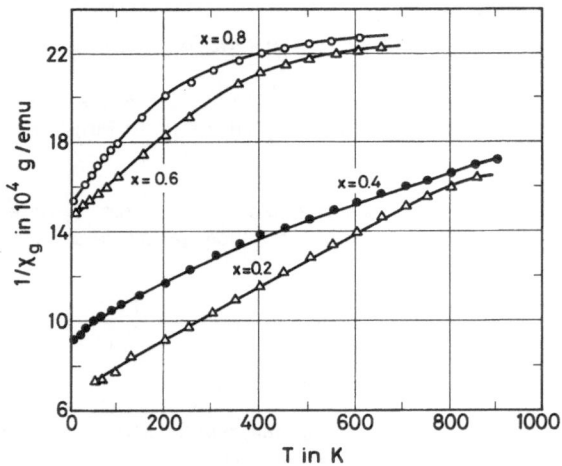

Fig. 3-88. Reciprocal susceptibilities versus temperature for $U(Al_{1-x}Mn_x)_2$ alloys with $0.2 < x < 0.8$ [4].

Chemical Reactions

U−Al−M alloys (with M = Mn, Ni, Co) were hydrogenated under a hydrogen pressure of up to 150 atm at temperatures cycling in the range 20 to 250°C after the samples had been degassed at 450°C in vacuum of 10^{-4} Torr for several hours. For UAlMn a maximum hydrogen absorption of 10.51 mL H_2/g was found at a synthesis pressure of 40 atm, resulting in a maximum hydride composition of $UAlMnH_{0.15}$ [8].

The oxidation behavior of $U(Al, Mn)_2$ (33.3 at% U, 33.3 at% Al, 33.3 at% Mn) in air was measured manometrically at 400°C and 500°C. The maximum speed of oxidation, calculated from linear regions of the oxygen take-up versus time diagrams, is 220 µg · cm^{-2} · min^{-1} at 400°C and 800 µg · cm^{-2} · min^{-1} at 500°C [7].

3.1.7.12.4 UAl_8Mn_4

A further ternary alloy with the composition of UAl_8Mn_4 was prepared by arc melting of stoichiometric amounts of the elements. UAl_8Mn_4 crystallizes with body-centered tetragonal structure ($ThMn_{12}$ type), the space group is $I4/mmm-D_{4h}^{17}$ (No. 139). The lattice parameters, based on X-ray diffraction determinations, are a = 8.849 Å, c = 5.104 Å with c/a = 0.577 and a unit cell volume of 399.67 Å^3. The calculated X-ray density is 5.60 g/cm^3 as compared to a measured value of 5.55 g/cm^3 [13].

The magnetic susceptibility of UAl_8Mn_4 was measured in the range of 4.2 to 300 K. A distinct anomaly was found close to 180 K, which is attributed to antiferromagnetic ordering. The low-temperature magnetization at 4.2 K shows a linear dependence on applied external fields, which was measured up to 100 kOe [13].

References for 3.1.7.12:

[1] Petzow, G.; Steeb, S. (Z. Metallk. **55** [1964] 456/9).
[2] Petzow, G.; Sampaio, A. O. (J. Less-Common Metals **13** [1967] 281/93).
[3] Dwight, A. E. (ANL-82-14 [1982] 1/44; C.A. **98** [1983] No. 186482).
[4] Burzo, E.; Gratz, E.; Lucaci, P. (Solid State Commun. **60** [1986] 241/4).
[5] Steeb, S.; Petzow, G. (Z. Metallk. **55** [1964] 453/6).
[6] Ondracek, G.; Petzow, G. (Z. Metallk. **56** [1965] 498/502).
[7] Thümmler, F.; Exner, H. E.; Petzow, G. (J. Nucl. Mater. **24** [1967] 328/39).
[8] Drulis, H.; Petrynski, W.; Stalinski, B.; Zygmunt, A. (J. Less-Common Metals **83** [1982] 87/93).
[9] Lam, D. J.; Darby, J. B., Jr.; Nevitt, N. V. (in: Freeman, A. J.; Darby, J. B., Jr., The Actinides: Electronic Structure and Related Properties, Vol. 2, Academic, New York − San Francisco − London 1974, pp. 119/84).
[10] Dwight, A. E. (in: Giessen, B. C., Developments in the Structural Chemistry of Alloy Phases, Plenum, New York 1969, pp. 181/226).
[11] Pearson, W. B. (A Handbook of Lattice Spacings and Structures of Metals and Alloys, Vol. 2, Pergamon, New York 1967).
[12] Trainor, R. J.; Brodsky, M. B.; Culbert, H. V. (Phys. Rev. Letters **34** [1975] 1019/22).
[13] Baran, A.; Suski, W.; Mydlarz, T. (J. Magn. Magn. Mater. **63/64** [1987] 196/8).

3.1.7.13   U−Al−Ni

3.1.7.13.1   Phase Relationships

The quasi-binary system UAl_2−NiAl and the Al-rich section Al−UAl_2−NiAl of the ternary system U−Al−Ni were investigated in detail using microscopic, X-ray, and thermal analysis [1 to 3].

UAl_2 and NiAl form a simple eutectic system with the eutectic point at 27.0 at% Ni at 1150 ± 10°C (see **Fig. 3-89**, p. 216). No solubility was observed for U in NiAl. UAl_2 dissolves

References for 3.1.7.13 on p. 223

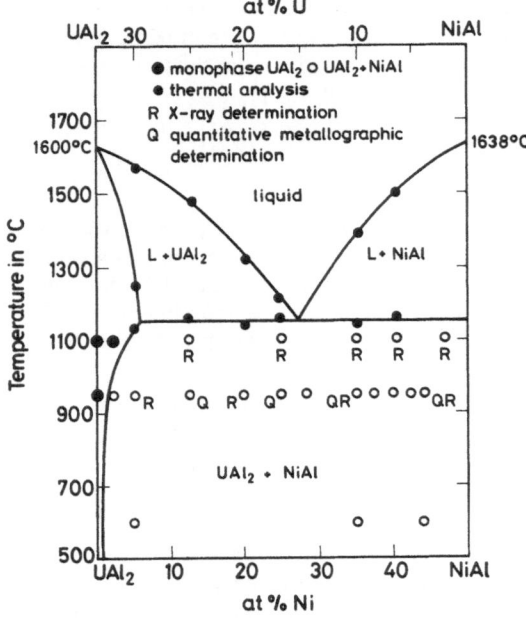

Fig. 3-89. Phase diagram of the quasi-binary system UAl_2-NiAl [1].

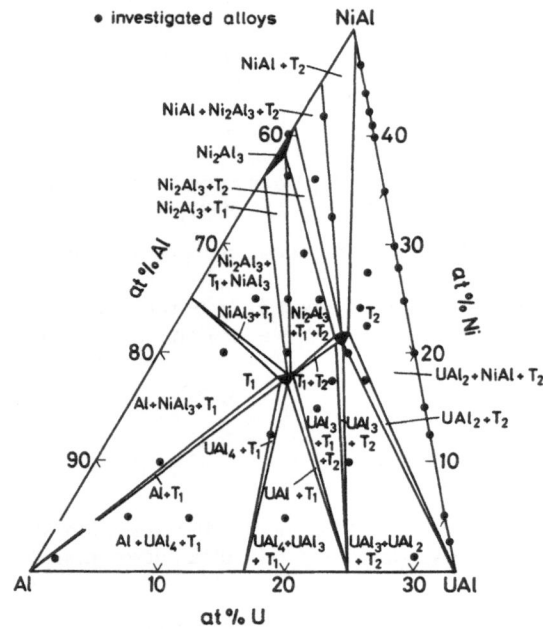

Fig. 3-90. Isothermal section of the ternary system $Al-UAl_2-NiAl$ at 600°C [2].

up to 5.5 at% Ni at the eutectic temperature, whereas a contraction of the unit cell occurs by about 0.3% [1].

The ternary system $Al−UAl_2−NiAl$, isothermal section at 600°C, is shown in **Fig. 3-90**. Two ternary intermetallic alloys were found, whereas in the more Al-rich corner a three-phase equilibrium exists together with Al and UAl_4 [2 to 4].

A ternary equiatomic alloy is reported, crystallizing with hexagonal structure (Fe_2P type), based on X-ray diffraction analysis [5,6]. Single-phase solid solutions with the composition of $UAl_xNi_{5−x}$ are formed in the range of $0 \leq x \leq 1$ as analyzed by thermal and X-ray determinations. The solid solutions crystallize cubically [7, 8] with $AuBe_5$-type structure [11].

3.1.7.13.2 Preparation and Properties

Specimens of ternary U−Al−Ni alloys were generally prepared by arc melting of the elements under argon (or helium [6]) with subsequent annealing for homogenization at different temperatures and times in evacuated quartz ampoules: 600°C for 14 d [1], 700 to 800°C for one week [9, 10], 800°C for 700 h and slowly cooled (50°C/d) [8], 950°C for 30 h [5], or 950 to 1100°C for 7 d. Solid solutions of $UAl_xNi_{5−x}$ were also prepared by hot-pressing under vacuum, subsequently sintered at 1200°C and annealed at 800°C for 700 h [7]. Larger amounts of the ternary alloys were prepared by melting the elements in an induction furnace under argon in Al_2O_3 crucibles [2, 4].

UAlNi crystallizes with hexagonal Fe_2P-type structure (C 22). The unit cell contains three molecules, the space group is $D_{3h}^3−P\bar{6}2m$ (No. 189) [5, 6, 11 to 14]. The structure of the unit cell is shown in **Fig. 3-91**, see also [6]. Lattice parameters, measured by X-ray diffraction determination, are summarized in Table 3/45.

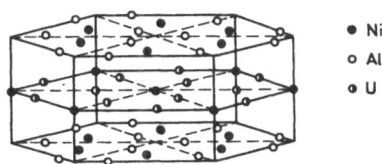

Fig. 3-91. Structure of the unit cell of UAlNi [13].

Table 3/45
Measured Lattice Parameters of UAlNi.

lattice parameter			Ref.	also given in
a	c	c/a		
6.703 Å	4.005 Å		[5]	
6.733 Å	4.035 Å	0.599	[6]	[10 to 12]
675.3 pm	408.8 pm		[15]	
675.1 pm	408.8 pm		[14]	

References for 3.1.7.13 on p. 223

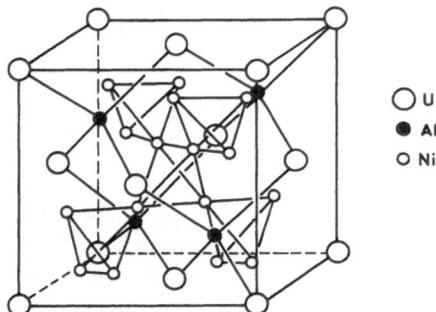

Fig. 3-92. Crystal structure of the unit cell of $UAlNi_4$ (UNi_5 type) [7].

The atomic positions of the Fe_2P-type structure are: (origin at $\bar{6}2m$) 3 Al in f (mm): x,0,0; 0,x,0; $\bar{x},\bar{x},0$. 3 U in g (mm): x,0,1/2; 0,x,1/2; $\bar{x},\bar{x},1/2$. 1 Ni in b ($\bar{6}mz$): 0,0,1/2. 2 Ni in c ($\bar{6}$): ±1/3,2/3,0 [6, 11]. The U—U distance is 351 pm [15] or 358 pm [14].

The UAl_xNi_{5-x} solid solutions with $0 \leqq x \leqq 1$ crystallize with cubic structure (UNi_5 type), the space group is $F\bar{4}3m - T_d^2$ (No. 216) [7, 8]. The structure of the unit cell is given in **Fig.** 3-**92**. The lattice parameters vary from 6.780 Å (UNi_5) to 6.810 Å ($UAlNi_4$), obeying Végard's law [7]. Measured lattice parameters, based on X-ray diffraction determination, are summarized in Table 3/46.

Table 3/46
Measured Lattice Parameters of UAl_xNi_{5-x} Solid Solutions.

x in UAl_xNi_{5-x}	lattice parameter a in Å	Ref.
0	6.780	[7]
0.2	6.787	[8]
0.4	6.793	[8]
0.6	6.797	[8]
0.8	6.805	[8]
1.0	6.810	[7,8]

The specific heat of UAlNi was measured at temperatures of up to 300 K [15, 16]. The low-temperature specific heat of UAlNi is shown in **Fig.** 3-**93**. A maximum was observed at 18 K, which is considerably lower than the maximum of the magnetic susceptibility at 23 K [15, 16]. This maximum at 18 K is shifted to approximately 17 K in a magnetic field of 5 T [15]. From C/T versus T^2 plots, a value for the coefficient of the electronic specific heat $\gamma = 160$ [16] or 167 [15] $mJ \cdot mol^{-1} \cdot K^{-2}$ is derived.

The electrical resistance of UAlNi is shown in **Fig.** 3-**94**. There is a maximum observed in the low-temperature region and a nearly temperature-independent behavior at higher temperatures [16].

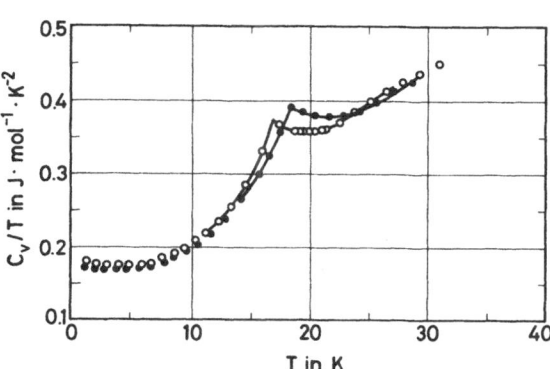

Fig. 3-93. Temperature dependence of the specific heat of UAlNi measured in zero field (●—●—●) and in an applied magnetic field of 5 T (○—○—○) [15].

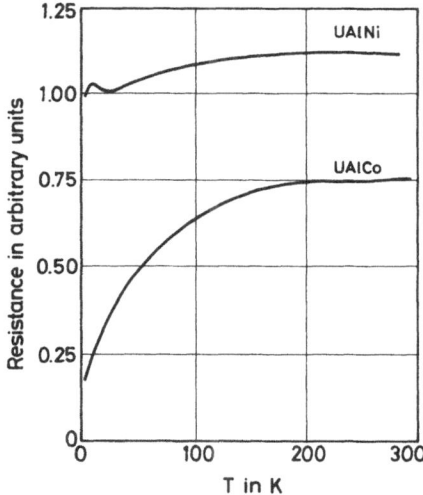

Fig. 3-94. Temperature dependence of the electrical resistance of UAlNi and UAlCo [16].

UAlNi is an antiferromagnet with a Néel temperature of $T_N = 23$ K [14 to 16]. The magnetism in UAlNi is mainly attributed to the 5f electrons of U. The magnetization curves in magnetic fields of up to 35 T (see **Fig.** 3-**95**, p. 220) are characterized by a rapid change in the magnetization between 10 and 12 T, resembling a metamagnetic transition. No complete magnetic saturation was observed in fields of up to 35 T. The spontaneous magnetization amounts to 0.3 μ_B/formula unit [14] or to 0.5 μ_B/formula unit for oriented powder samples and 0.27 μ_B/ formula unit for a tightly packed powder sample, suggesting the axis of magnetization to be oriented along the c axis [15, 16]. The magnetic susceptibility of UAlNi is given in **Fig.** 3-**96**, p. 220, see also [13].

References for 3.1.7.13 on p. 223

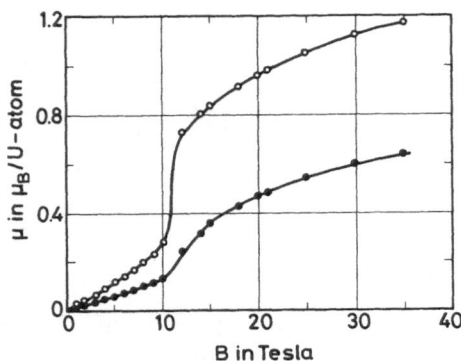

Fig. 3-95. Magnetization of UAlNi at 4.2 K for tightly packed powder (●—●—●) and for oriented powder (○—○—○) [15].

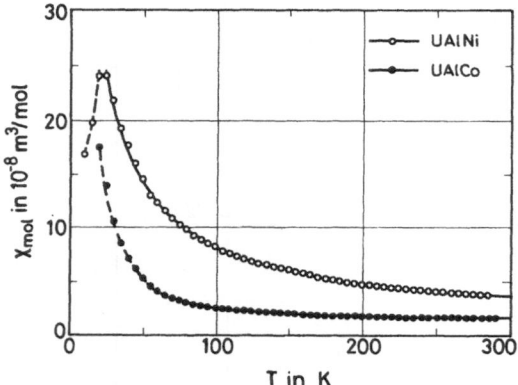

Fig. 3-96. Temperature dependence of the magnetic susceptibility of UAlNi and UAlCo [15].

At higher temperatures a Curie-Weiss law $\chi = \chi_0 + C/(T - \Theta)$ is obeyed with a Curie temperature of $\Theta_p = -7$ K and $\chi_0 = 1.24 \times 10^{-8}$ m³/mol. An effective magnetic moment is $\mu_{eff} = 2.2\ \mu_B$/formula unit [15, 16].

A comparison of the magnetic parameters and the values of the coefficient γ of the specific heat for different UTM alloys (with T = transition metal, M = Al, Ga, or Sn), using data from [14 to 18] is given in [19].

The UAl$_x$Ni$_{5-x}$ alloys order antiferromagnetically [8]. Néel temperatures and effective magnetic moments are summarized in Table 3/47.

It is postulated, that the magnetic cell orders antiferromagnetically as UCu$_5$ if Al is substituted for Ni in UNi$_5$, which has a nearly temperature-independent magnetic susceptibility. The magnetic lattice of UCu$_5$ consists of parallel sheets, (111) planes, with the magnetic moments of U coupled ferromagnetically, but with antiferromagnetic coupling between neighboring sheets. The magnetic moments are perpendicular to the ferromagnetic sheets [8].

Table 3/47

Magnetic Data for UAl_xNi_{5-x} Alloys [8].

x in UAl_xNi_{5-x}	Θ_N in K	μ_{eff} in μ_B
0.2	−619	3.70
0.4	−468	3.38
0.6	−329	1.96
0.8	−604	4.21
1.0	−472	2.92

The surface segregation of UAlNi exposed to air, residual gases, and oxygen was investigated by X-ray photoelectron spectroscopy (XPS). The analysis of the experimental data indicated that the driving force for the segregation is associated with the free energy of the pure metal constituents [19, 20].

3.1.7.13.3 UAlNiH Alloys

U—Al—M alloys (with M = Mn, Ni, Co) were hydrogenated under a hydrogen pressure of up to 150 atm at cycling temperatures in the range of 20 to 250 °C after the samples had been degassed at 450 °C in vacuum of 10^{-4} Torr for several hours. For UAlNi, a maximum hydrogen absorption of 189.73 mL H_2/g [9] or 2.5 hydrogen atoms per formula unit [10] was found at a synthesis pressure of 55 atm, resulting in a maximum hydride composition of $UAlNiH_{2.74}$ [9]. Desorption isotherms (see **Fig. 3-97**) revealed the existence of at least two hydride phases: $UAlNiH_{1.9}$ and $UAlNiH_{2.7}$ [10, 13]. Heats of formation were derived from the dissociation isotherms to be $\Delta H = -46.5$ kJ/mol H_2 and $\Delta H = -63.6$ kJ/mol H_2, respectively. Entropies of

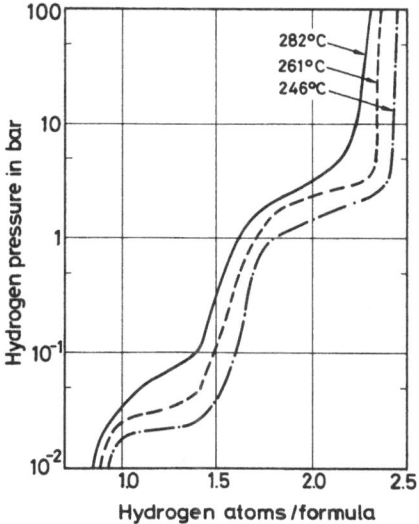

Fig. 3-97. Hydrogen dissociation isotherms for the UAlNi−H system [10].

References for 3.1.7.13 on p. 223

formation were obtained from pressure versus 1/T plots to be $\Delta S = -93.8 \, J \cdot K^{-1} \cdot mol^{-1} \, H_2$ and $\Delta S = 90 \, J \cdot K^{-1} \cdot mol^{-1} \, H_2$, respectively [10]. Further thermodynamic values for hydrogenated UAlNi alloys are summarized in Table 3/48.

Table 3/48
Thermodynamic Data for $UAlNiH_x$ Hydrides [9].

hydrogen content, x	ΔH in kcal/mol H_2	ΔS in cal $\cdot K^{-1} \cdot mol^{-1} \, H_2$	ΔG in kcal/mol H_2
2.30	− 9.8	−22.7	−3.0
2.20	−10.4	−22.9	−3.5
1.55	−11.4	−21.6	−5.0
1.35	−12.6	−21.0	−6.3

From X-ray diffraction measurement, it was found that the hydrogenated UAlNi alloys crystallize with the same hexagonal Fe_2P-type structure as the pure UAlNi alloys, although the volume of the unit cell increases by about 12%. Lattice constants are a = 7.223 ± 0.008 Å, c = 3.953 ± 0.005 Å for $UAlNiH_{2.74}$ [9], a = 7.182 Å, c = 4.012 Å for $UAlNiH_{2.5}$ [21]. Several diffraction lines of the Fe_2P-type structure are absent in the X-ray pattern of the hydrogenated samples [10]. From theoretical considerations the hydrogen atoms may be situated near or even at the 3 g (2/3, 1/3, 1/2) positions [10].

A continuous-wave (cw) and pulsed-proton nuclear magnetic resonance (NMR) study conducted on $UAlNiH_x$ (x = 1.9), together with measurements of the magnetic susceptibility, indicated the existence of two antiferromagnetic transition temperatures [13]. The second moment increases with decreasing temperature to 128 K, and only a narrow (1 to 2 Oe) resonance line of low intensity was found at 77 K and 87 K, which strongly suggests magnetic ordering below 128 K. The temperature dependence of the second moment, which is expected for paramagnetic materials, follows the equation $M_2 = M_2(0) + C/(T + \Theta)^2$, with $\Theta = 66$ K and $C_m = 7.4 \times 10^5 \, (Oe \cdot K)^2$ and $M_2(0) = 7.8 \, Oe^2$ [13]. The temperature dependence of $T_1 \cdot T$, where T_1 = spin-lattice relaxation time and T = absolute temperature, is linear for $T_1 \cdot T \gtrsim 1.3 \, s \cdot K$, which is also known for phosphides and hydrides of U and Pu [13], see also [22].

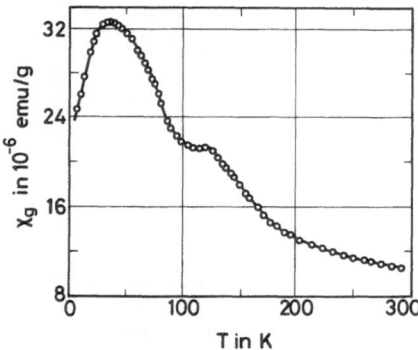

Fig. 3-98. Temperature dependence of the magnetic susceptibility in UAlNiH [13].

The magnetic susceptibility of UAl_xNi ($x = 1.9$) shows two anomalies near 34 K and 122 K (see **Fig.** 3-**98**), whereas for unhydrogenated UAlNi, a maximum was observed at 23 K. Possibly, $UAlNiH_x$ contains two phases. The maximum at 34 K may represent a hydrogen solid solution, in which the dissolved hydrogen would be responsible for shifting and broadening of the susceptibility maximum in UAlNi at 23 K. The maximum at 122 K and the disappearance of the NMR signal suggests magnetic ordering for the hydride phase [13].

References for 3.1.7.13:

[1] Petzow, G.; Exner, H. E.; Kiessler, G. (J. Less-Common Metals **14** [1968] 127/31).

[2] Exner, H. E.; Petzow, G. (Metall **23** [1969] 220/5).

[3] Petzow, G.; Exner, H. E. (BMwF-FBK-67-88 [1967] 1/69; C.A. **69** [1968] No. 38273).

[4] Exner, H. E.; Petzow, G. (Intern. Leichtmetalltag. **5** [1968/69] 77/83; C.A. **74** [1971] No. 37351).

[5] Steeb, S.; Petzow, G. (Trans. AIME **236** [1966] 1756/8).

[6] Dwight, A. E.; Mueller, M. H.; Conner, R. A., Jr.; Downey, J. W.; Knott, H. (Trans. AIME **242** [1968] 2075/80).

[7] Blazina, Z.; Ban, Z. (J. Less-Common Metals **33** [1973] 321/5).

[8] Blazina, Z.; Ban, Z.; Szytulla, A. (Fizika [Zagreb] Suppl. **12** No. 1 [1980] 200/4; C.A. **95** [1981] No. 34376).

[9] Drulis, H.; Petryński, W.; Staliński, B.; Zygmunt, A. (J. Less-Common Metals **83** [1982] 87/93).

[10] Jacob, I.; Hadari, Z.; Reilly, J. J. (J. Less-Common Metals **103** [1984] 123/7).

[11] Lam, D. J.; Darby, J. B., Jr.; Nevitt, N. V. (in: Freeman, A. J.; Darby, J. B., Jr., The Actinides: Electronic Structure and Related Properties, Vol. 2, Academic, New York — San Francisco — London 1974, pp. 119/84).

[12] Dwight, A. E. (in: Giessen, B. C., Developments in the Structural Chemistry of Alloy Phases, Plenum, New York 1969, pp. 181/226).

[13] Zogal, O. J.; Lam, D. J.; Zygmunt, A.; Drulis, H.; Petryński, W.; Staliński, S. (Phys. Rev. [3] B **29** [1984] 4837/42).

[14] Andreev, A. V.; Bartashevich, M. I. (Fiz. Metal. Metalloved. **62** No. 2 [1986] 266/8; Phys. Metallog. [USSR] **62** No. 2 [1986] 50/3; C.A. **105** [1986] No. 182534).

[15] Sechovsky, V.; Havela, L.; de Boer, F. R.; Franse, J. J. M.; Veenhuizen, P. A.; Sebek, J.; Stehno, J.; Andreev, A. V. (Physica B + C **142** [1986] 283/93).

[16] Sechovsky, V.; Havela, L.; Neuzil, L.; Andreev, A. V.; Hilscher, G.; Schmitzer, C. (J. Less-Common Metals **121** [1986] 169/74).

[17] Havela, L.; Neuzil, L.; Sechovsky, V.; Andreev, A. V.; Schmitzer, C.; Hilscher, G. (J. Magn. Magn. Mater. **54/57** [1986] 551/2).

[18] Sechovsky, V.; Havela, L.; Pillmayr, N.; Hilscher, G.; Andreev, A. V. (J. Magn. Magn. Mater. **63/64** [1987] 199/201).

[19] Jacob, I.; Fisher, M.; Hadari, Z. (Solid State Commun. **49** [1984] 1161/4).

[20] Jacob, I.; Fisher, M.; Hadari, Z. (Solid State Commun. **60** [1986] 401/5).

[21] Biderman, S.; Jacob, I.; Mintz, M. H.; Hadari, Z. (Trans. Joint Ann. Meeting Nucl. Soc. Israel **10** [1982] 129/32).

[22] Zogal, O. J.; Zygmunt, A.; Petrynski, W.; Drulis, H.; Lam, D. J. (13th Ogolnopol Semin. Magn. Rezon. Jad. Jego Zastosow., Krakow, Pol., 1980 [1981], pp. 68/72; C.A. **96** [1982] No. 45112).

3.1.7.14   U−Al−Co

3.1.7.14.1   Phase Relationships

The quasi-binary system UAl_2-UCo_2 and the U-rich section $U-UAl_2-UCo_2$ of the ternary system U−Al−Co have been investigated in detail based on microscopic, X-ray, and thermal analysis [1, 2].

The system UAl_2-UCo_2 (see **Fig. 3-99**) forms a peritectic three-phase equilibrium (liquid + UCo_2 = U_2AlCo_3) which extends at 990°C from 4 to 13 at% Al. The ternary Laves phase U_2AlCo_3, extending from 13 at% Al (at 990°C) to 17 at% Al (at 970°C), has a hexagonal $MgZn_2$-type structure. Two eutectic ternary equilibria were found in the range of 17 to 31 at% Al at 970°C (liquid = U_2AlCo_3 + $U(Al, Co)_2$) and in the range of 36 to 64 at% Al at 1150°C (liquid = UAl_2 + $U(Al, Co)_2$). The ternary Laves phase $U(Al, Co)_2$ extends from 31 at% Al (at 970°C) to 36 at% Al (at 1150°C), crystallizing hexagonally with $MgNi_2$-type structure, the melting point is 1180°C. UCo_2 dissolves up to 4 at% Al [1], see also [2]. Over this, an equiatomic alloy UAlCo was found crystallizing with hexagonal Fe_2P-type structure [3, 4].

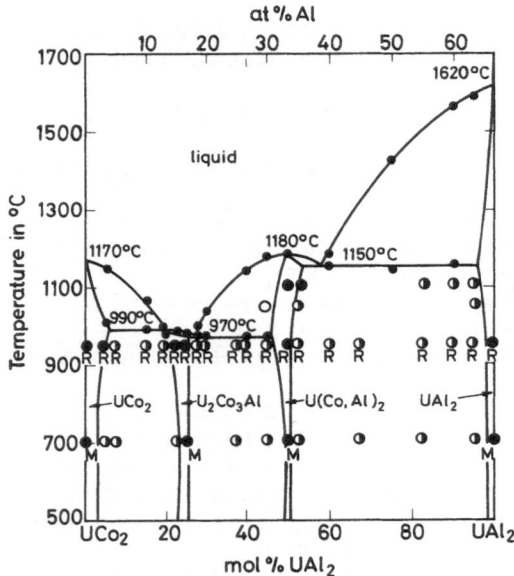

Fig. 3-99. Quasi-binary phase diagram of UAl_2-UCo_2 [1]. ● monophase, ◑ two-phase, ○ melt + one solid phase, • thermal analysis, M analysis by microhardness, R X-ray determination.

The ternary system $U-UAl_2-UCo_2$ shows eight melting surfaces of primary solidification. The compositions of the corners of the different four-phase equilibria are summarized in [2]. A reaction scheme of the phase equilibria is also given in [2]. A ternary phase U_2AlCo_2 was found [2], crystallizing with tetragonal U_3Si_2-type structure [5], which results from the ternary reaction at 910°C of liquid + $U(Al, Co)_2$ + U_2AlCo_3 = U_2AlCo_2 [2], see also [6]. Phase diagrams at different temperatures and temperature concentration sections are given in [2].

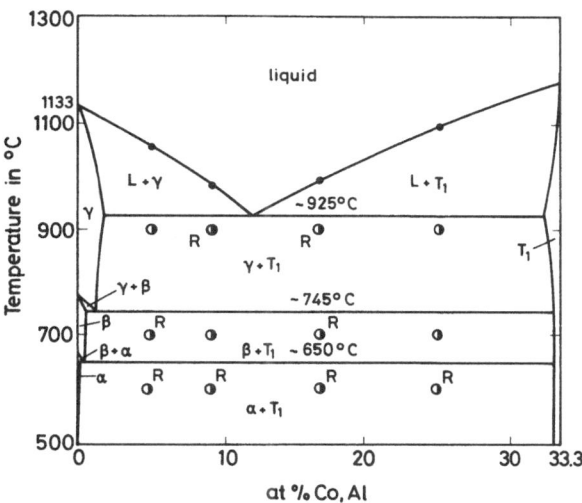

Fig. 3-100. The quasi-binary system placed between U and U(Al, Co)$_2$ [2].

The U-rich quasi-binary section U−U(Al, Co)$_2$ forms a simple eutectic system (see **Fig. 3-100**), separating the ternary system U−UAl$_2$−UCo$_2$ in two further sections: U−U(Al, Co)$_2$−UAl$_2$ and U−U(Al, Co)$_2$−UCo$_2$. The eutectic temperature is about 925°C [2].

3.1.7.14.2 Preparation and Properties

The U−Al−Co alloys were generally prepared by arc melting under argon or argon/ helium in water-cooled copper hearths [1, 3, 4, 7 to 10] or by usual melting techniques [11]. The samples were homogenized by remelting several times [4] or annealing in evacuated quartz ampoules at different temperatures and time: 700 to 725°C for 14 d and then water quenched [1, 4], 700 to 850°C for one week [9, 11], or 950°C for 30 h and then water quenched [1, 3]. The alloy U$_2$AlCo$_2$ was obtained by the peritectic reaction of UAlCo + U$_2$AlCo$_3$ + melt at 910°C on cooling [12]. The alloys obtained are generally brittle [1, 8].

Single crystals of several mm in length were obtained by remelting the prepared samples in an "alundum" crucible in He atmosphere and cooling at experimentally chosen temperature gradients and cooling rates [10, 13].

The brittle U(Al, Co)$_2$ alloys can easily be crushed to powders. Sintered specimen can be obtained by cold pressing (without binding agents) at 8 to 10 t/cm^2 and sintering in vacuum (10^{-4} to 10^{-5} Torr). Sintering starts at about 1243 K, which is $(0.86 \pm 0.02) \cdot T_M$ (with melting temperature T_M of U(Al,Co)$_2$ = 1453 K). This high sintering temperature is related to the Laves-phase structure. The sintering process is suggested to be based on solid state diffusion [8].

The U-rich U(Al$_{1-x}$Co$_x$)$_2$ solid solutions (with $0 \leq x \leq 1$) crystallize with cubic MgCu$_2$-type structure (C 15), obeying Végard's law [11].

The U(Al, Co)$_2$ alloys crystallize with hexagonal MgNi$_2$-type structure (C 36) [1], see also [2, 8]. The homogeneity range is given with 31 at% Al (at 970°C) to 36 at% Al (at 1 150°C) [1].

References for 3.1.7.14 on p. 230   15

It should be noted that the C 36-type structure was not confirmed by [4]. The lattice parameters are a = 6.501 kX, c = 21.371 kX [33], also reported in [2]. The measured density for a molten sample is 10.58 g/cm^3 [8].

UAlCo crystallizes hexagonally with Fe_2P-type structure (C 22) with 9 atoms per unit cell, the space group is $P\bar{6}2m-D_{3h}^3$ (No. 189) [3, 14 to 17]. Measured lattice parameters are given in Table 3/49. The atomic positions for the Fe_2P-type structure are given in [15]. The U−U distance is 349 pm [10] or 353 pm [17].

Table 3/49
Measured Lattice Parameters of the Ternary Alloy UAlCo.

lattice parameter		c/a	cell volume in Å^3	Ref.	also given in
a in Å	c in Å				
6.536	3.915	0.599	16.09	[3]	
6.686 + 0.001	3.966 + 0.001	0.593	17.06	[4]	[15, 16]
6.677	3.965			[14]	
666.7 (in pm)	3.965 (in pm)				[10, 17]

U_2AlCo_2 crystallizes with tetragonal U_3Si_2-type structure with Z = 10, the space group is $P_4/mbm-D_{4h}^5$ (No. 127) [12]. U_2AlCo_2 has no homogeneity range. The measured lattice parameters are a = 7.091 Å, c = 3.461 Å, c/a = 0.488 (or a = 7.1356 kX, c = 3.463 kX [2]) [12]. The calculated density is 11.85 g/cm^3 as compared to a pycnometrically obtained density of 11.3 g/cm^3 [12], see also [2]. The atomic positions are: 4 U in 4 (h): x, 1/2−x, 1/2 with x = 0.17, 4 Co in 4 (g): x, 1/2−x, 0 with x = 0.39, 2 Al in 2 (a): 0, 0, 0 [12].

U_2AlCo_3 crystallizes with hexagonal $MgZn_2$-type structure (C 14), the space group is $P6_3/mmc-D_{6h}^5$ (No. 194). The homogeneity range is 13 at% Al (at 990°C) to 17 at% Al (at 970°C) [1, 2, 7, 15, 16]. The measured lattice parameters are a = 5.110 kX, c = 7.677 kX [1], see also [2, 7], a = 5.120 Å, c = 7.693 Å, c/a = 1.503 [16], see also [5, 15]. The calculated density is 12.97 g/cm^3 as compared to a value for the measured density of 12.6 g/cm^3 [7], see also [5]. Concerning the AB_2 Laves-phase structure (with A = uranium) the Co and U atoms are statistically distributed at the B positions [7]. The atomic positions for the $MgZn_2$-type structure are given in [15]. The U−U distance is 3.10 kX and the B−B distance 2.41 kX [7].

Melting points of the different ternary U−Al−Co alloys are summarized in Table 3/50.

Table 3/50
Melting Points of Different U−Al−Co Alloys.

composition	melting point in °C	Ref.
U(Al,Co)$_2$	1180	[1,2,8]
U$_2$AlCo$_2$	peritectic at 910	[2,12]
U$_2$AlCo$_3$	peritectic at 990	[1,2]

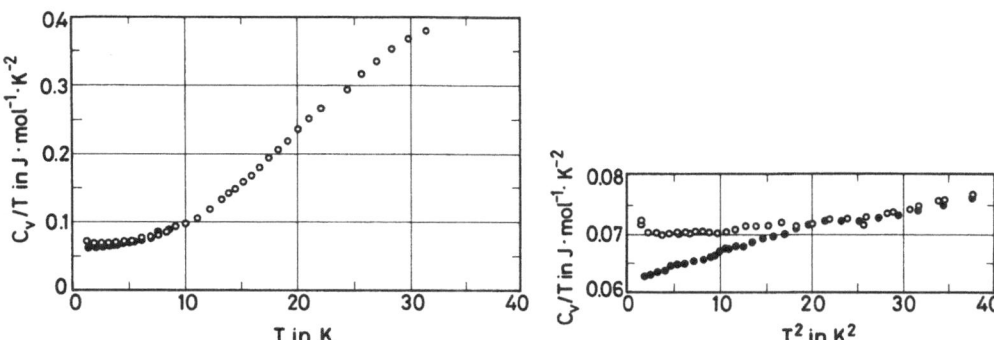

Fig. 3-101. Temperature dependence of the specific heat of UAlCo in zero-field (○) and in an applied field of 5T (●) (left figure); related low-temperature C/T versus T^2 plots (right figure) [10].

The thermal expansion coefficient of $U(AlCo)_2$ is given in Fig. 3-87, p. 214, as compared to UAl_2 and UCo_2 [8].

Measurements of the specific heat of UAlCo showed no anomalies at low temperatures (see **Fig. 3-101**). From C/T versus T^2 plots a value for the coefficient of the electronic specific heat, γ, is derived with $\gamma = 65$ [18, 19] or 68 (in $mJ \cdot mol^{-1} \cdot K^{-2}$) [10]. An observed low-temperature upturn of the specific heat at zero-field measurements is probably due to the nuclear specific heat contribution of Co [18, 19].

The temperature dependence of the electrical resistance of UAlCo is given in Fig. 3-94, p. 219. The tendency to saturation at higher temperatures indicates a considerable influence of spin fluctuation [18, 19].

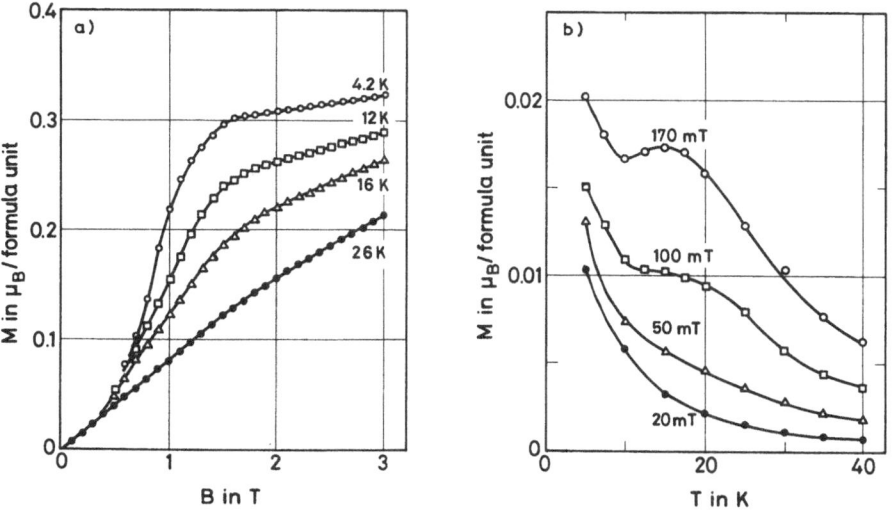

Fig. 3-102. a) Field dependence of the magnetization of UAlCo at various temperatures [19]. b) Temperature dependence of the low-field magnetization of UAlCo [19].

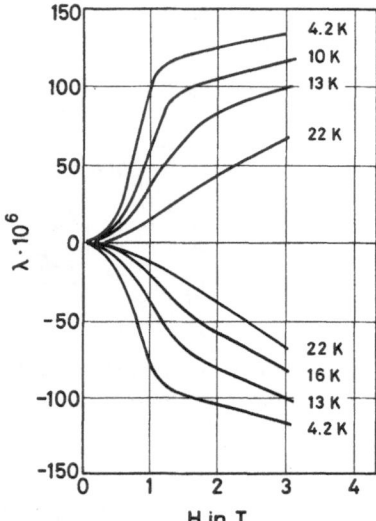

Fig. 3-103. Field dependence of the longitudinal (λ_{cc}, positive values) and transverse (λ_{ac}, negative values) magnetostriction $\lambda (= \Delta V/V)$ of a UAlCo single crystal at various temperatures [14].

The electrical resistivity of $U(Al,Co)_2$ for molten and sintered (88% th.d.) specimen is $2.3 \times 10^{-6}\ \Omega \cdot$ cm and $1.9 \times 10^{-2}\ \Omega \cdot$ cm, respectively [8].

The magnetization and magnetostriction of UAlCo was measured at temperatures of up to 40 K in applied fields of up to 3 T (see **Fig.** 3-**102**, p. 227, and **Fig.** 3-**103**) [10, 14, 19], see also [17]. From these observations UAlCo is deduced to be an antiferromagnet. In applied external magnetic fields of about 0.8 T, a metamagnetic transition from antiferromagnetism to ferromagnetism is induced below 16 K [19], see also [18], or 17 K [10], see also [17]. The magnetic moment, extrapolated from high-field measurements, amounts to 0.3 μ_B/formula unit at 4.2 K [19], see also [17, 18]. In measurements performed on a single crystal a strongly anisotropic behavior was observed, from which the c axis is shown to be the easy axis of magnetization (see **Fig.** 3-**104**) [10, 14, 19].

The magnetic susceptibility of UAlCo is given in Fig. 3-96, p. 220, see also [10, 18]. At higher temperatures a Curie-Weiss law is obeyed ($\chi = \chi_0 + C/(T - \Theta)$) with a Curie temperature of $\Theta_p = 25$ K and $\chi_0 = 1.28 \times 10^{-8}$ m³/mol. An effective magnetic moment was derived to be $\mu_{eff} = 0.8$ μ_B/formula unit [10, 18].

The occurrence of magnetism in UAlCo is mainly attributed to the 5f electrons of uranium. The results are discussed in terms of possible spin fluctuations [10, 14, 18, 19]. A comparison of the magnetic parameters and the values of the coefficient γ of the specific heat for different UMT alloys (with T = transition metal, M = Al, Ga, or Sn) from [10, 17, 18, 20] is given in [10].

The temperature dependence of the magnetic susceptibility of $U(Al_{1-x}Co_x)_2$ alloys (with $0 \leqq x \leqq 0.8$) is given in **Fig.** 3-**105**. The data are represented by the equation $\chi = \chi_0 + C/(T - \Theta)$. For the resulting Curie constants C, Pauli-type contribution X_0, and effective magnetic moments see the original paper. The magnetic behavior of the $U(Al_{1-x}Co_x)_2$ alloys may be described by a simple dilution model [11].

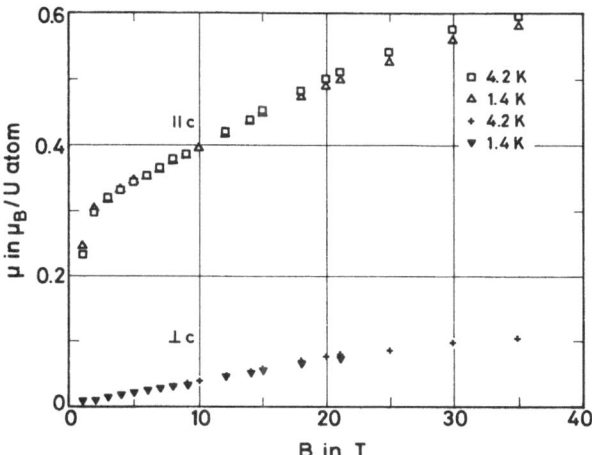

Fig. 3-104. Magnetization of a UAlCo single crystal ∥ and ⊥ to the easy axis of magnetization (c) at 4.2 and 1.4 K [10].

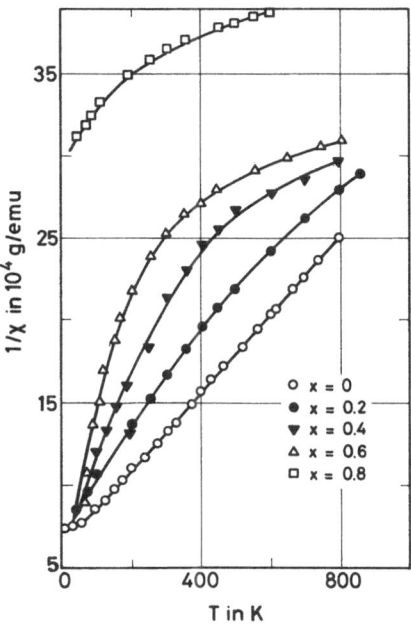

Fig. 3-105. Temperature dependence of the reciprocal susceptibility of $U(Al_{1-x}Co_x)_2$ alloys
(with $0 \leq x \leq 1$) [11].

UAlM alloys (with M = Mn, Ni, Co) were hydrogenated under a hydrogen pressure of up to 150 atm at cycling temperatures in the range of 20 to 250°C after the samples had been degassed at 450°C in vacuum of 10^{-4} Torr for several hours. For UAlCo, a maximum hydrogen absorption of 83.03 mL H_2/g was found at a synthesis pressure of 40 atm, resulting in a maximum hydride composition of $UAlCoH_{1.2}$ [9].

References for 3.1.7.14:

[1] Petzow, G.; Steeb, S.; Kiessler, G. (J. Nucl. Mater. **12** [1964] 271/6).
[2] Sampaio, A. O.; Marta, E. S.; Petzow, G. (Z. Metallk. **59** [1968] 118/24).
[3] Steeb, S.; Petzow, G. (Trans. AIME **236** [1966] 1756/8).
[4] Lam, D. J.; Darby, J. B., Jr.; Downey, J. W.; Norton, L. J. (J. Nucl. Mater. **22** [1977] 22/7).
[5] Pearson, W. B. (A Handbook of Lattice Spacings and Structures of Metals and Alloys, Vol. 2, Pergamon, New York 1967).
[6] Dwight, A. E. (ANL-82-14 [1982] 1/44; C.A. **98** [1983] No. 186482).
[7] Steeb, S.; Petzow, G. (Z. Metallk. **55** [1964] 453/6).
[8] Ondracek, G.; Petzow, G. (Z. Metallk. **56** [1965] 498/502).
[9] Drulis, H.; Petrynski, W.; Stalinski, B.; Zygmunt, A. (J. Less-Common Metals **83** [1982] 87/93).
[10] Sechovsky, V.; Havela, L.; de Boer, F. R.; Franse, J. J. M.; Veenhuizen, P. A.; Sebek, J.; Stehno, J.; Andreev, A. V. (Physica B+C **142** [1987] 283/93).

[11] Burzo, E.; Lucaci, P. (Solid State Commun. **56** [1985] 537/9).
[12] Sampaio, A. O.; Marta, E. S.; Lukas, H. L.; Petzow, G. (J. Less-Common Metals **14** [1968] 472/3).
[13] Andreev, A. V.; Deryagin, A. V.; Yumaguzhin, R. Yu. (Zh. Eksperim. Teor. Fiz. **86** [1984] 1862/9; Soviet Phys.-JETP **59** [1984] 1082/6).
[14] Andreev, A. V.; Levitin, R. Z.; Popov, Yu. F.; Yumaguzhin, R. Yu. (Fiz. Tverd. Tela [Leningrad] **27** [1985] 1902/4; Soviet Phys. Solid State **27** [1985] 1145/6; C.A. **103** [1985] No. 80639).
[15] Lam, D. J.; Darby, J.B., Jr.; Nevitt, N. V. (in: Freeman, A. J.; Darby, J. B., Jr., The Actinides: Electronic Structure and Related Properties, Vol. 2, Academic, New York − San Francisco − London 1974, pp. 119/84).
[16] Dwight, A. E. (in: Giessen, B. C., Developments in the Structural Chemistry of Alloy Phases, Plenum, New York 1969, pp. 181/226).
[17] Andreev, A. V.; Bartashevich, M. I. (Fiz. Metal. Metalloved. **62** No. 2 [1986] 266/8; Phys. Metals Metallog. [USSR] **62** No. 2 [1986] 50/3; C.A. **105** [1986] No. 182534).
[18] Sechovsky, V.; Havela, L.; Neuzil, L.; Andreev, A. V.; Hilscher, G.; Schmitzer, C (J. Less-Common Metals **121** [1986] 169/74).
[19] Havela, L.; Neuzil, L.; Sechovsky, V.; Andreev, A. V.; Schmitzer, C.; Hilscher, G. (J. Magn. Magn. Mater. **54/57** [1986] 551/2).
[20] Sechovsky, V.; Havela, L.; Pillmayr, N.; Hilscher, G.; Andreev, A. V. (J. Magn. Magn. Mater. **63/64** [1987] 199/201).

3.1.7.15   U−Al−Fe

3.1.7.15.1   Phase Relationships

The quasi-binary system UAl_2−UFe_2 and the U-rich section U−UAl_2−UFe_2 of the ternary system U−Al−Fe have been investigated in detail based on microscopic, X-ray, and thermal analysis [1 to 4].

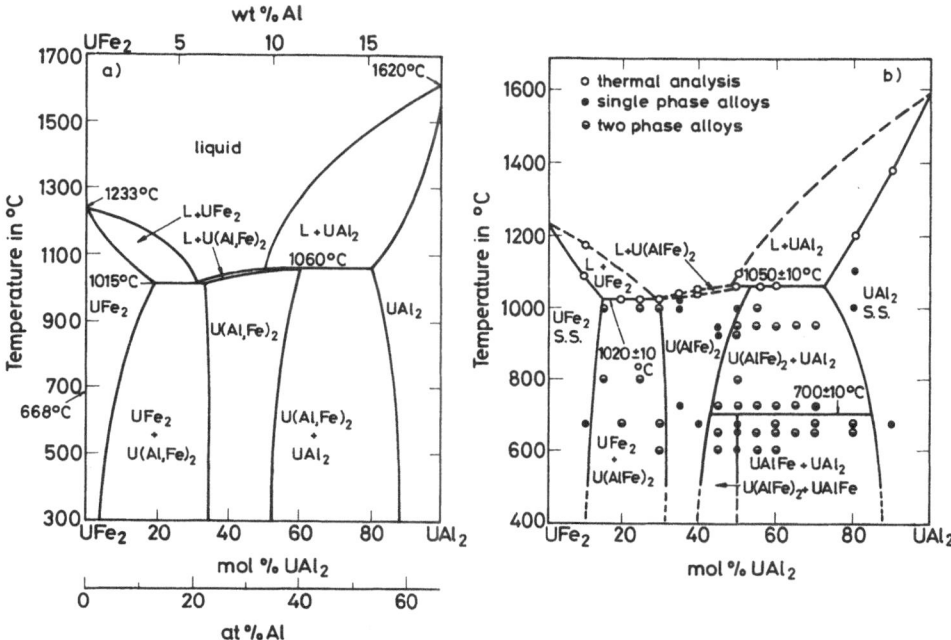

Fig. 3-106. Phase diagram of the quasi-binary system $UAl_2 - UFe_2$; a) from [3], b) from [1].

The system $UAl_2 - UFe_2$ (see **Fig. 3-106**a) forms a eutectic three-phase equilibrium (liquid = $UFe_2 + U(AlFe)_2$) which extends from 13 to 22 at% Al at 1015°C [2] or 1020 ± 10°C [1] with the eutectic point at 20.5 at% Al [2]. A peritectic three-phase equilibrium (liquid + UAl_2 = $U(Al,Fe)_2$) is placed between 33 and 53 at% Al at 1060°C [2] or 1050 ± 10°C [1]. The ternary Laves phase $U(Al,Fe)_2$ extends from 22 at% Al (at 1015°C) to 40 at% Al (at 1060°C), crystallizing hexagonally with $MgZn_2$-type structure. This phase field contains the compositions U_2AlFe_3 [5], $U_4Al_3Fe_5$, and $U_4Al_4Fe_4$ [2]. A new investigation of the quasi-binary system indicates the existence of a second ternary phase UAlFe, together with a concomitant narrowing of the $U(Al,Fe)_2$ phase field (see **Fig. 3-106**b) [1, 6, 7]. UAlFe is formed by a peritectic reaction at about 700°C, crystallizing with hexagonal structure, isostructurally with UAlCo and UAlIr [1]. The revised phase diagram of [1] has not been accepted by the authors of the first phase diagram [3].

The solubility of UFe_2 in UAl_2 is given with 19 mol% UFe_2 (at 1000°C), resulting in a contraction of the UAl_2 unit cell of about 2.6% of the cubic lattice parameter. UFe_2 dissolves up to about 19 mol% UAl_2 (at 1000°C), resulting in an expansion of the UFe_2 unit cell [2, 19], see also [1, 8].

The ternary system $U - UAl_2 - UFe_2$ shows five melting surfaces of primary solidification [3]. The compositions of the different four-phase equilibria are summarized in [3]. A reaction scheme of the phase equilibria and phase diagrams at different temperatures and temperature-concentration sections are also shown in the original paper. The ternary system is dominated by extended three-phase equilibria [3].

References for 3.1.7.15 on pp. 236/7

A phase diagram of the U-rich corner of the ternary system U—Al—Fe, summarizing the section of the so-called adjusted uranium, is given in [9], see also [4, 10 to 12].

A series of ternary alloys of the composition $UAl_{12-x}Fe_x$ have been found to crystallize body-centered tetragonally with $ThMn_{12}$-type structure [13 to 17]. The range of single-phase alloys in the series is not yet known exactly. Alloys with $x = 4$ [13 to 16], $x = 5$ [16], $x = 6$ [13, 16], and $3.2 \leqq x \leqq 4.8$ [17] were prepared for investigations of the magnetic behavior.

3.1.7.15.2 Preparation and Properties

The U—Fe—Al alloys were generally prepared by arc melting of the elements in argon or argon/helium atmosphere in water-cooled copper hearths with tungsten electrodes [1, 2, 18 to 21]. The preparation of the alloys was also carried out with U—Fe pre-alloys as starting material [2]. The samples obtained were remelted several times for homogenization or annealed at 1000°C for 8 d and then water quenched. The alloys are reported to be generally brittle [18]. Very pure samples were prepared by levitation melting in very pure argon atmosphere. The samples were held for 1 min in the levitation coil and then cast into a copper mold [1].

Especially the $UAl_{12-x}Fe_x$ alloys were prepared by melting of the elements in an induction furnace under argon in a tantalum crucible and then slowly cooled within 24 h or annealed at 700°C for 7 d for homogenization [13 to 15, 17].

The brittle $U(Al, Fe)_2$ alloys can easily be crushed to powders. Sintered specimen can be obtained by cold pressing (without binding agents) at 8 to 10 t/cm² and sintering in vacuum (10^{-4} to 10^{-5} Torr). Sintering starts at about 1173 K, which is $(0.89 \pm 0.02) \cdot T_M$ (with melting temperature T_M of $U(Al, Fe)_2 = 1333$ K). This high sintering temperature is related to the Laves-phase structure. The sintering process is suggested to be based on solid state diffusion [18].

The enthalpy of formation of UAlFe was measured by means of high-temperature calorimetry to be -5.9 kcal/g-atom [22].

The $U(Al, Fe)_2$ alloys crystallize with hexagonal $MgZn_2$-type structure (C 14) [1 to 3, 19] with $Z = 12$, the space group is $P6_3/mmc-D_{6h}^4$ (No. 194) [19], see also [25]. Measured lattice parameters are summarized in Table 3/51.

Table 3/51
Measured Lattice Parameters of $U(Al, Fe)_2$ Alloys.

composition in mol%	a	c	c/a	Ref.	also given in
65 $UFe_2 - 35\ UAl_2$	5.212 Å	8.036 Å	1.542	[1]	
62.5 $UFe_2 - 37.5\ UAl_2$	5.15 kX	7.98 kX	1.55	[19]	
62.4 $UFe_2 - 37.6\ UAl_2$	5.17 kX	8.06 kX	1.56	[19]	
	5.18 Å	8.08 Å	1.56		[23]
46 $UFe_2 - 54\ UAl_2$	5.22 kX	8.16 kX	1.565	[19]	
	5.23 Å	8.18 Å	1.565		[23]
$U_4Al_3Fe_5$	5.16 Å	8.00 Å	1.55	[24]	[23, 25]
41.6 Fe − 25 Al	5.145 kX	7.97 kX		[19]	
33 Fe − 33 Al	5.205 kX	8.11 kX		[19]	

Measured densities are D (in g/cm^3) = 11.50 (molten sample) [18], 11.21 (sample with 37.5 mol% UAl$_2$) and 10.21 (sample with 54 mol% UAl$_2$) [19], see also [23] as compared to calculated values of D (in g/cm^3) = 11.66 (sample with 37.5 mol% UAl$_2$) and 10.85 (sample with 54 mol% UAl$_2$) [19], see also [23].

A Mössbauer study, performed on the ternary phases U$_2$AlFe$_3$ and UAlFe (the phase boundaries of the U(Al, Fe)$_2$ alloys) resulted in a probable chemical ordering in the intermediate phases. The structure is suggested to be of Mg$_2$Cu$_3$Si type, which is an ordered variant of the MgZn$_2$ phase, in U$_2$AlFe$_3$ the U atoms would be placed in 4f sites, the Fe atoms in 6h, and the Al atoms in 2a. With increasing UAl$_2$ content, the Al atoms are expected to occupy U sites, which was confirmed by additional structure in the Mössbauer spectra [5].

UAlFe crystallizes hexagonally with Fe$_2$P-type structure with Z = 9, the space group is P$\bar{6}$2m−D$_{3h}^3$ (No. 189) [1, 24, 25]. Measured lattice parameters are a = 6.672 ± 0.001 Å, c = 3.981 ± 0.001 Å, c/a = 0.597 [1], also reported in [24, 25], or a = 667.0 pm, c = 399.2 pm [26]. The U−U distance is 353 pm [26].

The UAl$_{12-x}$Fe$_x$ alloys (with x = 4, 5, 6) crystallize with a body-centered tetragonal ThMn$_{12}$-type structure with Z = 2, the space group is I4/mmm−D$_{4h}^{17}$ (No. 139) [13 to 15]. Measured lattice parameters are summarized in Table 3/52.

Table 3/52
Measured Lattice Parameters of UAl$_{12-x}$Fe$_x$ Alloys.

composition	a in Å	c in Å	c/a	Ref.
UAl$_8$Fe$_4$	8.732 ± 0.001	5.039 ± 0.001	0.577	[13, 14]
	8.749	5.036	0.576	[15, 16]
UAl$_7$Fe$_5$	8.692	5.018		[16]
UAl$_6$Fe$_6$	8.638	4.987	0.577	[13, 16]

Measured densities are D = 5.85 [13], 5.83 [15] for UAl$_8$Fe$_4$ and 6.56 g/cm^3 for UAl$_6$Fe$_6$ [13], as compared to calculated densities of 5.85 g/cm^3 for UAl$_8$Fe$_4$ and 6.56 g/cm^3 for UAl$_6$Fe$_6$ [13, 15]. The crystal structure of UAl$_8$Fe$_4$ is given in **Fig. 3-107**.

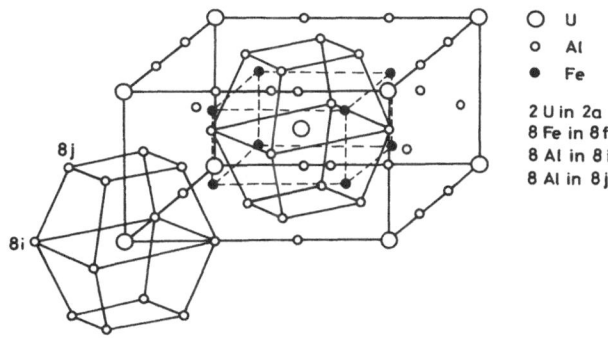

Fig. 3-107. The crystal structure of UAl$_8$Fe$_4$ [15].

References for 3.1.7.15 on pp. 236/7

The atomic positions for the $ThMn_{12}$-type structure are (origin at center 4/mmm): equivalent positions (0, 0, 0; 1/2, 1/2, 1/2) + 2 U in 2a (4/mmm): 0, 0, 0. 8 Fe in 8f (2/m): 1/4, 1/4, 1/4; 3/4, 3/4, 1/4; 1/4, 3/4, 1/4; 3/4, 1/4, 1/4. 8 Al in 8i (mm): ± (x, 0, 0; 0, x, 0) with x = 0.34399 ± 0.00046. 8 Al in 8j (mm): ± (x, 1/2, 0; 1/2, x, 0) with x = 0.28054 ± 0.00046 [25], see also [15]. The interatomic distances are given in Table 3/53.

The thermal expansion coefficient of $U(Al,Fe)_2$ is given in Fig. 3-87, p. 214, as compared to UAl_2 and UFe_2 [18].

Table 3/53
Interatomic Distances in the Crystal Structure of UAl_8Fe_4 [15].

atom		distances in Å	atom		distances in Å
U	2 U	5.036 ± 0.001	Al (8i)	1 U	3.009 ± 0.004
	8 Fe	3.340 ± 0.001		4 Fe	2.654 ± 0.001
	4 Al (8i)	3.009 ± 0.004		1 Al (8i)	2.730 ± 0.006
	8 Al (8j)	3.167 ± 0.002		4 Al (8i)	3.173 ± 0.002
				2 Al (8j)	2.744 ± 0.002
Fe	2 U	3.340 ± 0.001		2 Al (8j)	2.808 ± 0.004
	2 Fe	2.5181 ± 0.0002			
	4 Al (8i)	2.654 ± 0.001	Al (8j)	2 U	3.167 ± 0.002
	4 Al (8j)	2.5378 ± 0.0004		4 Fe	2.5378 ± 0.0004
				2 Al (8i)	2.744 ± 0.002
				2 Al (8i)	2.808 ± 0.004
				1 Al (8j)	3.840 ± 0.006
				2 Al (8j)	2.715 ± 0.004

Mössbauer effect measurements, performed on $U(Al_{1-x}Fe_x)_2$ alloys (with x = 0.08 to 0.97) in the paramagnetic state, showed two overlapping sub-spectra, which were analyzed according to the statistical distribution of nearest neighbors to iron. The results are discussed in view of charge transfers, and s- and d-electronic densities at the Fe nuclei [27].

From ferromagnetic resonance (FMR) and electron paramagnetic resonance (EPR) measurements, performed on $U(Al_{1-x}Fe_x)_2$ alloys (with x = 0.8 to 1.0), it is suggested that the electronic configuration of the transition metal is not much effected by alloying [28], see also [20, 21].

The electrical resistivity of a molten and a sintered (75% th.d.) specimen of $U(Fe,Al)_2$ alloys is $1.4 \times 10^{-6} \, \Omega \cdot cm$ and $12.5 \times 10^{-2} \, \Omega \cdot cm$, respectively [18].

The magnetic properties of $U(Al_{1-x}Fe_x)_2$ alloys were measured at temperatures of up to 300 K and in magnetic fields of up to 50 kG. The alloys are Pauli-type paramagnets in the composition range of x < 0.84. The magnetization increases with increasing external field. The saturation magnetization decreases nearly linearly with increasing substitution of Fe by Al. The magnetic susceptibility follows the equation $\chi = \chi_0 + C/(T - \Theta)$. The Curie temperatures also decrease linearly with increasing Al content. At x = 0.84 the Curie temperature is zero [20, 21], see also [8]. The derived Curie constants C and the effective magnetic moments supposed the U-effective moment to be negligible compared to that of Fe. The reduced magnetizations M(T)/M(0) as a function of reduced temperature are shown in **Fig. 3-108**.

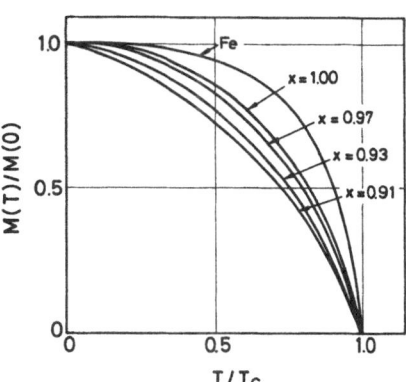

Fig. 3-108. Reduced magnetizations M(T)/M(0) as a function of reduced temperatures for
U(Al$_{1-x}$Fe$_x$)$_2$ alloys [21].

The magnetic behavior of UAl$_{12-x}$Fe$_x$ alloys (with x = 4, 5, 6, and 3.2 ≦ x ≦ 4.8) for powdered
samples was measured at temperatures of 4.2 to 300 K and in magnetic fields of up to 140 kOe.
The alloys show a typical ferromagnetic behavior but they do not reach magnetic saturation
even in a magnetic field of 140 kOe [13, 14, 16, 17]. For alloys with x ≦ 4.2 magnetic remanence
was observed below 70 kOe and was most pronounced for x = 4.2 (see **Fig. 3-109**). The

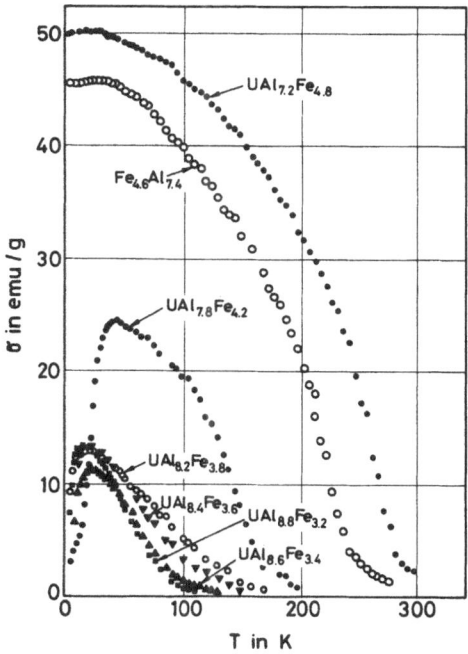

Fig. 3-109. Magnetization of UAl$_{12-x}$Fe$_x$ alloys versus temperature. Applied field 5 kOe [17].

$UAl_{12-x}Fe_x$ alloys with $x < 4.4$ show broad maxima in the magnetization plots which disappear when the samples are cooled in an external field prior to magnetization measurement. This is attributed to the formation of magnetic domains with narrow Bloch walls, characteristic of strong magnetocrystalline anisotropy [17]. Concerning the UAl_8Fe_4 lattice, it is assumed that the Fe atoms occupy only the 8f sites. A simple sublattice is formed with $a' = a/2 = 4.37$ Å and $c' = c/2 = 2.52$ Å. Each Fe atom then has two Fe atoms as neighbors at a distance of c', which is close to the nearest neighbor distance in α iron (2.48 Å) and much greater in the planes perpendicular to the c axis (4.37 Å). From this, ferromagnetic interactions along the c axis are expected, along with more complex ordering in the (004) and (00$\bar{4}$) planes [14]. A linear Curie-Weiss behavior is observed only at temperatures above 230 K; ordering temperatures are derived from ^{57}Fe Mössbauer measurements (see Table 3/54) [16].

Table 3/54
Magnetic Data of $UAl_{12-x}Fe_x$ Alloys with $x = 4, 5, 6$ [16].

composition	ordering temperature in K	estimated saturation moment at 4.2 K in μ_B/formula unit	Curie constant in $emu \cdot K \cdot mol^{-1}$
UAl_8Fe_4	160	—	16.7
UAl_7Fe_5	268	6.16	—
UAl_6Fe_6	355	6.84	—

References for 3.1.7.15:

[1] Lam, D. J.; Darby, J. B., Jr.; Downey, J. W.; Norton, L. J. (J. Nucl. Mater. **22** [1967] 22/7).
[2] Petzow, G.; Steeb, S.; Tank, R. (Z. Metallk. **53** [1962] 526/9).
[3] Petzow, G.; Tank, R. (Z. Metallk. **54** [1963] 91/8).
[4] Khakimova, D. K.; Ivanov, O. S.; Virgil'ev, Yu. A. (Stroenie Svoistva Splavov Urana Toriya i Tsirkoniya **1963** pp. 16/21; C.A. **59** [1963] 7207).
[5] Kimball, C. W.; Hannon, R. H.; Hummel, C. L.; Dwight, A. E.; Shenoy, G. K. (Conf. Dig. Inst. Phys. [London] No. 3 [1971] 105/8).
[6] Lam, D. J.; Darby, J. B., Jr.; Norton, L. J.; Downey, J. W. (TID-17624 [1962] 1/8; C.A. **60** [1964] 15543).
[7] Darby, J. B., Jr.; Lam, D. J. (ANL-6516 [1961] 213; N.S.A. **16** [1961] No. 30717).
[8] Hilscher, G. (J. Magn. Magn. Mater. **27** [1982] 1/31).
[9] Ostberg, G.; Haglund, B. O.; Lehtinen, B.; Storm, L. (J. Microsc. [Paris] **5** [1966] 21/30).
[10] Russell, R. B. (NMI-2813 [1964] 1/44; C.A. **62** [1965] 7460).

[11] Schierding, R. G.; Fergason, L. A. (MCW-1503 [1966] 1/20; C.A. **67** [1967] No. 6251).
[12] Kaderabek, E.; Haubelt, I.; Kraus, V. (UJV-2526 M [1971] 1/43; C.A. **76** [1972] No. 75661).
[13] Baran, A.; Suski, W.; Mydlarz, T. (in: Burzo, E.; Rogalski, M., Proc. Intern. Conf. Magn. Rare Earths Actinides, Bucharest, Rom., 1983, Vol. 1, pp. 224/6, C.A. **102** [1985] No. 16231).
[14] Baran, A.; Suski, W.; Mydlarz, T. (J. Less-Common Metals **96** [1984] 269/73).
[15] Stepien-Damm, J.; Baran, A.; Suski, W. (J. Less-Common Metals **102** [1984] L5/L8).
[16] Baran, A.; Suski, W.; Mydlarz, T. (Physica B + C **130** [1985] 219/21).
[17] Baran, A.; Suski, W.; Zogal, O. J.; Mydlarz, T. (J. Less-Common Metals **121** [1986] 175/80).
[18] Ondracek, G.; Petzow, G. (Z. Metallk. **56** [1965] 498/502).

[19] Steeb, S.; Petzow, G.; Tank, R. (Acta Cryst. **17** [1964] 90/5).

[20] Burzo, E.; Valeanu, M.; Ungur, D.; Lazar, D. P. (Proc. Intern. Conf. Magn. Rare Earths Actinides, Bucharest, Rom., 1983, Vol. 1, pp. 230/3; C.A. **102** [1985] No. 38190).

[21] Burzo, E.; Valeanu, M. (Appl. Phys. A **35** [1984] 79/85).

[22] Dannöhl, H. D.; Luikas, H. L. (Z. Metallk. **65** [1974] 642/9).

[23] Pearson, W. B. (A Handbook of Lattice Spacings and Structures of Metals and Alloys, Vol. 2, Pergamon, New York 1967).

[24] Dwight, A. E. (in: Giessen, B. C., Developments in the Structural Chemistry of Alloy Phases, Plenum, New York 1969, pp. 181/226).

[25] Lam, D. J.; Darby, J. B., Jr.; Nevitt, N. V. (in: Freeman, A. J.; Darby, J. B., Jr., The Actinides; Electronic Structure and Related Properties, Vol. 2, Academic, New York − San Francisco − London 1974, pp. 119/84).

[26] Andreev, A. V.; Bartashevich, M. I. (Fiz. Metal. Metalloved. **62** No. 2 [1986] 266/8; Phys. Metals Metallog. [USSR] **62** No. 2 [1986] 50/3; C.A. **105** [1986] No. 182534).

[27] Fayek, M. K.; Bahgat, A. A.; Eltawansi, A. (Appl. Phys. A **26** [1981] 175/8).

[28] Burzo, E. (in: Mueller, K. A.; Kind, R.; Roos, J., Magn. Resonance Relat. Phenom. Proc. 22nd Congr. AMPERE, Zürich, 1984, pp. 295/6; C.A. **102** [1985] No. 124262).

3.1.7.16   Quaternary U − Al − Fe − Mn Alloys

Phase Relationships

The quasi-ternary system $UAl_2 - UFe_2 - UMn_2$ and the U-rich section of the quaternary system $U - UAl_2 - UFe_2 - UMn_2$ have been investigated in detail, based on microscopic, X-ray, and thermal analysis. Samples of the investigated alloys were prepared by arc melting of the elements in argon atmosphere and were homogenized by remelting several times [1 to 4].

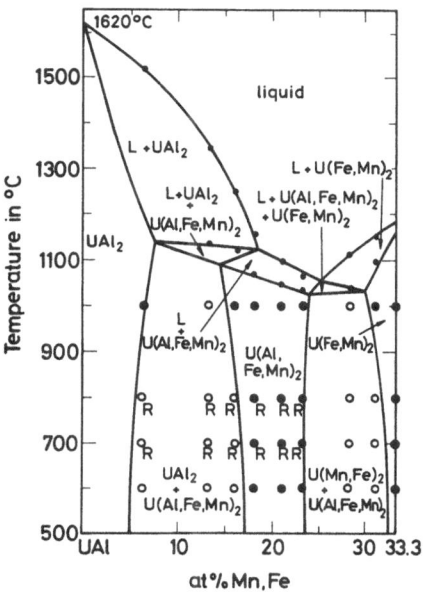

Fig. 3-110. Phase diagram of the quasi-ternary system $UAl_2 - UFe_2 - UMn_2$ at a constant ratio Mn:Fe = 1:1 [4]. ● monophase, ○ two phase, R = X-ray determination.

References for 3.1.7.16 on p. 238

The quasi-ternary system $UAl_2 - UFe_2 - UMn_2$ (see **Fig.** 3-**110**, p. 237) is characterized by extending two-phase areas in the solid state. The two ternary phases $U(Al,Fe)_2$ and $U(Al,Mn)_2$ form complete solid solutions: $U(Al,Mn)_2 \leftrightarrow U(Al,Mn,Fe)_2 \leftrightarrow U(Al,Fe)_2$. Non-variant phase equilibria are not present [4].

The quaternary system $U - UAl_2 - UFe_2 - UMn_2$ shows four non-variant five-phase equilibria, existing at 760, 715, 660, and 590°C. The five-phase equilibria at 760 and 715°C are transition equilibria involving the melt. The other two are solid state eutectoid reactions characterized by the decomposition of γ- and β-solid solutions [1]. The binary, ternary, and quaternary non-variant melting equilibria, a reaction scheme of the phase equilibria, and the composition of the corners of the five-phase areas are summarized in [1]. At 550°C the same phase relations are observed as at room temperature. The pertinent isothermal phase tetrahedron and different sections of this tetrahedron are also given in [1].

Further information on the quaternary system $U - UAl_2 - UFe_2 - UMn_2$ with a constant iron content of 10 and 50 at% are reported in [3] and [2], respectively.

References for 3.1.7.16:

[1] Petzow, G.; Sampaio, A. O. (Z. Metallk. **57** [1966] 625/32).
[2] Petzow, G.; Sampaio, A. O. (Z. Metallk. **57** [1966] 676/81).
[3] Petzow, G.; Sampaio, A. O. (Z. Metallk. **57** [1966] 741/6).
[4] Petzow, G.; Sampaio, A. O.; de Lourdes Pinto, M. (J. Nucl. Mater. **26** [1968] 331/7).

3.1.7.17   U — Al — Cu

3.1.7.17.1   Phase Relationships

Up to now, there is no detailed information on the system U — Al — Cu. Several single-phase alloys were prepared and determined mainly by X-ray analysis [1 to 3].

The following alloys exist: U_2AlCu_3, formed by substitution of Al for Cu in UCu_2, crystallizing hexagonally with $MgZn_2$-type structure [2]; UAl_2Cu, formed by substitution of Al for Cu in UCu_3, crystallizing tetragonally with $TiAl_3$-type structure [2]; U_2AlCu_9 and $U_2Al_3Cu_7$, formed by substitution of Al for Cu in UCu_5, crystallizing hexagonally with $MgZn_2$-type and $CaCu_5$-type structure, respectively [1]; $U_2Al_{10}Cu_7$, crystallizing trigonally with Th_2Zn_{17}-type structure [3, 4]; UAl_8Cu_4 and $UAl_{7.5}Cu_{4.5}$ ($= UAl_{12-x}Cu_x$), crystallizing tetragonally with $CeAl_8Mn_4$-type [3, 4] or $ThMn_{12}$-type [5] structure.

3.1.7.17.2   Preparation

U_2AlCu_3 and UAl_2Cu were prepared by arc melting of the elements in argon atmosphere in a titanium button which acted as a getter, followed by annealing at 800°C for 700 h in evacuated quartz tubes and slowly cooling at a rate of 50°C per day [2].

U_2AlCu_9 and $U_2Al_3Cu_7$ were prepared by hot-pressing of the powdered elements in vacuum (10^{-3} Torr), sintering at 1100°C, and subsequent annealing at 800°C for 700 h [1].

$U_2Al_{10}Cu_7$ and UAl_8Cu_4 were prepared by arc melting of the elements in argon atmosphere in water-cooled copper hearths [4] or by heating in an induction furnace in aluminium oxide crucibles in an argon atmosphere at 1000 K for 30 min and then slowly cooling at 5 K/min. Small single crystals were selected [3, 4].

$UAl_{7.5}Cu_{4.5}$ was prepared by arc melting in argon atmosphere and then cooling rapidly [5].

3.1.7.17.3 Crystallographic Properties

U_2AlCu_3 crystallizes with hexagonal $MgZn_2$-type structure; $Z = 2$. The space group is $P6_3/mmc - D_{6h}^4$ (No. 194). The lattice parameters are given in Table 3/55, p. 241. The calculated density is 12.48 g/cm³ as compared to a pycnometrically measured value of 12.19 g/cm³ [2].

The atomic positions are 4 U in 4 (f): $1/3, 2/3, z$; $2/3, 1/3, 1/2 + z$; $2/3, 1/3, z$; $1/3, 2/3, 1/2 - z$ with $z = 0.025$. 2 Al in 2 (a): $0, 0, 0$; $0, 0, 1/2$. 6 Cu in 6 (h): $x, 2x, 1/4$; $2\bar{x}, \bar{x}, 1/4$; $x, \bar{x}, 1/4$; $\bar{x}, 2\bar{x}, 3/4$; $2x, x, 3/4$; $\bar{x}, x, 3/4$ with $x = 0.840$ (see **Fig.** 3-**111**) [2].

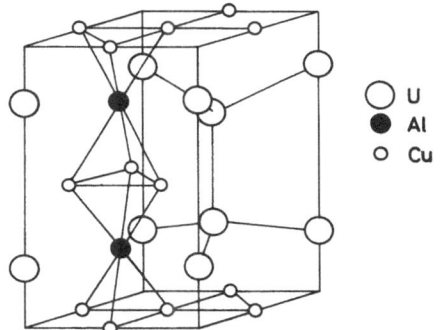

<table>
<tr><td>◯</td><td>U</td></tr>
<tr><td>●</td><td>Al</td></tr>
<tr><td>○</td><td>Cu</td></tr>
</table>

Fig. 3-111. The crystal structure of U_2AlCu_3 [2].

Interatomic distances are $U-U = 2.939$ Å, $U-Al = 2.932$ Å, $U-Cu = 2.743$ Å, $Al-Cu = 2.507$ Å, $Cu-Cu = 2.431$ Å [2].

UAl_2Cu crystallizes with tetragonal $TiAl_3$-type structure; $Z = 2$. The space group is $I4/mmm - D_{4h}^{17}$ (No. 139). The lattice parameters are given in Table 3/55, p. 241. The calculated density is 9.08 g/cm³ as compared to a pycnometrically measured value of 8.79 g/cm³. The atomic positions are 2 U in 2 (a): $0, 0, 0$; $1/2, 1/2, 1/2$. 2 Cu in 2 (b): $0, 0, 1/2$; $1/2, 1/2, 0$. 4 Al in 4 (d): $1/2, 0, 3/4$; $0, 1/2, 3/4$; $0, 1/2, 1/4$; $1/2, 0, 1/4$ (see **Fig.** 3-**112**, p. 240) [2].

Interatomic distances are $U-Cu = 2.727$ Å, $U-Al = 2.913$ Å, $Cu-Al = 2.913$ Å, $Al-Al = 2.727$ Å [2].

U_2AlCu_9 crystallizes with hexagonal $MgZn_2$-type structure, with $Z = 12$. The space group is $P6_3/mmc$ D_{6h}^4 (No. 194). The lattice parameters are given in Table 3/55, p. 241. The calculated density is 9.05 g/cm³ as compared to a pycnometrically measured value of 9.02 g/cm³. The atomic positions are 2 U in 2 (a): $0, 0, 0$; $0, 0, 1/2$. 4 Cu in 4 (f): $1/3, 2/3, z$; $2/3, 1/3, 1/2 + z$; $2/3, 1/3, z$; $1/3, 2/3, 1/2 - z$ with $z = 0.06$. 5 Cu + 1 Al in 6 (h) (statistically): $x, 2x, 1/4$; $2\bar{x}, \bar{x}, 1/4$; $x, \bar{x}$,

                    References for 3.1.7.17 on p. 242

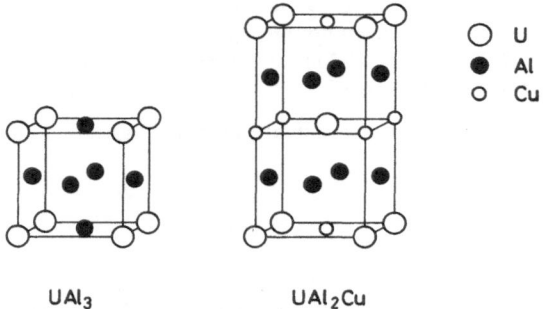

UAl₃   UAl₂Cu

Fig. 3-112. The crystal structure of UAl₂Cu as compared to the crystal structure of UAl₃ [2].

1/4; $\bar{x}, 2\bar{x}, 3/4$; $2x, x, 3/4$; $\bar{x}, x, 3/4$ with $x = 0.77$. Interatomic distances are Cu 4 (f) − Cu 4 (f) = 3.435 Å, Cu 4 (f) − Cu 4 (f) = 3.068 Å, Cu 4 (f) − U 2 (a) = 2.937 Å, Cu 4 (f) − Cu, Al 6 (h) = 2.774 Å, U 2 (a) − Cu 4 (f) = 2.937 Å, U 2 (a) − Cu, Al 6 (h) = 3.012 Å, Cu, Al 6 (h) − Cu 4 (f) = 2.774 Å, Cu, Al 6 (h) − U 2 (a) = 3.012 Å, Cu, Al 6 (h) − Cu, Al 6 (h) = 3.448 Å [1].

U₂Al₃Cu₇ crystallizes with hexagonal CaCu₅-type structure; Z = 6. The space group is P6/mmm − D_{6h}^1 (No. 191). The lattice parameters are given in Table 3/55. The calculated density is 8.94 g/cm³ as compared to a pycnometrically measured value of 8.85 g/cm³. The atomic positions are 1 U in 1 (a): 0, 0, 0; 2/3, 1/3, 0. 2 Cu in 2 (c): 1/3, 2/3, 0; 0, 1/2, 1/2. 1.5 Cu + 1.5 Al in 3 (g): 1/2, 0, 1/2; 1/2, 1/2, 1/2 (statistically) (see **Fig.** 3-**113**) [1].

U₂Al₁₀Cu₇ crystallizes with trigonal Th₂Zn₁₇-type structure; Z = 3. The space group is R̄3m − D_{3d}^5 (No. 166). The lattice parameters are given in Table 3/55. The calculated density is 6.99 g/cm³ as compared to a pycnometrically measured value of 7.0 g/cm³. Interatomic distances (main) are U − U = 404.7 ± 3 pm, U − Cu, Al = 309.2 ± 8 to 337.8 ± 3 pm, Cu, Al − Cu, Al = 250.7 ± 2 to 286.4 ± 5 pm [3].

UAl₈Cu₄ crystallizes with tetragonal CeMn₄Al₈-type structure; Z = 2. The space group is I4/mmm − D_{4h}^{17} (No. 139). The lattice parameters are given in Table 3/55. The calculated density is 6.00 g/cm³ as compared to a pycnometrically measured value of 5.89 g/cm³. Interatomic distances are U − U = 510.4 pm, U − Cu = 335.1 pm, U − Al = 303.7 ± 11 pm, U − Al = 319.6 ± 11 pm, Cu − Al = 267.3 ± 4 pm, Cu − Al = 254.9 ± 1 pm, Cu − Cu = 255.2 pm, Al − Al = 268.9 ± 22 pm, Al − Al = 279.9 ± 14 pm [3].

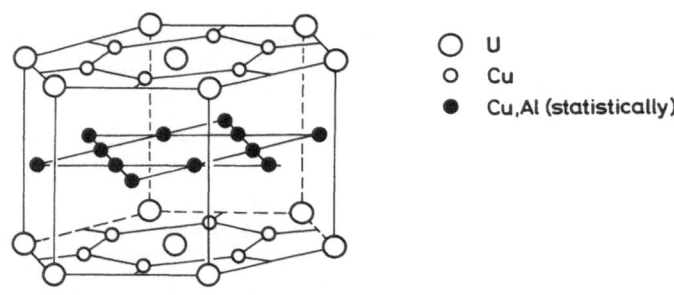

Fig. 3-113. The crystal structure of U₂Al₃Cu₇ [1].

$UAl_{7.5}Cu_{4.5}$ crystallizes with body-centered tetragonal $ThMn_{12}$-type structure. The space group is $I4/mmm - D_{4h}^{17}$ (No. 139) [5]. The lattice parameters are given in Table 3/55.

Table 3/55
Lattice Parameters of U – Al – Cu Alloys.

alloy composition	lattice parameter		c/a	Ref.
	a in Å	c in Å		
U_2AlCu_3	5.065 ± 0.005	8.307 ± 0.005	1.640	[2]
UAl_2Cu	3.857 ± 0.005	8.736 ± 0.005	2.265	[2]
U_2AlCu_9	4.988 ± 0.005	9.040 ± 0.005	1.809	[1]
$U_2Al_3Cu_7$	5.083 ± 0.005	4.155 ± 0.005	0.817	[1]
$U_2Al_{10}Cu_7$ (in pm)	875.1 ± 3	1280.0 ± 5	1.4633	[3,4]
UAl_8Cu_4 (in pm)	876.3 ± 1	510.4 ± 1	0.5825	[3,4]
$UAl_{7.5}Cu_{4.5}$	8.792	5.122	–	[5]

3.1.7.17.4 Electrical Properties

The temperature dependence of the electric resistivity for $U_2Al_{10}Cu_7$ and UAl_8Cu_4 is given in **Fig. 3-114** and **Fig. 3-115**, respectively. The resistivity of $U_2Al_{10}Cu_7$ shows a shallow minimum near 30 K. No superconductivity was detected for either alloy down to a temperature of 30 mK [4].

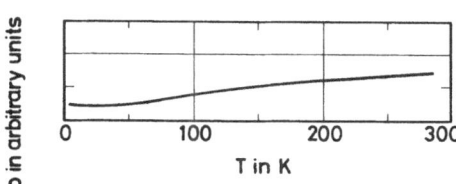

Fig. 3-114. Temperature dependence of the electrical resistivity of $U_2Al_{10}Cu_7$ [4].

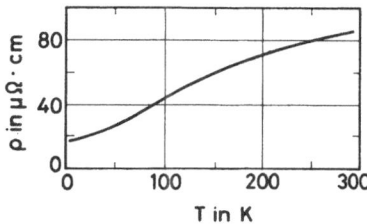

Fig. 3-115. Temperature dependence of the electrical resistivity of UAl_8Cu_4 [4].

3.1.7.17.5 Magnetic Properties

$U_2Al_{10}Cu_7$ and UAl_8Cu_4 show no sign of magnetic ordering down to a temperature of 30 mK [4].

The temperature dependence of the magnetic susceptibility of $UAl_{7.5}Cu_{4.5}$ (see **Fig.** 3-**116**) exhibits a pronounced maximum at 30 K, suggesting an onset of antiferromagnetic ordering. A Curie-Weiss law is obeyed at 40 to 180 K with a Curie temperature of $\Theta = -138$ K, from which an effective magnetic moment of $\mu_{eff} = 3.11$ μ_B was derived [5].

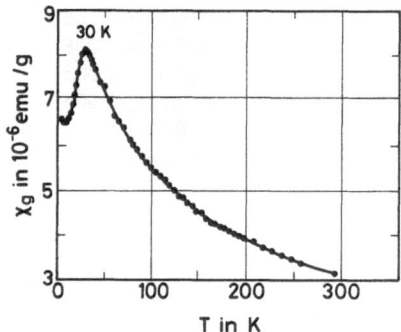

Fig. 3-116. Temperature dependence of the magnetic susceptibility of $UAl_{7.5}Cu_{4.5}$ [5].

No resonance was detected at ^{63}Cu NMR measurements [5].

References for 3.1.7.17:

[1] Blazina, Z.; Ban, Z. (Z. Naturforsch. **28b** [1973] 561/4).
[2] Blazina, Z.; Ban, Z. (Z. Naturforsch. **35b** [1980] 1162/5).
[3] Cordier, G.; Czech, E.; Schaefer, H.; Woll, P. (J. Less-Common Metals **110** [1985] 327/30).
[4] Rauchschwalbe, U.; Gottwick, U.; Ahlheim, U.; Mayer, H. M.; Steglich, F. (J. Less-Common Metals **111** [1985] 265/75).
[5] Baran, A.; Suski, W.; Zogl, O. J.; Mydlarz, T. (J. Less-Common Metals **121** [1986] 175/80).

3.1.7.18 U−Al−Ru. U−Al−Rh

Only the alloys UAlRu and UAlRh of the ternary systems U−Al−Ru and U−Al−Rh have been investigated.

Samples of UAlRu and UAlRh were prepared by arc melting in helium/argon [1] or argon [2] atmosphere and subsequently homogenized at temperatures of 700 to 900°C [1].

UAlRu and UAlRh both crystallize with hexagonal Fe_2-type structure; Z = 3. The space group is $P\bar{6}2m - D_{3h}^3$ (No. 189) [1, 3]. Measured lattice parameters are a = 6.895 ± 0.001 Å, c = 4.029 ± 0.001 Å, c/a = 0.584 for URuAl [1], see also [3, 4], a = 692.0 pm, c = 402.0 pm [2, 5]; a = 6.9647 ± 0.0006 Å, c = 4.0192 ± 0.0006 Å, c/a = 0.577 for UAlRh [1], see also [3, 4].

The U−U distance is reported to be 361 pm [2] or 367 pm [5].

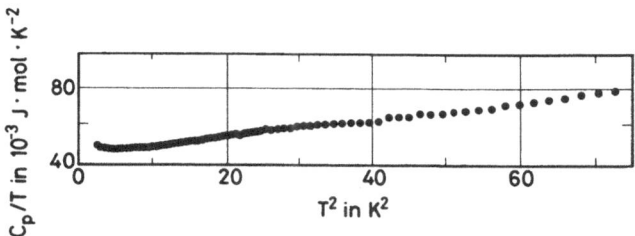

Fig. 3-117. Low-temperature specific heat of UAlRu as C_p/T versus T^2 [6].

The low-temperature specific heat of UAlRu is given in **Fig. 3-117** as a function of T^2. From this, a value of the electronic specific heat coefficient was derived with $\gamma = 45$ mJ · mol^{-1} · K^{-2} [6], see also [2].

The temperature dependence of the electrical resistance of UAlRu is given in **Fig. 3-118**. The tendency to saturation at higher temperatures, together with the susceptibility behavior indicates a considerable influence of spin fluctuation [6].

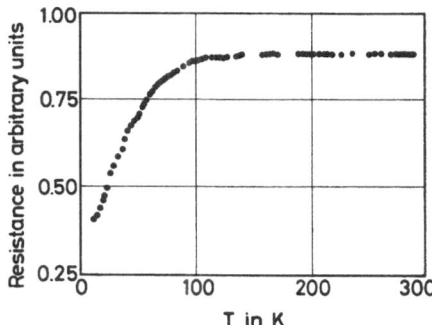

Fig. 3-118. Temperature dependence of the electrical resistance of UAlRu [6].

The field dependence of the paramagnetic susceptibility of UAlRu is given in **Fig. 3-119**, p. 244. There was no indication of any magnetic superconductive transition down to a temperature of 20 mK. The broad maximum observed in the susceptibility plots at about 50 K resembles the influence of magnetic impurities, which leads to a spin-fluctuation behavior [2, 6]. UAlRu remains paramagnetic down to a temperature of 2.5 K [6], and a modified Curie-Weiss law is obeyed ($\chi = \chi_0 + C/(T - \Theta)$) above 100 K with a Curie temperature of $\Theta_p = -95$ K, and $\chi_0 = 0.65 \times 10^{-8}$ m^3/mol, from which an effective moment was derived of $\mu_{eff} = 2.0 \, \mu_B$ [2, 6].

Further values are reported for ferromagnetic UAlRh with a transition temperature of $T_c = 35$ K and a spontaneous magnetic moment of $\mu_s = 0.15 \, \mu_B$ [5].

The magnetic parameters and the values of the coefficient γ of the specific heat for different UTM alloys (with T = transition metal, M = Al, Ga, or Sn) are compared in [2], using data from [2, 5 to 8].

References for 3.1.7.18 on p. 244   16*

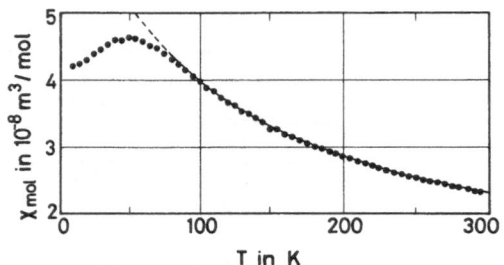

Fig. 3-119. Temperature dependence of the magnetic susceptibility of UAlRu [2]. Full line represents the modified Curie-Weiss fit.

References for 3.1.7.18:

[1] Dwight, A. E.; Mueller, M. H.; Conner, R. A., Jr.; Downey, J. W.; Knott, H. (Trans. AIME **242** [1968] 2075/80).

[2] Sechovsky, V.; Havela, L.; de Boer, F. R.; Franse, J. J. M.; Veenhuizen, P. A.; Sebek, J.; Stehno, J.; Andreev, A. V. (Physica B + C **142** [1986] 283/93).

[3] Lam, D. J.; Darby, J. B., Jr.; Nevitt, N. V. (in: Freeman, A. J.; Darby, J. B., Jr., The Actinides: Electronic Structure and Related Properties, Vol. 2, Academic, New York − San Francisco − London 1974, pp. 119/84).

[4] Dwight, A. E. (in: Giessen, B. C., Developments in the Structural Chemistry of Alloy Phases, Plenum, New York 1969, pp. 181/226).

[5] Andreev, A. V.; Bartashevich, M. J. (Fiz. Metal. Metalloved. **62** [1986] 266/8; Phys. Metals Metallog. [USSR] **62** No. 2 [1986] 50/3; C.A. **105** [1986] No. 182534).

[6] Sechowsky, V.; Havela, L.; Neuzil, L.; Andreev, A. V.; Hilscher, G.; Schmitzer, C. (J. Less-Common Metals **121** [1986] 169/74).

[7] Havela, L.; Neuzil, L.; Sechowsky, V.; Andreev, A. V.; Schmitzer, C.; Hilscher, G. (J. Magn. Magn. Mater. **54/57** [1986] 551/2).

[8] Sechowsky, V.; Havela, L.; Pillmayr, N.; Hilscher, G.; Andreev, A. V. (J. Magn. Magn. Mater. **63/64** [1987] 199/201).

3.1.7.19   U−Al−Os. U−Al−Ir. U−Al−Pt

Only the alloys U_2Al_3Os, UAlIr, and UAlPt of the ternary systems U−Al−Os, U−Al−Ir, and U−Al−Pt have been investigated.

U_2Al_3Os and UAlIr or UAlPt were prepared by arc melting in argon [1, 2] or helium/argon [3, 4] atmosphere, remelted several times, and homogenized by annealing in evacuated quartz tubes at 950°C for 30 h [1, 2], at 1000°C for one week [3], or at 700 to 900°C [4], and subsequently water-quenched. The alloys are reported to be bright and brittle [1].

U_2Al_3Os crystallizes with hexagonal $MgZn_2$-type structure (C 14); Z = 4. The space group is $P6_3/mmc−D_{6h}^4$ (No. 194) [2, 5, 6]. The measured lattice parameters are a = 5.369 kX, c = 8.465 kX [2]; a = 5.380 Å, c = 8.482 Å, c/a = 1.577 [5, 6]. The calculated density is 11.65 g/cm³, as compared to a pycnometrically measured value 11.2 g/cm³ [2]. Atomic distances are U−U = 3.27 kX and B−B = 2.61 kX, whereas the Os and Al atoms are statistically distributed in the B positions [2].

UAlIr and UAlPt crystallize with hexagonal Fe_2P-type structure (C 22); Z = 3. The space group is $P\bar{6}2m - D_{3h}^3$ (No. 189) [1, 3 to 6]. The measured lattice parameters are a = 6.893 Å, c = 3.987 Å, c/a = 0.577 [1]; a = 6.968 ± 0.001 Å, c = 4.030 ± 0.001 Å, c/a = 0.579 [3], see also [5, 6]; a = 695.0 pm, c = 401.0 pm [7]. A value for the U — U distance is reported with 371 pm for UAlIr [7].

UAlIr and UAlPt are reported to order ferromagnetically with a transition temperature of T_c = 64 K and 52 K, respectively [7] and spontaneous magnetic moments of μ_s = 0.4 μ_B and 0.8 μ_B, respectively [7].

The magnetic parameters and the values of the coefficient γ of the specific heat for different UTM alloys (with T = transition metal, M = Al, Ga, or Sn) are compared in [11] using data from [7 to 11].

References for 3.1.7.19:

[1] Steeb, S.; Petzow, G. (Trans. AIME **236** [1966] 1756/8).
[2] Steeb, S.; Petzow, G. (Z. Metallk. **55** [1964] 453/6).
[3] Lam, D. J.; Darby, J. B., Jr.; Downey, J. W.; Norton, L. J. (J. Nucl. Mater. **22** [1967] 22/7).
[4] Dwight, A. E.; Mueller, M. H.; Conner, R. A., Jr.; Downey, J. W.; Knott, H. (Trans. AIME **242** [1968] 2075/80).
[5] Dwight, A. E. (in: Giessen, B. C., Developments in the Structural Chemistry of Alloy Phases, Plenum, New York 1969, pp. 181/226).
[6] Lam, D. J.; Darby, J. B., Jr.; Nevitt, N. V. (in: Freeman, A. J.; Darby, J. B., Jr., The Actinides: Electronic Structure and Related Properties, Vol. 2, Academic, New York — San Francisco — London 1974, pp. 119/84).
[7] Andreev, A. V.; Bartashevich, M. J. (Fiz. Metal. Metalloved. **62** [1986] 266/8; Phys. Metals Metallog. [USSR] **62** No. 2 [1986] 50/3; C.A. **105** [1986] No. 182534).
[8] Havela, L.; Neuzil, L.; Sechovsky, V.; Andreev, A. V.; Schmitzer, C.; Hilscher, G. (J. Magn. Magn. Mater. **54/57** [1986] 551/2).
[9] Sechovsky, V.; Havela, L.; Neuzil, L.; Andreev, A. V.; Hilscher, G.; Schmitzer, C. (J. Less-Common Metals **121** [1986] 169/74).
[10] Sechovsky, V.; Havela, L.; Pillmayr, N.; Hilscher, G.; Andreev, A. V. (J. Magn. Magn. Mater. **63/64** [1987] 199/201).
[11] Sechovsky, V.; Havela, L.; de Boer, F. R.; Franse, J. J. M.; Veenhuizen, P. A.; Sebek, J.; Stehno, J.; Andreev, A. V. (Physica B + C **142** [1986] 283/93).

3.2. Uranium — Gallium

3.2.1 The U — Ga System

The solution of Ga in U leads to a variety of metastable structures at room temperature. Uranium alloys with 1.5 at% Ga crystallize with tetragonal β-uranium structure when rapidly cooled to room temperature from the γ-uranium or β-uranium phase. This β-uranium structure is metastable and undergoes a shear-like isothermal transformation into an α'-uranium structure, which is a distorted α-uranium-like structure. That structure decomposes after aging at 590°C for 150 h (upper α-uranium phase temperature range) into a two-phase structure consisting of Ga-free α uranium and the intermetallic compound U_3Ga_5 (small spheroidal δ-phase particles), whereas Widmanstatten platelets are formed by U_3Ga_5 at higher Ga contents and slower cooling rates of the initial alloys [1, 2]. U_3Ga_5 crystallizes with Pu_3Pd_5-type structure [2].

                    References for 3.2.1 on p. 248

A further U-rich alloy was produced for nuclear application and is referred to as U_3Ga [3], see also [4].

At the U-rich side of the phase diagram, a eutectic was found at 22 at% Ga with a eutectic temperature of 1030°C [5]. These data are in disagreement with a former value for the eutectic placed at about 15.5 at% Ga [6], see also [7 to 9].

With increasing Ga content, the alloy UGa is formed, crystallizing with orthorhombic structure [7], see also [10, 11]. A re-investigation of the phase diagram led to a two-phase composition for alloys with 50 at% Ga in the as-cast as well as in the as-annealed condition (850°C for 10 d) [5]. Single-phase alloys were obtained for a composition near 60 at% Ga from which this phase is reported to be U_2Ga_3 rather than UGa [5]. U_2Ga_3 crystallizes face-centered orthorhombic [5]. UGa is reported to melt congruently [9], whereas U_2Ga_3 is reported to decompose peritectically at 1260°C [5].

At higher Ga content, the alloy UGa_2 is formed, first mentioned by [12], see also [6, 8, 13]. The X-ray diagrams of UGa_2 were indexed hexagonally [5, 7], see also [9, 10], the crystal structure is body-centered tetragonal, isotypic with USi_2 [9, 14], see also [11]. UGa_2 melts congruently at 1355°C [5].

With still increasing Ga content in the system, the alloy UGa_3 is formed, first mentioned in [15, 16]. UGa_3 crystallizes simple cubic with $AuCu_3$-type structure, isomorphous with UAl_3, USn_3, UIn_3, USi_3, UGe_3, or UPb_3 [5, 15, 16], see also [8, 13]. UGa_3 decomposes peritectically at 1250°C [5], see also [10, 17]. UGa_3 was assumed to melt congruently at about 1300°C [8, 13].

A eutectic was found at the Ga-rich end of the phase diagram placed at 86 at% Ga with a eutectic temperature of about 28°C [13], see also [8], which does not agree with the liquidus measured by [18, 19].

The solubility of U in Ga is very restricted, the melting point of Ga is virtually unaffected by small additions of uranium [19], see also [8, 12, 13]. Measured values from [9, 11, 17] are summarized in Table 3/56, see **Fig. 3-120** [17]. The solubility of U in liquid Ga was expressed by the empirical equation $\log a$ (a in at% U) $= 3.571 - 3823 \cdot T^{-1}$ [17].

Table 3/56
Measured Liquidus Data for the System U — Ga.

temperature in °C	composition in at% U	Ref.	temperature in °C	composition in at% U	Ref.
1129	4.9	[9]	554	0.0860	[17]
875	3.2	[9]	522	0.0719	[17]
800	1.5	[9]	500	0.091	[9]
700	0.93	[9]	499	0.0345	[17]
649	0.26	[17]	474	0.025	[17]
600	0.177	[17]	448	0.0163	[17]
580	0.118	[17]	420	0.0140	[17]
560	0.13	[9]	343	0.00412[*]	[17]

[*] This does not appear to be an equilibrium value [11].

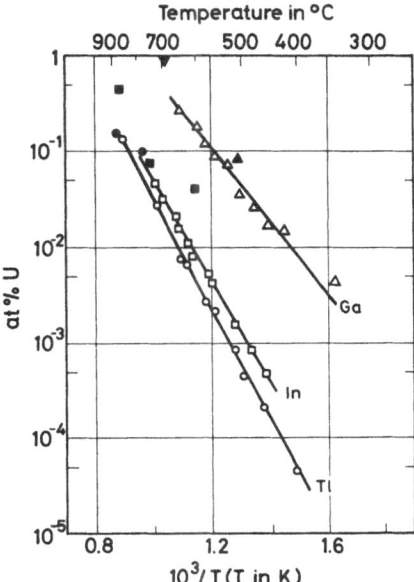

Fig. 3-120. Solubility of uranium in liquid gallium [17]. Measured values: △ from [17], ▲ from [20], ▼ from [21]. Solubilities of U in In and in Tl, respectively, are shown for comparison. In: □ [17], ■ [20]; Tl: ○ [17], ● [20].

A first phase diagram was constructed in 1958 by [8] from data summarized in [6]. A revised phase diagram based on X-ray determinations, metallography, and thermal analysis was established in 1973 by [5], see **Fig. 3-121.**

Vapor pressures of Ga in the system U−Ga are given in [22].

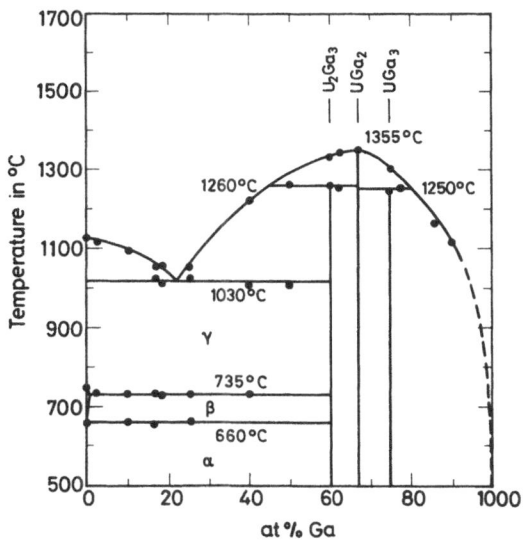

Fig. 3-121. Phase diagram of the U−Ga system [5].

References for 3.2.1 on p. 248

References for 3.2.1:

[1] Dayan, D.; Dariel, M. P.; Dapht, M. (J. Less-Common Metals **121** [1986] 399/404).

[2] Dayan, D.; Kimmel, G.; Dariel, M. P. (J. Nucl. Mater. **135** [1985] 40/5).

[3] Travelli, A. (U.S. Appl. 791235 [1986]; C.A. **106** [1987] No. 145796).

[4] Lazarev, B. G.; Ovcharenko, O. N.; Matsakova, A. A. (Metalloved. Fiz. Khim. Metallofiz. Sverkhprovodnikov **1967** 98/100; C.A. **70** [1969] No. 24269).

[5] Buschow, K. H. J. (J. Less-Common Metals **31** [1973] 165/8).

[6] Saller, H. A.; Rough, F. A. (BMI-1000 [1955] from [8]).

[7] Makarov, E. S.; Levdik, V. A. (Kristallografiya **1** [1956] 506/10; Soviet Phys.-Cryst. **1** [1956] 506/10; C.A. **1957** 7095).

[8] Hansen, M. (Constitution of Binary Alloys, McGraw-Hill, New York — Toronto — London 1958, pp. 143/4).

[9] Wilkinson, W. D. (Uranium Metallurgy, Vol. 2, Uranium Corrosion and Alloys, Interscience, New York 1962).

[10] Elliott, R. P. (Constitution of Binary Alloys, First Suppl., McGraw-Hill, New York — Toronto — London 1965).

[11] Shunk, F. A. (Constitution of Binary Alloys, Second Suppl., McGraw-Hill, New York — Toronto — London 1969).

[12] Dempster, A. J. (AL-4120 [1948] from [13]).

[13] Rough, F. A.; Bauer, A. A. (BMI-1300 [1958] 1/138; N.S.A. **12** [1958] No. 13935).

[14] Wilkinson, W. D.; Kelman, LeRoy R. (U.S. 3193380 [1965]; C.A. **63** [1965] 9408).

[15] Iandelli, A.; Ferro, R. (Ann. Chim. [Rome] **42** [1952] 598/608; C.A. **1953** 3165).

[16] Frost, B. R. T.; Maskrey, J. T. (J. Inst. Metals **82** [1953/54] 171/80; AERE-M-R-1027 [1952] 1/27; N.S.A. **7** [1953] No. 1547).

[17] Johnson, I.; Chasanov, M. G. (ASM [Am. Soc. Metals] Trans. Quart. **56** [1963] 272/7; C.A. **59** [1963] 4869).

[18] Hayes, E. E.; Gordon, P. (TID-65 [1948] 130/41 from [13]).

[19] Jaffee, R. I.; Evans, R. M.; Fromm, E. O.; Gonser, B. W. (BMI-T-20 [1950] from [13]).

[20] Hayes, E. E.; Gordon, P. (TID-2501 [1957] 115; N.S.A. **12** [1958] No. 17280).

[21] Wilkinson, W. D. (ANL-4150 [1948] 30 from [17]).

[22] Alcock, C. B.; Cornish, J. B.; Grieveson, P. (Thermodyn. Proc. Symp., Vienna 1965 [1966], Vol. 1, pp. 211/30, 367/8; C.A. **65** [1966] 8092).

3.2.2 Triuranium Pentagallide, U_3Ga_5

U_3Ga_5 was prepared as a nearly single-phase material by melting of an U — 62.5 at% Ga alloy and subsequent annealing at 1000°C for 7 d for homogenization [1], see also [2].

U_3Ga_5 crystallizes with orthorhombic Pu_3Pd_5-type structure, the space group is $Cmcm - D_{2h}^{17}$ (No. 63). The lattice parameters from the X-ray pattern are a = 9.396 ± 0.003, b = 7.575 ± 0.003, c = 9.387 ± 0.004 Å. Measured density 10.726 g/cm³ [1]. The atomic positions are summarized in Table 3/57. The X-ray pattern could also be indexed according to an orthorhombic unit cell for UGa [4, 5] or U_2Ga_3 [6]. For UGa the space group $Cmcm - D_{2h}^{17}$ (No. 63) is reported [4, 5] and the determined lattice parameters of U_2Ga_3 [6] are close to those given for UGa. But, it was not possible to achieve any reasonable agreement between the observed diffraction-line intensities and those calculated on the basis given by [4, 5], whereas excellent agreement was obtained for U_3Ga_5 [1].

Table 3/57
Atomic Positions for U_3Ga_5 [1].
They are those reported for Th_3In_5 [3].

atomic position	x	y	z
U(1) in 4 (c)	0.0	0.632	0.25
U(2) in 8 (c)	0.206	0.0	0.0
Ga(1) in 4 (c)	0.0	0.018	0.25
Ga(2) in 8 (f)	0.0	0.309	0.448
Ga(3) in 8 (g)	0.208	0.284	0.25

References for 3.2.2:

[1] Dayan, D.; Kimmel, G.; Dariel, M. P. (J. Nucl. Mater. **135** [1985] 40/5).
[2] Dayan, D.; Dariel, M. P.; Dapht, M. (J. Less-Common Metals **121** [1986] 399/404).
[3] Palenzona, A.; Manfrinetti, P.; Cirifaci, S. (J. Less-Common Metals **97** [1984] 231/36).
[4] Makarov, E. S.; Levdik, V. A. (Kristallografiya **1** [1956] 644/9; Soviet Phys.-Cryst. **1** [1956] 506/10; C.A. **1957** 7095).
[5] Makarov, E. S. (Kristallchimiya Prostejsich Soedineniy Urana, Toriya, Plutoniya i Neptuniya, Izd. Akad. Nauk SSSR, Moscow 1958; Crystal Chemistry of Simple Compounds of Uranium, Thorium, Neptunium, Plutonium, Consultant Bureau, New York 1959, pp. 83/4).
[6] Buschow, K. H. J. (J. Less-Common Metals **31** [1973] 165/8).

3.2.3   Uranium Monogallide, UGa, or Diuranium Trigallide, U_2Ga_3

Preparation

U – Ga alloys with the composition UGa were prepared from the powdered elements. The reaction was carried out at 230 to 280°C in an "alundum" or zirconia crucible encapsulated in an evacuated (10^{-4} Torr) quartz tube. After the reaction was completed, the samples were homogenized at 650°C for a long time [1], or at 800 to 900°C for 80 h [2], or at 1240 to 1250°C and subsequently at 650°C [3]. Lastly, the samples were cooled within 24 h. It is uncertain whether the samples really belong to the composition UGa or if the composition is U_2Ga_3 instead [4, 5].

U_2Ga_3 was prepared by arc melting of the elements followed by homogenizing by vacuum annealing in sintered Al_2O_3 crucibles [4].

Single crystals of UGa were obtained by cooling the prepared samples from the melting point to 150°C with a cooling rate smaller than 50°C/min [6].

Thermodynamic Data of Formation

An equation $\Delta G°$ (in cal/mol) $= -9800 + 2.05 \cdot T$ (at 820 to 1041 K) for the free enthalpy of formation of UGa according to $U(\alpha) + Ga(l) \rightarrow UGa(s)$ was obtained by combining the results for gallium vaporization, based on Knudsen-effusion measurements, with the data for the free energy of the $\alpha \rightarrow \gamma$ transformation in solid uranium [13], see also [14].

References for 3.2.3 on p. 252

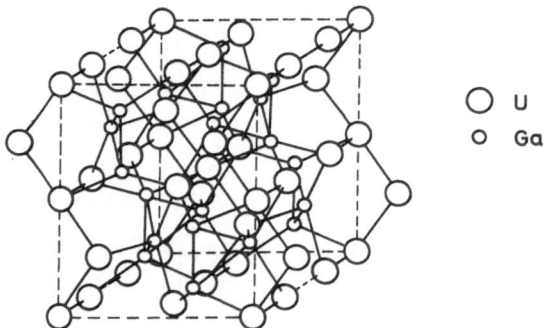

Fig. 3-122. Structure of the unit cell of UGa [6].

Crystallographic Properties

UGa crystallizes with orthorhombic structure with $Z = 16$; the space group is Cmcm $- D_{2h}^{17}$ (No. 63) [6], see also [7 to 10]. The lattice parameters, derived from Laue photographs, are: $a = 9.40$, $b = 7.60$, $c = 9.42$ Å [6], see also [7, 8, 10, 11]. A value $a = 7.40$ Å for the lattice parameter [9, 12], is assumed to be a printing mistake and should be written as $a = 9.40$ Å (footnote of the author). The X-ray density was calculated to be 12.1 g/cm³ [6]. The atomic positions are: origin at center (2/m), equivalent positions (0,0,0; 1/2, 1/2, 0) + 4 U(1) in 4 (a) (2/m): 0,0,0; 0,0,1/2. 4 U(2) in 4 (c) (mm): 0, y, 1/4; 0, $\bar{y}$, 3/4 with $y = 0.212$. 8 U(3) in 8 (e) (2): x,0,0; $\bar{x}$,0,0; x,0,1/2 $\bar{x}$,0,1/2 with $x = 0.300$. 8 Ga(1) in 8 (f) (m): 0, y, z; 0, $\bar{y}$, $\bar{z}$ with $y = 0.689$, 0, y, 1/2 − z; 0, y, 1/2 + z; $z = 0.118$. 8 Ga(2) in 8 (g) (m): x, y, 1/4; $\bar{x}$, y, 1/4 with $x = 0.260$, x, $\bar{y}$, 3/4; $\bar{x}$, $\bar{y}$, 3/4; $y = 0.354$ [8, 9]. The crystal structure of UGa is shown in **Fig. 3-122**. Atomic distances are U(1)−2 U(3) = 2.82 Å, U(1)−2 U(2) = 2.85 Å, U(1)−2 Ga(1) = 2.61 Å, U(2)−2 Ga(2) = 2.66 Å, U(3)−2 Ga(1) = 2.605 Å, U(3)−2 Ga(2) = 2.66 Å, Ga(1)−Ga(1) = 2.48 Å, Ga(1)− 2 Ga(2) = 2.86 Å, Ga(1)−2 U(1) = 2.61 Å, Ga(1)−2 U(3) = 2.605 Å, Ga(2)−2 Ga(1) = 2.86 Å, Ga(2)−2 U(2) = 2.6 Å, Ga(2)−2 U(3) = 2.66 Å [6, 8].

The structure of UGa is characterized by the formation of infinite broken chains of U(1)−U(2) atoms with a bond length of 2.85 Å, which is characteristic for α uranium, of strong U(1)−U(3) bonds, and of "molecules" of Ga(1)−Ga(1) (2.48 Å), which are also present in the structure of metallic gallium (2.45 Å) [6, 8].

U_2Ga_3 was indexed based on a b-face-centered orthorhombic structure. The lattice parameters are $a = 7.583$, $b = 9.398$, $c = 9.382$ Å [4], which are close to the values given for UGa by [6].

Electrical Properties

The room-temperature resistivity of UGa (U_2Ga_3) is 70 μΩ · cm [3]. The temperature dependence of the relative resistivity $\varrho(T)/\varrho_{300}$ for UGa (U_2Ga_3) is given in [3]. The relative resistivity of UGa (U_2Ga_3) increases sharply from 0.6 to 0.84 at about 27 K [3, 5], almost the same temperature at which a sharp increase of the initial magnetic susceptibility was observed [1], indicating an antiferromagnetic transition.

A weak hysteresis behavior was observed in the low-temperature region [3, 5]. In the region below 20 K, $\Delta\varrho$ is proportional to T^2 and at higher temperatures proportional to T, which is in accordance with the theory of a localized spin-fluctuation model [3].

Magnetic Properties

A magnetic transition was observed with UGa, as well as with U$_2$Ga$_3$, around a temperature of 20 K [1, 4, 5]. But the nature of this magnetic transition is still under discussion together with the crystallographic results. Both investigated materials are possibly of the same nature [4, 5].

An antiferromagnetic transition is assumed for UGa with a transition temperature at 27 K, based on measurements of the magnetic susceptibility performed by an ac bridge with 4 kHz and an applied field of about 0.8 kOe (see **Fig.** 3-**123**) [1], see also [5]. The susceptibility of UGa was observed to be field-independent up to 600 Oe, measured at 20, 88, and 225 K [1].

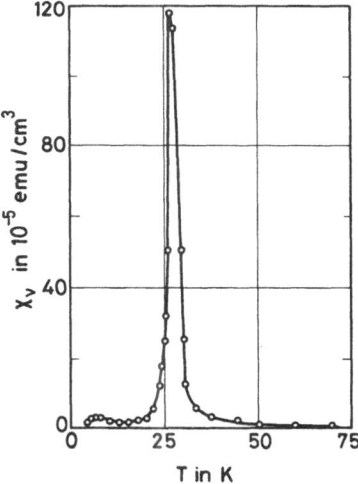

Fig. 3-123. Temperature dependence of the magnetic susceptibility of UGa [1].

A ferromagnetic transition is reported for U$_2$Ga$_3$ (see **Fig.** 3-**124**, p. 252) [4, 5]. A Curie-Weiss behavior was found below 250 K with a ferromagnetic Curie temperature T$_c$ of 23 K [4], slightly higher than the asymptotic Curie temperature Θ_p of 19 K [4] or 22.5 K [5]. At temperatures above 250 K a deviation from the Curie-Weiss law was observed [5]. A magnetization of 18.5 emu/g was found at 4.2 K and with 30 kOe, from which a ferromagnetic moment of 1.1 μ_B per U atom was derived [4]. From the susceptibility measurements, an effective moment of $\mu_{eff} = 3.1$ μ_B per U atom [4] or 3.0 μ_B per U atom [5] was derived.

References for 3.2.3 on p. 252

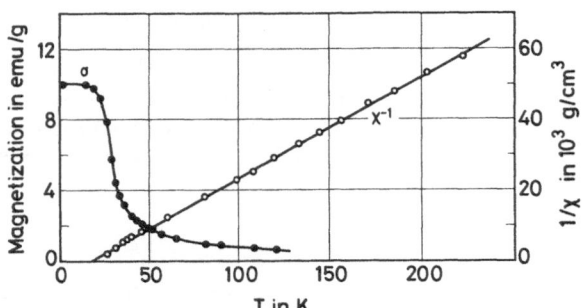

Fig. 3-124. Temperature dependence of the magnetization (left-hand scale) and the reciprocal susceptibility (right-hand scale) of U_2Ga_3 [4].

References for 3.2.3:

[1] Sternberk, J.; Menovsky, A.; Svec, T.; Zentko, A. (Phys. Status Solidi A **28** [1975] K45/K47).

[2] Ansorge, V.; Menovsky, A. (Phys. Status Solidi **30** [1968] K31/K32).

[3] Smetana, Z.; Houdek, V.; Menovsky, A.; Simsa, Z. (Phys. Status Solidi A **28** [1975] K103/K105).

[4] Buschow, K. H. J. (J. Less-Common Metals **31** [1973] 165/8).

[5] Sechovsky, V.; Smetana, Z.; Menovsky, A. (Plutonium 1975 Other Actinides Proc. 5th Intern. Conf., Baden-Baden, FRG, 1975 [1976], pp. 641/8).

[6] Makarov, E. S.; Levdik, V. A. (Kristallografiya **1** [1956] 644/9; Soviet Phys.-Cryst. **1** [1956] 506/10; C.A. **1957** 7095).

[7] Elliot, R. P. (Constitution of Binary Alloys, First Suppl., McGraw-Hill, New York — Toronto — London 1965).

[8] Makarov, E. S. (Kristallchimiya Prostejsich Soedineniy Urana, Toriya, Plutoniya i Neptuniya, Izd. Akad. Nauk SSSR, Moscow 1958; Crystal Chemistry of Simple Compounds of Uranium, Thorium, Neptunium, Plutonium, Consultant Bureau, New York 1959, pp. 83/4).

[9] Lam, D. J.; Darby, J. B., Jr.; Nevitt, N. V. (in: Freeman, A. J.; Darby, J. B., Jr., The Actinides: Electronic Structure and Related Properties, Vol. 2, Academic, New York — San Francisco — London 1974, pp. 119/84).

[10] Pearson, W. P. (A Handbook of Lattice Spacings of Metals and Alloys, Vol. 2, Pergamon Press, New York 1967).

[11] Sternberk, J.; Ansorge, V.; Hrebik, J.; Menovsky, A.; Smetana, Z. (Czech. J. Phys. B **21** [1971] 969/78).

[12] Dwight, A. E. (in: Giessen, B. C., Developments in the Structural Chemistry of Alloy Phases, Plenum, New York 1969, pp. 181/226).

[13] Alcock, C. B.; Cornish, J. B.; Grieveson, P. (Thermodyn. Proc. Symp., Vienna 1965 [1966], Vol. 1, pp. 211/30, 367/8; C.A. **65** [1966] 8092).

[14] Krivy, I. (UJV-1783 [1967] 1/48; N.S.A. **22** [1968] No. 59; ANL-Trans-584 [1968]; N.S.A. **22** [1968] No. 25335).

3.2.4 Uranium Digallide, UGa$_2$

Preparation

UGa$_2$ crystals were formed in a mixture of 26.4 wt% uranium and 73.6 wt% gallium by heating to 600°C for 2 d under argon, resulting in solid crystals inbedded in a liquid gallium phase. The crystals were separated from the liquid by filtration [1]. Samples of UGa$_2$ were prepared by reaction of stoichiometric amounts of the powdered elements at 600°C in an "alundum" crucible and encapsulated in an evacuated (10^{-4} Torr) silica tube [2, 3]. The multiphase mixture obtained was powdered under dry-box conditions in an argon atmosphere, pressed into pellets, and melted in an arc furnace. Finally, the samples were homogenized in evacuated silica tubes at 800°C for 150 h [3]. UGa$_2$ samples were also prepared without a melting step by direct homogenization after the reaction was completed (240 to 260°C [4]) at 600°C for 200 to 250 h [4] or at 800 to 900°C for 80 h in vacuum [2].

UGa$_2$ was also prepared by arc melting in sintered Al$_2$O$_3$ crucibles without subsequent annealing [5]. To increase the grain size after arc melting (under He in water-cooled copper hearths) the resultant ingots were heated to the melting point (1200°C) in an "alundum" crucible in a resistance furnace. Ingots of UGa$_2$ with fairly large grains were obtained by cooling with high-cooling gradients of 150°C/cm [6]. Samples with cylindrical shape were obtained after arc melting in argon atmosphere and remelting 5 times to increase the homogeneity by vacuum casting into Al$_2$O$_3$ crucibles. The cast samples, wrapped in tantalum foils, were annealed for homogenization at 800°C for 10 d in evacuated silica tubes [7].

Single crystals of UGa$_2$ were cut from polycrystalline samples and ground to spherical shape for measurement of Barkhausen noise spectra [8]. Single crystals of 300 mg (or 2 to 3 mm long with spherical or cubical shape [9]) were cut from large grain-size ingots [6]. Not more than 5% UGa$_3$ was observed from X-ray diffraction. Misorientation of the sub-grain boundaries did not exceed 2° [6, 9]. Single crystals of UGa$_2$ were also obtained by cooling the prepared samples from the melting point to 150°C with a cooling rate smaller than 50°C/min [10].

Thermodynamic Data of Formation

An equation for the free energy of formation, $\Delta G°$ (in cal/mol) $= -20050 + 7.45 \cdot T$ (at 820 to 1041 K), of UGa$_2$ according to U(α) + 2 Ga(l) $\rightarrow$ UGa$_2$(s) was obtained by combining the results for Ga vaporization, based on Knudsen-effusion measurements, with the data for the free energy of the $\alpha \rightarrow \gamma$ transformation in solid U [21], see also [22].

Crystallographic Properties

UGa$_2$ crystallizes with hexagonal AlB$_2$-type structure with Z = 1. The space group is P6/mmm$-$D$_{6h}^1$ (No. 191) [6, 11], see also [12] (formerly given as C6/mmm$-$D$_{6h}^1$ [10]). The lattice parameters (summarized in Table 3/58, p. 254) were derived from Laue photographs [10] or X-ray diffraction pattern. The X-ray density is 10.3 g/cm^3 [10]. The temperature dependence of the lattice parameters is given in [6]. The atomic positions are: 1 U in 1 (a) (6/mmm): 0,0,0. 2 Ga in 2 (d) ($\bar{6}$/m2): 1/3, 2/3, 1/2; 2/3, 1/3, 1/2 [4, 12]. The shortest atomic distances are: U$-$U = 4.01 Å, U$-$Ga = 3.15 Å, Ga$-$Ga = 2.43 Å [10], or U$-$U = 4.02 Å, U$-$Ga = 3.15 Å [3].

References for 3.2.4 on pp. 258/60

Table 3/58
Measured (Hexagonal) Lattice Parameters of UGa_2 at Room Temperature.
For tetragonal lattice parameters of UGa_2, see p. 253.

a in Å	c in Å	c/a	Ref.	also given in
4.21	4.01	0.954	[10]	[4, 12, 13, 15 to 18]
4.22	4.02	0.95	[3]	
4.212	4.024		[5]	
powder prepared from an ingot:				
4.2130	4.0171		[6]	[19]
powder synthesized by diffusion sintering:				
4.2103	4.0193		[6]	

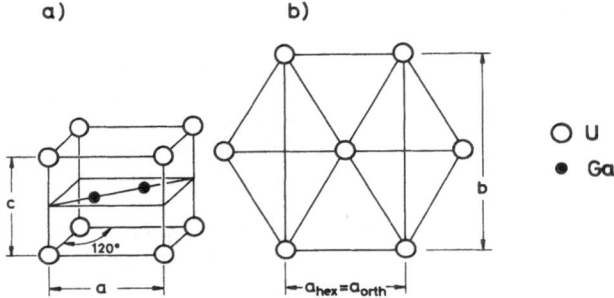

Fig. 3-125. a) The hexagonal unit cell of UGa_2. b) Relationship between the hexagonal and orthorhombic cell in the (001) plane [6].

The crystallographic structure of UGa_2 is also reported as body-centered tetragonal, isomorphous with $ThSi_2$, USi_2, $PuSi_2$, and $CeSi_2$, with $Z = 4$. The lattice constants are $a = 4.238 \pm 0.001$, $c = 4.664 \pm 0.003$ Å [1, 13], see also [14]. The calculated X-ray density is 9.45 g/cm³ [1]. According to the tetragonal interpretation of the structure, each U atom is bonded to 12 Ga atoms with the distance $U - Ga = 3.23$ Å, and each Ga atom is bonded to 6 U and 3 Ga atoms with the distance $Ga - Ga = 2.44$ Å [1].

The room-temperature hexagonal structure of the unit cell of UGa_2 (paramagnetic state) is shown in **Fig. 3-125**. With decreasing temperature, below 125 K (ferromagnetic ordered state), the structure of the unit cell changes to an orthorhombic structure with the space group $Cmmm - D_{2h}^{19}$ (No. 65), because of a large magnetostriction. The distortion was defined as $\Delta = b/\sqrt{3} - a$, where the value of Δ increases with decreasing temperature below 125 K [6, 9, 20]. Lattice parameters at 4.2 K, from measurements with a polarized neutron diffractometer, are $a = 4.190$, $b = 7.287$, $c = 4.011$ Å [19].

Thermal Properties

UGa_2 melts congruently at 1355 °C [5].

Electrical Properties

The room-temperature resistivity of UGa$_2$ is 450 $\mu\Omega \cdot$ cm [23]. The temperature dependence of the resistivity was measured in cooling and heating cycles up to 300 K, showing a temperature hysteresis in the range of 140 to 250 K. The resistivity increases steeply with temperature up to 123 K [24] or 127 K [23], see also [25], where a maximum in the temperature derivative dϱ/dT is observed, which is connected with the ferromagnetic ordering and represents the Curie temperature T$_c$ (T$_c$ = 129 K from magnetic measurements) [24].

In the low-temperature region (below 20 K) $\Delta\varrho$ is proportional to T^2 and in the temperature range of 20 K < T < 80 K proportional to T, which is in accordance with the theory of a localized spin-fluctuation model [23].

Emission Spectra

The observed 7 eV satellites in X-ray photoemission spectra (XPS) are interpreted as an indication of a weak f-d hybridization and, when compared to uranium-transition metal compounds, increased 5f localization. The following values are derived from the spectra: The binding energies relative to the Fermi energy E$_F$ are 377.5 $\pm$ 0.1 eV for U4f$_{1/2}$, 388.4 $\pm$ 0.1 eV for U4f$_{5/2}$, 18.4 $\pm$ 0.1 eV for Ga3d. The Fermi level is located at 0.3 eV lower binding energy relative to the 5f maximum. The position of the satellite is 7.0 $\pm$ 0.1 eV relative to the main line [26].

From resonant photoemission spectra, performed on α-U, UGa$_2$, and UGa$_3$, it is assumed that the 1.4 eV feature in the off-resonance spectra of UGa$_2$ and UGa$_3$ and the 2 eV feature in α-U are originated by U6d valence states and should not be interpreted as 5f-shake-up satellites [27].

Magnetic Properties

UGa$_2$ is a ferromagnet at temperatures below 125 K, whereas the crystallographic structure changes from hexagonal AlB$_2$-type structure in the paramagnetic state to an orthorhombic structure in the ferromagnetic state (compare "Crystallographic Properties", p. 253). The magnetic structure of UGa$_2$ was determined by X-ray and neutron diffraction studies [4, 6, 11, 20, 25, 28].

UGa$_2$ crystallizes in the ferromagnetic ordered state orthorhombically with the space group Cmmm as determined by X-ray diffraction, whereas the directions of easy magnetization are placed in the basal plane along the [100] directions [6]. The X-ray determinations are confirmed by neutron diffraction measurements leading to a co-linear structure with the magnetic space group P2'm [4, 25], see also [19]. The lattice parameters at 4.2 K, measured with a polarized neutron diffractometer (λ = 0.9 Å) at a single crystal, are a = 4.190, b = 7.287, c = 4.011 Å [19]. The temperature dependence of the magnetic contributions to the (100) peak leads to a Curie temperature of T$_c$ = 126 $\pm$ 3 K [4] or 130 $\pm$ 5 K [28]. Assuming "spin-only" magnetic contributions, a ferromagnetic moment of μ_f = 2.28 μ_B was derived, at 4.2 K [4] or 3.0 $\pm$ 0.2 μ_B (average value) at 27 K, indicating some delocalization of the 5f electrons [28].

Measured values for Curie temperatures T$_c$, paramagnetic Curie temperatures Θ_p, moments for magnetic saturation μ_s, and effective magnetic moments μ_{eff} are summarized in Table 3/59, p. 256.

 References for 3.2.4 on pp. 258/60

Table 3/59

Measured Values of Curie Temperatures T_c, Paramagnetic Curie Temperatures Θ_p, Moments for Magnetic Saturation μ_s, and Effective Magnetic Moments μ_{eff} for UGa_2.

Curie temperature T_c in K	magnetic saturation moment μ_s in μ_B	paramagnetic Curie temperature Θ_p in K	effective moment μ_{eff} in μ_B	Ref.
from neutron diffraction measurements:				
126 $\pm$ 3	2.28 (at 4.2 K)			[4, 25]
130 $\pm$ 5	3.0 $\pm$ 0.2 (at 27 K)			[28]
from magnetization and susceptibility measurements:				
		126	3.12	[2]
126 $\pm$ 2		126 $\pm$ 2	3.12	[18]
128	2.1	125	3.56	[5, 29]
125.5 $\pm$ 2	2.35 $\pm$ 5%	126 $\pm$ 2	3.60 $\pm$ 5%	[30]
133	1.8 (at 77 K)	133	3.2	[3]
129				[24]
125	2.71	127	3.0	[6, 11, 20]
120				[7]

The magnetization of UGa_2 shows large anisotropy effects [6, 19, 20]. The field and temperature dependence of the magnetization are shown in **Fig. 3-126** and **Fig. 3-127**. The magnetization of spherical samples in liquid nitrogen at a field of 9 kG yields 22.0 G $\cdot$ cm$^3 \cdot$ g^{-1} [2]. Magnetic moments per uranium atom for magnetic saturation (measured in the easy directions of magnetization) are μ_s (in μ_B) = 2.1 [29], 2.35 $\pm$ 5% (polycrystalline material) [30], 1.8 at 77 K [3], 2.71 at 4.2 K (single crystal) [6, 11], 2.28 at 4.2 K (from neutron-diffraction measurement) [4]. Values for the Curie temperature T_c, derived from the magnetization measurements are summarized in Table 3/59 together with observed values for the paramag-

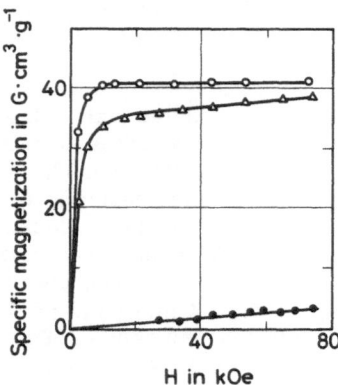

Fig. 3-126. The field dependence of the magnetization of UGa_2 at 4.2 K measured at a single crystal [6]. $\bigcirc$ Along the a axis [100], $\triangle$ along the b axis [120], $\bullet$ along the c axis [001] (hexagonal indices).

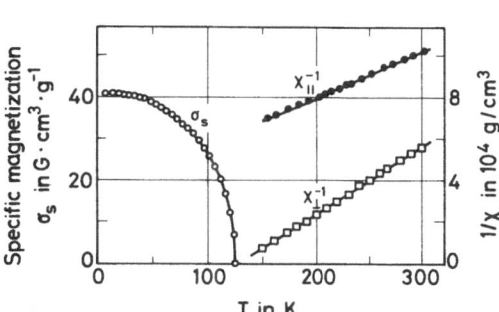

Fig. 3-127. The temperature dependence of the specific spontaneous magnetization σ_s and the reciprocal susceptibility χ^{-1} of UGa$_2$, measured at a single crystal [6]. □ Along the a axis [100], ● along the c axis [001].

netic Curie temperature Θ_p and for the magnetic moments μ_s and μ_{eff}. The pressure dependence of the Curie temperature was measured in fields of up to 10 T and pressures of up to 8 kbar, resulting in $\partial T_c/\partial p = 0.22 \pm 0.05$ K/kbar [7].

The temperature dependence of the magnetic susceptibility of UGa$_2$ is shown in [3]. The reciprocal susceptibility follows a Curie-Weiss law. Paramagnetic Curie temperatures Θ_p and effective moments μ_{eff} are derived from the susceptibility values and are listed in Table 3/59, together with measured values of the moments for magnetic saturation μ_s and the Curie temperatures T_c. A room temperature value of the susceptibility is $\chi = 8000 \times 10^{-6}$ emu/g at 290 K [3]. The pressure dependence of the magnetic susceptibility was measured in fields of up to 10 T and pressures of up to 8 kbar, resulting in $\partial\ln\chi/\partial_p = 1.8$ Mbar^{-1} [7].

The power spectrum of Barkhausen impulses, performed on a UGa$_2$ single crystal, was investigated in external fields of up to 480 kA/m (about 6 kOe). The observed power spectra of the Barkhausen noise reveal an apparent anisotropy in UGa$_2$ [8].

UGa$_2$ shows large magnetic anisotropy effects (see Fig. 3-126 and Fig. 3-127). The magnetic susceptibility, performed on UGa$_2$ single crystals, was measured along the c axis, $\chi_\parallel^{-1}$, and the a axis (basal plane), $\chi_\perp^{-1}$. In both cases a Curie-Weiss law is obeyed, resulting in different paramagnetic Curie temperatures, $\Theta_p^\parallel = -148$ K and $\Theta_p^\perp = 127$ K [6, 20] and effective magnetic moments of $\mu_{eff}^\parallel = 3.55\,\mu_B$ and $\mu_{eff}^\perp = 3.0\,\mu_B$ [6, 20]. The difference in the magnetic moments indicates a temperature-independent component χ_0 due to the polarization of conduction electrons: $\chi = \chi_0 + C/(T - \Theta_p)$. In this case (with $\chi_0 = 2.8 \times 10^{-6}$), the Curie-Weiss law is obeyed for the two different paramagnetic Curie temperatures $\Theta_p^\parallel = -64$ K and $\Theta_p^\perp = 134$ K, resulting in the same effective moment of $\mu_{eff} = 2.8\,\mu_B$ [6, 11, 20].

Constants of the magnetic anisotropy energy were derived from the equation $E_\sigma = K_1 \cdot \sin^2\vartheta + K_2 \cdot \sin^4\vartheta + K_3 \cdot \sin^6\vartheta \cdot \cos 6\varphi$ with ϑ = angle between magnetization and c axis, and φ = angle between magnetization and b axis in the basal plane, whereas the first two terms describe the uniaxial anisotropy and the third one the hexagonal anisotropy in the basal plane. The constants are (at 4.2 K) $K_1 + 2 K_2 = -2.10^7$ erg/g, $K_3 = -0.6 \times 10^6$ erg/g [6, 20], see also [9]. The corresponding anisotropy field H_A, derived from $H_A = 2(K_1 + 2K_2)/\sigma_s$, was calculated to be about 10^6 Oe, and from the equation $H_A = (\chi_\parallel^{-1} - \chi_\perp^{-1})/\sigma_s$ (if the anisotropy energy depends on temperature as $\sigma^2(T)$): $H_A = 3.10^6$ Oe at 4.2 K [6].

References for 3.2.4 on pp. 258/60

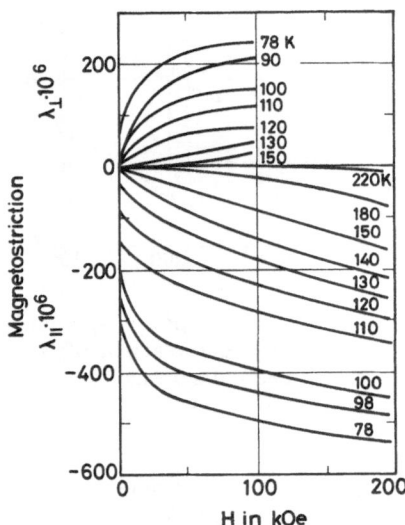

Fig. 3-128. The field dependence of the longitudinal ($\lambda_\parallel$) and transverse ($\lambda_\perp$) magnetostriction of UGa$_2$ at various temperatures [24].

The observed moment for magnetic saturation in UGa$_2$, $\mu_s = 2.71\ \mu_B$ or $2.8\ \mu_B$ [6], does not agree with the magnetic moments for U ions of different valence as calculated from $\mu_U = g \cdot J \cdot \mu_B$ to be $\mu_U = 3.20\ \mu_B$ (for U^{4+} ions), $\mu_U = 2.14\ \mu_B$ (for U^{5+} ions) [9], or calculated within the framework of Russell-Saunders coupling with $\mu_{eff} = g_J \cdot (J(J + 1))^{1/2}$ to be $\mu_U = 3.62\ \mu_B$ (for U^{3+} ions), $\mu_U = 3.58\ \mu_B$ (for U^{4+} ions) [19]. The differences are discussed as partial delocalization of the 5f electrons, corresponding to a larger hybridization with the conduction band, crystal field effects, and magnetocrystalline anisotropy [9, 19]. The determination of the magnetic form factors for UGa$_2$ leads to the conclusion of a larger delocalization of the 5f electrons than that calculated by the Russell-Saunders coupling approximation [19].

A hysteresis loop of a UGa$_2$ single crystal magnetized along the a axis at 4.2 K and the temperature dependence of the coercive force is shown in [6]. The coercive force in UGa$_2$ was found to be $H_c = 1.1$ Oe at 4.2 K [6].

The field dependence of the longitudinal, $\lambda_\parallel$, and the transverse, $\lambda_\perp$, magnetostriction of UGa$_2$, performed on a polycrystalline sample, at different temperatures above 78 K is shown in **Fig. 3-128**, from which $\lambda_{s\perp} = 1/2\ \lambda_{s\parallel}$ was observed for the spontaneous magnetostriction at low temperatures. This relation is not fullfilled in the vicinity of the Curie temperature, where the spontaneous transverse magnetostriction is equal to zero at about 130 K [24], see also [6, 11]. The magnetostriction of UGa$_2$ is very high, reaching a value of about -4×10^{-3} at 4.2 K, which is comparable to the magnetostriction of rare earths [6].

References for 3.2.4:

[1] Wilkinson, W. D.; Kelman, LeRoy R. (U.S. 3 193 380 [1965]; C.A. **63** [1965] 9 408).
[2] Ansorge, V.; Menovsky, A. (Phys. Status Solidi **30** [1968] K31/K32).
[3] Misiuk, A.; Mulak, J.; Czopnik, A.; Trzebiatowski, W. (Bull. Acad. Polon. Sci. Ser. Sci. Chim. **20** [1972] 337/41).

[4] Sechovsky, V.; Smetana, Z.; Menovsky, A. (Phys. Status Solidi A **28** [1975] K37/K40).

[5] Buschow, K. H. J. (J. Less-Common Metals **31** [1973] 165/8).

[6] Andreev, A. V.; Belov, K. P.; Deryagin, A. V.; Kazei, Z. A.; Levitin, R. Z.; Menovsky, A.; Popov, Yu. F.; Silantev, V. I. (Zh. Eksperim. Teor. Fiz. **75** [1978] 2351/61; Soviet Phys.-JETP **48** [1978] 1187/93; C.A. **90** [1979] No. 114061).

[7] Franse, J. J. M.; Frings, P. H.; de Boer, F. R.; Menovsky, A. (Phys. Solids High Pressure Proc. Intern. Symp., Bad Honnef, FRG, 1981, pp. 181/91; C.A. **96** [1982] No. 134510).

[8] Zentko, A.; Filka, S.; Deryagin, A. V.; Andreev, A. V. (Acta Phys. Slovaca **30** [1980] 319/22; C.A. **94** [1981] No. 40485).

[9] Andreev, A. V.; Below, K. P.; Deryagin, A. V.; Levitin, R. Z.; Popov, Yu. F. (Izv. Akad. Nauk SSSR Ser. Fiz. **44** [1980] 1352/5; Bull. Acad. Sci. USSR Phys. Ser. **44** No. 7 [1980] 22/5; C.A. **93** [1980] No. 124752).

[10] Makarov, E. S.; Levdik, V. A. (Kristallografiya **1** [1956] 644/9; Soviet Phys.-Cryst. **1** [1956] 506/10); C.A. **1957** 7095).

[11] Andreev, A. V.; Belov, K. P.; Deryagin, A. V.; Zeleny, M.; Levitin, R. Z.; Popov, Yu. F. (Proc. Conf. Phys. **1** [1981] 16/24; C.A. **95** [1981] No. 179843).

[12] Lam, D. J.; Darby, J. B., Jr.; Nevitt, N. V. (in: Freeman, A. J.; Darby, J. B., Jr., The Actinides: Electronic Structure and Related Properties, Vol. 2, Academic, New York — San Francisco — London 1974, pp. 119/84).

[13] Wilkinson, W. D. (Uranium Metallurgy, Vol. 2, Uranium Corrosion and Alloys, Interscience, New York 1962).

[14] Shunk, F. A. (Constitution of Binary Alloys, Second Suppl., McGraw-Hill, New York — Toronto — London 1969).

[15] Elliot, R. P. (Constitution of Binary Alloys, First Suppl., McGraw-Hill, New York — Toronto — London 1965).

[16] Pearson, W. B. (A Handbook of Lattice Spacings and Structures of Metals and Alloys, Vol. 2, Pergamon, New York 1967).

[17] Dwight, A. E. (in: Giessen, B. C., Developments in the Structural Chemistry of Alloy Phases, Plenum, New York 1969, pp. 181/226).

[18] Sternberk, J.; Ansorge, V.; Hrebik, J.; Menovsky, A.; Smetana, Z. (Czech. J. Phys. B **21** [1971] 969/78).

[19] Ballou, R.; Deryagin, A. V., Givord, F.; Lemaire, R.; Levitin, R. Z.; Tasset, F. (J. Phys. Colloq. [Paris] **40** [1979] C7-279/C7-285).

[20] Andreev, A. V.; Belov, K. P.; Deryagin, A. V.; Levitin, R. Z.; Menovsky, A. (J. Phys. Colloq. [Paris] **40** [1979] C4-82/C4-83).

[21] Alcock, C. B.; Cornish, J. B.; Grieveson, P. (Thermodyn. Proc. Symp., Vienna 1965 [1966], Vol. 1, pp. 211/30; 367/8; C.A. **65** [1966] 8092).

[22] Krivy, I. (UJV-1783 [1967] 1/48; N.S.A. **22** [1968] No. 59; ANL-Trans-584 [1968] 1/55; N.S.A. **22** [1968] No. 25335).

[23] Smetana, Z.; Houdek, V.; Menovsky, A.; Simsa, Z. (Phys. Status Solidi A **28** [1975] K103/K105).

[24] Levitin, R. Z.; Dmitrievskii, A. S.; Henkie, Z.; Misiuk, A. (Phys. Status Solidi A **27** [1975] K109/K112).

[25] Sechovsky, V.; Smetana, Z.; Menovsky, A. (in: Blank, H.; Lindner, R., Plutonium 1975 Other Actinides, Proc. 5th Intern. Conf., Baden-Baden, FRG, 1975 [1976], pp. 641/8; C.A. **85** [1976] No. 103097).

[26] Schneider, W. D.; Laubschat, C. (Phys. Rev. Letters **46** [1981] 1023/7).

[27] Reihl, B.; Domke, M.; Kaindl, G.; Kalkowski, G.; Laubschat, C.; Hulliger, F.; Schneider, W. D. (Phys. Rev. B [3] **32** [1985] 3530/3).

[28] Lawson, A. C.; Williams, R.; Smith, J. L.; Seeger, P. A.; Goldstone, J. A.; O'Rourke, J. A.; Fisk, Z. (J. Magn. Magn. Mater. **50** [1985] 83/7).

[29] Buschow, K. H. J.; van Daal, H. J. (AIP [Am. Inst. Phys.] Conf. Proc. No. 5 [1972] 1464/77; C.A. **77** [1972] No. 11324).

[30] Sternberk, J.; Hrebik, J.; Menovsky, A.; Smetana, Z. (J. Phys. Colloq. [Paris] **32** [1971] C1-744/C1-745).

3.2.5 Uranium Trigallide, UGa_3

Preparation

UGa_3 crystals were formed in a mixture of 23.9 wt% U and 76.1 wt% Ga by heating to 800°C for 2 d in a protective atmosphere (argon), resulting in solid crystals inbedded in a liquid Ga phase. The crystals were separated by filtration [1].

Samples of UGa_3 were prepared by reaction of stoichiometric amounts of the powdered elements at 230°C [2] or at 400°C for 100 h [3], in an "alundum" crucible and encapsulated in an evacuated (10^{-4} Torr) silica tube [2, 3]. The resulting multiphase mixture was homogenized in vacuum (10^{-4} Torr) at 800 to 900°C for 80 h [2] or powdered in an argon atmosphere, pressed into pellets, placed in a beryllium oxide crucible, encapsulated in an evacuated silica tube, and homogenized at 700°C for 150 h [3], see also [4, 5].

UGa_3 was also prepared by usual melting techniques with the metals molten in stoichiometric amounts in a BeO crucible in a vacuum furnace [6], or by arc melting in argon atmosphere [7 to 12] and subsequent vacuum annealing in an Al_2O_3 crucible [7].

Thermodynamic Data of Formation

From emf measurements (molten salt electrolytes): $\Delta H_f^\circ = -10.34$ kcal/g-atom, $\Delta G_f^\circ = -7.73$ kcal/g-atom, $\Delta S_f^\circ = -3.64$ cal $\cdot$ g-atom$^{-1} \cdot$ K^{-1} for the formation of UGa_3 at 450°C were derived [13], see also [14, 15]. The following functions were derived for the formation of UGa_3, according to $U(s) + 3\,Ga(l) \rightarrow UGa_3(s)$ at 370 to 740°C: ΔG_f° (in kcal/mol) $= -59.28 + 82.15 \times 10^{-3} \cdot T - 84.36 \times 10^{-6} \cdot T^2 + 34.65 \times 10^{-9} \cdot T^3$; ΔH_f° (in kcal/mol) $= -59.28 + 84.36 \times 10^{-6} \cdot T^2 - 69.30 \times 10^{-9} \cdot T^3$; ΔS_f° (in cal $\cdot$ mol$^{-1} \cdot$ K^{-1}) $= -82.15 + 168.7 \times 10^{-3} \cdot T - 104.0 \times 10^{-6} \cdot T^3$ [13], see also [14, 15].

For the formation reaction $U(s) + 3\,Ga(l) \rightarrow UGa_3(s)$ at 370 to 700 K the equation ΔG_f° (in kcal/mol) $= -41.42 + 14.51 \times 10^{-3} \cdot T$ is given in [26].

The equation ΔG_f° (in cal/mol) $= -27400 + 12.15 \cdot T$ for the free energy of formation of UGa_3, according to $U(\alpha) + 3\,Ga(l) \rightarrow UGa_3(s)$ at 820 to 1041 K, was obtained by combining the results for Ga vaporization, based on Knudsen effusion measurements, with the data for the $\alpha \rightarrow \gamma$ transformation in solid uranium [27], see also [14, 21].

From dynamic differential calorimetric (DDC) measurements the enthalpy of formation at room temperature ΔH_f (298 K) $= -6.1$ kcal/g-atom is derived [5].

Crystallographic Properties

UGa_3 crystallizes with simple cubic Cu_3Au-type structure with $Z = 1$, the space group is $Pm\bar{3}m - O_h^1$ (No. 221) [16], see also [4, 6]. Measured lattice parameters are summarized in Table 3/60. The calculated X-ray density is 9.686 [6, 16] or 9.67 g/cm^3 [1].

Table 3/60
Measured Lattice Parameters of UGa₃.

a in Å	Ref.	also given in
4.249	[4]	[16 to 19]
4.2475	[6]	[16 to 18, 20, 21]
4.24 ± 0.01	[1]	
4.248	[9, 22]	[23]
4.256 ± 0.003	[3, 10]	
4.2567	[7]	

The atomic positions are 1 U in: 0,0,0. 3 Ga in: 0, 1/2, 1/2; 1/2, 0, 1/2; 1/2, 1/2, 0 [4]. Each U atom is bonded to 12 Ga atoms with the distance $U-Ga = a/\sqrt{2} = 2.995$ Å [6], 2.99 Å [1], or 3.010 Å [3]; each Ga atom is bonded to 4 U atoms and eight Ga atoms with the distance $Ga-Ga = 2.99$ Å [1].

Melting Point

UGa₃ decomposes peritectically at 1250°C [7], see also [24, 25]. UGa₃ was assumed to melt congruently at about 1300°C [16, 18], see also [6].

Specific Heat

The low-temperature specific heat of UGa₃ was measured on compacted powders (see **Fig. 3-129**). Following the equation $C = \gamma \cdot T + \alpha \cdot T^3$, with γ = coefficient of the electronic specific heat and α = coefficient of the lattice contribution to the specific heat, the following values are calculated: $\gamma = 52.0$ mJ·mol⁻¹·K⁻² and $\alpha = 0.321$ mJ·mol⁻¹·K⁻⁴ [8]. An approximation of the Debye temperature was calculated to be Θ_D (at 0 K) = 288 K [8].

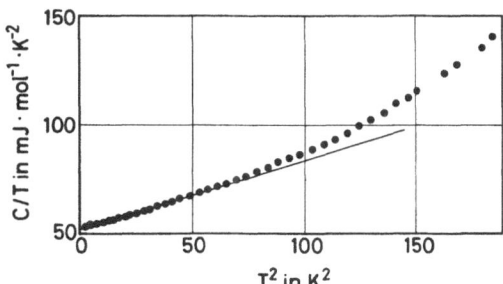

Fig. 3-129. The heat capacity of UGa₃ given as C/T versus T^2 [8]. The straight line corresponds to $C = \gamma \cdot T + \alpha \cdot T^3$ [8].

Photoemission Spectra

From resonant photoemission spectra, performed on α-U, UGa₂, and UGa₃, it is assumed that the 1.4 eV feature in the off-resonance spectra of UGa₂ and UGa₃, and the 2 eV feature in α-U are originated by the U 6d valence states and should not be interpreted as 5f shake-up satellites [28].

Electrical Properties

The temperature dependence of the electrical resistivity of UGa$_3$ was reported in [34]. In contrast to this observation, the resistivity was found to decrease very slowly on cooling in the temperature range of 100 to 67 K. At 62 K the resistivity increases with decreasing temperature and passes a maximum at 62 K, and finally reaches a rather large residual resistivity at liquid-helium temperature (see **Fig.** 3-**130**) [10]. The anomaly below 67 K indicates a phase transition, which was established by neutron diffraction experiments [29, 30] to be of antiferromagnetic nature [10].

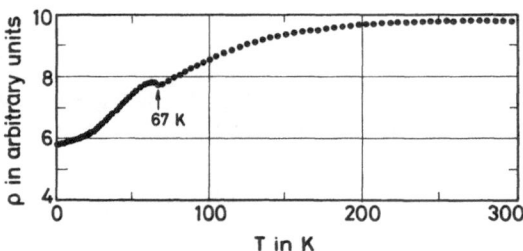

Fig. 3-130. The temperature dependence of the electrical resistivity of UGa$_3$ [10].

The resistivity of UGa$_3$ shows a variety of different temperature dependences as predicted by the theory of Rivier and Zlatic [31, 32] for localized spin fluctuations in dilute alloys [33, 34]. Different spin-fluctuation temperatures, analyzed from different temperature regions are summarized in Table 3/61.

Table 3/61
Calculated Spin-Fluctuation Temperatures T_{SF} at Different Temperature Regions for UGa$_3$ [33].

temperature region	T_{SF}
from low-temperature end of $\varrho \sim \ln T$ region, $T = T_{SF}$	50 K
from high-temperature end of $\varrho \sim T^2$ region, $T = T_{SF}/2\pi$	310 K
from the slope of the $\varrho \sim T^2$ region, $d\varrho/d(T^2) = 3\,\pi^2/4\,T_{SF}^2$	192 K
from the slope of the $\varrho \sim T^{-1}$ region, $d\varrho/dT = 1.12\,T_{SF}$	94 K
from the approach to the high temperature limit of the $\varrho \sim T^{-1}$ region, $-d\varrho/d(1/T) = T_{SF}$	29 K
from susceptibility measurements; beginning of the Curie-Weiss law temperature range	200 K

Following the theory of Kaiser and Doinach [35], the estimated spin-fluctuation temperatures T_{SF} are: from the high-temperature end of $\varrho \sim T^2$ region: $T_{SF} = 50$ K, from the high-temperature end of $\varrho \sim T$ region: $T_{SF} = 60$ K [34].

Calculations within the framework of the Friedel-Anderson model led to a calculated spin-fluctuation temperature for UGa₃ of $T_{SF} = 63.9$ K [33].

Superconductivity with transition temperatures of about 1 K was observed in uranium compounds with partly disordered Cu₃Au-type structure. A possible transition temperature for UGa₃ was found to be less than 20 mK from measurements of the magnetic susceptibility at 0.02 to 4 K [10].

The thermoelectric power of UGa₃ was measured at temperatures of up to 300 K, showing a large increase with temperature of up to about 60 µV/K (see **Fig.** 3-**131**). The thermoelectric power of UGa₃ clearly tends to saturation at higher temperatures [36].

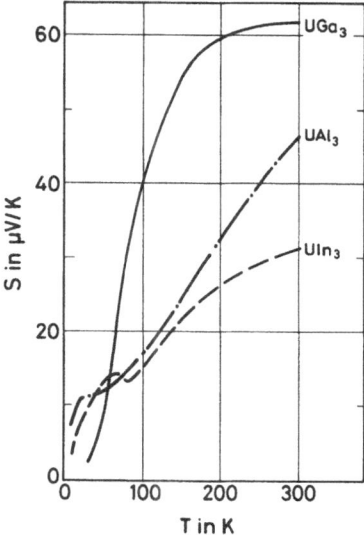

Fig. 3-131. The temperature dependence of the thermopower of UGa₃ [36]. The curves for UAl₃ and for UIn₃ are shown for comparison.

Magnetic Properties

UGa₃ is an antiferromagnet with a transition temperature of about 70 K [11, 29, 30]. The magnetic structure was observed from neutron diffraction studies. The magnetic unit cell is doubled in three directions as compared to the crystallographic unit cell, indicating antiferromagnetic ordering with antiparallel spins on adjacent (111) planes [29, 30], see also Fig. 3-143, p. 289, showing the magnetic structure of UIn₃. The Néel temperature and the magnetic moment of UGa₃ was derived from the temperature dependence of the magnetic reflections to be $T_N = 70 \pm 3$ K [29, 30], and from the intensities $\mu = 0.72 \pm 0.05\,\mu_B$ at 4.2 K [29] or $0.95 \pm 0.07\,\mu_B$ at 27 K [30], respectively. The temperature dependence of the magnetic moment is shown in [28].

References for 3.2.5 on pp. 264/5

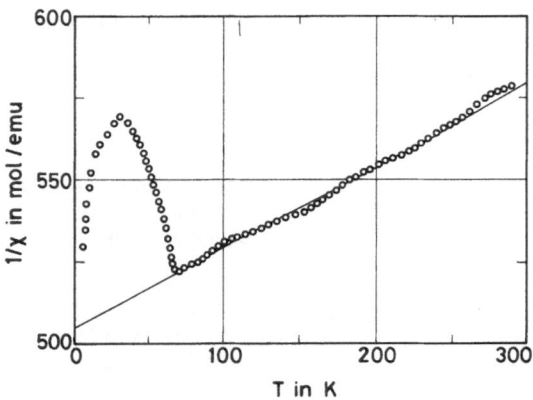

Fig. 3-132. The temperature dependence of the inverse susceptibility of UGa_3. The solid line is a fit to $\chi(T) = C/(T + \Theta_p)$ with $\Theta_p = 2080$ K and $\mu_{eff} = 1.4$ μ_B [11].

The temperature dependence of the magnetic susceptibility is shown in **Fig.** 3-**132**, see also [2, 3, 34]. Single values of the susceptibility are $\chi = 2340 \times 10^{-6}$ emu/mol at 0 K [8], $\chi = 1950 \times 10^{-6}$ emu/mol at 4.2 K [3], $\chi = 1700 \times 10^{-6}$ emu/mol at 290 K [3], $\chi = 1150 \times 10^{-6}$ emu/mol at 900 K [3].

The Néel temperature, derived from the susceptibility measurements, is $T_N = 67$ K [11]. In the temperature range of 67 to 300 K, a modified Curie-Weiss law is obeyed, $\chi = C/(T + \Theta_p)$, with a paramagnetic Curie temperature of $\Theta_p = 2080$ K and a paramagnetic moment of $\mu_{eff} = 1.4$ μ_B, which is much smaller than the moment for localized U 5f electrons. The large value of the paramagnetic Curie temperature and the reduced paramagnetic effective moment both indicate strong hybridization [11].

References for 3.2.5:

[1] Wilkinson, W. D.; Kelman, LeRoy R. (U.S. 3193380 [1965]; C.A. **63** [1965] 9408).

[2] Ansorge, V.; Menovsky, A. (Phys. Status Solidi **30** [1968] K31/K32).

[3] Misiuk, A.; Mulak, J.; Czopnik, A. (Bull. Acad. Polon. Sci. Ser. Sci. Chim. **20** [1972] 891/6).

[4] Iandelli, A.; Ferro, R. (Ann. Chim. [Rome] **42** [1952] 598/608; C.A. **1953** 3165).

[5] Palenzona, A.; Cirafici, S. (Thermochim. Acta **13** [1975] 357/60).

[6] Frost, B. R. T.; Maskrey, J. T. (J. Inst. Metals **82** [1953/54] 171/80; AERE-M/R-1027 [1952] 1/27; N.S.A. **7** [1953] No. 1547).

[7] Buschow, K. H. J. (J. Less-Common Metals **31** [1973] 165/8).

[8] van Maaren, M. H.; van Daal, H. J.; Buschow, K. H. J.; Schinkel, C. J. (Solid State Commun. **14** [1974] 145/7).

[9] Lawrence, J. M.; den Boer, M. L.; Parks, R. D.; Smith, J. L. (Phys. Rev. [3] B **29** [1984] 568/75).

[10] Ott, H. R.; Hulliger, F.; Rudigier, H.; Fisk, Z. (Phys. Rev. [3] B **31** [1985] 1329/33).

[11] Zhou, L. W.; Jee, C. S.; Lin, C. L.; Crow, J. E.; Bloom, S.; Guertin, R. P. (J. Appl. Phys. **61** Pt. 2A [1987] 3377/9).

[12] Lin, C. L.; Zhou, L. W.; Jee, C. S.; Wallash, A.; Crow, J. E. (J. Less-Common Metals **133** [1987] 67/75).

[13] Johnson, I.; Feder, H. M. (Thermodyn. Nucl. Mater. Proc. Symp., Vienna 1962 [1963], pp. 319/28; C.A. **62** [1965] 15498).

[14] Lebedev, V. A.; Seregin, V. M.; Poyarkov, A. M.; Nichkov, I. F.; Raspopin, S. P. (Zh. Fiz. Khim. **47** [1973] 712/14; Russ. J. Phys. Chem. **47** [1973] 462/3; C.A. **79** [1973] No. 10554).

[15] Lebedev, V. A.; Cherkezov, V. A. (Zh. Fiz. Khim. **49** [1975] 2154; Russ. J. Phys. Chem. **49** [1975] 1266/7; C.A. **83** [1975] No. 185460).

[16] Rough, F. A.; Bauer, A. A. (BMI-1300 [1958] 1/138; N.S.A. **12** [1958] No. 13935).

[17] Saller, H. A.; Rough F. A. (BMI-1100 [1955] from [18]).

[18] Hansen, M. (Constitution of Binary Alloys, McGraw-Hill, New York — Toronto — London 1958).

[19] Sternberk, J.; Ansorge, V.; Hrebik, J.; Menovsky, A.; Smetana, Z. (Czech. J. Phys. B **21** [1971] 969/78).

[20] Pearson, W. B. (A Handbook of Lattice Spacings of Metals and Alloys, Vol. 2, Pergamon Press, New York 1967).

[21] Krivy, I. (UJV-1783 [1967] 1/60; N.S.A. **22** [1968] No. 59; ANL-Trans-584 [1968] 1/55; N.S.A. **22** [1968] No. 25335).

[22] Dwight, A. E. (in: Giessen, B. C., Developments in the Structural Chemistry of Alloy Phases, Plenum, New York 1969, pp. 181/226).

[23] Lam, D. J.; Darby, J. B., Jr.; Nevitt, N. V. (in: Freeman, A. J.; Darby, J. B., Jr., The Actinides: Electronic Structure and Related Properties, Vol. 2, Academic, New York — San Francisco — London 1974, pp. 119/84).

[24] Johnson, I.; Chasanov, M. G. (ASM [Am. Soc. Metals] Trans. Quart. **56** [1963] 272/7; C.A. **59** [1963] 4869).

[25] Elliott, R. P. (Constitution of Binary Alloys, First Suppl., McGraw-Hill, New York — Toronto — London 1965).

[26] Johnson, I. (Intern. Symp. Compounds Interest Nucl. Reactor Technol., Boulder 1964, pp. 171/92).

[27] Alcock, C. B.; Cornish, J. B.; Grieveson, P. (Thermodyn. Nucl. Mater. Proc. Symp., Vienna 1965 [1966], Vol. 1, pp. 211/30, 367/8; C.A. **65** [1966] 8092).

[28] Reihl, B.; Domke, M.; Kaindl, G.; Kalkowski, G.; Laubschat, C.; Hulliger, F.; Schneider, W. D. (Phys. Rev. [3] B **32** [1985] 3530/3).

[29] Murasik, A.; Leciejewicz, J.; Ligenza, S.; Zygmunt, A. (Phys. Status Solidi A **23** [1974] K147/K149).

[30] Lawson, A. C.; Williams, A.; Smith, J. L.; Seeger, P. A.; Goldstone, J. A.; O'Rourke, J. A.; Fisk, Z. (J. Magn. Magn. Mater. **50** [1985] 83/7).

[31] Rivier, N.; Zlatik, V. (J. Phys. F **2** [1972] L87/L92).

[32] Rivier, N.; Zlatik, V. (J. Phys. F **2** [1972] L99/L104).

[33] Brosky, M. B. (Phys. Rev. [3] B **9** [1974] 1381/7).

[34] Buschow, K. H. J.; van Daal, H. J. (AIP [Am. Inst. Phys.] Conf. Proc. No. 5 Pt. 2 [1972] 1464/77; C.A. **77** [1972] No. 11324).

[35] Kaiser, A. B.; Doniach, S. (Intern J. Magn. **1** [1970] 11/22).

[36] van Daal, H. J.; Buschow, K. H. J.; van Aken, P. B. (Thermoelectr. Met. Conduct. Proc. 1st Intern. Conf., East Lansing, Mich., 1977 [1978], pp. 107/15; C.A. **92** [1980] No. 32756).

3.2.6 Ternary Uranium Gallides

3.2.6.1 U — Ga — Y. U — Ga — Gd. U — Ga — Ho

Phase Relationships

The ternary systems U — Ga — M (with M = Y, Gd, Ho) have not yet been investigated in detail, but a series of samples of the quasi-binary systems $UGa_2 - UM_2$ were reported to form monophase solid solutions, which crystallize with hexagonal AlB_2-type structure: $(U_xY_{1-x})Ga_2$ (with $0 \leqq x \leqq 1$), $(U_xGd_{1-x})Ga_2$ (with $0 \leqq x \leqq 1$), $(U_xHo_{1-x})Ga_2$ (with $x = 0.5$) [1 to 3].

Preparation

Mixed crystals of $(U_xY_{1-x})Ga_2$, $(U_xGd_{1-x})Ga_2$, and $U_{0.5}Ho_{0.5}Ga_2$ were prepared by arc melting of stoichiometric amounts of the elements, followed by an annealing step for homogenization [1, 3].

Crystallographic Properties

The solid solutions of the quasi-binary systems $UGa_2 - UY_2$, $UGa_2 - UGd_2$, and $U_{0.5}Ho_{0.5}Ga_2$ crystallize with hexagonal AlB_2-type structure [1 to 3]. The dependences of the lattice parameters on the uranium content are shown in [1] for the alloys of the composition $(U_xY_{1-x})Ga_2$ and $(U_xGd_{1-x})Ga_2$. The lattice parameters of $U_{0.5}Ho_{0.5}Ga_2$ are a = 0.4205 nm, c = 0.3985 nm [3].

Electrical Resistivity

The temperature dependence of the electrical resistivity of $(U_{1-x}Y_x)Ga_2$ alloys are shown in **Fig. 3-133**. There is a remarkable shift to low temperatures of the minima in the ϱ versus temperature curves observed from values above room temperature for UGa_2 to about 20 K for $U_{0.01}Y_{0.99}Ga_2$. The results are discussed in terms of a localized spin-fluctuation system [2].

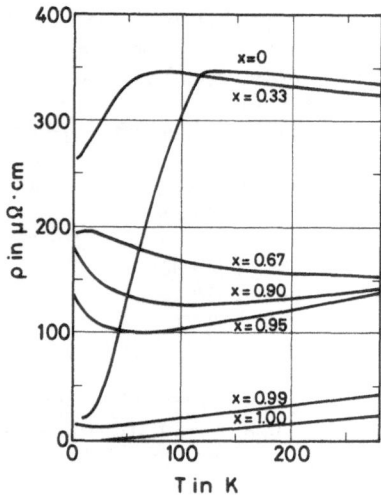

Fig. 3-133. The temperature dependence of the electrical resistivity of $(U_{1-x}Y_x)Ga_2$ alloys [2].

Magnetic Properties

UGa$_2$ is a magnetically-ordered ferromagnet at temperatures below 125 K (see Section 3.2.4, Magnetic Properties, p. 255), whereas YGa$_2$ is a temperature-independent paramagnet with a susceptibility of about 7×10^{-10} m^3/mol [1]; GdGa$_2$ and HoGa$_2$ are antiferromagnets with transition temperatures of $T_N = 12$ K for GdGa$_2$ [4] and $T_N = 8$ K for HoGa$_2$ [3]. In general, a loss of ferromagnetism was observed with increasing substitution of Y, Gd, or Ho for U in the quasi-binary solid solutions [1 to 3]. Magnetic parameters of the solid solutions, derived from modified Curie-Weiss law, are summarized in Table 3/62.

U$_x$Y$_{1-x}$Ga$_2$

The substitution of Y for U in the solid solutions results in a disappearance of the ferromagnetic behavior at about 50 at % Y. A further increase of the Y content leads to negative values of the paramagnetic Curie temperature [2]. The observed effective moments are roughly constant over the total range of solid solutions (see Table 3/62) [1, 2]. The temperature-independent contribution to the susceptibility is of the order of 10^{-8} m^3/mol [1].

Table 3/62
Measured Paramagnetic Parameters of (U$_x$M$_{1-x}$)Ga$_2$ Alloys (with M = Y, Gd, Ho) Derived from Modified Curie-Weiss Fits.

composition	x	Curie temperature Θ_p in K	Curie constant C in m$^3 \cdot$ mol$^{-1} \cdot$ K	effective moment μ_{eff} in μ_B/formula unit	Ref.
U$_x$Y$_{1-x}$Ga$_2$	0.9	105	13.8×10^{-6}	3.1	[1]
	0.8	68	9.99×10^{-6}	2.8	[1]
	0.7	59	8.78×10^{-6}	2.8	[1]
	0.6	37	7.11×10^{-6}	2.7	[1]
	0.4	13	4.93×10^{-6}	2.8	[1]
U$_x$Gd$_{1-x}$Ga$_2$	1.0	126	1.92×10^{-5}	3.5	[1]
	0.9	95	1.80×10^{-5}	3.4	[1]
	0.8	54	1.81×10^{-5}	3.4	[1]
	0.7	34	2.61×10^{-5}	4.1	[1]
	0.6	31	3.85×10^{-5}	5.0	[1]
	0.5	9	4.09×10^{-5}	5.1	[1]
	0.4	4	5.08×10^{-5}	5.7	[1]
	0.0	−24	9.99×10^{-5}	7.9	[1]
U$_x$Ho$_{1-x}$Ga$_2$	1.0	125.5	—	3.56	[3]
	0.5	15.5	—	7.92	[3]
	0.0	− 1	—	11.03	[3]

U$_x$Gd$_{1-x}$Ga$_2$

The temperature dependence of the magnetization of the U$_x$Gd$_{1-x}$Ga$_2$ solid solutions and its field dependence at 4.1 K are shown in [1]. The ferromagnetic ordering of these alloys disappears with Gd contents of x = 0.7. This effect is attributed to the antiferromagnetic

        References for 3.2.6.1 on p. 268

exchange interactions of Gd—Gd (and probably U—Gd), leading to a spin-glass-like region for higher Gd contents. The susceptibilities of the solid solutions follow a modified Curie-Weiss law from which paramagnetic Curie temperatures and effective moments were derived (see Table 3/62, p. 267). A rapid decrease of the Curie temperatures was observed with increasing Gd content, which was found to be more pronounced than for Y substitution, whereas an increase of the effective moments was observed. The temperature-independent contribution to the susceptibility is of the order of 10^{-8} m^3/mol [1].

$U_xHo_{1-x}Ga_2$

$U_{0.5}Ho_{0.5}Ga_2$ was found to be an antiferromagnet as well as $HoGa_2$ [3]. The field dependence of the magnetization is shown in [3]. The magnetic moment at 4.2 K at an applied field of 5 T is $\mu = 5.05$ μ_B/formula unit [3], as compared to $\mu = 2.35$ μ_B/formula unit for UGa_2 [5] and $\mu = 8.25$ μ_B/formula unit for $HoGa_2$ [3]. A Néel temperature of $T_N = 16 \pm 2$ K was derived from ac susceptibility measurements [3], as compared to $T_N = 125.5$ K for UGa_2 [5] and $T_N = 8 \pm 2$ K for $HoGa_2$ [3]. In the paramagnetic region, a Curie-Weiss law is obeyed from which paramagnetic parameters were derived (see Table 3/62, p. 267). The values for the effective moments follow the equation $\mu_{eff}^2 [1/2(U + Ho)] = 1/2 [\mu_{eff}^2 (U) + \mu_{eff}^2 (Ho)]$, which is consistent with a localized character of both the U and the Ho magnetic moments [3].

References for 3.2.6.1:

[1] Sechovsky, V.; Havela, L.; Svoboda, P.; Andreev, A. V. (J. Less-Common Metals **121** [1986] 163/7).

[2] Buschow, K. H. J.; van Daal, H. J. (AIP Conf. Proc. No. 5 Pt. 2 [1971/72] 1464/77; C.A. **77** [1972] No. 11324).

[3] Smetana, Z.; Sima, V.; Burianek, J.; Sebek, J. (Acta Phys. Slovaca **31** [1981] 149/51).

[4] Tsai, T. H.; Gerber, J. A.; Weymouth, J. W.; Sellmyer, D. J. (J. Appl. Phys. **49** [1978] 1507/9).

[5] Sternberk, J.; Hrebik, J.; Menovsky, A.; Smetana, Z. (J. Phys. Colloq. [Paris] **32** [1971] C1-744/C1-745).

3.2.6.2   U—Ga—Ge

3.2.6.2.1   Phase Relationships

The ternary system U—Ga—Ge has not yet been investigated in detail. A series of alloys was prepared with the composition $U_{75}Ga_xGe_{25-x}$, which are isostructural with U_3Si [1]. Samples of the quasi-binary system UGa_3—UGe_3 were prepared, forming a complete series of solid solutions, $U(Ga_{1-x}Ge_x)_3$ with $0 \leq x \leq 1$, crystallizing in an ordered Cu_3Au-type structure [2, 3]. The existence of a further ternary compound, UGa_2Ge_2, with a cubic structure (a = 4.218 ± 0.003 Å), is reported in [4].

3.2.6.2.2 $U_{75}Ga_{10}Ge_{15}$

The ternary alloys $U_{75}Ga_xGe_{25-x}$ were prepared by arc melting of the elements in an inert atmosphere. The most interesting feature of $U_{75}Ga_{10}Ge_{15}$ or similar alloys in this series is their extremely high room temperature malleability. An arc melted button of $U_{75}Ga_{10}Ge_{15}$, annealed at 800°C, was cold hammered into a thin disk without an intermediate annealing step [1].

3.2.6.2.3 U(Ga$_{1-x}$Ge$_x$)$_3$

The ternary U(Ga$_{1-x}$Ge$_x$)$_3$ alloys were prepared by arc melting of the elements in an inert atmosphere [2, 3].

The specific heat of the alloys is reported to follow the equation C = $\gamma \cdot$ T + $\beta \cdot$ T^3 (with γ = electronic coefficient and β = lattice contribution of the specific heat) with low values for γ of about 20 mJ $\cdot$ mol$^{-1} \cdot$ K^{-2} in the total range of compositions [2, 3].

The temperature dependence of the electrical resistivity of U(Ga$_{1-x}$Ge$_x$)$_3$ alloys in the range of 0 < x < 0.15 and of dϱ/dT is drastically altered from that obtained for UGa$_3$ (antiferro-magnetic, T$_N$ = 67 K). With increasing Ge content (UGe$_3$ is paramagnetic with a nearly temperature-independent susceptibility), in the range of x = 0.15 to 0.18, an extremely high depression rate of the Néel temperature was observed. This may reflect such a large hybridization that a narrow f-band formation is not allowed, a condition which is necessary for a heavy-fermion behavior [2, 3].

The temperature dependence of the magnetic susceptibility of U(Ga$_{1-x}$Ge$_x$)$_3$ alloys is shown in [3]. The observed Néel temperature is rapidly depressed by small substitutions of Ge for Ga, and tends to zero when x is about 0.18. This implies an extremely rapid increase of f-spd hybridization with increasing Ge concentration. The low-temperature limit of the susceptibility χ(0) is small and nearly independent of x in the paramagnetic region [2, 3].

3.2.6.2.4 UGa$_2$Ge$_2$

UGa$_2$Ge$_2$ was prepared by arc melting of the elements in an inert atmosphere. The pycnometrically measured density is 9.13 ± 0.06 g/cm^3. UGa$_2$Ge$_2$ crystallizes with cubic struc-ture (a = 4.218 ± 0.003 Å) but the nature of the structure is not clear. The interpretation of the structure as a Cu$_3$Au cell with one quarter of the Au sites empty, (U$_{0.75}\square_{0.25}$)Ga$_{1.5}$Ge$_{1.5}$, leads to a calculated density of 8.706 g/cm^3, and the interpretation as a perovskite-type distribution, UGa(Ge$_2$Ga), leads to 11.61 g/cm^3. Neither of these values agrees with the observed density, which, instead, corresponds to a composition of (U$_{0.87}$Ga$_{0.13}$)Ga$_{1.6}$Ge$_{1.4}$ + 0.35 Ge [4].

From measurements of the ac magnetic susceptibility UGa$_2$Ge$_2$ was found to be a super-conductor with a transition temperature of T$_c$ = 0.87 ± 0.08 K [4].

References for 3.2.6.2:

[1] Dwight, A. E. (ANL-82-14 [1982] 1/44; C.A. **98** [1983] No. 186482).

[2] Lin, C. L.; Zhou, L. W.; Jee, C. S.; Wallash, A.; Crow, J. E. (J. Less-Common Metals **133** [1987] 67/75).

[3] Zhou, L. W.; Lee, C. S.; Lin, C. L.; Crow, J. E.; Bloom, S.; Guertin, R. P. (J. Appl. Phys. **61** Pt. 2A [1987] 3377/9).

[4] Oh, H. R.; Hulliger, F.; Rudigier, H.; Fisk, Z. (Phys. Rev. [3] B **31** [1985] 1329/33).

3.2.6.3 U – Ga – Ni

The ternary alloys UGaNi, crystallizing with hexagonal Fe$_2$P-type structure [1 to 4], UGa$_5$Ni, crystallizing with hexagonal HoCoGa$_5$-type structure [5], and U$_4$Ga$_{20}$Ni$_{11}$, crystallizing with monoclinic structure [6], of the system U – Ga – Ni were investigated for crystallographic and magnetic properties.

                   References for 3.2.6.3 to 3.2.6.5 on p. 280

3.2.6.3.1 UGaNi

Preparation

Samples of UGaNi were prepared by arc melting of stoichiometric amounts of the elements in a protective atmosphere (helium and/or argon). They were remelted at least three times [4], or were annealed at 700 to 900 °C [1], see also [7, 8], at 800 °C for 2 to 3 weeks in vacuum [9], or at 1300 K for 1 week [10, 11]. Small single crystals of UGaNi were obtained from the prepared samples by temperature gradient annealing [8, 12].

Crystallographic Properties

UGaNi crystallizes in hexagonal Fe_2P-type structure with $Z = 3$; the space group is $P\bar{6}2m - D_{3h}^3$ (No. 189) [1, 3, 4]. The measured lattice parameters are summarized in Table 3/63. The crystal structure is shown in **Fig.** 3-**134**. The atomic positions are: Ga in 3 (f) (mm): x, 0, 0; 0, x, 0.; $\bar{x}, \bar{x}, 0$. U in 3 (g) (mm): x, 0, 1/2; 0, x, 1/2; $\bar{x}, \bar{x}$, 1/2. Ni in 1 (b) ($\bar{6}$mz): 0, 0, 1/2. Ni in 2 (c) ($\bar{6}$): $\pm$ 1/3, 2/3, 0 [1, 3, 4].

The interatomic distances in UGaNi are (number of neighbors in parenthesis) [4]:

Ni(1) − Ni(1)	= 4.02 Å (2)	Ga − Ni(1)	= 2.65 Å (2)
Ni(1) − Ni(2)	= 4.32 Å (12)	Ga − Ni(2)	= 2.54 Å (2)
Ni(1) − Ga	= 2.65 Å (6)	Ga − Ga	= 3.03 Å (2)
Ni(1) − U	= 2.80 Å (3)	Ga − Ga	= 4.02 Å (1)
Ni(1) − U	= 3.93 Å (3)	Ga − Ga	= 4.38 Å (4)
Ni(2) − Ni(1)	= 4.38 Å (6)	Ga − U	= 2.97 Å (2)
Ni(2) − Ni(2)	= 3.89 Å (3)	Ga − U	= 3.18 Å (4)
Ni(2) − Ni(2)	= 4.02 Å (2)	U − Ni(1)	= 2.80 Å (1)
Ni(2) − Ga	= 2.54 Å (3)	U − Ni(1)	= 3.93 Å (1)
Ni(2) − U	= 2.85 Å (6)	U − Ni(2)	= 2.85 Å (4)
		U − Ga	= 2.97 Å (2)
		U − Ga	= 3.18 Å (4)
		U − U	= 3.49 Å (4)
		U − U	= 4.02 Å (2)

Table 3/63
Measured Lattice Parameters of UGaNi at Room Temperature.

a	c	Ref.	also given in
6.7328 ± 0.0006 Å	4.0218 ± 0.0003 Å	[1]	[2, 3]
0.6715 nm	0.4012 nm	[11]	[7]
6.7251 Å	4.0162 Å	[4]	
0.6725 nm	0.4016 nm	[13]	[12]
0.6733 nm	0.4022 nm	[9]	

At low temperatures the lattice parameters are nearly temperature-independent, whereas at higher temperatures a linear dependence on temperature was observed (see **Fig.** 3-**135**). There were no anomalies found in the temperature region of magnetic ordering [11]. The lattice parameters at 5 K are a = 6.7020 Å, c = 4.0115 Å [4].

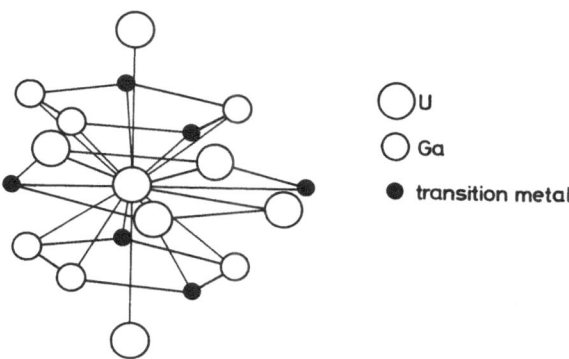

Fig. 3-134. The crystal structure of U−Ga−T alloys [4]. T = transition metal Ni, Co, Fe, Ru, Rh, Ir, Pt.

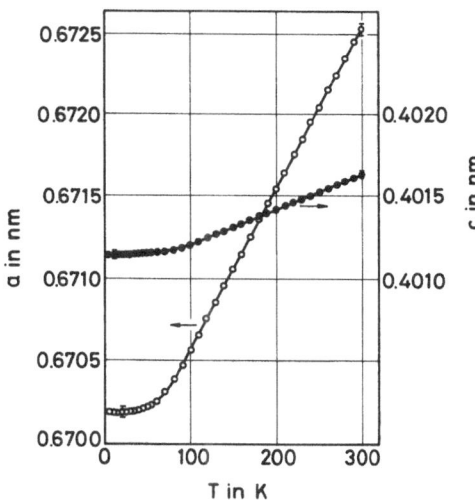

Fig. 3-135. Temperature dependence of the lattice parameters of UGaNi measured at a single crystal [11].

Mechanical and Thermal Properties

The velocity of the longitudinal sound propagation in UGaNi was measured along the c axis (see **Fig. 3-136**, p. 272). An anomaly was observed at 41 K, corresponding to the Curie temperature T_c [11].

The thermal expansion coefficients of UGaNi were derived from X-ray diffraction measurements for the high-temperature region to be: $\alpha_a = 1.44 \times 10^{-5}\,K^{-1}$ and $\alpha_c = 0.52 \times 10^{-5}\,K^{-1}$ [4, 7, 11].

References for 3.2.6.3 to 3.2.6.5 on p. 280

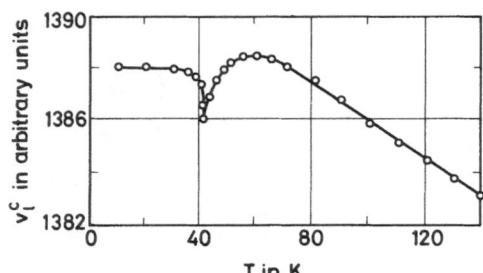

Fig. 3-136. Velocity of the longitudinal sound propagation in UGaNi along the c axis in zero field [11].

The temperature dependence of the specific heat of UGaNi is shown in **Fig.** 3-**137**. A λ-shaped anomaly was observed at a temperature of 36 K, which is connected with the ferromagnetic transition in UGaNi. The coefficient γ of the specific heat is γ = 59 mJ·mol^{-1}·K^{-2} [8, 9].

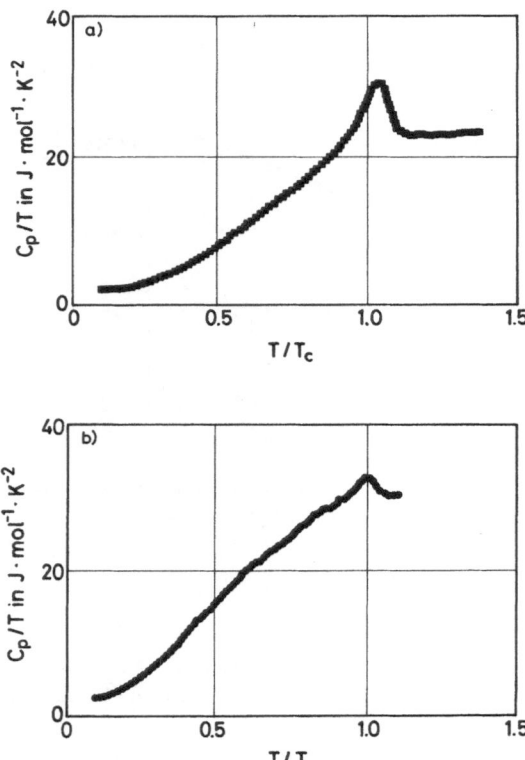

Fig. 3-137. Temperature dependence of the specific heat given as C_p/T versus T/T_c (with T_c = Curie temperature) for UGaNi (a) and UGaCo (b) [8].

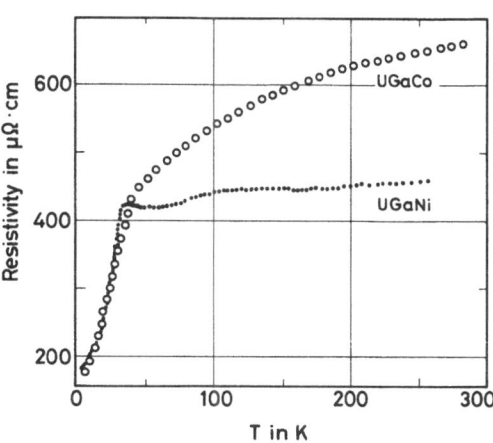

Fig. 3-138. Temperature dependence of the electrical resistivity of UGaNi and UGaCo [7].

Electrical Properties

The temperature dependence of the electrical resistivity of UGaNi is characterized by a steep increase of the resistivity up to the ferromagnetic transition temperature of $T_c = 41$ K [7] or 38 K [9]. A saturation effect was observed with further increasing temperature (see **Fig. 3-138**). The temperature behavior of the resistivity indicates a considerable influence of spin fluctuations [7, 9, 14]. Single values of the resistivity are $\varrho = 95\ \mu\Omega \cdot$ cm at 4 K [9], $\varrho = 470\ \mu\Omega \cdot$ cm [7], or 325 $\mu\Omega \cdot$ cm [9] at 300 K.

Magnetic Properties

From measurements on polycrystalline material and single crystals, UGaNi was found to be a highly anisotropic uniaxial ferromagnet with a Curie temperature of $T_c = 41$ K [4, 11], see also [7, 13], or 36 K [12, 14], or 38 K [9].

The temperature and field dependence of the magnetization is shown in **Fig. 3-139**, p. 274. In the magnetically ordered state, UGaNi shows magnetic relaxation following the equation $\sigma(t) = \sigma_0 - \sigma_1 \cdot \exp(-t/\tau)$ with the parameters $\sigma_0 = 12.3\ A \cdot m^2 \cdot kg^{-1}$, $\sigma_1 = 10.3\ A \cdot m^2 \cdot kg^{-1}$, and $\tau = 6$ min measured at 9 K [4, 11], see also [15]. In the easy direction of magnetization (c axis) UGaNi reaches a stabilized value of 18 $A \cdot m^2 \cdot mg^{-1}$ (corresponding to $\mu_s = 1.3\ \mu_B/$ formula unit [4, 11] or 0.7 μ_B/formula unit for polycrystalline samples [8]), whereas the magnetization in the basal plane reaches only a value of 0.25 $A \cdot m^2 \cdot kg^{-1}$ [11], see also [7]. The value for μ_0 is 1.19 μ_B [4, 12, 14], see also [13]. From these results, the anisotropy field $\mu_0 \cdot H_0$ was estimated to be higher than 100 T [8, 11], or at least 2000 kOe at 4.2 K [4], see also [13], with an anisotropy constant $K_1 = \sigma_s \cdot \varrho \cdot H_A/2 = 22 \times 10^7$ erg/cm^3 (ϱ (in g/cm^3) = 11.70 at 5 K and 11.60 at 300 K) [4]. From the observed hysteresis loop, a coercive force of about 7 kOe (or $\mu_0 \cdot H_c(0) = 0.7$ T [4, 8], see also [13]) was derived. The temperature dependence of H_c follows the equation $H_c(T) = H_c(0) \cdot \exp(-Q/kT)$ with $H_c(0) = 12$ kOe and an activation energy of $Q = 11 \times 10^{-16}$ erg [4], see also [15]. No magnetostriction was observed in fields of up to 300 kOe [4].

  References for 3.2.6.3 to 3.2.6.5 on p. 280 18

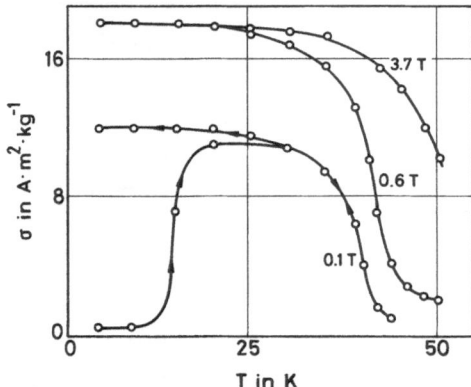

Fig. 3-139. Temperature dependence of the magnetization of a single crystal of UGaNi in the easy direction in different fields after cooling in zero field: $\mu_0 \cdot H = 3.7$ T, 0.6 T, and 0.1 T. Arrows denote the direction of temperature change [11].

The magnetic parameters along with the coefficients γ of the specific heat for different UTX alloys (with T = transition metal, X = Al, Ga, or Sn) are compared in a schematic plot in [12]. The data are summarized from [8, 12 to 14, 16].

The susceptibility of UGaNi was measured on polycrystalline samples in magnetic fields of up to 2 T. The susceptibility was observed to follow a modified Curie-Weiss law $\chi = \chi_0 + C/(T - \Theta_p)$ [9, 11, 14, 15], see also [8], with the parameters given in Table 3/64.

Table 3/64
Paramagnetic Parameters of UGaNi Derived from a Modified Curie-Weiss Law $\chi = \chi_0 + C/(T - \Theta_p)$.

Curie constant Θ_p in K	effective moment μ_{eff} in μ_B/formula unit	$\chi_0 \cdot 10^8$	Ref.	also given in
40.5	1.9	3.0 m³/kg	[11]	[7]
39	2.0	1.17 m³/mol	[14]	[12]
28	2.71 (μ_B/U)		[9]	

Concerning the anisotropy in UGaNi, the measured average susceptibility might be written as $\chi = (\chi_{\parallel} + 2 \cdot \chi_{\perp})$ [3], whereas $\chi_{\parallel}$ and $\chi_{\perp}$ are the values along the easy axis of magnetization (c axis) and in the basal plane, respectively. From this assumption the parameters $\Theta_{p\parallel} = 39$ K, $\Theta_{p\perp} = 0$ K, $\mu_{eff\parallel} = \mu_{eff\perp} = 1.7$ μ_B, $\chi_{0\parallel} = \chi_{0\perp} = 4.10^{-8}$ m³/kg are derived [15].

3.2.6.3.2 UGa₅Ni

Samples of UGa₅Ni were prepared by arc melting of the elements in water-cooled copper hearths in an argon atmosphere, followed by an annealing step at 600°C for 150 h for homogenization in an evacuated silica tube [5].

UGa_5Ni crystallizes with hexagonal $HoGa_5Co$-type structure, isotypic with UGa_5Co, with $Z = 1$, the space group is $P4/mmm−D_{4h}^1$ (No. 123). The measured lattice parameters are $a = 4.237 \pm 0.001$ Å, $c = 6.785 \pm 0.001$ Å, $c/a = 1.601$ [5]. The crystal structure is the same as that of UGa_5Co (see p. 278).

A weak temperature dependence of the magnetic susceptibility was observed, indicating a small magnetic moment on the U atom [5].

3.2.6.3.3 $U_4Ga_{20}Ni_{11}$

Samples of $U_4Ga_{20}Ni_{11}$ were prepared by arc melting of the elements in water-cooled copper hearths in an argon atmosphere, followed by an annealing step at 600 °C for 150 h in an evacuated silica tube, and finally quenching in water. A small single crystal was obtained from an arc-melted sample by mechanical fragmentation [6].

$U_4Ga_{20}Ni_{11}$ crystallizes with monoclinic structure with $Z = 2$, the space group is $C2/m−C_{2h}^3$ (No. 12). The lattice parameters are $a = 2.0734 \pm 0.0004$ nm, $b = 0.4119 \pm 0.0001$ nm, $c = 1.5338 \pm 0.0004$ nm, $\beta = 124.71 \pm 0.01°$. The calculated X-ray density is 9.26 g/cm^3 [6]. The atomic positions in $U_4Ga_{20}Ni_{11}$ are given in Table 3/65. Interatomic distances up to 0.42 nm are tabulated in [6].

Table 3/65
Atomic Positions for $U_4Ga_{20}Ni_{11}$ [6] (deviations in parentheses).

atom	site	x	y	z	atom	site	x	y	z
U(1)	4i	0.1187(1)	0	0.3323(1)	Ga(8)	4i	0.3990(2)	0	0.0028(3)
U(2)	4i	0.7732(1)	0	0.1781(1)	Ga(9)	4i	0.5410(3)	0	0.3783(3)
Ga(1)	4i	0.0057(2)	0	0.0891(3)	Ga(10)	4i	0.6275(2)	0	0.1944(3)
Ga(2)	4i	0.0647(2)	0	0.7957(3)	Ni(1)	4i	0.0243(3)	0	0.6111(4)
Ga(3)	4i	0.1109(3)	0	0.5501(3)	Ni(2)	4i	0.1335(3)	0	0.1053(4)
Ga(4)	4i	0.1976(2)	0	0.0019(3)	Ni(3)	4i	0.2476(3)	0	0.6138(4)
Ga(5)	4i	0.2509(2)	0	0.2968(3)	Ni(4)	4i	0.3458(3)	0	0.1139(4)
Ga(6)	4i	0.2983(2)	0	0.4980(3)	Ni(5)	4i	0.5041(3)	0	0.1902(4)
Ga(7)	4i	0.3911(2)	0	0.3095(3)	Ni(6)	2d	0	0.5	0.5

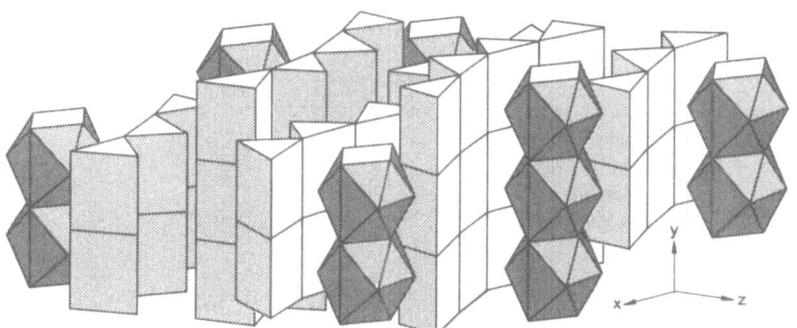

Fig. 3-140. Space model of the crystal structure of $U_4Ga_{20}Ni_{11}$ [6].

References for 3.2.6.3 to 3.2.6.5 on p. 280
18*

A space model of the structure of $U_4Ga_{20}Ni_{11}$ is shown in **Fig.** 3-**140**, p. 275. The crystal structure is an occupation variant of the $Ho_4Ga_{21}Ni_{10}$-type structure. 90% of the Ni atoms are placed in trigonal prismatic coordination: $Ni[Ga_4U_2]$ or $Ni[Ga_2Ni_2]$; the remaining Ni atoms are located in the center points of distorted icosahedra (or cuboctahedra): $Ni[Ga_8Ni_4]$; infinite columns of triangular prisms face-connected via the triangular face run parallel to the [010] direction and each four of the columns are edge-connected, forming a "fence-like" structural unit; two of these units, which are related by the two-fold axis, form the "side-walls" partly framing the inner icosahedral column [6].

3.2.6.4 U — Ga — Co

The ternary alloys UGaCo, crystallizing with hexagonal Fe_2P-type structure [1 to 3, 17], and UGa_5Co, crystallizing with hexagonal $HoGa_5Co$-type structure [5], of the system U — Ga — Co were investigated for crystallographic and magnetic properties.

3.2.6.4.1 UGaCo

Preparation

Samples of UGaCo were prepared by arc melting of stoichiometric amounts of the elements in a protective atmosphere (He and/or Ar) and were remelted at least three times for homogenization [4] or were annealed for homogenization: at 700 to 900 °C [1], see also [7, 8], at 800 °C for 2 to 3 weeks in vacuum [9], or at 1300 K for 1 week [10, 17]. Small single crystals of UGaCo were obtained from the prepared samples by gradient annealing [8, 12].

Crystallographic Properties

UGaCo crystallizes with hexagonal Fe_2P-type structure with Z = 3; the space group is $P\bar{6}2m - D_{3h}^3$ (No. 189) [1, 3, 4]. The measured lattice parameters are summarized in Table 3/66. The crystal structure, which is identical with that of UGaNi, is shown in Fig. 3-134, p. 271. The atomic distances are: Ga in 3 (f) (mm): x,0,0; 0,x,0; $\bar{x},\bar{x},0$. U in 3 (g) (mm): x,0,1/2; 0,x,1/2; $\bar{x},\bar{x},1/2$. Co in 1 (b) ($\bar{6}$mz): 0,0,1/2. Co in 2 (c) ($\bar{6}$): $\pm$ 1/3, 2/3, 0 [1,3,4].

Table 3/66
Measured Lattice Parameters at Room Temperature for UGaCo.

a	c	Ref.	also given in
6.6925 $\pm$ 0.0006 Å	3.9333 $\pm$ 0.0006 Å	[1]	[2, 3]
0.6661 nm	0.3939 nm	[17]	[7]
6.6910 Å	3.9370 Å	[4]	
0.6691 nm	0.3937 nm	[13]	[12]

The temperature dependence of the lattice parameters of UGaCo is graphically shown in [7]. No anomalies were found in the temperature region of magnetic ordering [7]. The lattice parameters at 5 K are a = 6.6657 Å, c = 3.9322 Å [4].

Thermal Expansion

The thermal expansion coefficients for UGaCo were derived from X-ray diffraction measurements for the high-temperature region to be $\alpha_a = 1.55 \times 10^{-5}\ K^{-1}$, $\alpha_c = 0.53 \times 10^{-5}\ K^{-1}$ [7].

Specific Heat

The temperature dependence of the specific heat of UGaCo is shown in Fig. 3-137, p. 272, together with that of UGaNi. A λ-shaped anomaly was observed at 48 K, which is connected with the ferromagnetic transition in UGaCo. The coefficient γ of the specific heat is $\gamma = 40\ mJ \cdot mol^{-1} \cdot K^{-2}$ [8, 12, 14, 16].

Electrical Resistivity

The temperature dependence of the electrical resistivity of UGaCo is characterized by a steep increase of the resistivity up to the ferromagnetic transition temperature of $T_c = 51$ K [7] or 48 K [16]. Contrary to UGaNi, no saturation effect was observed. The increasing resistivity reaches an extraordinary high value of $\varrho_{300K} = 660\ \mu\Omega \cdot cm$ at 300 K [7].

Magnetic Properties

From measurements on polycrystalline material and single crystals, UGaCo was found to be a highly anisotropic uniaxial ferromagnet with a Curie temperature of $T_c = 51$ K [4, 7, 13, 17], or 47 K [12, 14], or 48 K [16].

The temperature dependence of the magnetization is shown in **Fig. 3-141**. The field dependence is graphically shown in [17]. In contrast to UGaNi, there was no magnetic relaxation effect observed for UGaCo [17]. In the easy direction of magnetization (c axis) the spontaneous magnetization value corresponds to $\mu_s = 0.56\ \mu_B$/formula unit at 4.2 K [7, 17], see also [12, 14, 16] or 0.35 μ_B/formula unit for polycrystalline samples [8], whereas the magnetization in the basal plane does not depend on the magnetic field up to 2.2 T [17] or is less than 0.01 μ_B/formula unit in a field of 2 T [16]. The value of μ_0 is also given with 0.56 μ_B/formula unit [4], see also [13]. From these results, the anisotropy field $\mu_0 \cdot H_A$ at 4.2 K was estimated to be

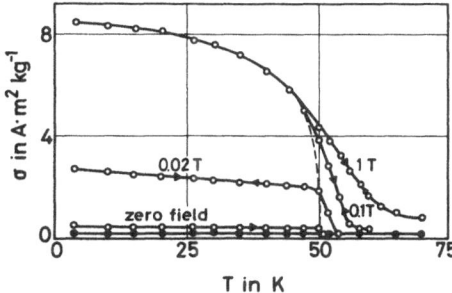

Fig. 3-141. Temperature dependence of the magnetization of UGaCo at different applied fields [17]. Easy magnetization along the c axis at 1 T, 0.1 T, 0.02 T, and zero field; hard magnetization along the a axis at 2.2 T (●).

References for 3.2.6.3 to 3.2.6.5 on p. 280

about 60 T [17] or 1000 kOe [4, 8] with an anisotropy constant of $K_1 = \sigma_s \cdot \varrho \cdot H_A/2 = 5 \times 10^7$ erg/cm^3 ($\varrho = 12.07$ g/cm^3 at 5 K and 11.96 g/cm^3 at 300 K) [4]. From the observed hysteresis loop, a coercive force at 4.2 K was about 1 kOe [4] or $\mu_0 \cdot H_0 = 0.15$ T [16, 17], or 0.1 T [8]. No magnetostriction was observed in fields of up to 300 kOe [4].

In [8], the magnetic parameters and the coefficients of γ of the specific heat for different UTM alloys (with T = transition metal, M = Al, Ga, or Sn) are compared. The data are summarized from [8, 12 to 14, 16].

The susceptibility of UGaCo was measured on polycrystalline samples in magnetic fields of up to 2 T. The susceptibility was observed to follow a modified Curie-Weiss law $\chi = \chi_0 + C/(T - \Theta_p)$ [14, 17], see also [8], with the parameters of $\Theta_p = 55$ T, $\mu_{eff} = 1.9$ μ_B/formula unit, $\chi_0 = 2.6$ m^3/kg [17], see also [7]; or $\Theta_p = 53$ K, $\mu_{eff} = 1.6$ μ_B/formula unit, $\chi_0 = 1.03$ m^3/mol [14], see also [12].

3.2.6.4.2 UGa$_5$Co

Samples of UGa$_5$Co were prepared by arc melting of the elements in water-cooled copper hearths in an argon atmosphere, followed by an annealing step at 600°C for 150 h for homogenization in an evacuated silica tube. A single crystal specimen was isolated by mechanical fragmentation for X-ray diffraction measurements [5].

UGa$_5$Co crystallizes with hexagonal HoGa$_5$Co-type structure with Z = 1; the space group is P4/mmm−D$_{4h}^1$ (No. 123) [5]. The measured lattice parameters are a = 4.2357 $\pm$ 0.0005 Å, c = 6.7278 $\pm$ 0.0009 Å, c/a = 1.588 [5]. The calculated X-ray density is 8.88 g/cm^3 [5]. The crystal structure of UGa$_5$Co, which is similar to that of UGa$_5$Ni, is shown in **Fig. 3-142**. The atomic positions are U in 1 (a): 0, 0, 0. Co in 1 (b): 0, 0, 1/2. Ga in 1 (c): 1/2, 1/2, 0. Ga in 4 (i): 0, 1/2, z; with z = 0.3082 $\pm$ 0.0001 [5].

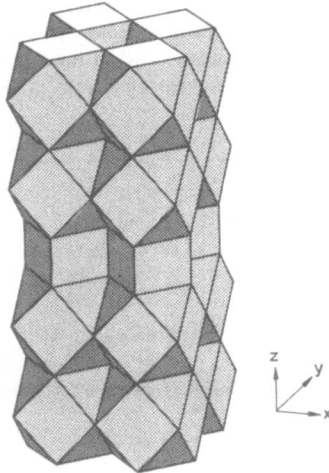

Fig. 3-142. Space model of the crystal structure of UGa$_5$Co [5]. Center points of cuboctahedra are occupied by U atoms, cubes are centered by Co atoms.

The interatomic distances are (deviations in parentheses): U−4 Ga(1) = 2.9951(2) Å, U−8 Ga(2) = 2.9638(7) Å, Co−8 Ga(2) = 2.4801(5) Å, Ga(1)−4 U = 2.9951(2) Å, Ga(1)−8 Ga(2) = 2.9638(7) Å, Ga(2)−2 U = 2.9638(7) Å, Ga(2)−2 Ga(1) = 2.9638(7) Å, Ga(2)−4 Ga(2) = 2.9951(2) Å, Ga(2)−1 Ga(2) = 2.5810(14) Å, Ga(2)−2 Co = 2.4801(5) Å [5].

A weak temperature dependence of the magnetic susceptibility was observed, similar to those of the rare earth aluminides [5].

3.2.6.5 U−Ga−Fe

Similar to the quasi-binary system UAl_2-UFe_2, a field of compositions is formed in the system U−Ga−Fe, ranging from U_2GaFe_3 to UGaFe. U_2GaFe_3 crystallizes with hexagonal $MgZn_2$-type structure [18] and UGaFe with hexagonal Fe_2P-type structure [1 to 3, 10]. A further single-phase alloy, UGa_5Fe, was found to exist, crystallizing with hexagonal $HoGa_5Co$-type structure [5]. The alloys were investigated for crystallographic and magnetic properties.

3.2.6.5.1 UGaFe

Samples of UGaFe were prepared by arc melting of the elements in a He and/or Ar atmosphere, followed by an annealing step at 700 to 900 °C [1] or at 1300 K for 1 week [10] for homogenization.

UGaFe crystallizes with hexagonal Fe_2P-type structure with $Z = 3$; the space group is $P\bar{6}2m − D_{3h}^3$ (No. 189) [1, 3]. The measured lattice parameters are $a = 6.731 \pm 0.001$ Å, $c = 3.903 \pm 0.001$ Å [1], see also [2, 3], or $a = 0.6725$ nm, $c = 0.3901$ nm [13]. The crystal structure is shown in Fig. 3-134, p. 271. The atomic positions are Ga in 3 (f) (mm): x,0,0; 0,x,0; $\bar{x},\bar{x},0$. U in 3 (g) (mm): x, 0, 1/2; 0, x, 1/2; $\bar{x},\bar{x}$, 1/2. Fe in 1 (b) ($\bar{6}mz$): 0, 0, 1/2. Fe in 2 (c) ($\bar{6}$): ± 1/3, 2/3, 0 [1,3].

The magnetic properties of UGaFe depend strongly on the stoichiometry and on the content of intermediate phases [10]. The magnetization behavior of a UGaFe sample is graphically shown in [10]. A value of $\sigma \approx 3$ A·m²·kg^{-1} at 4.2 K was obtained in a field of 2.2 T. The susceptibility follows a modified Curie-Weiss law, $\chi = \chi_0 + C/(T − \Theta_p)$ with the parameters $\chi_0 = 5.1 \times 10^{-8}$ m³/kg, C = 4.10^{-6} m³/kg^{-1}·K^{-1}, and $\Theta_p = 73.6$ K [10].

3.2.6.5.2 U_2GaFe_3

A Mössbauer study was performed on the alloy U_2GaFe_3. The spectrum at 4.2 K is broadened with no resolved structure. This broadening was attributed to magnetic ordering, and a maximum magnetic hyperfine field of about 50 kG at the ^{57}Fe nucleus was deduced. This corresponds roughly to a magnetic moment of 0.3 μ_B on Fe atoms [18].

3.2.6.5.3 UGa_5Fe

Samples of UGa_5Fe were prepared by arc melting of the elements in water-cooled copper hearths in an argon atmosphere, followed by an annealing step at 600 °C for 150 h for homogenization in an evacuated silica tube [5].

UGa_5Fe crystallizes with hexagonal $HoGa_5Co$-type structure, isotypic with UGa_5Co, with $Z = 1$; the space group is $P4/mmm − D_{4h}^1$ (No. 123) [5]. The measured lattice parameters are

References for 3.2.6.3 to 3.2.6.5 on p. 280

a = 4.261 ± 0.001 Å, c = 6.734 ± 0.003 Å, c/a = 1.580 [5]. The crystal structure, given for UGa_5Co, is shown in Fig. 3-142, p. 278.

The alloy UGa_5Fe revealed a nearly temperature-independent thermomagnetic behavior, suggesting a diamagnetic state for the Fe atoms [5].

References for 3.2.6.3 to 3.2.6.5:

[1] Dwight, A. E.; Mueller, M. H.; Conner, R. A., Jr.; Downey, J. W.; Knott, H. (Trans. AIME **242** [1968] 2075/80).
[2] Dwight, A. E. (in: Giessen, B. C., Developments in the Structural Chemistry of Alloy Phases, Plenum, New York 1969, pp. 181/226).
[3] Lam, D. J.; Darby, J. B., Jr.; Nevitt, N. N. (in: Freeman, A. J.; Darby, J. B., Jr., The Actinides, Electronic Structure and Related Properties, Vol. 2, Academic, New York — San Francisco — London 1974, pp. 119/84).
[4] Andreev, A. V.; Deryagin, A. V.; Yumaguzhin, R. Yu. (Zh. Eksperim. Teor. Fiz. **86** [1984] 1862/9; Soviet Phys.-JETP **59** [1984] 1082/6; C.A. **101** [1984] No. 32102).
[5] Grin, Yu. N.; Rogl, P.; Hiebl, K. (J. Less-Common Metals **121** [1986] 497/505).
[6] Grin, Yu. N.; Rogl, P. (J. Nucl. Mater. **137** [1986] 89/93).
[7] Havela, L.; Hrebik, J.; Zellny, M.; Andreev, A. V. (Acta Phys. Polon. A **68** [1985] 493/8).
[8] Sechovsky, V.; Havela, L.; Pillmayr, P.; Hilscher, G.; Andreev, A. V. (J. Magn. Magn. Mater. **63/64** [1987] 199/201).
[9] Palstra, T. T. M.; Nieuwenhuys, G. J.; Vlasturin, R. F. M.; van den Berg, J.; Mydosh, J. A.; Buschow, K. H. J. (J. Magn. Magn. Mater. **67** [1987] 331/42).
[10] Zeleny, M.; Sternberk, J.; Andreev, A. V.; Snegirev, V. V. (Acta Phys. Slovaca **34** [1984] 358/62; C.A. **102** [1985] No. 54864).

[11] Andreev, A. V.; Zeleny, M.; Havela, L.; Hrebik, J. (Phys. Status Solidi A **81** [1984] 301/11).
[12] Sechovsky, V.; Havela, L.; de Boer, F. R.; Franse, J. J. M.; Veenhuizen, P. A.; Sebek, J.; Stehno, J.; Andreev, A. V. (Physica B + C **142** [1986] 283/93).
[13] Andreev, A. V.; Bartashevich, M. I. (Fiz. Metal. Metalloved. **62** [1986] 266/8; Phys. Metals Metallog. [USSR] **62** No. 2 [1986] 50/3; C.A. **105** [1986] No. 182534).
[14] Sechovsky, V.; Havela, L.; Neuzil, L.; Andreev, A. V.; Hilscher, G.; Schmitzer, C. (J. Less-Common Metals **121** [1986] 169/74).
[15] Zeleny, M.; Schreiber, J.; Kobe, S. (J. Magn. Magn. Mater. **50** [1985] 27/31).
[16] Havela, L.; Neuzil, L.; Sechovsky, V.; Andreev, A. V.; Schmitzer, C.; Hilscher, G. (J. Magn. Magn. Mater. **54/57** [1986] 551/2).

3.2.6.6   U—Ga—Ru

The ternary alloys UGaRu, crystallizing with hexagonal Fe_2P-type structure [1 to 4], and UGa_5Ru, crystallizing with hexagonal $HoGa_5Co$-type structure [5], of the system U—Ga—Ru were investigated for crystallographic and magnetic properties.

3.2.6.6.1   UGaRu

Samples of UGaRu were prepared by arc melting of stoichiometric amounts of the elements in a protective atmosphere (He and/or Ar) followed by an annealing step for homogenization at 700 to 900°C [1], see also [4, 6, 7].

UGaRu crystallizes with hexagonal Fe_2P-type structure with $Z = 3$; the space group is $P\bar{6}2m - D^3_{3h}$ (No. 189) [1, 3, 4]. The measured lattice parameters are a = 7.076 ± 0.001 Å, c = 3.818 ± 0.001 Å [1], see also [2, 3, 6, 8], or a = 0.7076 nm, c = 0.381 nm [4]. The temperature dependence of the lattice parameters is shown in [4]. The crystal structure is shown in Fig. 3-134, p. 271. The atomic positions are Ga in 3 (f) (mm): x,0,0; 0,x,0; $\bar{x},\bar{x}$,0. U in 3 (g) (mm): x, 0, 1/2; 0, x, 1/2; $\bar{x}$, $\bar{x}$, 1/2. Ru in 1 (b) ($\bar{6}$mz): 0, 0, 1/2. Ru in 2 (c) ($\bar{6}$): ± 1/3, 2/3, 0 [1,3].

The thermal expansion coefficients for UGaRu were derived from X-ray diffraction measurements for the high-temperature region to be: $\alpha_a = 1.07 \times 10^{-5} K^{-1}$ and $\alpha_c = 1.10 \times 10^{-5} K^{-1}$ [4].

The coefficient of the electronic contribution to the specific heat of UGaRu is $\gamma = 52$ mJ·mol^{-1}·K^{-2} [6, 9].

The temperature dependence of the electrical resistivity of UGaRu shows a broad maximum at $T_{max} = 70$ to 80 K with a resistivity of $\varrho_{max} = 272$ μΩ·cm. Above T_{max} the resistivity decreases slightly towards 300 K, reaching a value of $\varrho_{300K} = 259$ μΩ·cm [4].

There is no indication for any magnetic or superconductive transition with UGaRu down to 20 mK [6], see also [4]. The paramagnetic susceptibility follows a modified Curie-Weiss law $\chi = \chi_0 + C/(T - \Theta_p)$ with the parameters $\Theta_p = -455$ K, $\mu_{eff} = 3.0$ μ$_B$/formula unit [4], or $\Theta_p = -192$ K, $\mu_{eff} = 1.7$ μ$_B$/formula unit, $\chi_0 = 1.02 \times 10^{-8}$ m^3/mol [6, 9].

The magnetic parameters and the coefficient γ of the specific heat for different UTM alloys (with T = transition metal, M = Al, Ga, or Sn) are compared in [7], the data are summarized from [6 to 10].

3.2.6.6.2 UGa$_5$Ru

Samples of UGa$_5$Ru were prepared by arc melting of the elements in water-cooled copper hearths in an argon atmosphere, followed by an annealing step at 600°C for 150 h for homogenization in an evacuated silica tube [5]. UGa$_5$Ru crystallizes with hexagonal HoGa$_5$Co-type structure, isotypic with UGa$_5$Co, with Z = 1; the space group is P4/mmm − D$^1_{4h}$ (No. 123). The measured lattice parameters are a = 4.3412 ± 0.001 Å, c = 6.800 ± 0.002 Å, c/a = 1.577 [5]. The crystal structure, given for UGa$_5$Co, is shown in Fig. 3-142, p. 278.

UGa$_5$Ru reveals a nearly temperature-independent thermomagnetic behavior [5].

3.2.6.7 U−Ga−Rh

The ternary alloys of U with Ga and Rh, UGaRh, crystallizing with hexagonal Fe$_2$P-type structure [1 to 3, 7], and UGa$_5$Rh, crystallizing with hexagonal HoGa$_5$Co-type structure [5] were investigated for crystallographic and magnetic properties.

3.2.6.7.1 UGaRh

Samples of UGaRh were prepared by arc melting of stoichiometric amounts of the elements in a protective atmosphere (He and/or Ar) followed by an annealing step for homogenization at 700 to 900°C [1], see also [7].

          References for 3.2.6.6 to 3.2.6.8 on p. 283

UGaRh crystallizes with hexagonal Fe_2P-type structure with $Z = 3$; the space group is $P\bar{6}2m - D_{3h}^3$ (No. 189) [1, 3]. The measured lattice parameters are a = 7.0064 ± 0.0006 Å, c = 3.9449 ± 0.0006 Å [1], see also [2, 3], or a = 0.7002 nm, c = 0.3950 nm [8]. The crystal structure is the same as that of UGaNi. It is shown in Fig. 3-134, p. 271. The atomic positions are Ga in 3 (f) (mm): x, 0, 0; 0, x, 0; $\bar{x}, \bar{x}, 0$. U in 3 (g) (mm): x, 0, 1/2; 0, x, 1/2; $\bar{x}, \bar{x}$, 1/2. Rh in 1 (b) ($\bar{6}mz$): 0, 0, 1/2. Rh in 2 (c) ($\bar{6}$): ± 1/3, 2/3, 0 [1, 3].

The temperature dependence of the specific heat, given as C_p/T versus T/T_c, is shown in [7]. The coefficient of the electronic contribution to the specific heat of UGaRh is $\gamma = 40$ mJ · mol^{-1} · K^{-2} [7].

UGaRh is a ferromagnet with a Curie temperature of $T_c = 40$ K [7, 8]. The spontaneous moment at 4.2 K, measured at polycrystalline samples, is $\mu_s = 0.4$ μ_B/formula unit [7, 8]. A coercive force of $\mu_0 \cdot H_c = 0.2$ T [7] ($H_c = 160$ kA/m [8]) was derived from the hysteresis loop. The temperature dependence of the magnetic susceptibility is graphically shown in [7].

The magnetic parameters and the coefficient γ of the specific heat for different UTM alloys (with T = transition metal, M = Al, Ga, or Sn) are graphically compared in [7]. The data are taken from [6 to 10].

3.2.6.7.2 UGa₅Rh

Samples of UGa₅Rh were prepared by arc melting of the elements in water-cooled copper hearths in an argon atmosphere, followed by an annealing step at 600°C for 150 h in an evacuated silica tube for homogenization [5].

UGa₅Rh crystallizes with hexagonal HoGa₅Co-type structure, isotypic with UGa₅Co, with $Z = 1$; the space group is $P4/mmm - D_{4h}^1$ (No. 123). The measured lattice parameters are a = 4.299 ± 0.001 Å, c = 6.800 ± 0.001 Å, c/a = 1.582 [5].

The crystal structure, given for UGa₅Co, is shown in Fig. 3-142, p 278.

A weak temperature dependence of the magnetic susceptibility was observed, similar to those of the rare earth aluminides [5].

3.2.6.8 U−Ga−Pd

Only the ternary alloys UGa₅Pd, crystallizing with hexagonal HoGa₅Co-type structure [5] and $U_4Ga_{20}Pd_{11}$, crystallizing with monoclinic structure [11], of the system U−Ga−Pd were investigated for crystallographic and magnetic properties.

3.2.6.8.1 UGa₅Pd

Samples of UGa₅Pd were prepared by arc melting of the elements in water-cooled copper hearths in an argon atmosphere, followed by an annealing step at 600°C for 150 h in an evacuated silica tube for homogenization [5].

UGa₅Pd crystallizes with hexagonal HoGa₅Co-type structure, isotypic with UGa₅Co, with $Z = 1$; the space group is $P4/mmm - D_{4h}^1$ (No. 123). The measured lattice parameters are a = 4.321 ± 0.001 Å, c = 6.862 ± 0.001 Å, c/a = 1.588 [5]. The crystal structure is shown in Fig. 3-142, p 278.

A weak temperature dependence of the magnetic susceptibility was observed, indicating a small magnetic moment on the U atom [5].

3.2.6.8.2 $U_4Ga_{20}Pd_{11}$

Samples of $U_4Ga_{20}Pd_{11}$ were prepared by arc melting of the elements in water-cooled copper hearths in an argon atmosphere, followed by an annealing step at 600°C for 150 h in an evacuated silica tube for homogenization and finally quenching in water [11].

$U_4Ga_{20}Pd_{11}$ crystallizes with monoclinic structure with Z = 2, isotypic with $U_4Ga_{20}Ni_{11}$; the space group is $C2/m-C_{2h}^3$ (No. 12). The measured lattice parameters are a = 2.1255 ± 0.0010 nm, b = 0.4295 ± 0.0001 nm, c = 1.5765 ± 0.0008 nm, β = 124.72° ± 0.03° [11]. A space model of the structure, given for $U_4Ga_{20}Ni_{11}$, is shown in Fig. 3-140, p. 275.

References for 3.2.6.6 to 3.2.6.8:

[1] Dwight, A. E.; Mueller, M. H.; Conner, R. A., Jr.; Downey, J. W.; Knott, H. (Trans. AIME **242** [1968] 2075/80).

[2] Dwight, A. E. (in: Giessen, B. C., Developments in Structural Chemistry of Alloy Phases, Plenum, New York 1969, pp. 181/226).

[3] Lam, D. J.; Darby, J. B., Jr.; Nevitt, N. V. (The Actinides, Electronic Structure and Related Properties, Vol. 2, Academic, New York − San Francisco − London 1974, pp. 119/84).

[4] Havela, L.; Hrebik, J.; Zeleny, M.; Andreev, A. V. (Acta Phys. Polon. A **68** [1985] 493/8).

[5] Grin, Yu. N.; Rogl, P.; Hiebl, K. (J. Less-Common Metals **121** [1986] 497/505).

[6] Sechovsky, V.; Havela, L.; de Boer, F. R.; Franse, J. J. M.; Veenhuizen, P. A.; Sebek, J.; Stehno, J.; Andreev, A. V. (Physica B + C **142** [1986] 283/93).

[7] Sechovsky, V.; Havela, L.; Pillmayr, N.; Hilscher, G.; Andreev, A. V. (J. Magn. Magn. Mater. **63/64** [1987] 199/201).

[8] Andreev, A. V.; Bartashevich, M. I. (Fiz. Metal. Metalloved. **62** [1986] 266/8; Phys. Metals Metallog. [USSR] **62** No. 2 [1986] 50/3; C.A. **105** [1986] No. 182534).

[9] Sechovsky, V.; Havela, L.; Neuzil, L.; Andreev, A. V.; Hilscher, G.; Schmitzer, C. (J. Less-Common Metals **121** [1986] 169/74).

[10] Havela, L.; Neuzil, L.; Sechovsky, V.; Andreev, A. V.; Schmitzer, C.; Hilscher, G. (J. Magn. Magn. Mater. **54/57** [1986] 551/2).

[11] Grin, Yu. N.; Rogl, P. (J. Nucl. Mater. **137** [1986] 89/93).

3.2.6.9 U−Ga−Os

The ternary alloys UGaOs, crystallizing with hexagonal $MgZn_2$-type structure [1], and UGa_5Os, crystallizing with hexagonal $HoGa_5Co$-type structure [2], of the system U−Ga−Os were investigated for crystallographic and magnetic properties.

3.2.6.9.1 UGaOs

No method of preparation was reported for UGaOs.

UGaOs crystallizes with hexagonal $MgZn_2$-type structure, with Z = 4; the space group is $P6_3/mmc-D_{6h}^4$ (No. 194). The lattice parameters are a = 5.398 Å, c = 8.485 Å, c/a = 1.57 [1].

References for 3.2.6.9 to 3.2.6.11 on p. 286

3.2.6.9.2 UGa$_5$Os

Samples of UGa$_5$Os were prepared by arc melting of the elements in water-cooled copper hearths in an argon atmosphere, followed by an annealing step at 600°C for 150 h in an evacuated silica tube for homogenization. UGa$_5$Os crystallizes with hexagonal HoGa$_5$Co-type structure, isotypic with UGa$_5$Co, with Z = 1; the space group is P4/mmm — D$_{4h}^1$ (No. 123). The measured lattice parameters are a = 4.318 ± 0.004 Å, c = 6.813 ± 0.007 Å, c/a = 1.577 [2].

The crystal structure, given for UGa$_5$Co, is shown in Fig. 3-142, p. 278. UGa$_5$Os reveals a nearly temperature-independent thermomagnetic behavior [2].

3.2.6.10   U — Ga — Ir

The ternary alloys of the system U — Ga — Ir, UGaIr, crystallizing with hexagonal Fe$_2$P-type structure [1, 3 to 5], and UGa$_5$Ir, crystallizing with hexagonal HoGa$_5$Co-type structure [2], were investigated for crystallographic and magnetic properties.

3.2.6.10.1   UGaIr

Samples of UGaIr were prepared by arc melting of stoichiometric amounts of the elements in a protective atmosphere (He and/or Ar) followed by an annealing step for homogenization at 700 to 900°C [3], see also [5].

UGaIr crystallizes with hexagonal Fe$_2$P-type structure, with Z = 3; the space group is P$\bar{6}$2m — D$_{3h}^3$ (No. 189) [1, 3]. The measured lattice parameters are a = 7.0330 ± 0.0001 Å, c = 3.9444 ± 0.0001 Å [3], see also [1, 4], or a = 0.7030 nm, c = 0.3935 nm [6]. The crystal structure is shown in Fig. 3-134, p. 271. The atomic positions are: Ga in 3 (f) (mm): x, 0, 0; 0, x, 0; $\bar{x}$, $\bar{x}$, 0. U in 3 (g) (mm): x, 0, 1/2; 0, x, 1/2; $\bar{x}$, $\bar{x}$, 1/2. Ir in 1 (b) ($\bar{6}$mz): 0, 0, 1/2. Ir in 2 (c) ($\bar{6}$): ± 1/3, 2/3, 0 [1, 3].

The temperature dependence of the specific heat of UGaIr is shown in [5]. The coefficient of the electronic contribution to the specific heat is γ = 41 mJ · mol^{-1} · K^{-2} [5].

UGaIr is a ferromagnet with a Curie temperature of T$_c$ = 64 K [5] or 63 K [6]. The spontaneous moment at 4.2 K, measured at polycrystalline samples, is μ_s = 1.0 μ_B/formula unit [5, 6]. A coercive force of μ_0 · H$_c$ = 0.45 T [5] (H$_c$ = 360 kA/m [6]) was derived from the hysteresis loop. The temperature dependence of the magnetic susceptibility is shown in [5].

The magnetic parameters and the coefficient γ of the specific heat for different UTX alloys (with T = transition metal, X = Al, Ga, or Sn) are compared in [5]; data are summarized from [5 to 9].

3.2.6.10.2   UGa$_5$Ir

Samples of UGa$_5$Ir were prepared by arc melting of the elements in water-cooled copper hearths in an argon atmosphere, followed by an annealing step at 600°C for 150 h in an evacuated silica tube for homogenization. UGa$_5$Ir crystallizes with hexagonal HoGa$_5$Co-type structure, isotypic with UGa$_5$Co, Z = 1; the space group is P4/mmm — D$_{4h}^1$ (No. 123). The measured lattice parameters are a = 4.317 ± 0.001 Å, c = 6.745 ± 0.003 Å, c/a = 1.563 [2].

A weak temperature dependence of the magnetic susceptibility was observed, similar to those of the rare earth aluminides [2].

3.2.6.11   U — Ga — Pt

The ternary alloys of the system U — Ga — Pt, UGaPt, crystallizing with hexagonal Fe_2P-type structure [1, 3 to 5], UGa_5Pt, crystallizing with hexagonal $HoGa_5Co$-type structure [2], and $U_4Ga_{20}Pt_{11}$, crystallizing with monoclinic structure [10], were investigated for crystallographic and magnetic properties.

3.2.6.11.1   UGaPt

Samples of UGaPt were prepared by arc melting of stoichiometric amounts of the elements in a protective atmosphere (He and/or Ar) followed by an annealing step for homogenization at 700 to 900°C [3], see also [5].

UGaPt crystallizes with hexagonal Fe_2P-type structure with $Z = 3$; the space group is $P\bar{6}2m - D_{3h}^3$ (No. 189) [1, 3]. The measured lattice parameters are $a = 7.063 \pm 0.001$ Å, $c = 4.065 \pm 0.001$ Å [3], see also [1, 4], or $a = 0.7057$ nm, $c = 0.4068$ nm [6]. The crystal structure is shown in Fig. 3-134, p. 271. The atomic positions are: Ga in 3 (f) (mm): x, 0, 0; 0, x, 0; $\bar{x}, \bar{x}, 0$. U in 3 (g) (mm): x, 0, 1/2; 0, x, 1/2; $\bar{x}, \bar{x}$, 1/2. Pt in 1 (b) ($\bar{6}mz$): 0, 0, 1/2. Pt in 2 (c) ($\bar{6}$): $\pm$ 1/3, 2/3, 0 [1, 3].

The temperature dependence of the specific heat of UGaPt is graphically shown in [5]. The coefficient of the electronic contribution to the specific heat is $\gamma = 72$ mJ $\cdot$ mol$^{-1} \cdot$ K^{-2} [5].

UGaPt is a ferromagnet with a Curie temperature of $T_c = 68$ K [5] or 79 K [6]. In the magnetically ordered state, UGaPt shows magnetic relaxation, as also observed for the alloy UGaNi (see Section 3.2.6.3.1, p. 273). The spontaneous moment at 4.2 K, measured at polycrystalline samples, is $\mu_s = 1.2 \mu_B$/formula unit [5, 6]. An extremely high value for the coercive force of $\mu_0 \cdot H_c = 2.4$ T [5] ($H_c = 2000$ kA/m [6]) was derived from the hysteresis loop. The temperature dependence of the magnetic susceptibility is graphically shown in [5]. The magnetic parameters and the coefficient γ of the specific heat for different UTM alloys (with T = transition metal, M = Al, Ga, or Sn) are compared in [5]; the data are taken from [5 to 9].

3.2.6.11.2 UGa_5Pt

Samples of UGa_5Pt were prepared by arc melting of the elements in water-cooled copper hearths in an argon atmosphere, followed by an annealing step at 600°C for 150 h in an evacuated silica tube for homogenization [2]. UGa_5Pt crystallizes with a hexagonal $HoGa_5Co$-type structure, isotypic with UGa_5Co, with $Z = 1$; the space group is $P4/mmm - D_{4h}^1$ (No. 123). The measured lattice parameters are $a = 4.341 \pm 0.001$ Å, $c = 6.813 \pm 0.003$ Å, $c/a = 1.570$ [2]. The crystal structure, given for UGa_5Co, is shown in Fig. 3-142, p. 278. A weak temperature dependence of the magnetic susceptibility was observed, indicating a small magnetic moment in the U atom [2].

3.2.6.11.3 $U_4Ga_{20}Pt_{11}$

Samples of $U_4Ga_{20}Pt_{11}$ were prepared by arc melting of the elements in water-cooled copper hearths in an argon atmosphere, followed by an annealing step at 600°C for 150 h for homogenization in an evacuated silica tube and quenching in water [10]. $U_4Ga_{20}Pt_{11}$ crystallizes with monoclinic structure, with $Z = 2$; isotypic with $U_4Ga_{20}Ni_{11}$, the space group is $C2/m - C_{2h}^3$ (No. 12). The measured lattice parameters are $a = 2.1150 \pm 0.0010$ nm, b = 0.4272 $\pm$

References for 3.2.6.9 to 3.2.6.11 on p. 286

0.0001 nm, c = 1.5691 $\pm$ 0.0009 nm, β = 124.71° $\pm$ 0.02° [10]. A space model of the crystal structure, given for $U_4Ga_{20}Ni_{11}$, is shown in Fig. 3-140, p. 275.

References for 3.2.6.9 to 3.2.6.11:

[1] Lam, D. J.; Darby, J. B., Jr.; Nevitt, N. V. (in: Freeman, A. J.; Darby, J. B., Jr., The
Actinides, Electronic Structure and Related Properties, Vol. 2, Academic, New York —
San Francisco — London 1974, pp. 119/84).

[2] Grin, Yu. N.; Rogl, P.; Hiebl, K. (J. Less-Common Metals **121** [1986] 497/505).

[3] Dwight, A. E.; Mueller, M. H.; Conner, R. A., Jr.; Downey, J. W.; Knott, H. (Trans. AIME
242 [1968] 2075/80).

[4] Dwight, A. E. (in: Giessen, B. C., Developments in the Structural Chemistry of Alloy
Phases, Plenum, New York 1969, pp. 181/226).

[5] Sechovsky, V.; Havela, L.; Pillmayr, N.; Hilscher, G.; Andreev, A. V. (J. Magn. Magn.
Mater. **63/64** [1987] 199/201).

[6] Andreev, A. V.; Bartashevich, M. I. (Fiz. Metal. Metalloved. **62** [1986] 266/8; Phys. Metals
Metallog. [USSR] **62** No. 2 [1986] 50/3; C.A. **105** [1986] No. 182534).

[7] Havela, L.; Neuzil, L.; Sechovsky, V.; Andreev, A. V.; Schmitzer, C.; Hilscher, G. (J. Magn.
Magn. Mater. **54/57** [1986] 551/2).

[8] Sechovsky, V.; Havela, L.; Neuzil, L.; Andreev, A. V.; Hilscher, G.; Schmitzer, C. (J. Less-
Common Metals **121** [1986] 169/74).

[9] Sechovsky, V.; Havela, L.; de Boer, F. R.; Franse, J. J. M.; Veenhuizen, P. A.; Sebek, J.;
Stehno, J.; Andreev, A. V. (Physica B + C **142** [1986] 283/93).

[10] Grin, Yu. N.; Rogl, P. (J. Nucl. Mater. **137** [1986] 89/93).

3.3 Uranium — Indium

3.3.1 Phase Relationships

UIn_3 is the only compound known to exist in the binary system U — In and was first
mentioned by [1, 2]. It crystallizes with simple cubic $AuCu_3$-type structure, isomorphous with
UAl_3, USn_3, UGa_3, USi_3, UGe_3, or UPb_3 [1, 2], see also [3, 4].

The solubility of U in In is very restricted, measured values are summarized in Table 3/67, see Fig. 3-120, p. 247. The solubility of U in liquid In was expressed by the empirical
equation log a (a in at % U) = $3.781 - 5146 \cdot T^{-1}$ [5], see also [6].

Table 3/67
Measured Liquidus Data for the System U — In.
The values of [9] are also given in [3, 6, 7], those of [5] in [8].

temperature in °C	composition in wt %	composition in wt %	temperature in °C	composition in wt %	composition in wt %
900	1.05 [9]	1.03 [6]	710	0.15 [10]	6.5×10^{-2} [5]
883	0.93 [10]		700	0.15 [9]	0.18 [6]
850	0.70 [9]	0.71 [6]	663		4.2×10^{-2} [5]
800	0.45 [9]	0.48 [6]	650	0.10 [6, 9]	3.2×10^{-2} [5]
750	0.25 [9]	0.30 [6]	618		2.11×10^{-2} [5]
720		9.3×10^{-2} [5]	611		1.60×10^{-2} [5]

Table 3/67 (continued)

tempera-ture in °C	composition in wt%	composition in wt%		tempera-ture in °C	composition in wt%	composition in wt%
605	0.08 [10]			470		1.70×10^{-3} [5]
600	0.08 [9]	0.06 [6]		455		$9.4 \ \times 10^{-4}$ [5]
567		1.06×10^{-2} [5]		374		1.88×10^{-4} [5]
557		$8.6 \ \times 10^{-3}$ [5]		339		1.56×10^{-4} [5]
509		3.12×10^{-3} [5]				

References for 3.3.1:

[1] Iandelli, A.; Ferro, R. (Ann. Chim. [Rome] **42** [1952] 598/608; C.A. **1953** 3165).

[2] Frost, B. R. T.; Maskrey, J. T. (J. Inst. Metals **82** [1953/54] 171/80, AERE-M-R-1027 [1952] 1/27; N.S.A. **7** [1953] No. 1547).

[3] Rough, F. A.; Bauer, A. A. (BMI-1300 [1958] 1/138; N.S.A. **12** [1958] No. 13935).

[4] Hansen, M. (Constitution of Binary Alloys, McGraw-Hill, New York — Toronto — London 1958).

[5] Johnson, I.; Chasanov, M. G. (ASM [Am. Soc. Metals] Trans. Quart. **56** [1963] 272/7).

[6] Lebedev, V. A.; Seregin, V. M.; Poyarkov, A. M.; Nichkov, I. F.; Raspopin, S. P. (Zh. Fiz. Khim. **48** [1974] 542/5; Russ. J. Phys. Chem. **48** [1974] 314/9; C.A. **81** [1974] No. 30526).

[7] Elliott, R. P. (Constitution of Binary Alloys, 1st Suppl., McGraw-Hill, New York — Toronto — London 1965).

[8] Shunk, F. A. (Constitution of Binary Alloys, 2nd Suppl., McGraw-Hill, New York — Toronto — London 1969).

[9] Hayes, E. E.; Gordon, P. (TID-65 [1948] 130/41, from [3]).

[10] Hayes, E. E.; Gordon, P. (TID-2501 [1957] 115; N.S.A. **12** [1958] No. 17280).

3.3.2   Uranium Triindide, UIn$_3$

Preparation and Formation

Samples of UIn$_3$ were prepared by reaction of stoichiometric amounts of the powdered elements at 600°C within 100 to 150 h in an evacuated silica tube [1, 2] or at 1000°C within 20 h [3], annealed in an inert atmosphere (argon) [3 to 5]. After the reaction was completed the samples were slowly cooled to room temperature within 24 h [1] or 70 h [2]. The reaction is exothermic and starts at 430°C [5].

UIn$_3$ was also prepared by usual melting [6] or arc melting techniques [7, 8].

UIn$_3$ forms during the electrodeposition of U from melts of UCl$_3$ in LiCl—KCl on molten In. Parameters for the nucleation from experiments at 400, 506, 586, and 745°C and current densities of 0.1 and 0.2 A/cm^2 are given in [10].

Thermodynamic Data of Formation

From emf measurements (molten salt electrolytes) the following thermodynamic values were derived for the formation of UIn$_3$ at 450°C: $\Delta H_f^\circ = -6.58$ kcal/g-atom, $\Delta G_f^\circ = -4.23$ kcal/

References for 3.3.2 on pp. 290/1

g-atom, $\Delta S_f^\circ = -3.26$ cal·g-atom^{-1}·K^{-1} [9]. The enthalpy of formation at room temperature is derived from dynamic differential calorimetry (DDC) as ΔH_f (at 298 K) = -3.8 kcal/g-atom [5].

The following equations, according to U(s) + 3 In(l) → UIn$_3$(s) at 353 to 676°C, for the formation of UIn$_3$ were derived from emf measurements: ΔG_f° (in kcal/mol) = $-24.28 + 7.439 - 10^{-3} \cdot T + 3.875 \times 10^{-6} \cdot T^2$; ΔH_f° (in kcal/mol) = $-24.28 - 3.875 \times 10^{-6} \cdot T^2$; ΔS_f° (in cal·mol^{-1}·K^{-1}) = $7.439 - 7.750 \times 10^{-3} \cdot T$ [9, 17, 18].

The free energy for the reaction 1/3 UIn$_3$(s) = 1/3 U(γ) + In(s) is ΔG_f° (in cal/mol) = $68300 - 32.2 \cdot T$ at 820 to 1041°C. From this, an analytical relation for the free energy of formation according to U(α) + 3 In(l) = UIn$_3$(s) was derived to be ΔG_f° (in cal/mol) = $-23050 + 11.32 \cdot T$ [14].

Crystallographic Properties

UIn$_3$ crystallizes with simple cubic AuCu$_3$-type structure (Ll$_2$) with Z = 1; the space group is Pm$\bar{3}$m−O$_h^1$ (No. 221) [11], see also [4, 6]. Measured lattice parameters are summarized in Table 3/68. The calculated X-ray density is 10.12 g/cm^3 [6], see also [12]. The atomic positions are: 1 U in: 0,0,0. 3 In in: 0,1/2,1/2; 1/2,0,1/2; 1/2,1/2,0 [4]. Each U atom is bonded to 12 In atoms and each In atom is bonded to 4 U atoms and 8 In atoms with the distance $a/\sqrt{2} = 3.25$ Å [6], 3.24 Å [4], or 3.258 Å [1].

Table 3/68
Measured Lattice Parameters of UIn$_3$.

a in Å	Ref.	also given in
4.588	[4]	[13]
4.6013	[6]	[11, 12, 14 to 16]
4.606 ± 0.002	[1]	[2]
4.601	[3]	

Specific Heat

The low-temperature specific heat of UIn$_3$ was measured at compacted powders. Following the equation $C = \gamma \cdot T + \alpha \cdot T^3$, with γ = coefficient of the electronic specific heat and α = coefficient of the lattice contribution to the specific heat, the values $\gamma = 49.9$ mJ·mol^{-1}·K^{-2} and $\alpha = 1.47$ mJ·mol^{-1}·K^{-4} are reported. The Debye temperature was calculated to be approximately $\Theta_D = 174$ K (at 0 K). No correction was made for the contribution due to antiferromagnetic ordering [7].

Electrical Properties

The electric resistivity of UIn$_3$ increases steeply with temperature up to about 80 K and then tends to saturate at a level of about 100 $\mu\Omega \cdot$ cm [19]. In the low-temperature region, UIn$_3$ shows a variety of different temperature dependences, as predicted by the theory of Rivier and Zlatic [20, 21] for localized spin fluctuations in dilute alloys [19, 22]. Different spin-fluctuation temperatures, analyzed from different temperature regions, are summarized in Table 3/69.

Table 3/69

Spin-Fluctuation Temperatures T_{SF} Calculated by Different Methods in Different Temperature Regions for UIn₃ [22].

temperature region	T_{SF}
low-temperature end of $\varrho \sim \ln T$ region	40 K
high-temperature end of $\varrho \sim T^2$ region	250 K
from the slope of the $\varrho \sim T^2$ region	138 K
from the slope of the $\varrho \sim T^{-1}$ region	86 K
from the approach to the high-temperature limit of the $\varrho \sim T^{-1}$ region	12 K
begin of the temperature region where the Curie-Weiss law in susceptibility measurements is obeyed	150 K

Estimated spin-fluctuation temperatures T_{SF} according to the theory of Kaiser and Doinach [23] are: $T_{SF} = 30$ K from the high-temperature end of $\varrho \sim T^2$ region, $T_{SF} = 80$ K from the high-temperature end of $\varrho \sim T$ region [19]. Calculations within the framework of the Friedel-Anderson model led to a calculated spin-fluctuation temperature of $T_{SF} = 36.9$ K [22].

The thermoelectric power of UIn₃ was measured at temperatures of up to 300 K, showing an increase with temperature of up to about 30 µV/K. At about 80 K an onset of the thermopower is observed, which is connected with the antiferromagnetic transition. The thermoelectric power of UIn₃ tends to saturation at higher temperatures [24].

Magnetic Properties

UIn₃ is antiferromagnetic with a transition temperature of $T_N = 95$ K [2] or 108 K [8, 25]. The magnetic structure, as derived from a neutron diffraction study, shows a magnetic cell which is doubled in three directions, indicating antiferromagnetic ordering with antiparallel spins on adjacent (111) planes (see **Fig. 3-143**). The magnetic moment of UIn₃ was derived from the intensity of the M(111) magnetic reflections to be $\mu \approx 1$ μ_B at 4.2 K [2]. A value for the Néel temperature was derived from the exchange integrals calculated on the basis of molecular field theory to be $T_N = 95$ K [2].

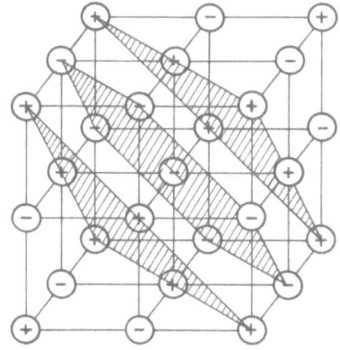

Fig. 3-143. The magnetic structure of UIn₃ [2]. (|1/2, 1/2, 1/2| mode).

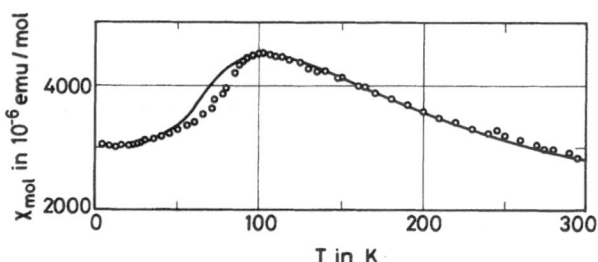

Fig. 3-144. Temperature dependence of the magnetic susceptibility of UIn_3 [1].

The temperature dependence of the magnetic susceptibility is shown in **Fig.** 3-**144**. Single values of the susceptibility are $\chi = 3240 \times 10^{-6}$ emu/mol at 0 K [7], $\chi = 3050 \times 10^{-6}$ emu/mol at 4.2 K [1], $\chi = 2900 \times 10^{-6}$ emu/mol at 290 K [1].

The Curie-Weiss law is approximately obeyed above 200 K; an effective magnetic moment of $\mu_{eff} = 3.6\ \mu_B$ was derived. The broad maximum of the susceptibility at 95 to 105 K is not quite typical for antiferromagnetic transition. Thus, the observed maximum in Fig. 3-144 was discussed in terms of a paramagnetic susceptibility maximum model, resulting in the van Vleck equation in the form of $\chi_m = \dfrac{0.003 + \dfrac{4}{T}\, e^{-250/T}}{1 + 5\, e^{-250/T}}$ [1].

References for 3.3.2:

[1] Misiuk, A.; Mulak, J.; Czopnik, A. (Bull. Acad. Polon. Sci. Ser. Sci. Chim. **20** [1972] 891/6).

[2] Murasik, A.; Leciejewicz, J.; Ligenza, S.; Misiuk, A. (Phys. Status Solidi A **20** [1973] 395/401).

[3] Lawrence, J. M.; de Boer, M. L.; Parks, R. D.; Smith, J. L. (Phys. Rev. [3] B **29** [1984] 568/75).

[4] Iandelli, A.; Ferro, R. (Ann. Chim. [Rome] **42** [1952] 598/608; C.A. **1953** 3165).

[5] Palenzona, A.; Cirafici, S. (Thermochim. Acta **13** [1975] 357/60).

[6] Frost, B. R. T.; Maskrey, J. T. (J. Inst. Metals **82** [1953/54] 171/80; AERE-M-R-1027 [1952] 1/27; N.S.A. **7** [1953] No. 1547).

[7] van Maaren, M. H.; van Daal, H. J.; Buschow, K. H. J.; Schinkel, C. J. (Solid State Commun. **14** [1974] 145/7).

[8] Lin, C. L.; Zhou, L. W.; Jee, C. S.; Wallash, A.; Crow, J. E. (J. Less-Common Metals **133** [1987] 67/75).

[9] Johnson, I.; Feder, H. M. (Thermodyn. Nucl. Mater. Proc. Symp., Vienna 1962 [1963], pp. 319/28; C.A. **62** [1965] 15498).

[10] Lebedev, V. A.; Cherkezov, V. A. (Zh. Fiz. Khim. **49** [1975] 1853; Russ. J. Phys. Chem. **49** [1975] 1092; C.A. **83** [1975] No. 15456).

[11] Rough, F. A.; Bauer, A. A. (BMI-1300 [1958] 1/138; N.S.A. **12** [1958] No. 13935).

[12] Pearson, W. B. (A Handbook of Lattice Spacings and Structures of Metals and Alloys, Vol. 2, Pergamon, New York 1967).

[13] Hansen, M. (Constitution of Binary Alloys, McGraw-Hill, New York — Toronto — London 1958).

[14] Krivy, I. (UJV-1783 [1967] 1/48; N.S.A. **22** [1968] No. 59; ANL-Trans-584 [1968] 1/55; N.S.A. **22** [1968] No. 25335).

[15] Dwight, A. E. (in: Giessen, B. C., Developments in the Structural Chemistry of Alloy Phases, Plenum, New York 1969, pp. 181/226).

[16] Lam, D. J.; Darby, J. B., Jr.; Nevitt, N. V. (in: Freeman, A. J.; Darby, J. B., Jr., The Actinides: Electronic Structure and Related Properties, Vol. 2, Academic, New York — San Francisco — London 1974, pp. 119/84).

[17] Johnson, I. (Met. Soc. AIME Inst. Metals Div. Spec. Rept. Ser. **10** No. 13 [1964] 171/92).

[18] Lebedev, V. A.; Seregin, V. M.; Poyarkov, A. M.; Nichkov, I. F.; Raspopin, S. P. (Zh. Fiz. Khim. **48** [1974] 542/5; Russ. J. Phys. Chem. **48** [1974] 317/9; C.A. **81** [1974] No. 30526).

[19] Buschow, K. H. J.; van Daal, H. J. (AIP Conf. Proc. No. 5, Pt. 2 [1972] 1464/77; C.A. **77** [1972] No. 11324).

[20] Rivier, N.; Zlatic, V. (J. Phys. F **2** [1972] L87/L92).

[21] Rivier, N.; Zlatic, V. (J. Phys. F **2** [1972] L99/L104).

[22] Brodsky, M. B. (Phys. Rev. [3] B **9** [1974] 1381/7).

[23] Kaiser, A. B.; Doinach, S. (Intern. J. Magn. **1** [1970] 11/22).

[24] van Daal, H. J.; Buschow, K. H. J.; van Aken, P. B. (Thermoelectr. Met. Conduct. Proc. 1st Intern. Conf., East Lansing, Mich., 1977 [1978], pp. 107/15; C.A. **92** [1980] No. 32756).

[25] Lin, C. L.; Zhou, L. W.; Crow, J. E.; Mihalisin, T.; Brooks, J.; Guertin, R. P. (J. Less-Common Metals **127** [1987] 273/9).

3.3.3   Ternary Uranium — Indium Compounds

3.3.3.1   U — In — Sn

The ternary alloys $U(In_{1-x}Sn_x)_3$ (with $0 \leq x \leq 1$) form a complete series of solid solutions, which were investigated for electrical and magnetic properties [1 to 4].

The specific heat of $U(In_{1-x}Sn_x)_3$ compounds with $0.5 \leq x \leq 1$ shows a large enhancement of the C/T values for compositions in the vicinity of $x = 0.6$, reaching 530 mJ · mol^{-1} · K^{-2} at 1.35 K. The coefficient γ of the electronic contribution decreases with decreasing values for x prior to the onset of long range magnetic ordering. A maximum of the specific heat was observed at 1.4 K with a sample of $x = 0.6$. This maximum vanishes in applied magnetic fields of 5 and 10 T, which is not connected with magnetic ordering. The results indicate a very narrow f-band formation near the Fermi level, i.e., enhanced heavy fermion behavior within the paramagnetic region [1], see also [2 to 4].

The electric resistivity of alloys with $x \leq 0.4$ is temperature-independent for temperatures below T_N, but at temperatures about the Néel temperature an anomaly is observed [1, 3].

$U(In_{1-x}Sn_x)_3$ alloys with $x \leq 0.4$ are antiferromagnets. Alloys with tin concentrations of $x > 0.5$ are paramagnets. The field dependence of the magnetization, which is linear for USn_3 up to 9 T, was also observed for the antiferromagnetic $U(In_{1-x}Sn_x)_3$ samples with $x \leq 0.4$ [2].

For more details on $U(In_{1-x}Sn_x)_3$ alloys see Section 4.1.10.2, p. 308.

References for 3.3.3.1:

[1] Lin, C. L.; Zhou, L. W.; Crow, J. E.; Guertin, R. P.; Stewart, G. R. (J. Magn. Magn. Mater. **54/57** [1986] 391/2).

[2] Zhou, L. W.; Lin, C. L.; Crow, J. E.; Bloom, S.; Guertin, R. P.; Foner, S. (Phys. Rev. [3] B **34** [1986] 483/6).

[3] Lin, C. L.; Zhou, L. W.; Jee, C. S.; Wallash, A.; Crow, J. E. (J. Less-Common Metals **133**
     [1987] 67/75).
[4] Lin, C. L.; Zhou, L. W.; Crow, J. E.; Mihalisin, T.; Brooks, J.; Guertin, R. P. (J. Less-
     Common Metals **127** [1987] 273/9).

3.3.3.2 U−In−Ni

Ternary alloys of UIn_xNi_{5-x}, crystallizing with UNi_5-type structure, are solid solutions of
$UNi_5 - UNi_4In$. The alloys were investigated for crystallographic and magnetic behavior [1, 2].

Preparation

Samples of UIn_xNi_{5-x} were prepared by arc melting of stoichiometric amounts of the
elements in an argon atmosphere, followed by an annealing step at 800°C for 700 h in an
evacuated silica tube for homogenization, and then slowly cooled to room temperature with
a rate of 50°C per day [1, 2].

As or Sb were substituted for In in $UInNi_4$, forming alloys of compositions $U(In_{0.5}As_{0.5})Ni_4$
or $U(In_{0.5}Sb_{0.5})Ni_4$ [1, 2].

Crystallographic Properties

The UIn_xNi_{5-x} and $U(In_{0.5}As_{0.5})Ni_4$ or $U(In_{0.5}Sb_{0.5})Ni_4$ alloys crystallize in cubic UNi_5-type
structure with $Z = 4$; the space group is $F\bar{4}3m - T_d^2$ (No. 216) [1, 2]. Measured lattice parameters
are summarized together with measured values of the Vickers hardness in Table 3/70. The
lattice parameters obey Végard's law. The crystal structure is similar to that of $USnNi_4$, which
is shown in Fig. 4-15, p. 314. It is considered to be a superlattice structure of the $MgSnCu_4$
type (space group $F\bar{4}3m - T_d^2$ (No. 216)) [1]. The atomic positions are: origin at $\bar{4}3m$; equivalent
positions: (0,0,0; 1/2,1/2,1/2; 1/2,0,1/2; 1/2,1/2,0) + 4 U in 4(a): 0,0,0. 4 In in 4(c): 1/4,1/4,1/4.
16 Ni in 16(e): x,x,x; x,$\bar{x}$,$\bar{x}$; $\bar{x}$,x,$\bar{x}$; $\bar{x}$,$\bar{x}$,x with x = 5/8 [1].

Table 3/70
Measured Lattice Parameters and Vickers Hardness H_v for UIn_xNi_{5-x} Alloys [1], $U(In_{0.5}As_{0.5})$-
Ni_4, and $U(In_{0.5}Sb_{0.5})Ni_4$ Alloys [2].

composition	lattice parameter a (in pm) ± 0.5	Vickers Hardness H_v in kg/mm^2
$UIn_{0.2}Ni_{4.8}$	680.7	587
$UIn_{0.4}Ni_{4.6}$	682.4	596
$UIn_{0.6}Ni_{4.4}$	684.9	638
$UIn_{0.8}Ni_{4.2}$	687.0	658
$UInNi_4$	689.0	708
$U(In_{0.5}As_{0.5})Ni_4$	6.810 Å	
$U(In_{0.5}Sb_{0.5})Ni_4$	6.884 Å	

Magnetic Susceptibility

As observed from magnetic susceptibility measurements in the temperature range 87 to
290 K, the UIn_xNi_{5-x}, $U(In_{0.5}As_{0.5})Ni_4$, and $U(In_{0.5}Sb_{0.5})Ni_4$ alloys are antiferromagnets [1].

$U(In_{0.5}As_{0.5})Ni_4$ and $U(In_{0.5}Sb_{0.5})Ni_4$ follow a Curie-Weiss law $\chi = C/(T - \Theta_p)$ with the parameters: Curie temperature $\Theta_p = 153$ K and 208 K, respectively, effective moment $\mu_{eff} = 2.56\ \mu_B$ and $2.64\ \mu_B$, respectively [2].

References for 3.3.3.2:

[1] Blazina, Z.; Drasner, A.; Ban, Z. (J. Nucl. Mater. **96** [1981] 141/6).
[2] Blazina, Z.; Ban, Z.; Szytula, A. (Fizika [Zagreb] **12** [1980] Suppl. 1, pp. 200/4; C.A. **95** [1981] No. 34376).

3.3.3.3 U — In — Pd. U — In — Pt

The ternary alloys UInPd and UInPt, crystallizing with hexagonal Fe_2P-type structure, were investigated for crystallographic and magnetic properties [1 to 4].

Preparation

Samples of UInPd and UInPt were prepared by arc melting of stoichiometric amounts of the elements in a protective atmosphere (He and/or Ar), followed by an annealing step for homogenization at 700 to 900 °C [1].

Crystallographic Properties

UInPd and UInPt crystallize with hexagonal Fe_2P-type structure with $Z = 3$; the space group is $P\bar{6}2m - D_{3h}^3$ (No. 189) [1, 3]. The measured lattice parameters are for UInPd: $a = 7.414 \pm 0.001$ Å, $c = 4.0963 \pm 0.0009$ Å, $c/a = 0.553$ [1], see also [2, 3], or $a = 0.7421$ nm, $c = 0.4093$ nm [4]; for UInPt: $a = 7.4134 \pm 0.0004$ Å, $c = 4.0581 \pm 0.0004$ Å, $c/a = 0.547$ [1], see also [2, 3], or $a = 0.7405$ nm, $c = 0.4055$ nm [4]. The atomic positions are: In in 3 (f) (mm): $x, 0, 0$; $0, x, 0$; $\bar{x}, \bar{x}, 0$. U in 3 (g) (mm): $x, 0, 1/2$; $0, x, 1/2$; $\bar{x}, \bar{x}, 1/2$. Pd (Pt) in 1 (b) ($\bar{6}mz$): $0, 0, 1/2$. Pd (Pt) in 2 (c) ($\bar{6}$): $\pm\ 1/3, 2/3, 0$ [1,3].

Magnetic Properties

UInPd and UInPt are ferromagnetically ordered below the Curie temperatures of $T_c = 22$ K and about 30 K, respectively. The spontaneous moment at 4.2 K, measured at polycrystalline samples, is $0.15\ \mu_B$ for UInPd and $0.25\ \mu_B$ for UInPt. A coercive field of $H_c = 400$ kA/m was observed for UInPt from the magnetic measurements [4].

References for 3.3.3.3:

[1] Dwight, A. E.; Mueller, M. H.; Conner, R. A., Jr.; Downey, J. W.; Knott, H. (Trans. AIME **242** [1968] 2075/80).
[2] Dwight, A. E. (in: Giessen, B. C., Developments in the Structural Chemistry of Alloy Phases, Plenum, New York 1969, pp. 181/226).
[3] Lam, D. J.; Darby, J. B., Jr.; Nevitt, N. V. (in: Freeman, A. J.; Darby, J. B., Jr., The Actinides: Electronic Structure and Related Properties, Vol. 2, Academic, New York — San Francisco — London 1974, pp. 119/84).
[4] Andreev, A. V.; Bartashevich, M. I. (Fiz. Metal. Metalloved. **62** [1986] 266/8; Phys. Metals Metallog. [USSR] **62** No. 2 [1986] 50/3; C.A. **105** [1986] No. 182534).

3.4 Uranium — Thallium

3.4.1 Phase Relationships

UTl$_3$ is the only compound known to exist in the binary system U — In. It was first mentioned by [1] and crystallizes with simple cubic AuCu$_3$-type structure, isomorphous with UAl$_3$, USn$_3$, UGa$_3$, USi$_3$, UGe$_3$, or UPb$_3$ [1, 2], see also [3].

The solubility of U in Tl is very restricted; measured values are summarized in Table 3/71, see Fig. 3-120, p. 247. The solubility of U in liquid Tl was expressed by the empirical equation log (at% U) = 4.184 − 5728 · T^{-1} [4].

Table 3/71
Measured Liquidus Data for the System U — Tl [4], see also [2, 5 to 7].
Values from [8] are separately indicated.

temperature in °C	wt% U	temperature in °C	wt% U
890	0.16 [8]	574	3.1×10^{-3}
854	0.155	554	2.6×10^{-3}
779	0.11 [8]	509	9.8×10^{-4}
722	3.1×10^{-2}	496	5.2×10^{-4}
643	8.5×10^{-3}	453	2.4×10^{-4}
630	7.8×10^{-3}	400	5.4×10^{-5}

References for 3.4.1:

[1] Iandelli, A.; Ferro, R. (Ann. Chim. [Rome] **42** [1952] 598/608; C.A. **1953** 3165).
[2] Rough, F. A.; Bauer, A. A. (BMI-1300 [1958] 1/138; N.S.A. **12** [1958] No. 13935).
[3] Hansen, M. (Constitution of Binary Alloys, McGraw-Hill, New York — Toronto — London 1958).
[4] Johnson, I.; Chasanov, M. G. (ASM [Am. Soc. Metals] Trans. Quart. **56** [1963] 272/7).
[5] Shunk, F. A. (Constitution of Binary Alloys, 2nd Suppl., McGraw-Hill, New York — Toronto — London 1969).
[6] Hayes, E. E.; Gordon, P. (TID-65 [1948] 130/41 from [2]).
[7] Elliott, R. P. (Constitution of Binary Alloys, 1st Suppl., McGraw-Hill, New York — Toronto — London 1965).
[8] Hayes, E. E.; Gordon, P. (TID-2501 [1957] 115; N.S.A. **12** [1958] No. 17280).

3.4.2 Uranium Trithallide, UTl$_3$

Preparation

Samples of UTl$_3$ were prepared by reaction of stoichiometric amounts of the powdered elements at 600°C within 100 to 150 h in an evacuated silica tube [1, 2], or annealed in an inert atmosphere (argon) [3, 4]. After the reaction was completed the samples were slowly cooled to room temperature within 24 h [1] or 70 h [2]. UTl$_3$ was also prepared by usual melting techniques [5].

Thermodynamic Data of Formation

Thermodynamic functions for the formation of UTl$_3$ according to U(s) + 3 Tl(l) → Tl(s) at 385 to 673°C were derived from emf measurements: ΔG_f° (in kcal/mol) = −12.00 + 8.498 × 10^{-3} · T + 2.208 × 10^{-6} · T^2, ΔH_f° (in kcal/mol) = −12.00 − 2.208 × 10^{-6} · T^2, ΔS_f° (in cal · mol^{-1} · K^{-1}) = 8.498 − 4.416 × 10^{-3} · T [6], see also [11]. From emf measurements (molten salt electrolytes), the following thermodynamic values were derived for the formation of UTl$_3$ at 450°C: ΔH_f° = −3.29 kcal/g-atom, ΔG_f° = −1.18 kcal/g-atom, ΔS_f° = −2.92 cal · g-atom^{-1} · K^{-1} [6]. The enthalpy of formation at room temperature, ΔH_f = −3.6 kcal/g-atom at 298 K was determined by dynamic differential calorimetry (DDC) [4].

Crystallographic Properties

UTl$_3$ crystallizes with simple cubic AuCu$_3$-type structure (L1$_2$) with Z = 1; the space group is Pm$\bar{3}$m − O$_h^1$ (No. 221) [7], see also [3]. Measured lattice parameters are: a = 4.675 Å [3], see also [7 to 10], or a = 4.688 ± 0.002 Å [1], see also [2].

The atomic positions are: 1 U in: 0,0,0. 3 Tl in: 0,1/2,1/2; 1/2,0,1/2; 1/2,1/2,0 [3].

Each U atom is bonded to 12 Tl atoms and each Tl atom is bonded to 4 U atoms and 8 Tl atoms with the distance a/$\sqrt{2}$ = 3.31 Å [3], or 3.315 Å [1].

Magnetic Properties

UTl$_3$ is an antiferromagnet with a transition temperature of T$_N$ = 90 K [2] or 35 K [12]. The magnetic structure, which was observed from a neutron diffraction study, shows a magnetic cell which is doubled in three directions, indicating antiferromagnetic ordering with antiparallel spins on adjacent (111) planes (compare Fig. 3-143, p. 289, showing the magnetic structure of UIn$_3$). The magnetic moment of UTl$_3$ was derived from the intensity of the M(111) and M(311) magnetic reflections to be μ = 1.60 ± 0.09 μ_B at 4.2 K [2]. A value for the Néel temperature was derived from the temperature dependence of the intensity of the M(111) reflection to be T = 90 ± 5 K [2].

The temperature dependence of the magnetic susceptibility of UTl$_3$ is graphically shown in [1]. Single values of the susceptibility are χ = 4750 × 10^{-6} emu/mol at 4.2 K, χ = 3600 × 10^{-6} emu/mol at 290 K.

The Curie-Weiss law is approximately obeyed above 200 K, from which an effective magnetic moment of μ_{eff} = 3.6 μ_B was derived. The broad maximum of the susceptibility at 80 to 90 K is not quite typical for antiferromagnetic transition. Thus, the observed maximum was discussed in terms of a paramagnetic susceptibility maximum model [1].

References for 3.4.2:

[1] Misiuk, A.; Mulak, J.; Czopnik, A. (Bull. Acad. Polon. Sci. Ser. Sci. Chim. **20** [1972] 891/6).

[2] Murasik, A.; Leciejewicz, J.; Ligenza, S.; Misiuk, A. (Phys. Status Solidi A **20** [1973] 395/401).

[3] Iandelli, A.; Ferro, R. (Ann. Chim. [Rome] **42** [1952] 598/608; C.A. **1953** 3165).

[4] Palenzona, A.; Cirafici, S. (Thermochim. Acta **13** [1975] 357/60).

[5] Frost, B. R. T.; Maskrey, J. T. (J. Inst. Metals **82** [1953/54] 171/80; AERE-M-R-1027 [1952] 1/27; N.S.A. **7** [1953] No. 1547).

[6] Johnson, I.; Feder, H. M. (Thermodyn. Nucl. Mater. Proc. Symp., Vienna 1962 [1963], pp. 319/28; C.A. **62** [1965] 15498).

[7] Rough, F. A.; Bauer, A. A. (BMI-1300 [1958] 1/138; N.S.A. **12** [1958] No. 13935).

[8] Hansen, M. (Constitution of Binary Alloys, McGraw-Hill, New York — Toronto — London 1958).

[9] Lam, D. J.; Darby, J. B., Jr.; Nevitt, N. V. (in: Freeman, A. J.; Darby, J. B., Jr., The Actinides: Electronic Structure and Related Properties, Vol. 2, Academic, New York — San Francisco — London 1974, pp. 119/84).

[10] Dwight, A. E. (in: Giessen, B. C., Developments in the Structural Chemistry of Alloy Phases, Plenum, New York 1969).

[11] Johnson, I. (Met. Soc. AIME Inst. Metals Div. Spec. Rept. Ser. **10** No. 13 [1964] 171/92).

[12] Lin, C. L.; Zhou, L. W.; Crow, J. E.; Mihalisin, T.; Brooks, J.; Guertin, R. P. (J. Less-Common Metals **127** [1987] 273/9).

4 With Metals of the 4th Main Group

Hans Ulrich Borgstedt,
Institut für Material- und Festkörperforschung,
Kernforschungszentrum Karlsruhe,
Karlsruhe, Federal Republic of Germany

4.1 Uranium – Tin

4.1.1 General

The alloying of U with Sn is favored by energy effects as is shown by means of a cellular model, which relates these effects to the atomic cells of the pure constituents [1]. The alloying effects originate from the change in boundary conditions when an atom is transferred from the pure metal to the alloy. The energy effects are determined by two terms. The first one represents the difference in the electronegativity between the two types of atoms in an alloy. The second one reflects the discontinuity in the density of electrons at the boundary between dissimilar Wigner-Seitz atomic cells. The electronegativity is similar to the experimental work functions of pure metallic elements. The heat of solution, $\Delta \overline{H}^\circ_M$, of U in liquid Sn is calculated using a model [2]. The value of $\Delta \overline{H}^\circ_M = -13$ kcal/g-atom indicates that there is a clear tendency to form alloys in the liquid state. The same authors applied the model to predict the alloying behavior in the solid state [3]. They found that the model predictions of values for the heat of formation agree fairly well with much recent experimental data of intermetallic compounds, including some U compounds [4]. The model is able to predict the thermodynamic stability of several equiatomic compounds of a transition metal and Sn [5]. Tables for a variety of transition and nontransition elements have been established by applying a computer program in Algol 60. The value for the infinitely dilute solution of U in Sn is $\Delta \overline{H}^\circ = -99$ kJ/g-atom (solute) and $\Delta \overline{H}^\circ = -84$ kJ/g-atom for Sn in liquid U [4, 6].

Thus, the U — Sn alloys and compounds are analogous to alloys and intermetallics generally formed in the systems of U and group III and IV metals or elements. With some of them there is a more or less complete miscibility [7].

4.1.2 Preparation of the U – Sn Alloys

U — Sn alloys were prepared from the pure components of not less than 99.9% purity. Mixtures were melted under argon cover gas in an induction furnace. The alloys prepared in this way must be annealed for 200 h at 500°C and then quenched [19]. Alloys with low U contents were annealed at 250°C and slowly cooled to room temperature. The same alloys were prepared in dry argon glove boxes with drops of metallic Sn of 99.999% purity and U turnings of 99.94% purity by means of arc melting. Graphite crucibles were used for the melting procedure [20].

U — Sn alloys were also prepared for physical measurements in small BeO crucibles by fusing ultrapure Sn with 99.6% pure U in the desired weight relations in a pure Ar atmosphere [68]. In another example, mixtures of U (99.94% purity) with Sn of highest purity (99.999%) were treated in an arc melting process or in an induction-heated graphite crucible [26].

The compound USn_3 was obtained by direct synthesis from powdered elements of unspecified purity at 900°C in vacuum, followed by the homogenization at the same temperature for 100 h [41].

References for 4.1 on pp. 316/9

4.1.3 Phase Diagram

4.1.3.1 General

The U — Sn phase diagram is based on the results of metallographic examination, thermal analysis, and X-ray diffraction. It was first published in a compilation of data by Saller, Rough [8], and this phase diagram was incorporated into Hansen, Anderko [9]. It is also the basis for more recent phase-diagram compilations such as those of Rough, Bauer [69], Chiotti et al. [42], Ivanov et al. [65], and Massalski [37].

4.1.3.2 Phase Relations

The phase diagram published in [8] and compiled in [9, 37, 42, 65, 69] is shown in **Fig. 4-1**. The diagram is based on experimental work of Wilhelm, Treick [10]. The compounds which exist in the U — Sn system are highly brittle and pyrophoric. Their accurate composition is, therefore, difficult to determine [10].

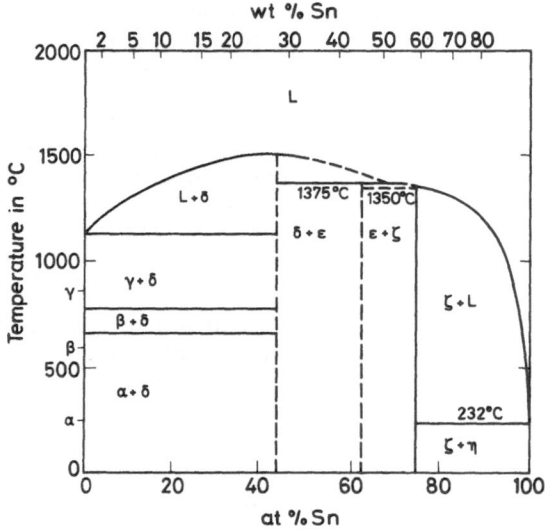

Fig. 4-1. The U — Sn phase diagram according to [8].

The solubility of Sn in solid α-U (and β-U) is negligible, as should be the solubility of U in solid Sn. Several sources give information on the solubility of U in liquid Sn. A saturation concentration $c_{sat} < 0.02$ at% U at 600°C is reported in [11]. Another early source presented values of <0.01 wt% or <0.005 at% U at the same temperature [12]. The solubility is expressed by the equation $\ln x_U = 6.693 - 13550/T$, where x_U is the mole fraction of U, and T is the temperature in K [42, 67]. Some single values are listed in [68]:

temperature in K	700	800	900	1000
concentration in at-fraction U	3.1×10^{-4}	3.0×10^{-3}	1.7×10^{-2}	7.1×10^{-2}

The equation, $\log x_U = 2.383 - 5530/T$, fits to the listed solubilities, but there is only poor agreement with the previously given equation [68]. The activity coefficient of U in liquid Sn is given as $\log \gamma_U = 1.9713 - 4555/T$ [67].

Three compounds occur in the U−Sn system. U_5Sn_4 melts congruently at 1500°C and U_3Sn_5 peritectically at 1380°C. The third compound USn_3 seems to melt peritectically at 1350°C [12]. X-ray diffraction studies gave useful results only for USn_3 [13]. Very broad X-ray reflections of the other compounds indicate high stresses in the crystals. The compounds with 44.4 and 62 at% Sn are U_5Sn_4 and U_3Sn_5, respectively. Alcock, Grieveson [14] formulate the first of these two compounds with a composition of between 36 and 45 at% Sn as U_3Sn_2 from measurements of the dissociation pressures of the U−Sn alloys.

U is completely soluble in liquid Sn at temperatures above 1670 K. Metallographic examinations of specimens with more than 75 at% Sn show dendritic precipitates of the USn_3 intermetallic in a Sn matrix. The U content was shown by autoradiographic techniques to be completely in the dendrites [26]. Electron microprobe analyses of the U−Sn system with concentrations of more than 40 at% Sn identify the compounds USn_3, U_3Sn_7, USn_2, U_4Sn_5, USn, and U_5Sn_3. Some of these compounds are not included in the phase diagram. The compositions of these compounds are shown in Table 4/1.

Table 4/1
Theoretical and Observed U Contents of U−Sn Compounds [26].

U−Sn compound	theoretical U content in at%	observed U content in at%
USn_3	25	24.9 ± 0.3
U_3Sn_7	30	29.5 ± 0.5
USn_2	33.3	33.8 ± 0.3
U_4Sn_5	44.4	44.4 ± 0.5
USn	50	50.7 ± 0.6
U_5Sn_3	62.5	62.0 ± 0.3

These results indicate that the phase diagram shown in Fig. 4-1 may still be incomplete [26].

4.1.4 Crystallographic Properties

USn_3 has the $L1_2$ structure and the space group $Pm3m - O_h^1$ (No. 221) in analogy with the isomorphous $AuCu_3$ compound. The simple face-centered cubic structure with the lattice constant a = 4.62 Å has a unit cell with one U atom and three Sn atoms in the positions U at 0,0,0 and 3 Sn at 0,1/2,1/2; 1/2,0,1/2; 1/2,1/2,0 [13].

The lattice constant a = 4.626 Å is reported in [15]. The U−Sn distance is 3.00 Å. The lattice constant, redetermined by X-ray diffraction, has range between the limits 4.603 to 4.609 Å [36]; a = 4.6089 ± 0.0023 Å [26]. The higher constant published by Frost, Maskrey [15] may be due to impurity effects.

Intensities of weak superlattice reflections were measured in a series of compounds UM_3 (M = Ru, Rh, Ge, and Sn) by means of neutron diffraction. Excellent fits to the observed

References for 4.1 on pp. 316/9

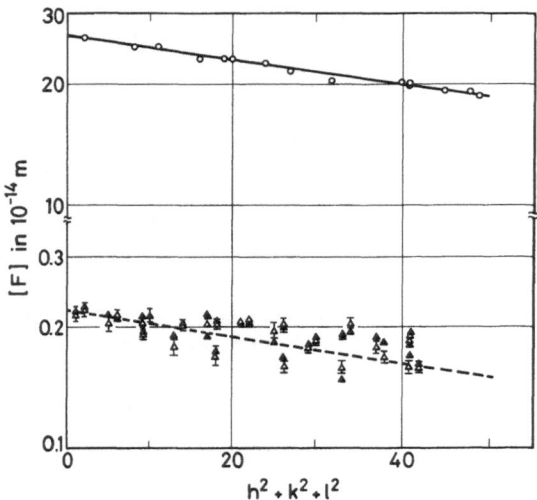

Fig. 4-2. Observed structure factor, [F], (open points) and calculated values (solid line for fundamental and closed triangles for superlattice) for USn_3 as a function of $h^2 + k^2 + l^2$. The dashed line indicates the expected values for the superlattice reflections if the thermal motion of the Sn atom is isotropic [28].

intensities were obtained by introducing anisotropic second-order and fourth-order (anharmonic) terms to describe the probability distribution at the Sn site. The ratio of the magnitude of the mean-square thermal vibration parallel and perpendicular to the unique tetragonal axis of the M atoms appears to depend on whether p bonding (UGe_3, USn_3) or d bonding (URu_3, URh_3) occurs with the U electrons. Values of structural dimensions of USn_3, evaluated from such measurements, are: lattice constant a = 4.626 Å, nuclear scattering potential (related to the structure factor) $b_{Sn} = 0.622 \times 10^{-14}$ m, temperature factor $B_U = 0.71 \pm 0.03$ Å^2. The observed structure factor of USn_3 as a function of $h^2 + k^2 + l^2$ is shown in **Fig.** 4-2 [28].

The compound of the composition U_2Sn is claimed to have the crystal structure of Cu_2Sb. The lattice constants for U_2Sn are a = 4.470 Å and c = 8.933 Å [58]. This compound is not indicated in any of the phase diagrams and its existence is doubtful.

4.1.5 Thermodynamic Data

Free energies of formation for three U — Sn compounds have been estimated from dissociation pressures measured by a Knudsen-cell technique [14]. The results, combined with the free enthalpy data for the $\alpha \rightarrow \beta$ transformation of the solid U, are given by the following equations [42]:

$\Delta G° = -32000 + 1.25 \cdot T$ (in cal)   for $3\,U(\alpha) + 2\,Sn(s) \rightarrow U_3Sn_2(s)$
$\Delta G° = -52000 + 2.24 \cdot T$ (in cal)   for $3\,U(\alpha) + 5\,Sn(s) \rightarrow U_3Sn_5(s)$
$\Delta G° = -23000 + 0.60 \cdot T$ (in cal)   for $U(\alpha) + 3\,Sn(s) \rightarrow USn_3(s)$

The heat of formation for USn_3 was estimated by means of aqueous solution calorimetry. The mean value of $\Delta G° = -5400$ cal/g-mol is in agreement with the value from the vapor-pressure measurements [14]. The enthalpy of formation of USn_3 at 298 K, derived from solution-calorimetric measurements, is $\Delta H = -34.95$ kJ/mol [72].

Free energies, enthalpies, and entropies of formation of USn_3 according to $U(s) + 3Sn(l) \rightleftarrows USn_3(s)$ have been determined by measurement of the electromotive force (emf) of high-temperature galvanic cells (360 to 680°C) in which USn_3 is formed from solid U and liquid Sn. A molten salt mixture was used as electrolyte. The applied galvanic cell is of the type $U // UCl_3 + KCl - LiCl$ (eutectic) $// U - Sn$ (2 phase alloy) in which the alloys are mixtures of the saturated liquids and the corresponding intermetallic compounds [16]. For experimental details see [18].

The free energy of formation of USn_3 according to $U(s) + 3 Sn(l) \rightleftarrows USn_3(s)$ is given by the equation $\Delta G_f°$ (in kcal/mol) $= a + bT + cT^2$ with $a = -39.18$, $b \times 10^3 = 9.502$, $c \times 10^6 = 3.062$ in the temperature range from 363 to 677°C [16].

The enthalpy of formation is $\Delta H_f°$ (in kcal/mol) $= a + bT^2$ with the coefficients $a = -39.18$ and $b = 3.062 \times 10^{-6}$. The entropy of formation is given by $\Delta S_f°$ (in cal $\cdot$ mol$^{-1} \cdot$ K^{-1}) $= a + bT$. The coefficients are $a = -9.502$ and $b = -6.124 \times 10^{-3}$. At 450°C, the values for USn_3 are $\Delta G_f° = -7.68$ kcal/mol, $\Delta H_f° = -10.20$ kcal/mol, $\Delta S_f° = -3.48$ cal $\cdot$ mol$^{-1} \cdot$ K^{-1} [16].

The estimated free energy of formation from the dissociation pressures of $U - Sn$ alloys indicates the compound U_3Sn_2 to be more likely than U_5Sn_4, as postulated by Rundle, Wilson [13], see [16].

The excess partial molal free energy $\overline{G}_U^{xs} = \overline{L}_U - T \cdot \overline{S}_U^{xs}$ of U for saturated solutions of U in Sn is determined by $\overline{L}_U = -13.8$ kcal/mol and $\overline{S}_U^{xs} = -0.7$ cal $\cdot$ K$^{-1} \cdot$ mol^{-1}. $\overline{L}_U$ is the partial molal free energy of solution, $\overline{S}_U^{xs}$ is the entropy [16]. The excess free energy (solid U reference state) in saturated solutions of U in Sn is given by the equation $\overline{G}_U^{xs} = -14.1 + 1 \times 10^{-1}$ T kcal/mol. Thus, at 500°C the activity coefficient of U in Sn is approximately 1.3×10^{-3} [27].

Chemical activities of U in the alloys are deduced from the equation $\Delta G° = +35350 - 20.0$ T. They are slightly decreased owing to the formation of USn_3 in the tin-rich alloys [38, 39].

Johnson [17] pointed out that the two thermodynamic studies differ by almost 10 kcal/mol at 1000 K. The emf data yield the more negative free energy of formation. The differences have not been resolved by the two groups of workers, [14] and [16].

It might be that a low-temperature eutectic at the U-rich side of the $U - Sn$ phase diagram influences the interpretation of the effusion results in the $U_3Sn_2 - U$ region. Such a eutectic is likely since it appears in several phase diagrams of U-low-melting B subgroup metals. Therefore, equations were computed for the free energy of formation of the three intermediate phases in the $U - Sn$ system based on the assumptions that the emf data may be correct for USn_3 and the effusion data for the $USn_3 - U_3Sn_5$ and $U_3Sn_5 - U_3Sn_2$ regions:

USn_3: $\Delta G_f°$ (in kcal/mol) $= -41.03 + 14.03 \times 10^{-3} \cdot$ T
U_3Sn_5: $\Delta G_f°$ (in kcal/mol) $= -99.41 + 30.12 \times 10^{-3} \cdot$ T
U_3Sn_2: $\Delta G_f°$ (in kcal/mol) $= -74.40 + 19.20 \times 10^{-3} \cdot$ T

Recommended data for 298.15 K are $\Delta H_f° = -21.6$ kcal/mol, $\Delta G_f° = -21.6$ kcal/mol, $S° = 49.0$ cal $\cdot$ mol$^{-1} \cdot$ K^{-1} [18], see also [19].

The heat of evaporation of USn_3 is $\Delta H_{evap}° = 72.0$ kcal/g-atom [25].

References for 4.1 on pp. 316/9

The assessment in [71] recommends the following values of entropy and enthalpy (Table 4/2):

Table 4/2
Recommended Values of Entropy (in $J \cdot mol^{-1} \cdot K^{-1}$) and Enthalpy (in kJ/mol) [71].

	U_3Sn_2	U_3Sn_5	USn_3
S_{298}	247.0	397.5	201.5
H_{298}	-136.3	-217.8	-96.6

The stability of UM_3 compounds with group IVB metals decreases with the atomic weight of M. The IVB-element compounds are more stable than the IIIB-element compounds [16].

4.1.6 Mechanical and Thermal Properties

4.1.6.1 Density

The density of USn_3 estimated from X-ray diffraction measurement is $D = 9.95$ g/cm^3 [13]; another value based on X-ray measurement is $D = 10.00$ g/cm^3 [15]; compilations of data in [8, 11] support the first mentioned value.

The density of USn and the other compounds have not yet been measured.

4.1.6.2 Hardness

The average microhardness of U — Sn compounds as measured by [26] increases with decreasing tin content, the highest value being observed with the composition USn, see Table 4/3.

4.1.6.3 Melting Temperature

The melting temperatures of the U — Sn compounds based on the phase diagram are given in [12]. U_5Sn_4 melts congruently at 1500°C, U_3Sn_5 peritectically at 1380°C, and USn_3 peritectically at 1350°C [12]. Katz, Rabinowitch [11] report the same value of the peritectic melting point of USn_3, which is also confirmed by more recent measurements [26].

4.1.6.4 Specific Heat

The electronic specific heat coefficient of USn_3 is $\gamma = 16.9$ mJ $\cdot$ mol$^{-1} \cdot$ K^{-2}, and the lattice specific heat coefficient is $\alpha = 1.04$ mJ $\cdot$ mol$^{-1} \cdot$ K^{-4}. The heat capacity C/T versus T^2 shows a straight line corresponding to $C = T + \alpha \cdot T^3$ [30]. These data were confirmed in a later study [53, 74].

Table 4/3
Microhardness of U—Sn Compounds [26].

compound	average hardness HV in kg/mm^2
U	210
Sn	9
USn$_3$	124
U$_3$Sn$_7$	195
USn$_2$	210
U$_4$Sn$_5$	290
USn	295

4.1.6.5 Thermal Conductivity

The thermal conductivity of U—Sn compounds as USn, USn$_3$, and alloys of undefined compositions have not yet been measured owing to the brittle and chemically reactive character of these materials [13].

4.1.7 Electrical Properties of USn$_3$

A survey of the low-temperature resistivity of UM$_3$ compounds including USn$_3$ is given in [56]. USn$_3$ shows a ϱ versus T behavior which conforms with Doniach's localized spin-fluctuation model. Below $T_1 = 10$ K the resistivity ϱ is proportional to T^2, above $T = 40$ K proportional to T. **Fig. 4-3** shows that the temperature dependence of ϱ for USn$_3$ differs from that of the other cubic UM$_3$ compounds (M = Al, Ga, In, Si, and Ge). The plots of $\Delta\varrho$ versus T in logarithmic scale can be used to define the two regimes of temperature dependence [32].

The resistivity of USn$_3$ at room temperature is in the order $\varrho \approx 55$ $\mu\Omega \cdot$ cm [32].

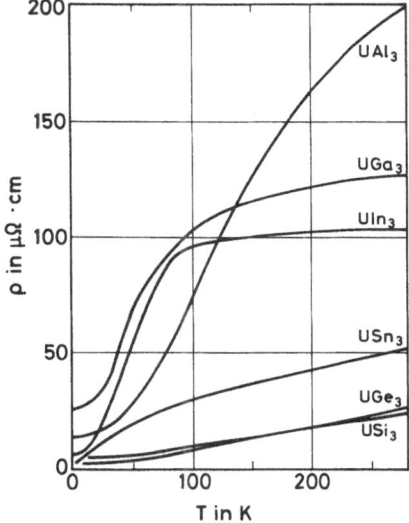

Fig. 4-3. Temperature dependence of the resistivity of the cubic compounds UM$_3$ below room temperature (M = Al, Ga, In, Si, Ge, and Sn) [32].

References for 4.1 on pp. 316/9

4.1.8 Magnetic Properties of USn_3

The literature on the magnetic susceptibility of USn_3 at low temperatures was reviewed in [23]. The susceptibility of USn_3 between 100 and 400 K, measured using a Gouy balance, is shown in **Fig. 4-4**. The susceptibility does not obey a Curie-Weiss law [31]. The magnetic susceptibility of the compound is linearly related to the ^{119}Sn Knight shift. The relationship differs from that of the rare earth compounds RSn_3 in that the K versus χ_{mol} curve intersects the χ-axis at a positive value when extrapolated to 0 K. The properties of USn_3 are related to the large U — U distance in the lattice. A survey is given in [61].

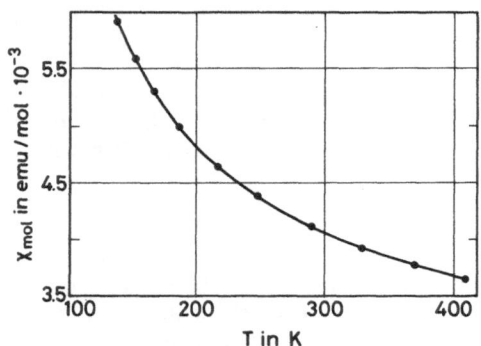

Fig. 4-4. Susceptibility of USn_3 in dependence of the temperature [31].

The susceptibility of USn_3 in the temperature range 80 to 900 K is discussed on the basis of the localization of the 5f electrons of U [29]. Susceptibility values between 10 and 900 K are given in [58]. The reciprocals χ_{mol}^{-1} from these values show a similar temperature dependence as that of the isotypic UPb_3 (see p. 327). The contributions of the conduction electrons, as well as of completed shells, to χ are negligible in these estimations [41].

Magnetic measurements on USn_3 were extended down to 4.2 K. **Fig. 4-5** exhibits the temperature function of χ_{mol}^{-1} for USn_3. The gradient becomes flat when the temperature falls below 20 K, indicating a nonmagnetic ground level. The results can be related to the negative ligand charge model. χ_{mol} reaches a value of 9.5×10^{-3} at 4.2 K. USn_3 does not show a magnetic transition at low temperatures above 4.2 K [41].

The flat gradient of χ_{mol}^{-1} versus T below 20 K is also related to a localized spin-fluctuation model, however, USn_3 was the only one of the cubic UM_3 compounds tested (M = Al, Ga, In, Si, Ge, and Sn) which obeys this relationship [32].

A systematic of the localization of f electrons in relation to the heavy-fermion behavior of UM_3 compounds (M = Ge, Sn, Pb, Tl, Ir, Rh, and Pd) is based upon the character of the f-ligand hybridization. Thus, the hybridization may produce "wide-band" transition metal-like behavior (e.g., UGe_3, UIr_3, URh_3), narrow-band behavior (USn_3), and local-moment behavior (UPd_3, UPb_3, UTl_3). These compounds are characterized by very large electronic specific heats corresponding to electronic specific masses, which are several hundred times the free electron mass [40].

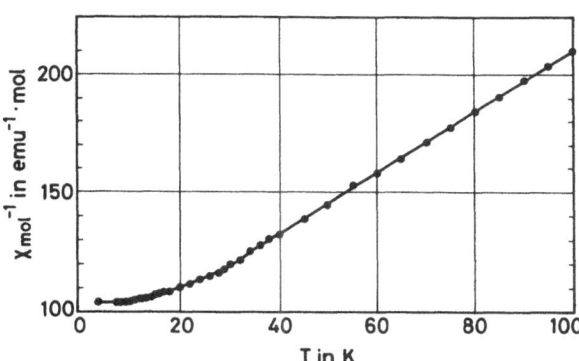

Fig. 4-5. Reciprocal susceptibility of USn₃ as a function of the temperature [41], using values
of [58].

The very large electronic specific heat of $\gamma = 169 \text{ mJ} \cdot \text{mol}^{-1} \cdot \text{K}^{-2}$ of USn₃, which may have
the maximum density of states, may be due to strong many-body effects as well as one-
electron hybridization in analogy to the behavior of CeSn₃. The study of pseudo-binary systems
is recommended in order to get more insight into the relationships. Compounds of the
compositions $U(Sn_xPb_{1-x})_3$ (see p. 309) or $U(Sn_xGe_{1-x})_3$ cover the range of behavior from
itinerant electrons, through spin fluctuation with large effective masses, to local-moment
behavior.

The heavy-fermion systems CeSn₃, USn₃, and NpSn₃, all of them with the cubic Au₃Cu
structure, form an excellent set of materials for the study of the representatives of strongly
enhanced systems [43]. Whereas USn₃ does not undergo a transition, the replacement of 8%
of the Sn by Pb changes the behavior; the ternary compound becomes antiferromagnetic
[43, 72]. The density of states of USn₃ at the Fermi energy, calculated by means of a fully
relativistic method, is $n(E_F) = 22.6 \text{ mJ} \cdot \text{mol}^{-1} \cdot \text{K}^{-2}$, which is compared to the experimental
electronic specific heat coefficient of $\gamma = 169 \text{ mJ} \cdot \text{mol}^{-1} \cdot \text{K}^{-2}$. The theoretical results are in
accordance with the near-instability of the compound. The f-states grade from very itinerant
behavior in USi₃ to rather local and magnetic behavior in UPb₃. USn₃ with its nearly-heavy-
fermion properties has a position near a transition point [70].

Another comparison of UM₃ compounds (M = group IIIB and IVB elements Al, Ga, In, Tl,
Si, Ge, Sn, and Pb), in relation to the heavy-electron character, shows that USn₃ has an extremely high value of the electronic specific heat compared to the other compounds, indicating the pronounced heavy-electron behavior of the U−Sn compound [59].

The properties of ternary compounds with Au₃Cu structure are treated in Sections 4.1.10.1
to 4.1.10.3.

The proposed compound U₂Sn, which is not in agreement with the phase diagram of the
U−Sn system, is reported to be ferromagnetic. Its Curie temperature is 200 K and its Weiss
constant $\Theta \approx 5$ K. The effective paramagnetic moment is 2.60 μ_B [58].

4.1.9 Chemical Reactions

Fig. 4-6, p. 306, shows the dissociation pressures of U−Sn alloys in logarithmic scale as
functions of the reciprocal absolute temperature. The mixtures containing the highest amounts
of Sn show the highest dissociation pressures.

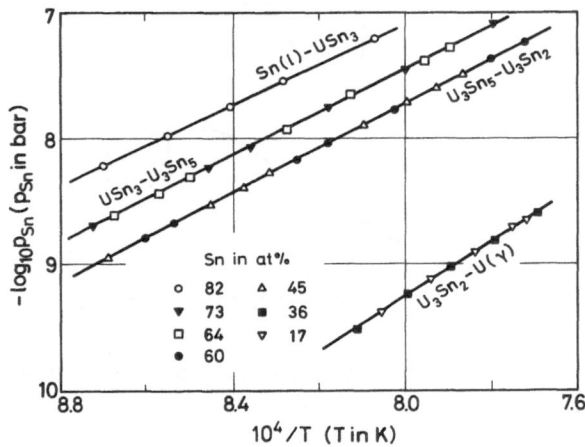

Fig. 4-6. The dissociation pressures of U — Sn alloys [16].

The alloys of U with Sn in the composition range 46 to 67 at% Sn are pyrophoric. The pyrophoric range includes the compounds U_5Sn_4 and U_3Sn_5. Even the preparation of the compound USn_3 must be performed under protective atmosphere because of its sensitivity to oxygen [13, 42].

The reaction of U — Sn alloys containing 2 to 18 wt% U with N_2 was measured between 600 and 1600°C. U_2N_3 is formed in a lower temperature region, whereas UN is the reaction product at high temperature (1475 to 1600°C). From volumetric measurements, equilibrium data of the reactions leading to UN and U_2N_3 and the standard free energies of these compounds were determined (see "Uranium" Suppl. Vol. C7, 1981, pp. 20, 60).

Fig. 4-**7** shows the absorption of N_2 by liquid U — Sn alloys at 1567°C.

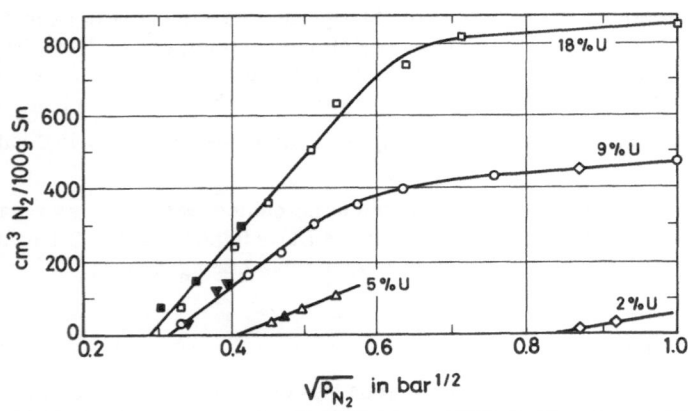

Fig. 4-7. Absorption of N_2 by liquid U — Sn alloys at 1567°C, U concentrations in wt% [39].
Nitriding: $\diamond$, $\triangle$, $\bigcirc$, $\square$; denitriding: $\blacktriangle$, $\blacktriangledown$, $\blacksquare$.

The nitrogen – nitride reaction of U – Sn alloys is useful for the preparation and purification of nuclear fuels [38, 39].

4.1.10 Ternary Systems

One group of ternary alloys U – Sn – M is based upon the compound USn_3 in which Sn is partly replaced by a ternary metallic element, see [36]. Other ternary compounds of U, Sn, and a third element belonging to the transition metals were recently discovered. These alloys are related to the hexagonal Fe_2P structure with the space group $P\bar{6}2m - D_{3h}^3$ (No 189). The resulting intermetallics USnM are members of a large group of ternary actinide phases, in which the transition metal atoms M occupy the P(1) and P(2) sites, whereas the actinide atoms are found to occupy the Fe(2), and the group IV (Sn) elements the Fe(1) sites [46].

The U – Sn compounds also have the tendency to partly replace U atoms by another actinide element. The ternary compounds mentioned here are interesting because of their uncommon properties at low temperatures.

4.1.10.1 U – Sn – Al

The phase equilibria in solid U – Sn – Al ternary alloys were studied by means of metallographic and X-ray analyses. The compounds USn_3 and UAl_3 form continuous solid solutions (cubic, Cu_3Au type) which show a narrow range of homogeneity for the U content. **Fig. 4-8** shows the solid state compatibility fields in the U – Sn – Al system [36].

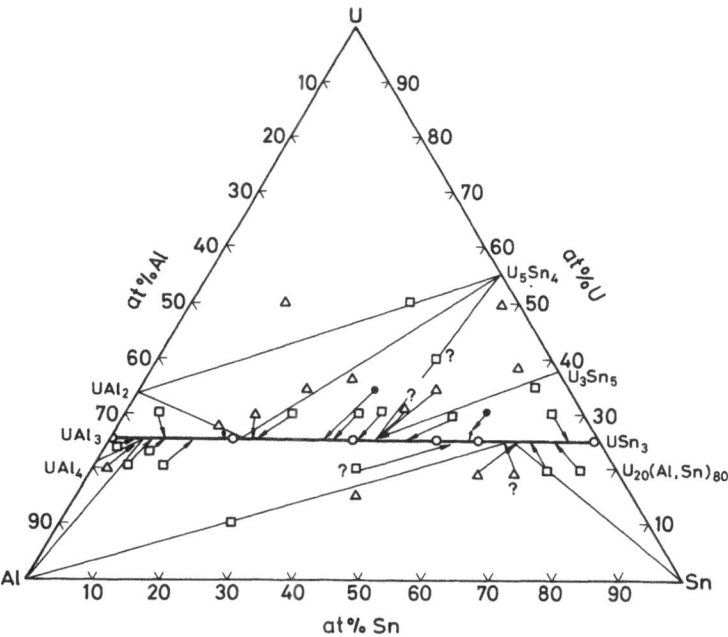

Fig. 4-8. Solid state compatibility fields in the Al – Sn – U system according to [36]; ○ homogeneous alloys, □ two-phase alloys, △ three-phase alloys. The arrows indicate the compositions of the solid solutions.

References for 4.1 on pp. 316/9

20·

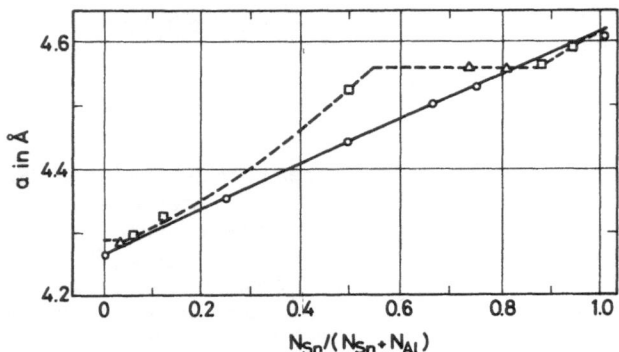

Fig. 4-9. Lattice parameters of solid solutions of U(Al, Sn)$_3$ as a function of $N_{Sn}/(N_{Sn} + N_{Al})$ [36];
——— homogeneous samples with composition U(Al, Sn)$_3$, – – – – heterogeneous samples of
composition U$_{20}$(Al, Sn)$_{80}$; $\bigcirc$ homogeneous samples, $\square$ two-phase alloys, $\triangle$ three-phase
alloys.

The lattice parameters of the solid solutions U(Al, Sn)$_3$, as determined by means of the
X-ray diffraction method, vary with the ratio $N_{Sn}/(N_{Sn} + N_{Al})$ as shown in **Fig.** 4-9. The lattice
parameter increases nearly linearly with the Sn content of the ternary compound [36].

The alloys were prepared from metals of not less than 99.9% purity by melting under
argon in an induction furnace. The samples were annealed for about 200 h at 500°C and then
quenched. Alloys with very low U contents were annealed at only 250°C and cooled slowly to
room temperature [36].

All mixtures, the compositions of which differ from U(Al, Sn)$_3$, are two- or three-phase
alloys. The alloys of the composition U$_{20}$(Al, Sn)$_{80}$, for instance, are mixtures of Al + U(Al, Sn)$_3$,
Sn + U(Al, Sn)$_3$, Al + UAl$_4$ + approximately U$_{25}$Al$_{70}$Sn$_5$ (saturated solution) or Al + Sn +
U$_{25}$Al$_{13}$Sn$_{62}$ (saturated solid solution) [36].

4.1.10.2   U—Sn—In

Polycrystalline samples of the compound U(In, Sn)$_3$ were prepared in the same way as the
Al-containing compound. The samples were annealed, after preparation, in vacuum at 600°C
for 4 days. Losses of In and Sn due to vaporization were compensated by means of addition
of these elements in proportion to their relative vapor pressures [53].

The ternary compound U(In, Sn)$_3$ crystallizes in the ordered Cu$_3$Au structure, and the room
temperature X-ray measurements indicate that all samples are single phase. The relation
between the lattice constant and the composition given by x follows Végard's law with a =
4.601 Å for UIn$_3$ increasing linearly to a = 4.610 Å for USn$_3$ [53].

The specific heat C/T versus T^2 for U(In$_{1-x}$Sn$_x$)$_3$ shows an influence of the composition (x).
While the compound with x = 0.8 exhibits behavior similar to USn$_3$, the compound with
x = 0.6 undergoes a steep decrease of C/T at low values of T^2 (T < 25 K) [53, 54, 60, 74].

The low-temperature magnetic susceptibility is similarly influenced by the composition of
the ternary compound [53, 54, 74]. The composition U(In$_{0.4}$Sn$_{0.6}$)$_3$ shows the largest difference
from the behavior of the two binary compounds with a characteristically large decrease of χ

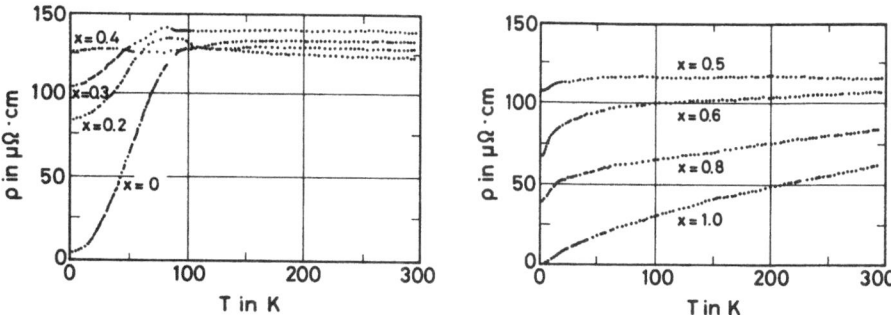

Fig. 4-10. Resistivity ϱ versus temperature for $U(In_{1-x}Sn_x)_3$ with $0 \leq x \leq 1$ [60].

with temperature at $T < 50$ K. The sharp maximum of χ at $x = 0.6$ ($\chi(0) \sim 450$ memu/mol) is also described in [60]. The Néel temperature of this ternary system decreases as χ approaches zero at $x \sim 0.4$ [45, 48, 60]. The low-temperature electronic specific heat coefficient γ (at $T = 1.35$ K) exhibits a sharp maximum at $x = 0.6$ [45, 48, 60, 74]. This behavior indicates magnetic ordering for compositions with $x < 0.45$ [54].

The very high electronic specific heat $C(T)/T$ of USn_3 is significantly enhanced in the ternary compound $U(In_{1-x}Sn_x)_3$, with the strongest effect at $x = 0.6$ [45]. At 1.5 K the electronic contribution $\gamma \cdot T$ to $C(T)$ is nearly three times as high as in the compound USn_3, the value of γ decreases markedly when the amount of Sn is lowered to $x = 0.55, 0.50$, and 0.45. Studies of the $U(In, Sn)_3$ system explicitly indicate that the heavy-fermion behavior is a result of a hybridization-driven mechanism. The heavy-fermion behavior is also observed in the magnetic to nonmagnetic transition of this pseudo-binary compound [54]. The heavy-fermion region is between $x = 0.45$ and $x = 0.80$. The system $U(In, Sn)_3$ evolves from the enhanced paramagnetic regime, through a strongly enhanced heavy-fermion regime, and finally to an antiferromagnetically ordered regime [48, 53].

The low-temperature resistivity of $U(In_{1-x}Sn_x)_3$ in the temperature range up to 300 K was studied for $0 \leq x \leq 1$. For $x \leq 0.4$, the ϱ versus T curve is surprisingly temperature-independent at T above the Néel temperature T_N and $x > 0$, whereas an anomaly occurs at $T \sim T_N$. This behavior of ϱ is typical for several U-based intermetallics or alloys. A gradual increase of ϱ with T was measured for the paramagnetic regime ($x \geq 0.5$), however, there is no strong enhancement at $x = 0.6$ as for the electronic specific heat [48]. The temperature function of the resistivity of $U(In_{1-x}Sn_x)_3$ is shown in [60] for $0 \leq x \leq 1$, as illustrated in **Fig. 4-10**.

4.1.10.3   U−Sn−Pb

The compound of composition $U(Sn_{1-x}Pb_x)_3$ was prepared in an inert atmosphere arc furnace using U of 99.99% purity and Sn and Pb of 99.999% purity. Losses of Sn and Pb occur because of their higher vapor pressures. Larger amounts of these elements were therefore added in proportion to their relative vapor pressures. The compounds were then annealed at 600°C for 7 days. Determination of room temperature lattice constants were measured by X-ray diffraction. The results indicate complete solid solubility across the pseudobinary alloy. The lattice constants follow Végard's law and the constants of the terminal systems (USn_3, a = 4.626 Å; UPb_3, a = 4.791 Å) are consistent with reported values. There is no evidence in the lattice constant versus x plot to indicate a significant change in the location of the 5f electrons in the alloy [44].

References for 4.1 on pp. 316/9

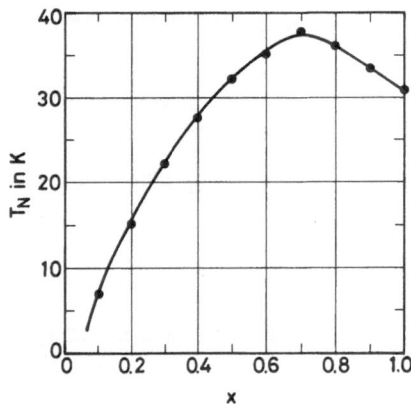

Fig. 4-11. The Néel temperature versus x for $U(Sn_{1-x}Pb_x)_3$ [44].

The magnetic susceptibility of the alloy $U(Sn_{0.3}Pb_{0.7})_3$ shows a sharp cusp in the $\chi(T)$ versus T curve at $T = 38.2 \pm 0.2$ K, similar to UPb_3, which shows this antiferromagnetic transition at $T = 31.0 \pm 0.2$ K (see p. 326). This temperature is equal to the Néel temperature T_N, which varies with x as is shown in **Fig. 4-11**. The $\chi(T)$ values were fitted to a modified Curie-Weiss law (i.e., $\chi(T) = \chi_0 + C/(T + T^*)$) for $T > T_N$. The term T^* decreases with x from 0 to ~ 0.5 and increases again approaching the composition of UPb_3 [44].

4.1.10.4 U—Sn—Bi

The isothermal section at 350°C of the phase diagram of this ternary system is given in [65]. The ternary diagram shown in **Fig. 4-12** is based on the work of Teitel [66]. At 350°C

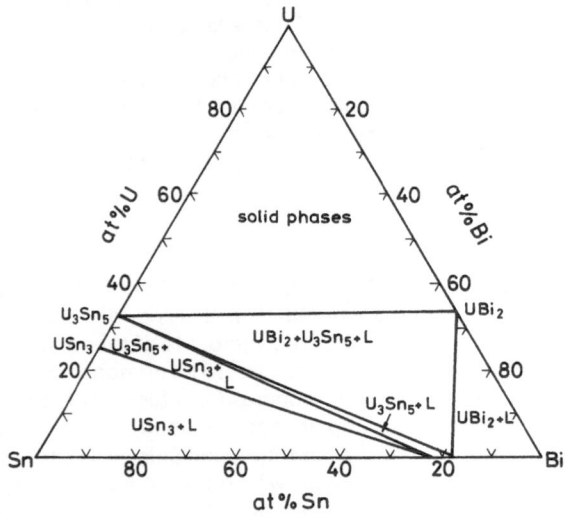

Fig. 4-12. Isothermal section at 350°C of the U—Sn—Bi system [65, 66].

the compounds USn_3 and U_3Sn_5 are in equilibrium with liquid phases containing up to 82 at% Bi. A two-phase region $UBi_2 + L$ also exists along with two regions of three-phase states of the alloys: $USn_3 + U_3Sn_5 + L$ and $U_3Sn_5 + UBi_2 + L$. Tin lowers the solubility of U in the liquid phase.

4.1.10.5 Uranium − Tin − Transition Metal Systems

4.1.10.5.1 Compounds of General Composition U − Sn − Tr

Binary compounds of uranium with metals of group VIII will be described in a later volume of the present series, according to the Gmelin system of last position. The Sn-containing systems with group VIII elements are described here owing to their close relation to the U − Sn system.

U and Sn form ternary intermetallics with several transition metals of group VIII, namely Co, Ni, Ru, Rh, Pd, Ir, and Pt. The structures of these compounds are related to three different types. Most of the compounds were examined by X-ray diffraction analyses. Table 4/4 gives an overview on the structure and the lattice constants of these intermetallics.

Table 4/4
The Structure of USnTr Compounds.

compound	structure type	space group	lattice constants a in pm	c in pm	Ref.
USnNi	MgAgAs	$F\bar{4}3m - T_d^2$ (No. 216)	638.5	−	[49, 62]
USnCo	Fe_2P	$P\bar{6}2m - D_{3h}^3$ (No. 189)	714.5	400.1	[47, 52, 62]
USnRu	Fe_2P	$P\bar{6}2m - D_{3h}^3$ (No. 189)	736.9	396.1	[47, 52]
USnRh	Fe_2P	$P\bar{6}2m - D_{3h}^3$ (No. 189)	736.5	399.3	[62]
USnPd	$CaIn_2$	$P6_3/mmc - D_{6h}^4$ (No. 194)	460.8	731.0	[62]
USnIr	Fe_2P	$P\bar{6}2m - D_{3h}^3$ (No. 189)	737.5	401.0	[46]
USnPt	MgAgAs	$F\bar{4}3m - T_d^2$ (No. 216)	661.7	−	[49, 62]

The Fe_2P-type structure is most common in this group of compounds. The constituents of USnCo are placed in the Fe_2P-type structure as follows (origin at $\bar{6}2$ m) [46]:

Sn at Fe(1)	3 f mm	x, 0, 0; 0, x, 0; x̄, x̄, 0
U at Fe(2)	3 g mm	x, 0, 1/2; 0, x, 1/2; x̄, x̄, 1/2
Co at P(1)	1 b $\bar{6}$mz	0, 0, 1/2
Co at P(2)	2 c $\bar{6}$	± 1/3, 2/3, 0

The Fe_2P structure representing the structure types of USnCo, USnRu, USnRh, and USnIr is isotype with that of UGaCo, in which 7 Co atoms and 12 U atoms form the plane in z = 0 and 6 Co atoms with 12 Sn atoms are placed in the plane at z = 0.5 (see **Fig. 4-13**, p. 312) [57]. The MgAgAs structure to which USnNi and USnPt belong is cubic with 4 formula units per cell and belongs to space group $F\bar{4}3m$. Atomic positions are given. This structure is an ordered derivative of the fluorite structure with Ag and Mg ordered on the F sites such that they occupy one set of tetrahedral holes in the fcc array of Ag atoms [46].

References for 4.1 on pp. 316/9

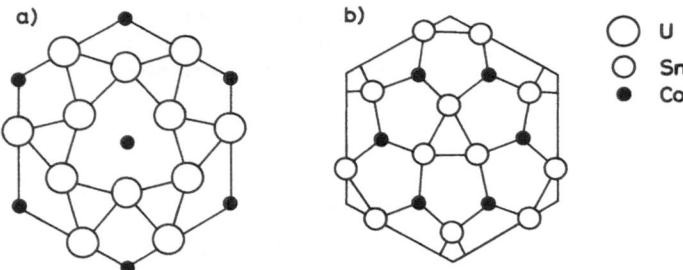

Fig. 4-13. USnTr (Tr = transition metal) structure, Fe$_2$P type, space group P$\bar{6}$2m − T$_d^2$ (No. 216), see Table 4/4, p. 311 [57]; a) z = 0 and b) z = 0.5 (z is the coordinate along the c axis).

Polycrystalline samples of these compounds were prepared by arc-furnace melting of stoichiometric amounts of the constituents in an Ar atmosphere. Small monocrystalline samples were obtained by a temperature-gradient annealing treatment [47].

The properties of this group of compounds are related to the U 5f states. USnCo and USnRu are ferromagnetic. The magnetic ordering temperature for USnCo is T$_C$ = 80.5 K, and for USnRu T$_C$ = 53 K [47, 52]. Magnetic ordering temperatures are also given for USnPd, T$_N$ = 29 K; USnCo, T$_C$ = 85 K; USnRh, T$_C$ = 25 K; USnIr, T$_C$ = 25 K; USnNi, T$_N$ = 40 K. The values of the asymptotic Curie temperatures are listed in Table 4/5 [62].

Table 4/5
Asymptotic Curie Temperatures of USnTr Compounds [62].

compound	U−U distance in nm	Curie temperature Θ_p in K
USnPd	0.365	− 10
USnCo	0.372	+ 25
USnRh	0.383	+ 20
USnIr	0.384	+ 20
USnNi	0.451	− 75
USnPt	0.468	− 100

The U−U distances are all greater than 0.36 nm, and magnetic moments in all of these compounds are in agreement with Hill's criterion [62]. The structural features seem to be rather unimportant in this respect.

Some other structural and physical properties are listed in the Table 4/6, among them the nearest neighbor U separation d (in nm), the Curie-Weiss temperature T$_C$ (in K), the Debye temperature Θ_D (in K), and the linear specific heat coefficient γ (in mJ · mol^{-1} · K^{-2}).

Fig. 4-14 shows the dependence of the electrical resistivity ϱ on the temperature from zero to 1000 K for the compounds USnTr with Tr = Ni, Co, Pd, and Pt [50]. Though the resistivity of USnCo is the lowest one, it is considerably larger than that of the similar compounds UGaCo (see pp. 273, 277) and UAlCo (see pp. 219, 227) [55]. The temperature dependence of the

Table 4/6
Physical Properties of USnTr Compounds.

compound	d in nm	T_C in K	Θ_D in K	γ in mJ·mol^{-1}·K^{-2}	Ref.
USnNi	0.451	47	215	28.1	[50]
USnCo	0.372	85	194	52.8	[50]
USnRu	0.399				[47]
USnRh	0.383	25			[62]
USnPd	0.365	29	−	4.3	[50]
USnIr	0.384	25			[62]
USnPt	0.468	75	185	10.9	[50]

Other values for γ are $\gamma = 61$ for USnCo and $\gamma = 50$ (in mJ·mol^{-1}·K^{-2}) for USnRu [52].

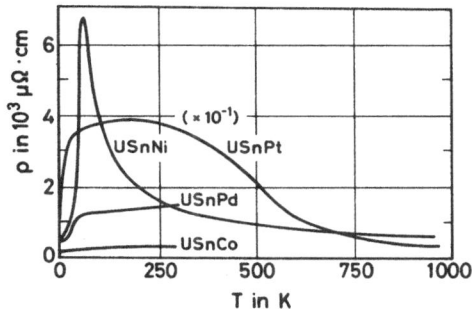

Fig. 4-14. Temperature dependence of the electrical resistivity ϱ for the compounds USnTr (Tr = Ni, Co, Pd, Pt) [50].

resistivities of USnNi and USnPt is exponential [49]. The resistivities at 4.2 K of some of these compounds are relatively large, 170 μΩ·cm for USnCo and 72000 μΩ·cm for USnRh [62]. The resistivity behavior of the USnTr compounds does not seem to be linked with the magnetic properties.

Photoemission spectra with photon energies between $h\nu = 22$ and 130 eV are used to locate the U 5f states of this group of compounds. They are close to the Fermi energy with a width ranging from ~1.1 eV for USnNi to ~1.5 eV for USnPt [49].

Whereas all USnTr compounds are ferromagnetic, USnPt is the only antiferromagnet in this group with the Néel temperature of 7 K [51].

USnAu has the same structure as the USnTr equiatomic ternary compounds. It crystallizes in the hexagonal CaIn$_2$ structure with the space group P6$_3$/mmc−D$_{6h}^4$ (No. 144), where the U atoms form trigonal prisms. The lattice parameters are a = 4.717 Å, c = 7.208 Å, and the U−U distance is d = 3.60 Å. The compound was prepared in the same way as all the USnTr compounds [64].

                   References for 4.1 on pp. 316/9

USnAu orders antiferromagnetically at 35 K and shows magnetic properties similar to USnPd. The electrical resistivity has a different temperature dependence than that of other $CaIn_2$-type compounds as USnPd. The resistivity is $\varrho = 650~\mu\Omega \cdot cm$ at 4 K and 610 $\mu\Omega \cdot cm$ at 300 K; the curve has a monotonic character [64].

4.1.10.5.2 USn_xTr_{5-x} Compounds

Sn can partly substitute for Ni in the compound UNi_5, the single phase region exists up to the composition UNi_4Sn [63]. Samples of this group of compounds were prepared by arc melting of appropriate quantities of the constituents in argon atmosphere using Ti as getter metal. An annealing procedure of 700 h at 800°C in evacuated quartz-glass ampoules and cooling at a rate of 50 K per day followed. The unit cell parameters of $USnNi_4$ with the superlattice of the $MgCu_4Sn$ type (space group $F\bar{4}3m - T_d^2$ (No. 216)) and the microhardness as a function of the composition are listed in Table 4/7.

Table 4/7.
Unit Cell Parameters a and Microhardness HV of USn_xNi_{5-x} Compounds [63].

composition	a in Å $\pm$0.5	HV in kg/mm^2
UNi_5	6.780	350
$USn_{0.2}Ni_{4.8}$	6.823	568
$USn_{0.4}Ni_{4.6}$	6.854	588
$USn_{0.6}Ni_{4.4}$	6.903	609
$USn_{0.8}Ni_{4.2}$	6.941	645
$USnNi_4$	6.984	697

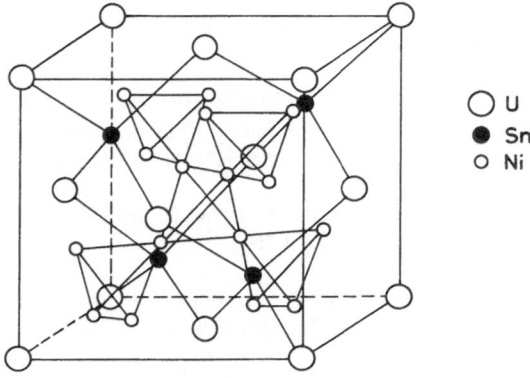

O U
● Sn
○ Ni

Fig. 4-15. The crystal structure of $USnNi_4$ [63].

The variation of the lattice constants obeys Végard's rule in the single phase region. The microhardness is strongly influenced by the substitution of Ni by Sn. The arrangement of the atoms in the unit cell of $USnNi_4$ is given by the following scheme [63]:

4 U in 4 (a)	0, 0, 0; 0, 0, 0; 0, 1/2, 1/2; 1/2, 0, 1/2; 1/2, 1/2, 0
4 Sn in 4 (c)	1/4, 1/4, 1/4; 0, 0, 0
16 Ni in 16 (e)	x, x, x; x, x̄, x̄; x̄, x, x̄; x̄, x̄, x; + (0, 0, 0; 0, 1/2, 1/2; 1/2, 0, 1/2; 1/2, 1/2, 0)

Fig. 4-15 shows this crystal structure of the composition $USnNi_4$.

Measurements of the magnetic susceptibilities of USn_xNi_{5-x} indicate antiferromagnetic ordering which does not exist in UNi_5 [63].

4.1.10.6   (U, Pu) − Sn Compounds

The U−Sn−Pu system was studied in the tin-rich region by Sari et al. [26]. The alloys, which are radioactive, toxic, and pyrophoric, were prepared and handled in glove boxes under an inert atmosphere. Samples of ∼1.5 g mass were prepared by means of arc melting or in an induction-heated graphite crucible [26].

Metallographic, autoradiographic, X-ray diffraction, electron microprobe analysis, differential thermal analysis, and microhardness techniques were applied to characterize the alloys. In mixtures with more than 75 at% Sn, intermetallic compounds were precipitated in the form of dendrites in an Sn matrix. All U and Pu is present in the dendrite precipitates. The solidified Sn contains only traces of U (<0.03 at%) and Pu (<0.02 at%) as detected by electron microprobe. The dendrites are MSn_3 (M = U + Pu) [26].

Several intermetallic compounds occur at higher concentrations of U and Pu in the ternary alloys. The analysis of the alloys indicates the formation of $(U, Pu)Sn_3$, $(U, Pu)_3Sn_7$, $(U, Pu)Sn_2$, $(U, Pu)_4Sn_5$, $(U, Pu)Sn$, $(U, Pu)_5Sn_4$, and $(U, Pu)_5Sn_3$. The binary compound U_5Sn_4 is not detected among the compounds [26].

Table 4/8 shows the influence of the Pu content of the compound $(U, Pu)Sn_3$ on the lattice constant a exhibiting a linear relationship.

Table 4/8
Lattice Parameter a of the Intermetallic Compounds $(U, Pu)Sn_3$ [26].

composition	a in nm
USn_3	0.46089 ± 0.00029
$U_{0.5}Pu_{0.5}Sn_3$	0.46181 ± 0.00010
$U_{0.3}Pu_{0.7}Sn_3$	0.46243 ± 0.00027
$U_{0.12}Pu_{0.88}Sn_3$	0.46293 ± 0.00023
$PuSn_3$	0.4630 ± 0.0001

References for 4.1 on pp. 316/9

The influence of the Pu content of some $(U, Pu)Sn_3$ alloys is shown in Table 4/9.

Table 4/9
Influence of the Pu Content on the Microhardness of $(U, Pu)Sn_3$.

composition	microhardness HV in kg/mm^2
USn_3	124
$PuSn_3$	130
$U_{0.5}Pu_{0.5}Sn_3$	150

The $(U, Pu)Sn_3$ compounds also show an influence of the $U/(U + Pu)$ ratio on the peritectic melting temperature of the compound, as is seen in **Fig. 4-16**.

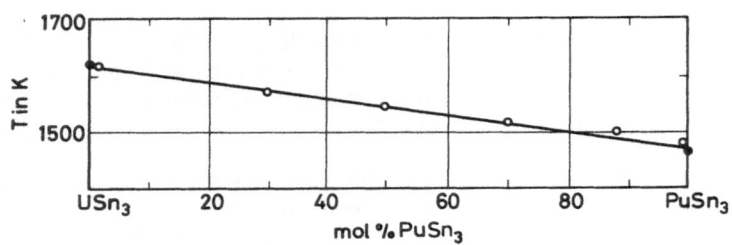

Fig. 4-16. Peritectic melting temperatures in the system $USn_3 - PuSn_3$ [26].

References for 4.1:

[1] Miedema, A. R. (J. Less-Common Metals **32** [1973] 117/36).
[2] Boom, R.; de Boer, F. R.; Miedema, A. R. (J. Less-Common Metals **46** [1976] 271/84).
[3] Miedema, A. R.; Boom, R.; de Boer, F. R. (J. Less-Common Metals **41** [1975] 283/98).
[4] Miedema, A. R.; de Boer, F. R.; Boom, R. (CALPHAD **1** [1977] 341/59).
[5] Miedema, A. R.; de Châtel, P. F.; de Boer, F. R. (Physica B + C **100** [1980] 1/28).
[6] Niessen, A. K.; de Boer, F. R.; Boom, R., de Châtel, P. F.; Mattens, W. C. M.; Miedema, A. R. (CALPHAD **7** [1983] 51/70).
[7] Cordfunke, E. H. P. (The Chemistry of Uranium, Including Applications in Nuclear Technology, Elsevier, Amsterdam 1969, pp. 45/8).
[8] Saller, H. A.; Rough, F. A. (BMI-1000 [1955] 1/141; N.S.A. **9** [1955] No. 5349).
[9] Hansen, M.; Anderko, K. (Constitution of Binary Alloys, McGraw-Hill, New York 1958, pp. 1215/6).
[10] Wilhelm, H. A.; Treick, D. A. (unpublished information [1944] cited in [8]).

[11] Katz, J. J.; Rabinowitch, E. (Natl. Nucl. Energy Ser. Div. VIII **5** I [1951] 174/9.
[12] Shunk, F. A. (Constitution of Binary Alloys, 2nd Suppl., McGraw-Hill, New York 1969).
[13] Rundle, R. E.; Wilson, A. S. (Acta Cryst. **2** [1949] 148/50).

[14] Alcock, C. B.; Grieveson, P. (J. Inst. Metals **90** [1962] 304/10).

[15] Frost, B. R. T.; Maskrey, J. T. (J. Inst. Metals **82** [1953] 177/80).

[16] Johnson, I.; Feder, H. M. (Thermodyn. Nucl. Mater. Proc. Symp., Vienna 1962, pp. 319/29).

[17] Johnson, I. (IMD-Spec. Rept. No. 13 [1964] 171/92).

[18] Wagman, D. D.; Evans, W. H.; Parker, V. P.; Schumm, R. H.; Nuttall, R. L. (Selected Values of Chemical Thermodynamic Properties — Compounds of Uranium, Protactinium, Thorium, Actinium, and the Alkali Metals, Natl. Bur. Std., Washington 1981, p. 6).

[19] Dwight, A. E. (Develop. Struct. Chem. Alloy Phases **1969** 181/226).

[20] Bader, S. D.; Knapp, G. S.; Culbert, H. V. (in: Graham, C. D., Jr.; Lander, G. H.; Rhine, J. J., Magnetism and Magnetic Materials — 1974, Am. Inst. Phys., New York 1975, p. 222).

[21] Fournier, J. M.; Troć, R. (in: Freeman, A. J.; Lander, G. H., Handbook of the Physics and Chemistry of the Actinides, Vol. 2, North-Holland, Amsterdam 1985, pp. 70/2).

[22] Buyers, W. J. L.; Holden, T. M. (in: Freeman, A. J.; Lander, G. H., Handbook of the Physics and Chemistry of the Actinides, Vol. 2, North-Holland, Amsterdam 1985, pp. 287/8).

[23] Lam, D. J.; Aldred, A. T. (in: Freeman, A. J.; Darby, J. B., The Actinides, Electronic Structure and Related Properties, Academic, New York 1974, pp. 167/8).

[24] Fradin, F. Y. (in: Freeman, A. J.; Darby, J. B., The Actinides, Electronic Structure and Related Properties, Academic, New York 1974, pp. 226/7).

[25] Alcock, C. B.; Cornish, J. B.; Grieveson, P. (Thermodyn. Proc. Symp., Vienna 1965 [1966], Vol. 1, pp. 211/30; C.A. **65** [1966] 8092).

[26] Sari, C.; Vernazza, F.; Müller, W. (J. Less-Common Metals **92** [1983] 301/6).

[27] Johnson, I. (IMD-Spec. Rept. No. 13 [1964] 171/92).

[28] Faber, J., Jr.; Lander, G. H.; Brown, P. J.; Delapalme, A. (Acta Cryst. A **37** [1981] 558/65).

[29] Mulak, J.; Misiuk, A. (Bull. Acad. Polon. Sci. Ser. Sci. Chim. **19** [1971] 207/13).

[30] van Maaren, M. H.; van Daal, H. J.; Buschow, K. H. J. (Solid. State Commun. **14** [1974] 145/7).

[31] Udaya Shankar Rao, V.; Vijayaraghavan, R. (J. Phys. Chem. Solids **29** [1968] 123/7).

[32] Buschow, K. H. J.; van Daal, H. J. (AIP Conf. Proc. No. 5 [1971] 1464/77).

[33] Loewenhaupt, M.; Horn, S.; Steglich, F.; Holland-Moritz, E.; Lander, G. H. (J. Phys. Colloq. [Paris] **40** [1979] C4-142/C4-144).

[34] Anderson, R. N.; Parlee, N. A. D.; Gallagher, J. M. (Nucl. Technol. **13** [1972] 29/35).

[35] Norman, M. R.; Bader, S. D.; Kierstead, H. A. (Phys. Rev. B **33** [1986] 8035/8).

[36] Marazza, R.; Ferro, R.; Rambaldi, G.; Mazzone, D. (J. Less-Common Metals **51** [1977] 51/4).

[37] Massalski, T. B. (Binary Alloy Phase Diagrams, Vol. 2, Am. Soc. Metals, Metals Park, Ohio, 1986, pp. 2079, 2081).

[38] Anderson, R. N. (Diss. Stanford Univ. 1969).

[39] Anderson, R. N.; Parlee, N. A. D. (Met. Trans. **2** [1971] 1599/604).

[40] Koelling, D. D.; Dunlap, B. D.; Crabtree, G. W. (Phys. Rev. [3] B **31** [1985] 4966/71).

[41] Misiuk, A.; Mulak, J.; Czopnik, A. (Bull. Acad. Polon. Sci. Ser. Sci. Chim. **20** [1972] 459/61).

[42] Chiotti, P.; Akhachinskij, V. V.; Ansara, I.; Rand, M. H. (The Chemical Thermodynamics of Actinide Elements and Compounds, Pt. 5, The Actinide Binary Elements, Intern. At. Energy Agency, Vienna 1981, pp. 177/81).

[43] Norman, M. R.; Koelling, D. D. (Physica B + C **135** [1985] 95/8).

[44] Lin, C. L.; Zhou, L. W.; Crow, J. E.; Guertin, R. P. (J. Appl. Phys. **57** [1985] 3146/8).

[45] Lin, C. L.; Zhou, L. W.; Crow, J. E.; Mihalisin, T.; Brooks, J.; Guertin, R. P. (J. Less-Common Metals **127** [1987] 273/9).

[46] Lam, D. J.; Darby, J. B., Jr.; Nevitt, M. V. (in: Freeman, A. J.; Darby, J. B., The Actinides; Electronic Structure and Related Properties, Academic, New York 1974, Vol. II, pp. 175/84).

[47] Sechovský, V.; Havela, L.; de Boer, F. R.; Franse, J. J. M.; Veenhuizen, P. A.; Sebek, J.; Stehno, J.; Andreev, A. V. (Physica B + C **142** [1986] 283/93).

[48] Lin, C. L.; Zhou, L. W.; Crow, J. E.; Guertin, R. P.; Stewart, G. R. (J. Magn. Magn. Mater. **54/57** [1986] 391/2).

[49] Höchst, H.; Tan, K.; Buschow, K. H. J. (J. Magn. Magn. Mater. **54/57** [1986] 545/6).

[50] Palstra, T. T. M.; Nieuwenhuys, G. J.; Mydosch, J. A.; Buschow, K. H. J. (J. Magn. Magn. Mater. **54/57** [1986] 549/50).

[51] Troć, R.; Nowak, L.; Krawczyk, L. (J. Less-Common Metals **121** [1986] 648).

[52] Sechovský, V.; Havela, L.; Neužil, L.; Andreev, A. V.; Hilscher, G.; Schmitzer, C. (J. Less-Common Metals **121** [1986] 169/74).

[53] Zhou, L. W.; Lin, C. L.; Crow, J. E.; Bloom, S.; Guertin, R. P.; Foner, S. (Phys. Rev. [3] B **34** [1986] 483/6).

[54] Zhou, L. W.; Lin, C. L.; Crow, J. E.; Bloom, S.; Guertin, R. P.; Foner, S.; Stewart, G. (Physica B + C **135** [1985] 99).

[55] Havela, L.; Neužil, L.; Sechovský, V.; Andreev, A. V.; Schmitzer, C.; Hilscher, G. (J. Magn. Magn. Mater. **54/57** [1986] 551/2).

[56] Brodsky, M. B.; Arko, A. J.; Harvey, A. R.; Nellis, W. J. (in: Freeman, A. J.; Darby, J. B., Jr., The Actinides, Electronic Structure and Related Properties, Vol. II, Academic, New York 1974, pp. 185/264).

[57] Andreev, A. V.; Havely, L.; Zelený, M.; Hrebík, J. (Phys. Status Solidi A **82** [1984] 191/4).

[58] Trzebiatowski, W. (in: Ringpfeil, H., Magnetismus, VEB Deut. Verlag Grundstoffindustrie, Leipzig 1967, pp. 88/101).

[59] Fisk, Z.; Ott, H. R.; Smith, J. L. (J. Less-Common Metals **133** [1987] 99/106).

[60] Lin, C. L.; Zhou, L. W.; Jee, C. S.; Wallash, A.; Crow, J. E. (J. Less-Common Metals **133** [1987] 67/75).

[61] Brodsky, M. B. (Rept. Progr. Phys. **41** [1978] 1547/602).

[62] Buschow, K. H. J.; De Mooij, D. B.; Palstra, T. T. M.; Nieuwenhuys, G. J.; Mydosh, J. A. (Philips J. Res. **40** [1985] 313/22).

[63] Blazina, Z.; Drasner, A.; Ban, Z. (J. Nucl. Mater. **96** [1981] 141/6).

[64] Palstra, T. T. M.; Nieuwenhuys, G. J.; Vlastuin, R. F. M.; van den Berg, J.; Mydosh, J. A. (J. Magn. Magn. Mater. **67** [1987] 331/42).

[65] Ivanov, O. S.; Badaeva, T. A.; Sofronova, R. M.; Kishenevskii, V. B.; Kushnir, N. P. (Phase Diagrams of Uranium Alloys, Amerind Publ., New Delhi, India, 1983, transl. from: Diagrammy Sostoyaniya i Fazovye Prevashcheniya Splavov Urana, Nauka, Moscow 1972).

[66] Teitel, R. J. (in: Rough, F. A.; Bauer, A. A., Constitutional Diagrams of Uranium and Thorium Alloys, Addison-Wesley, Reading, Mass., 1958).

[67] Sheldon, R. I.; Foltyn, E. M.; Peterson, D. E. (Bull. Alloy Phase Diagrams **8** [1987] 347/52, 413/4).

[68] Kadochnikov, V. A.; Lebedew, V. A.; Nichkov, I. F.; Raspopin, S. P. (Izv. Akad. Nauk SSSR Metally **1976** No. 4, pp. 67/9; Russ. Met. **1976** No. 4, pp. 61/3).

[69] Rough, F. A.; Bauer, A. A. (Constitutional Diagrams of Uranium and Thorium Alloys, Addison-Wesley, Reading, Mass., 1958).

[70] Koelling, D. D.; Norman, M. R.; Arko, A. J. (J. Magn. Magn. Mater. **63/64** [1987] 638/44).

[71] Rand, M. H.; Kubaschewski, O. (The Thermochemical Properties of Uranium Compounds, Oliver and Boyd, Edinburgh — London 1963, p. 47).

[72] Lin, C. L.; Zhou, L. W.; Crow, J. E. (Bull. Am. Phys. Soc. [2] **30** [1985] 407).

[73] Colinet, C.; Bessoud, A.; Pasturel, A.; Müller, W. (J. Less-Common Metals **143** [1988] 265/78).

[74] Ott, H. R.; Fisk, Z. (in: Freeman, A. J.; Lander, G. H., Handbook on the Physics and Chemistry of the Actinides, Vol. 5, Elsevier, Amsterdam 1987, pp. 85/225).

4.2 Uranium and Lead

4.2.1 General

The disparity in the sizes of the solvent and solute atoms emerges as an important factor in the alloying behavior in liquid metals. The temperature coefficients of the solubility are related to the size factor $S = r_B/r_A$, where r_B and r_A are the atomic radii of the solvent and solute atoms, respectively. Very large coefficients are found around $S \approx 0.7$ and $S \approx 1.35$, whereas very low coefficients are at $S \approx 1.0$ and $S \geq 1.45$. The value for the solvent Pb and the solute U is given with $S = r_{Pb}/r_U = 1.15$, indicating a temperature coefficient of the solubility of $-d(R \cdot \ln N_{(h)})/d(1/T) = 14.05$, about three times higher than the values of systems with very low miscibilities [43].

More recently, a cellular model was developed which allows an estimation of the heat of formation of binary alloys [41, 42]. The energy effects are described by two terms. The first represents the difference in electronegativity between the solvent and solute atoms, and the second reflects the discontinuity in the density of electrons at the boundaries between the Wigner-Seitz atomic cells. In later publications, the authors developed the basis for the calculation of heats of solution [40] and of the enthalpy of formation at infinite dilution. The result of such computations of enthalpies of solution at infinite dilution for the system U in Pb is $\overline{\Delta H}_0 = -32$ kJ/g-atom solute. The final values were recalculated using the experience of experimental work. These considerations, based on elemental properties of the pair of metals, clearly show that a certain miscibility and even a tendency for the formation of intermetallic compounds of the two elements has to be expected [38, 39].

4.2.2 Preparation of the Alloys

The materials used to prepare U–Pb alloys must be of high purity, U of 99.96 to 99.98% purity with Fe, C, Si, Ni, and O as main impurities, and Pb of analytical quality, the purity of which is 99.97% with Ag, Cu, Fe, As, and SO_4^{2-} as main contaminants [6].

In the direct method, weighed lumps of U and Pb were melted together in graphite crucibles in a vacuum furnace. The crucible, with a thermocouple sheath in the base, was surrounded by an aluminium oxide tube. To avoid the evaporation of a large part of the alloy, an argon pressure of ca. 0.1 bar was maintained in the tube. Insufficient stirring caused the formation of segregated products. The use of rotating crucibles solved this problem. The alloy-graphite interface showed evidence of some reaction [6].

A "soaking" method was used for the preparation of alloys with high Pb contents, which avoided raising the temperatures to the neighborhood of the melting point of uranium. U turnings were cleaned in an electropolishing bath or in dilute HNO_3, the U and lumps of Pb were placed in an aluminium oxide crucible, which was inserted in a Pyrex glass tube. After reaching a vacuum of 10^{-6} to 10^{-7} bar, the tube was sealed and heated to 500°C for about five days. The resulting ingots of the alloys were fairly homogeneous and did not contain free U [6].

References for 4.2 on pp. 330/1

A modified technique was used for alloys of the composition around UPb_3, the formation of which tends to be strongly exothermic. The elements were mixed together as fine powders, and a pellet formed under compression. This pellet was then heated inside a silica tube evacuated to ca. 10^{-6} bar. Heating to the melting point of Pb led to a reaction which proceeded quietly. Finally the samples were annealed at any temperature up to 1100°C [6].

A further refinement of this method is the hydride process which was developed for preparation of U — Hg alloys. Cleaned U turnings were heated to 250°C in a pure H_2 atmosphere to form UH_3. Evacuation at 300°C for 30 min converted the UH_3 into finely divided active U. After mixing with Pb, the same procedure was applied as for the "soaking" method. The apparatus had provisions to directly transfer specimens into capillaries for Debye-Scherrer powder diffraction studies. In order to get homogeneous Pb-rich samples, the ingots were heavily deformed and then annealed in vacuum. At the Pb-rich end of the system annealing must be performed at 290°C [6].

UPb_3

The compound UPb_3 was prepared by means of direct synthesis from the elements in the form of metal powders. The reaction was carried out in an evacuated quartz ampoule at 800°C, followed by annealing for 100 h at the same temperature to homogenize the polycrystalline samples [28]. The same method was proposed, suggesting 1200 to 1280°C as the most suitable temperature for the preparation [23].

UPb

The compound UPb was prepared in the same way as UPb_3, using powder mixtures of equiatomic composition. The annealing temperature must be in the range 500 to 600°C [16].

4.2.3 Phase Relations

4.2.3.1 General

The phase diagram of the U — Pb system is based upon two studies [6, 23], which agree in their results as far as the two intermetallics of the system, UPb and UPb_3, are concerned. There is, however, disagreement in the region with 37 to 90 at% U. One of the sources claims a miscibility gap in this region [23], whereas the other interpretes the findings in terms of a flat liquidus [6]. Thermal, microscopic, and structural methods, among them X-ray powder and neutron diffraction, were used to establish the phase diagrams. A high-temperature centrifuge and a resistivity method, were also applied in the Pb-rich part of the system [6]. Although it is not yet clearly decided whether or not the miscibility gap exists, the results of further studies [22] show more evidence for the miscibility gap, in contradiction with thermodynamic considerations [21]. Several compilations of the phase diagram have been published so far [1, 2, 8, 30, 44, 49, 52], all of them preferring Teitel's version with the miscibility gap [22, 23].

4.2.3.2 Phase Diagram

The eutectic of U and UPb is reported to be located very close to pure U (99.6 wt% U). The addition of Pb to U has no effect on the allotropic transformation temperatures of U. The solid solubility of Pb in U is, therefore, concluded to be very slight. The thermal analysis indicates

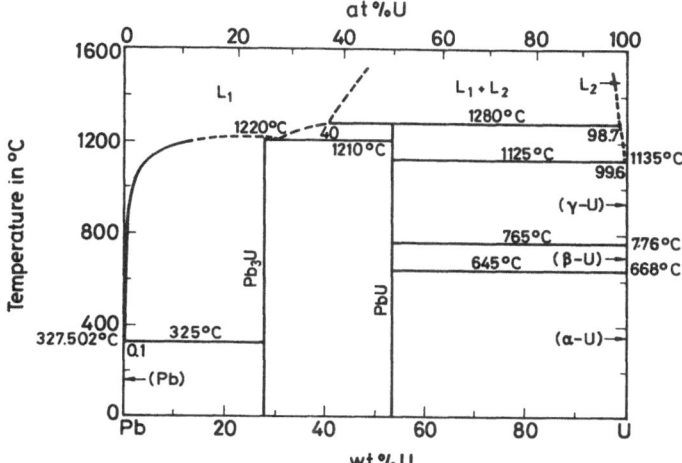

Fig. 4-17. Phase diagram of the U−Pb system according to Teitel [22, 23].

the α-β transformation at 640°C, the β-γ transformation at 765°C and the melting point at 1130°C [23].

Alloys containing more than 40 wt% U, which are prepared at a temperature above 1280°C, segregate into two phases, a U-rich one and a Pb-rich section. This is in agreement with the phase diagram proposed by Teitel [22, 23], shown in **Fig. 4-17**. Alloys, which contain more than 50 at% U and are molten below 1280°C, are mixtures of the compound UPb and its solid solution in U. The liquids of this composition range are two different phases above 1280°C [22, 23].

A eutectic of the two compounds UPb and UPb_3 contains 30 wt% of U and melts at 1210°C. At 25 at% U the intermetallic UPb_3 exists with a melting point of around 1220°C. The eutectic of UPb_3 with Pb melts at 325°C and contains less than 0.1 wt% U. The solid solubility of U in Pb seems to be close to zero. The metallographic results confirm this phase diagram. At different compositions of the alloys, which are prepared at a temperature of 1250°C, the micrographs show the expected phases. Thermal analyses show the ranges of compound formation, the eutectic mixtures, and the borders of the ranges of immiscibility. The syntectic with the melting point of 1280°C consists of L_1 (40 wt% U) + L_2 (98.7 wt% U). The exact melting temperature of UPb_3 is not yet determined, but it is probably between 1210°C and 1250°C [22, 23].

The phase diagram, originated by Teitel [22, 23], is cited in several compilations [1, 2, 8] of U−Pb phase diagrams. A different diagram was drawn by Frost, Maskrey [6] based on their microscopic, X-ray, and thermal-analysis studies. This phase diagram is shown in **Fig. 4-18**, p. 322. The phase diagram is in excellent agreement with Teitel's work [22, 23] as far as the composition of the compounds UPb and UPb_3 is concerned, however, the interpretation of the range of U contents, 30 to 95 at%, deviates from the earlier one. Frost, Maskrey [6] argue that a miscibility gap might be due to a dissimilarity between the two species of atoms. This is not the case in the U−Pb system, since a compound UPb is existent. Their argument is supported by thermodynamic considerations on the basis of the heat of mixing

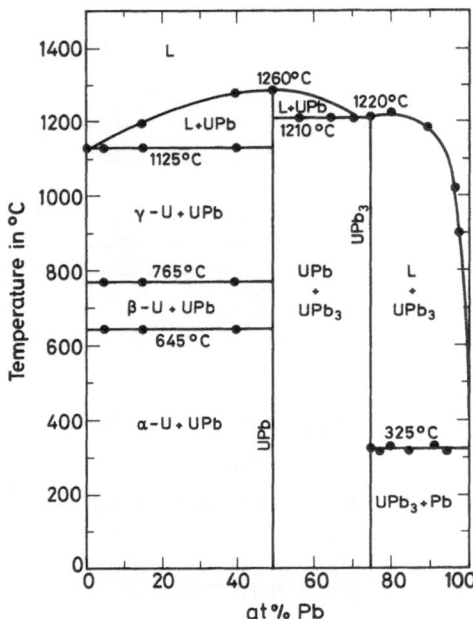

Fig. 4-18. Phase diagram of the U—Pb system according to Frost, Maskrey [6].

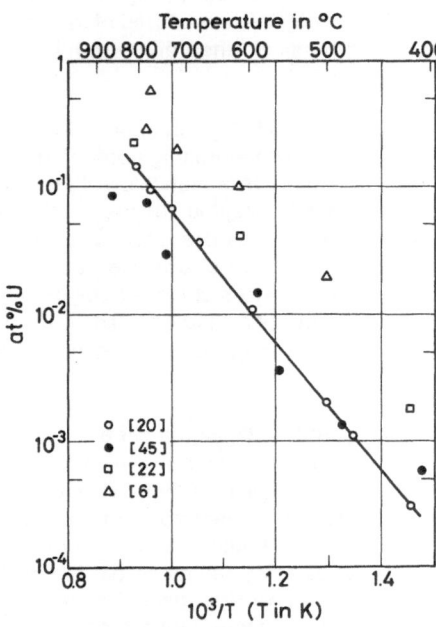

Fig. 4-19. Solubility of uranium in liquid lead [20].

of components with $r_A/r_B > 1$. The formation of liquid miscibility gaps in such systems is stated to be highly improbable, and this statement is brought into the discussion of the U—Pb phase diagram [21].

According to [52], the thermodynamic data predict a miscibility gap even when the data used in the optimization are restricted to temperatures below the syntectic reaction.

The solubility of U in liquid Pb in the temperature range 416 to 802°C is given by the equation $\log [U]_{sat}^{Pb} = 3.921 - 5121/T$ (with [U] in at% and T in K).

Fig. 4-19 shows the results of this study [20] to be very close to this relationship, and within the scatter band of solubility data published earlier [6, 22, 45]. High-purity U in the form of 20-mesh spheres was dissolved in Pb (99.997%). The enthalpy of solution is $\Delta H_{sol} =$ 23.4 kcal/g-atom ($= 97.8$ kJ/g-atom). The entropy of solution is $\Delta S_{sol} = 8.79$ cal·g-atom^{-1}·K^{-1} ($= 36.74$ J·g-atom^{-1}·K^{-1}) [20].

The solubility of U in liquid Pb is negligibly small in the temperature range 600 to 724°C, as is the solubility of Pb in solid U [15].

4.2.4 Crystallographic Data

UPb

The crystal structure of the compound UPb was evaluated from neutron-diffraction techniques. It has a body-centered tetragonal structure with c = 10.6 Å, a = 11.04 Å (c/a = 0.961), the unit cell contains 48 atoms; calculated density $D_{calc} = 13.7$ g/cm³ [22]. A later study reports c = 5.259 ± 0.001 Å and a = 4.579 ± 0.001 Å (c/a = 1.15) from X-ray measurements. The face-centered tetragonal structure has a close relationship to the cubic structure of ThPb₃ and UPb₃. The unit cell contains four atoms at the positions 0, 0, 0; 0, 1/2, 1/2; 1/2, 1/2, 0; and 1/2, 0, 1/2; alternative space groups are P4/mmm, respectively, I4/amd. The atomic arrangement and the distances and radii favor the space group $I4_1/amd-D_{4h}^{19}$ (No. 141) for the proposed structure (the original paper prints I4/amd) [16]. The structural data of UPb are reviewed in [5, 25].

UPb₃

The compound UPb₃ has a cubic structure with the lattice constant a = 4.791 Å and a calculated density $D_{calc} = 12.98$ g/cm³. The unit cell contains 4 atoms [23]. Another X-ray study resulted in the lattice constant a = 4.787 Å [19]. The value a = 4.7834 Å (with $D_{calc} = 13.25$ g/cm³) [6] is regarded as compatible when compared with a = 4.7915 ± 0.0002 of [16]. The difference between the value reported in [6] and that of Teitel [23] (above) is explained in [22]. Structure considerations led to the conclusion that UPb₃ has a simple face-centered cubic structure of A₁ type and the space group $Fm3m-O_h^5$ (No. 225) [16]. The reviews on the structural data of UPb₃ are based on these values. UPb₃ has the same structure as some other UX₃ compounds isotypic with the compound AuCu₃ [5, 25].

4.2.5 Thermodynamic Data

The free-energy equations $\Delta G° = 228330 - 105.76 \cdot T$ (in J/mol) for UPb(s) → U(β) + Pb(g) and $\Delta G° = 200397 - 99.79 T$ (in J/mol) for 1/2 UPb₃(s) → 1/2 UPb(s) + Pb(g) were calculated from dissociation pressures of U—Pb mixtures measured using the Knudsen effusion method [15], see p. 328.

The free energies of formation of the two compounds were calculated as $\Delta G° = -37245 + 2.47\,T$ (in J/mol) for $U(\alpha) + Pb(s) \rightarrow UPb(s)$ and $\Delta G° = -61635 + 1.65 \cdot T$ (in J/mol) for $U(\alpha) + 3\,Pb(s) \rightarrow UPb_3(s)$ [15].

The heat of formation of the compound UPb_3 was also deduced from aqueous solution calorimetric measurements. The value $\Delta G° = -16480 \pm 2060$ J/g-atom is in fair agreement with the value obtained from the Knudsen cell experiments (-15410 J/g-atom) [15].

The free energy of formation of UPb_3 in the temperature range 375 to 954 °C is given by the equation $\Delta G_f° = -96.66 + 89.03 \times 10^{-3} \cdot T - 78.24 \times 10^{-6} \cdot T^2 + 40.059 \times 10^{-9} \cdot T^3$ (in kJ/mol). These values are based on a high-temperature study using galvanic cells of the type $U\,//\,UCl_3 + KCl-LiCl(eutectic)\,//\,U-Pb$ alloys [14].

The enthalpy of formation of UPb_3 related to the equation $U(s) + 3\,Pb(l) \rightarrow UPb_3(s)$ is represented by the equation $\Delta H_f°$ (in kcal/mol) $= -23.46 + 18.99 \times 10^{-6}\,T^2 - 19.45 \times 10^{-9}\,T^3$ and the entropy of formation is then $\Delta S_f°$ (cal $\cdot$ mol$^{-1} \cdot$ K^{-1}) $= -21.61 + 37.98 \times 10^{-3}\,T - 29.17 \times 10^{-6}\,T^2$ [14].

The values from [14] for 723 K ($= 450$ °C) are listed in Table 4/10.

Table 4/10
Thermodynamic Data of Formation of UPb_3 at 723 K [14].

$\Delta G_f°$ in kJ/g-atom	$\Delta H_f°$ in kJ/g-atom	$\Delta S_f°$ in J $\cdot$ g-atom$^{-1} \cdot$ K^{-1}
-14.54	-21.51	-9.68

$\Delta G_f°$ values from emf measurements of the above-mentioned galvanic cell between 933 and 1143 K [53] are shown in **Fig. 4-20** together with the values from [14] and [15]. They are

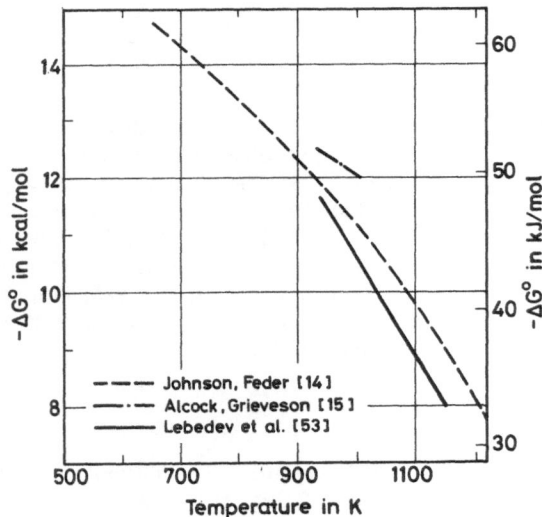

Fig. 4-20. Standard Gibbs energy of formation of $UPb_3(c)$ as a function of the temperature [53].

roughly 4 kJ/g-atom more positive than those of [14] at the upper end of the temperature range. $\Delta H_f^\circ = -65.92$ kJ/mol at 298.15 K (25°C) is tabulated in [13].

The enthalpy of evaporation ΔH_{evap}° of Pb from UPb_3, based on Knudsen cell studies, is 181.28 kJ/g-atom [12], the enthalpy of formation based on this study is $\Delta H_f^\circ = -18.95$ kJ/g-atom.

The temperature dependence of the free energy of formation of UPb from solid U and liquid Pb, based on an effusion study, fits the equation ΔG_f° (in kJ/mol) $= -42.02 + 10.3 \times 10^{-3}$ T [10].

The thermodynamic data for the two U−Pb compounds are reviewed in the compilations of [49, 52]. Enthalpy and entropy of UPb and UPb_3 are assessed in [54]. The standard values recommended in this publication are:

	UPb	UPb_3
S_{298} in $J \cdot mol^{-1} \cdot K^{-1}$	112.4	231.2
H_{298} in kJ/mol	−37.6	−71.9

4.2.6   Mechanical and Thermal Properties

4.2.6.1   Density. Atomic Volume

Although the pyrophoric character of UPb makes an adequate X-ray analysis and the estimation of the density difficult [23], a density D = 13.7 g/cm^3 based on X-ray measurements at room temperature is reported [22]. This density is included in several phase diagram compilations (see p. 320) [30, 44].

The density of UPb_3, based on an X-ray diffraction study at room temperature, is D = 12.98 g/cm^3 [23]. This value is preferably used in phase diagrams (see p. 320) [30, 44]. The value reported by Frost, Maskrey [6], D = 13.24 g/cm^3, may be incorrect.

The average atomic volume (volume of the unit cell in Å^3 divided by the number of atoms in the cell) in the intermetallic compounds UPb and UPb_3 deviates from the straight line between the values of the two components, U and Pb. The average atomic volume in UPb is given as ∼27.8 Å^3 above the straight line, the value of UPb_3 is ∼27.5 Å^3 below that line. The atomic volume in UPb is larger than in the mixture of the elements, a deviation common for intermetallic compounds of U [5].

4.2.6.2   Hardness

The compounds UPb and UPb_3 are very hard [16], but the results of quantitative measurements have not been published so far.

4.2.6.3   Melting Temperature

It was assumed that UPb melts congruently at 1280°C [6, 30, 44] (see phase diagrams, Figs. 4-17 and 4-18, pp. 321/2). The data are imprecise because of the sensitivity of UPb towards oxidation at high temperatures even at very low partial pressures of oxygen. The oxidation shifts the composition to higher Pb contents. The melting of this compound is critically discussed in [52].

                     References for 4.2 on pp. 330/1

The melting point of UPb$_3$ is not well defined; most sources give 1220°C [6, 22, 23, 30, 44]. The most recent compilation supports this value [52].

4.2.6.4 Thermal Conductivity

The thermal conductivity of the compounds UPb and UPb$_3$ and of U — Pb alloys of undefined compositions has not yet been reported. The measurement of such properties may be difficult, owing to the brittle and chemically reactive character of these materials. Since they are not used as fuel element materials, there has not been much interest in their thermal conductivity.

4.2.6.5 Diffusion

Information concerning the diffusion of Pb in solid U, or on the diffusion of any of the constituents of U — Pb alloys is not available. The work of Teitel [22] indicates that the diffusion in the multiphase alloys, which are formed by means of the dismixture at a temperature above 1280°C, might be very slow. The very low solubility of Pb in solid U may be the reason for the lack of diffusion data. The chemical reactivity of U-rich alloys makes the measurements of diffusion coefficients difficult at high temperatures.

4.2.7 Magnetic Properties

A survey on the magnetic properties of UM$_3$ compounds having the AuCu$_3$ structure is given in [34]. The observed magnetic behavior was qualitatively explained by assuming a 5f^2 electronic configuration of the U ion and a negative charge on the Pb ligands [33]. This crystal-field model is applicable to compounds in which the actinide ions are largely separated. As with compounds of the AuCu$_3$ structure, UPb$_3$ has a larger U — U separation than the critical Hill separation (i.e., 3.4 to 3.6 Å) beyond which f — f overlap is no longer a major factor influencing the magnetic behavior [50]. It was also suggested that the magnetic properties of the UX$_3$ compounds can be interpreted in terms of the localized spin-fluctuation model [46].

The magnetic susceptibility of a sample of powdered UPb$_3$ (a = 4.79 Å from 78 to 900 K) is given in [29, 33]. The plot of χ^{-1} versus T contains a linear region above 400 K with a Curie-Weiss temperature $\Theta \approx -170$ K and effective moment $\mu_{eff} \approx 3.2$ μ_B per formula unit. The plot is slightly curved between the temperatures of 78 K and 400 K. The values of the magnetic susceptibility have a pronounced temperature dependence, $\chi_M = 6700 \times 10^{-6}$ at 80 K, $\chi_M = 2840 \times 10^{-6}$ at 290 K, and $\chi_M = 1180 \times 10^{-6}$ at 900 K [33]. Measurements in the low-temperature range 4.2 to 100 K [26, 31] show an antiferromagnetic transition at $T_N = 32$ K, which does not occur with other UM$_3$ compounds. **Fig. 4-21** shows the temperature function of the reciprocal magnetic susceptibility indicating this transition. The ground state of U in UPb$_3$ (Γ_3 or Γ_5) still remains unknown because of the appearance of the antiferromagnetic range [31].

A powder sample of UPb$_3$ was also studied by means of neutron diffraction using profile analysis. Below T_N the magnetic structure is described by k = (0 0 1/2). The ordered magnetic moment ranges from 1.58 to 1.69 μ_B depending on the assumed 5f^n configuration. The measured reduced magnetization as a function of the temperature agrees reasonably well with calculated values based on a crystal-field model with Γ_5 ground state, including exchange in the molecular field approximation. T_N was redetermined, and its exact value is 30 $\pm$ 0.2 K [28].

Neutron diffraction patterns taken at 4.2 K give, apart from strong nuclear peaks, two superstructure reflections due to magnetic ordering. Both are indexed on a unit cell of tetragonal symmetry with a' = a (a is the lattice constant of UPb$_3$) and c' = 2a. The proposed

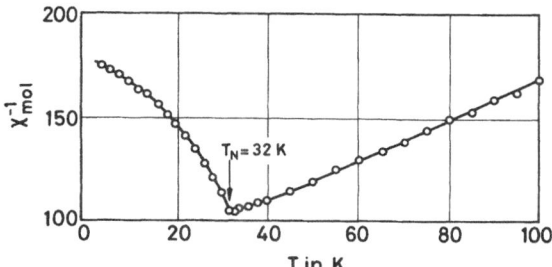

Fig. 4-21. The reciprocal magnetic susceptibility of UPb$_3$ as a function of the temperature [31].

magnetic structure consists of U magnetic moments aligned oppositely in adjacent (001) ferromagnetic places. The configurational symmetry is P4/mmm with 1 U (+) at 1 (a) 0,0,0; 1 U ($-$) at 1 (b) 0,0,1/2 [26].

The crystal field energy levels in UPb$_3$ were also studied by neutron spectroscopy at 50 K. The measurements indicate that the Γ_5 level is the ground state of the U ions in UPb$_3$. The overall splitting is 11.5 MeV, and the crystal field parameters are χ = 0.61 MeV and W = 0.224 MeV. These parameters give an excellent description of the magnetic properties of UPb$_3$ if 5f^4 configuration and ^{5}I$_4$ ground multiplet is assumed. On the basis of this neutron spectroscopy study, the multiaxial magnetic structure in UPb$_3$ was postulated [24].

The pressure dependence of the Néel temperature T$_N$ for UPb$_3$ is (dT$_N$/dP) = 0.26 K/kbar [48].

Since UPb$_3$ belongs to the group of U intermetallics in which the U$-$U distance is large (4.792 Å), the strength of the f-ligand hybridization controls the itinerant electron bandwidth. This is the reason for a local moment behavior of UPb$_3$ similar to UTl$_3$ [32]. There is no high electronic specific-heat coefficient γ as in USn$_3$. Compounds of the composition U(Pb$_x$Sn$_{1-x}$)$_3$ completely cover the range of behavior from itinerant electrons, through spin fluctuation with large effective masses, to local moment behavior depending on the value of x. The fact that UPb$_3$ does not possess a high electronic specific-heat coefficient means that the compound does not belong to the heavy-electron U compounds [46, 47].

The density of states at the Fermi energy is n(e$_F$) = 26.6 [47].

4.2.8 Electrical Properties

The electrical resistivity at high temperatures of Pb$-$UPb$_3$ mixtures was measured to support the determination of the liquidus curve in this part of the phase diagram (see p. 320). The resistivity versus temperature curve of the alloy with 0.25 at% U is shown in **Fig. 4-22**, p. 328. A change of the slope occurs at temperatures between 720°C and 735°C (993 to 1008 K). There is no information on the relation of resistivity and the concentration of U in the solutions. The measured resistivity is not related to the dimension of the alloy sample [6].

A superconducting transition of UPb$_3$ was not observed [46].

                    References for 4.2 on pp. 330/1

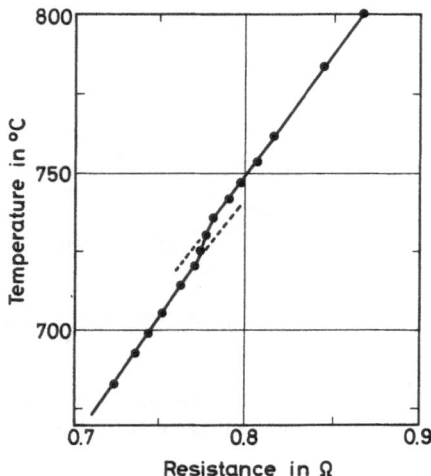

Fig. 4-22. Resistivity versus temperature plot for the U — Pb alloy containing 0.25 at% U [6].

4.2.9 Chemical Reactions

The dissociation pressures of U — Pb alloys were measured by means of a Knudsen effusion method [15]. The results of the pressure measurements of Pb of the mixtures $UPb_3 - UPb$ and $UPb - U(\beta)$ are shown in **Fig. 4-23**. These data were used to calculate the free-energy equations (see p. 323).

The compounds UPb and UPb_3 are highly pyrophoric, the compound UPb being the greatest [6, 8]. The extremely pyrophoric character of UPb causes difficulties in the handling of this compound and precludes an adequate X-ray analysis [23].

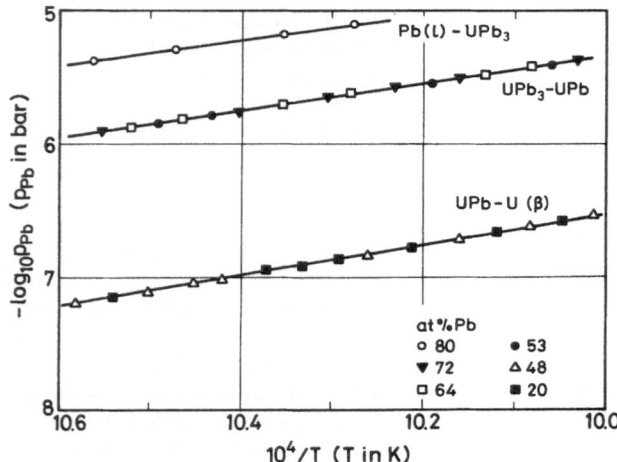

Fig. 4-23. The dissociation pressures of U — Pb alloys [15].

UPb_3 reacts with some liquid metals at high temperatures. Its reaction with liquid Sn leads to the formation of isostructural alloys $U(Pb, Sn)_3$ even at a temperature of 350 °C (623 K). The reaction with Bi does not form a compound of this type. UPb_3 might be partly decomposed while a compound UBi is formed [30, 44].

The preparation of U – Pb alloys in graphite crucibles at temperatures above 1000 °C shows some evidence for chemical reactions of the molten alloys with graphite, but the reaction products are not known [6].

4.2.10 Ternary Systems

No information is published on compounds of the types (U, X)Pb or $(U, X)Pb_3$. Compounds in which Pb is substituted by third elements are rare in comparison to systems containing Sn. The composition $U(Pb, X)_3$, crystallizing in the $CuAu_3$ structure, has only one example, $U(Pb, Sn)_3$, which is treated in the U – Sn system (see p. 309). Ternary alloys U – Pb – Bi (see below) are multi-phase systems.

4.2.10.1 U – Pb – Bi

Compounds of the binary system U – Bi as UBi and UBi_2 are in equilibrium with liquid phases containing up to 95 at% Pb in the system U – Pb – Bi at 800 °C. **Fig. 4-24** shows the isothermal section of the phase diagram at this temperature. The occurrence of one two-phase region, $UPb_3 + L$, and three regions of three-phase states is indicated: $UPb_3 + UBi + L$; $UBi + U_3Bi_4 + L$; and $U_3Bi_4 + UBi_2 + L$. The phase UPb_3 exists in equilibrium with UBi and the liquid containing less than 5 at% Bi [30, 44].

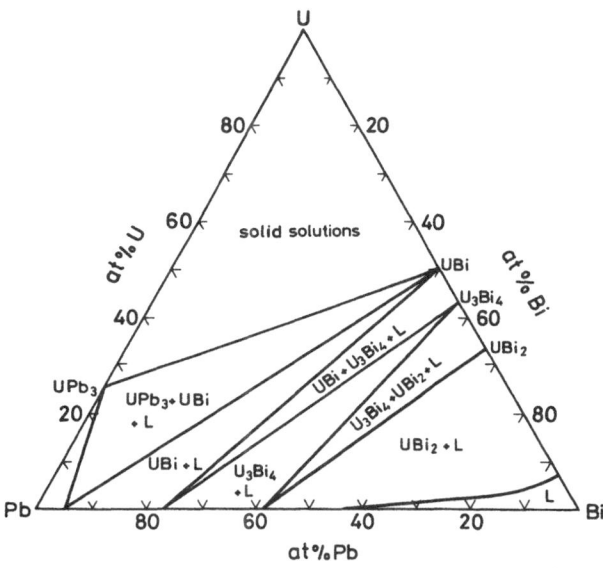

Fig. 4-24. The isothermal section of the U – Pb – Bi system at 800 °C [30].

References for 4.2 on pp. 330/1

References for 4.2:

[1] Massalski, T. B. (Binary Alloy Phase Diagrams, Vol. 2, Am. Soc. Metals, Metals Park, Ohio, 1985, pp. 1854/6).

[2] Hansen, M.; Anderko, K. (Constitution of Binary Alloys, McGraw-Hill, New York 1958, pp. 1116/7).

[3] Elliott, R. P. (Constitution of Binary Alloys, 1st Suppl., McGraw-Hill, New York 1965, pp. 726/7).

[4] Shunk, F. A. (Constitution of Binary Alloys, 2nd Suppl., McGraw-Hill, New York 1969, p. 605).

[5] Dwight, A. E. (in: Giessen, B. C., Developments in the Structural Chemistry of Alloy Phases, Plenum, New York 1969, pp. 181/226).

[6] Frost, B. R. T.; Maskrey, J. T. (J. Inst. Metals **82** [1953/54] 171/80).

[7] Katz, J. J.; Rabinowitch, E. (The Chemistry of Uranium, Pt. I, The Element, its Binary and Related Compounds, Chapter VII, McGraw-Hill, New York 1951, pp. 174/9).

[8] Saller, H. A.; Rough, F. A. (BMI-1000 [1955] 1/141; N.S.A. **9** [1955] No. 5349).

[9] Cordfunke, E. H. P. (The Chemistry of Uranium Including its Applications in Nuclear Technology, Chapter IV, Elsevier, Amsterdam 1969, pp. 39/58).

[10] Johnson, I. (IMD Spec. Rept. No. 13 [1964] 171/92).

[11] Hume-Rothery, W. (J. Inst. Metals **83** [1954/55] 535/6).

[12] Alcock, C. B.; Cornish, J. B.; Grieveson, P. (Thermodyn. Proc. Symp., Vienna 1965 [1966], pp. 211/30).

[13] Wagman, D. D.; Evans, W. H.; Parker, V. B.; Schumm, R. H.; Nuttall, R. L. (Selected Values of Chemical Thermodynamic Properties, Compounds of Uranium, Protactinium, Thorium, Actinium, and the Alkali Metals, Natl. Bur. Std., Washington 1981).

[14] Johnson, I.; Feder, H. M. (Thermodyn. Nucl. Mater. Proc. Symp., Vienna 1962, pp. 319/28).

[15] Alcock, C. B.; Grieveson, P. (J. Inst. Metals **93** [1961/62] 304/10).

[16] Brown, A. (Acta Cryst. **14** [1961] 856/60).

[17] Axon, H. J. (J. Inst. Metals **83** [1954/55] 537/8).

[18] Frost, B. R. T. (J. Inst. Metals **83** [1954/55] 539).

[19] Iandelli, A.; Ferro, R. (Ann. Chim. [Rome] **42** [1952] 598/606).

[20] Johnson, I.; Chasanov, M. G. (ASM [Am. Soc. Metals] Trans. Quart. **56** [1963] 272/7).

[21] Kubaschewski, O. (Thermodyn. Nucl. Mater. Proc. Symp., Vienna 1962, pp. 219/41).

[22] Teitel, R. J. (J. Inst. Metals **85** [1956/57] 409/12).

[23] Teitel, R. J. (Trans. AIME **194** [1952] 397/400).

[24] Murasik, A.; Zolierek, Z. (Physica B + C **98** [1980] 306/10).

[25] Lam, D. J.; Darby, J. B., Jr.; Nevitt, M. V. (in: Freeman, A. J.; Darby, J. B., Jr., The Actinides; Electronic Structure and Related Properties, Vol. 2, Academic, New York 1974, pp. 119/84).

[26] Leciejewicz, J.; Misiuk, A. (Phys. Status Solidi A **13** [1972] K79/K81).

[27] Wood, D. H.; Cramer, E. M.; Wallace, P. L.; Ramsay, W. J. (J. Nucl. Mater. **32** [1969] 193/207).

[28] Murasik, A.; Fischer, P.; Zolnierek, Z. (Physica B + C **102** [1980] 188/91).

[29] Brodsky, M. B. (Rept. Progr. Phys. **41** [1978] 1547/603).

[30] Rough, F. A.; Bauer, A. A. (Constitutional Diagrams of Uranium and Thorium Alloys, Addison-Wesley, Reading, Mass., 1958, p. 90).

[31] Misiuk, A.; Mulak, J.; Czopnik, A. (Bull. Acad. Polon. Sci. Ser. Sci. Chim. **20** [1972] 459/61).

[32] Koelling, D. D.; Dunlap, B. D.; Crabtree, G. W. (Phys. Rev. [3] B **31** [1985] 4966/71).

[33] Mulak, J.; Misiuk, A. (Bull. Acad. Polon. Sci. Ser. Sci. Chim. **19** [1971] 207/13).

[34] Lam, D. J.; Aldred, A. T. (in: Freeman, A. J.; Darby, J. B., Jr., The Actinides; Electronic Structure and Related Properties, Vol. 1, Academic, New York 1974, pp. 109/79).

[35] Fournier, J.-M.; Troć, R. (in: Freeman, A. J.; Lander, G. H., Handbook on the Physics and Chemistry of the Actinides, Vol. 2, North-Holland, Amsterdam 1985, pp. 70/3).

[36] Buyers, W. J. L.; Holden, T. M. (in: Freeman, A. J.; Lander, G. H., Handbook on the Physics and Chemistry of the Actinides, Vol. 2, North-Holland, Amsterdam 1985, pp. 288/9).

[37] Miedema, A.; de Châtel, P. F.; de Boer, F. R. (Physica B + C **100** [1980] 1/28).

[38] Niessen, A. K.; de Boer, F. R.; Boom, R.; de Châtel, P. F.; Mattens, W. C. M.; Miedema, A. R. (CALPHAD **7** [1983] 51/70).

[39] Miedema, A. R.; de Boer, F. R.; Boom, R. (CALPHAD **1** [1977] 341/59).

[40] Miedema, A. R.; Boom, R.; de Boer, F. R. (J. Less-Common Metals **41** [1975] 283/98).

[41] Boom, R.; de Boer, F. R.; Miedema, A. R. (J. Less-Common Metals **46** [1976] 271/84).

[42] Miedema, A. R. (J. Less-Common Metals **32** [1973] 117/36).

[43] Strauss, S. W.; White, J. L.; Brown, B. F. (Acta Met. **6** [1958] 604/6).

[44] Ivanov, O. S.; Badaeva, T. A.; Sofronova, R. M.; Kishenevskii, V. B.; Kushnir, N. P. (Phase Diagrams of Uranium Alloys, Amerind Publ., New Delhi 1983, transl. from: Diagrammy Sostoyaniya i Fazovye Prevrashcheniya Splavov Urana, Nauka, Moscow 1972).

[45] Petit, J. F.; Paoli, P. (unpublished data, included in [20]).

[46] Fisk, Z.; Ott, H. R.; Smith, J. L. (J. Less-Common Metals **133** [1987] 99/106).

[47] Koelling, D. D.; Norman, M. R.; Arko, A. J. (J. Magn. Magn. Mater. **63/64** [1987] 638/44).

[48] Lin, C. L.; Zhou, L. W.; Crow, J. E.; Guertin, R. P. (J. Appl. Phys. **57** [1985] 3146/8).

[49] Lin, C. L.; Zhou, L. W.; Crow, J. E.; Mihalisin, T.; Brooks, J.; Guertin, R. P. (J. Less-Common Metals **127** [1987] 273/9).

[50] Chiotti, P.; Akhachinskij, V. V.; Ansara, I.; Rand, M. H. (The Chemical Thermodynamics of Actinide Elements and Compounds, Pt. 5, The Actinide Binary Alloys, Intern. At. Energy Agency, Vienna 1981).

[51] Hunter Hill, H. (in: Miner, W. N., Plutonium 1970 and Other Actinides, Met. Soc. AIME, Metals Park, Ohio, 1970, Pt. I, pp. 2/19).

[52] Sheldon, R. I.; Foltyn, E. M.; Peterson, D. E. (Bull. Alloy Phase Diagrams **8** [1987] 536/41, 587/8).

[53] Lebedev, V. A.; Poyarkov, A. M.; Nichkov, I. F.; Raspopin, S. P. (At. Energiya SSSR **31** [1971] 621/2; Soviet At. Energy **31** [1971] 1408/9).

[54] Rand, M. H.; Kubaschewski, O. (The Thermochemical Properties of Uranium Compounds, Oliver & Boyd, Edinburgh—London 1963, p. 48).

Physical Constants and Conversion Factors

Avogadro constant N_A (or L) = 6.02214×10^{23} mol^{-1}
Faraday constant F = 9.64853×10^4 C/mol
molar gas constant R = 8.31451 J·mol^{-1}·K^{-1}
molar volume (ideal gas) V_m = 2.24141×10^1 L/mol
(273.15 K, 101325 Pa)

Planck constant h = 6.62608×10^{-34} J·s
elementary charge e = 1.60218×10^{-19} C
electron mass m_e = 9.10939×10^{-31} kg
proton mass m_p = 1.67262×10^{-27} kg

1 kg = 2.205 pounds
1 m = 3.937×10^1 inches = 3.281 feet
1 m^3 = 2.642×10^2 gallons (U.S.)
1 m^3 = 2.200×10^2 gallons (Imperial)

Force	N	dyn	kp
1 N	1	10^5	1.019716×10^{-1}
1 dyn	10^{-5}	1	1.019716×10^{-6}
1 kp	9.80665	9.80665×10^5	1

Pressure	Pa	bar	kp/m²	at	atm	Torr	lb/in²
1 Pa = 1N/m²	1	10^{-5}	1.019716×10^{-1}	1.019716×10^{-5}	9.86923×10^{-6}	7.50062×10^{-3}	1.450378×10^{-4}
1 bar = 10^6 dyn/cm²	10^5	1	1.019716×10^4	1.019716	9.86923×10^{-1}	7.50062×10^2	1.450378×10^1
1 kp/m²=1 mm H$_2$O	9.80665	9.80665×10^{-5}	1	10^{-4}	9.67841×10^{-5}	7.35559×10^{-2}	1.422335×10^{-3}
1 at (technical)	9.80665×10^4	9.80665×10^{-1}	10^4	1	9.67841×10^{-1}	7.35559×10^2	1.422335×10^1
1 atm = 760 Torr	1.01325×10^5	1.01325	1.033227×10^4	1.033227	1	7.60×10^2	1.469595×10^1
1 Torr = 1mmHg	1.333224×10^2	1.333224×10^{-3}	1.359510×10^1	1.359510×10^{-3}	1.315789×10^{-3}	1	1.933678×10^{-2}
1 lb/in² = 1psi	6.89476×10^3	6.89476×10^{-2}	7.03069×10^2	7.03069×10^{-2}	6.80460×10^{-2}	5.17149×10^1	1

Work, Energy, Heat	J	kW·h	kcal	Btu	eV
1 J = 1 W·s = 1 N·m = 10^7 erg	1	2.778×10^{-7}	2.39006×10^{-4}	9.4781×10^{-4}	6.242×10^{18}
1 kW·h	3.6×10^6	1	8.604×10^2	3.41214×10^3	2.247×10^{25}
1 kcal	4.1840×10^3	1.1622×10^{-3}	1	3.96566	2.6117×10^{22}
1 Btu (British thermal unit)	1.05506×10^3	2.93071×10^{-4}	2.5164×10^{-1}	1	6.5858×10^{21}
1 eV	1.602×10^{-19}	4.450×10^{-26}	3.8289×10^{-23}	1.51840×10^{-22}	1

1 cm^{-1} = 1.239842×10^{-4} eV
1 hartree = 27.2114 eV

1 Hz = 4.135669×10^{-15} eV
1 eV ≙ 23.0578 kcal/mol

Power	kW	hp	kp·m·s^{-1}	kcal/s
1 kW = 10^3 J	1	1.35962	1.01972×10^2	2.39006×10^{-1}
1 hp (horsepower, metric)	7.3550×10^{-1}	1	7.5×10^1	1.7579×10^{-1}
1 kp·m·s^{-1}	9.80665×10^{-3}	1.333×10^{-2}	1	2.34384×10^{-3}
1 kcal/s	4.1840	5.6886	4.26650×10^2	1

References:

International Union of Pure and Applied Chemistry, Manual of Symbols and Terminology for Physicochemical Quantities and Units, Pergamon, London 1979; Pure Appl. Chem. **51** [1979] 1/41.

The International System of Units (SI), National Bureau of Standards Spec. Publ. 330 [1972].

Landolt-Börnstein, 6th Ed., Vol. II, Pt. 1, 1971, pp. 1/14.

ISO Standards Handbook 2, Units of Measurement, 2nd Ed., Geneva 1982.

Cohen, E. R., Taylor, B. N., Codata Bulletin No. 63, Pergamon, Oxford 1986.